AUTOMOTIVE
AIR CONDITIONING

AUTOMOTIVE AIR CONDITIONING

8TH EDITION

BOYCE H. DWIGGINS

DELMAR

THOMSON LEARNING

Australia • Canada • Mexico • Singapore • Spain • United Kingdom • United States

DELMAR

THOMSON LEARNING

Business Unit Director:
Alar Elken

Executive Editor:
Sandy Clark

Acquisitions Editor:
Sanjeev Rao

Development Editor:
Christopher Shortt

Editorial Assistant:
Matthew Seeley

Executive Marketing Manager:
Maura Theriault

Marketing Coordinator:
Brian McGrath

Executive Production Manager:
Mary Ellen Black

Production Editor:
Barbara L. Diaz

Art and Design Coordinator:
Cheri Plasse

Library of Congress Cataloging-in-Publication Data:

Dwiggins, Boyce H.
Automotive air conditioning / Boyce Dwiggins.-- 8th ed.
   p. cm.
   Includes index.
   ISBN 0-7668-0788-6 (alk. paper)
   1. Automomobiles --Air conditioning--Maintenance and repair.   I.
Title.
TL271.5 .D9 2001
629.2'772--dc21

2001028775

Asia (including India):
Thomson Learning
60 Albert Street, #15-01
Albert Complex
Singapore 189969
Tel 65 336-6411
Fax 65 336-7411

Ausrtalia/New Zeland:
Nelson
102 Dodds Street
South Melbourne, Victoria 3205
Australia
Tel 61 (0)3 9685-4111
Fax 61 (0)3 9685-4199

Latin America:
Thomson Learning
Seneca 53
Colonia Polanco
11560 Mexico D. F. Mexico
Tel (525) 281-2906
Fax (525) 281-2656

Canada:
Nelson
1120 Birchmount Road
Toronto, Ontario
Canada M1K 5G4
Tel (416) 752-9100
Fax (416) 752-8102

UK/Europe/Middle East:
Thomson Learning
Berkshire House
168-173 High Holborn
London WC1V 7AA
United Kingdom
Tel 44 (0)171 497-1422
Fax 44 (0)171 497-1426

Business Press
Berkshire House
168-173 High Holborn
London WC1V 7AA
United Kingdom
Tel 44 (0)171 497-1422
Fax 44 (0)171 497-1426

Spain:
Paraninfo
Calle Magallanes 25
28015 Madrid
España
Tel 34 (0)91 446-3350
Fax 34 (0)91 445-6218

Distribution Services:
ITPS
Cheriton House
North Way
Andover,
Hampshire SP10 5BE
United Kingdom
Tel 44 (0)1264 34-2960
Fax 44 (0) 1264 34-2759

International Headquarters
Thomson Learning
International Division
290 Harbor Drive 2nd Floor
Stamford, CT 06902-7477
USA
Tel (203) 969-8700
Fax (203) 969-8751

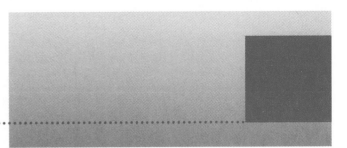

QUICK REFERENCE CONTENTS

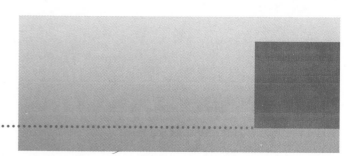

CONTENTS

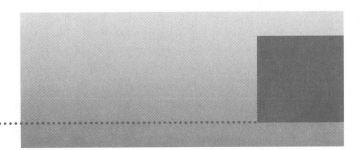

PREFACE

Automotive Air-Conditioning, 8th edition, continues in the best-selling tradition established in 1967 by the first edition. The most up-to-date state-of-the-art technology for motor vehicle air-conditioning (MVAC) systems has been included: manual, semiautomatic and automatic electronic single- and dual-zone temperature controls, domestic and foreign fixed and variable displacement compressors, thermostatic expansion valve and orifice tube systems.

The information provided in this textbook is a balanced introduction to motor vehicle air conditioning. The student will develop a basic understanding of the theory, diagnostic practices, and service procedures that are essential to successfully service air-conditioning systems. At the same time, the student will develop habits of sound practice and good judgment in the performance of all mobile air-conditioning diagnostic and repair procedures. The instructional units are regarded as entry level for those who immediately apply the basic skills developed in the class and shop. The units are preparatory for those who plan to continue studies in advanced phases of refrigeration, including systems not related to motor vehicle applications. The text is arranged in the natural order of dependence of one principle, law, or set of conditions upon another. The material follows an organized pattern that helps the student to see relationships.

Practices guide the student in the performance of system diagnostic and repair procedures. This is accomplished by stressing diagnosis through the use of a manifold and gauge set. Fifteen color plates are included for interpretation of various system functions and malfunctions.

A comprehensive glossary is provided to aid in identifying component parts, phrases, and terms relevant to motor vehicle air-conditioning systems.

It is suggested that each topic in this text be considered as an assignment to be carried out by the student.

An Instructor's Guide provides solutions to all of the objective questions and problems. Suggested answers are given wherever there may be variations in the responses given by the students.

ABOUT THE AUTHOR

In the mid 1960s the author, Boyce H. Dwiggins, organized one of the first courses for vocational education for Automotive Air Conditioning. He was in charge of automotive classes as a county-level administrator and was a consultant for the writing of educational specifications for a five-shop automotive complex in an area vocational center.

Mr. Dwiggins has served as an examiner, administering the "Automotive Excellence" test for the International Garage Owner's Association (IGOA) for the certification of auto mechanics. He has served on the automotive air-conditioning writing team of the National Institute for Automotive Service Excellence (ASE) to revise the certification test for automotive air conditioning. He was awarded ASE's coveted Blue Seal of Excellence in 1994 for his efforts.

Mr. Dwiggins holds patents on teaching devices and copyrights on teaching material in the automotive and refrigeration fields. He has conducted workshops for automotive and refrigeration teachers throughout the eastern United States. Until his retirement in 1994, he was chairperson of the industrial department of a large vocational-technical center in South Florida.

It should be recognized that this text would not have been possible without the generous cooperation of the many manufacturers of motor vehicle air-conditioning systems, equipment, and components. Their contributions, for over thirty-three years, have been most helpful in providing the latest information available. They include:

Best Educational Trainers, Inc.
Controls Company of America
Daimler Chrysler Corporation
Everhot Products Company
Ford Motor Company
General Motors Corporation: All divisions
Mapco
Murray Corporation
Ritchie Engineering
Robinair Manufacturing Company
Sankyo
Sears, Roebuck and Company
Tecumseh Products Company
Thermal Industries
T.I.F. Instruments
Uniweld Products
Warner Electric
York Corporation

Thanks to the many instructors who reviewed the manuscript for this edition. A special thanks to the memory of Mr. Richard G. Herd, former associate, Vocational-Industrial Education, State of New York, for his initial assistance and encouragement. Last, but by no means least, thanks to the great team at Delmar Publishers for steering this text through all eight editions.

This edition is lovingly dedicated to Janie and Lalani.

Boyce H. Dwiggins

CHAPTER 1

HISTORY AND DEVELOPMENT

OBJECTIVES

On completion and review of this chapter, you should be able to:

❑ Understand the history and development of the motor vehicle air-conditioning industry.
❑ Understand and discuss the popularity of the motor vehicle air-conditioning system.
❑ Be aware of the environmental problems that are associated with motor vehicle air conditioning.
❑ Be able to identify the technician certification requirements for servicing domestic and motor vehicle air-conditioning systems.
❑ Be aware of professional organizations and associations concerned with the motor vehicle air conditioning-industry.
❑ Develop an understanding of the role of the automotive technician relative to today's environmental concerns.

INTRODUCTION

Refrigeration and **air conditioning** are not discoveries of the twentieth century. Simple forms of refrigeration and air conditioning were in use more than twelve thousand years ago. Although these early systems were crude by today's standards, they served the same purpose as modern units.

Many aspects of modern life were made possible only after sophisticated air-conditioning systems were developed. Many components that are vital to the National Aeronautics and Space Administration (NASA) space exploration programs could not have been manufactured without the use of air conditioning. For example, many precision mechanical and electrical parts must be manufactured and assembled under very strict tolerances requiring the control of **temperature** and **humidity** within a range of a few degrees. For example, the microprocessor chip shown in Figure 1–1, which only measures about 1/16 inch (1.59 mm) square, is a miniaturized electronic circuit etched into a base of silicon. It was manufactured in

a temperature/humidity-controlled **environment** made possible by modern refrigeration techniques.

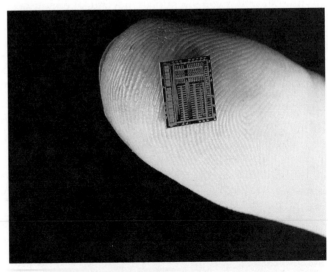

FIGURE 1–1 The actual size of this integrated circuit (chip) is less than 1/16" square. (*Courtesy of Texas Instruments*)

According to *Automotive News,* there were 142,600 new car dealer-installed air-conditioning systems in 1958. In 1962, slightly more than 11 percent of all cars sold were equipped with air conditioners. By 1965 there were 400,000 units installed and, just two years later, over 800,000 air-conditioning units were installed. The total number of vehicles equipped with air-conditioning systems was estimated to be 3,546,255 in 1967. In 1968 the average cost of an air-conditioning system was about 12 percent of the total cost of the vehicle, whereas today the cost is about 3 percent. Over 90 percent of all automobiles sold in the United States are equipped with air-conditioning systems. This means that more than 90 out of every 100 cars on the road today are equipped with a factory- or dealer-installed air-conditioning system. Trucks lag cars by about ten years in air conditioning, however, only becoming popular around 1970. Percentages of factory- and dealer-installed systems on trucks are on the rise at about the same rate as for cars.

Mobile air conditioning in Europe is becoming a popular **option** and is expected to increase at about the same rate as in the United States. When mobile air conditioning was first introduced, it was considered a luxury. Its usefulness, however, soon made it a necessity.

AIR CONDITIONING

The definition of air conditioning should be reviewed before tracing its history and its application to the automobile. *Air conditioning* is the process by which air is cooled or heated, cleaned or filtered, and circulated or recirculated. In addition, the quantity and quality of the conditioned air is controlled. This means that the temperature, humidity, and volume of air can be controlled at any time in any given situation. Under ideal conditions, air conditioning can be expected to accomplish all of these tasks at the same time. It should also be recognized that the air-conditioning process includes the process of *refrigeration* (cooling by removing heat).

HISTORICAL DEVELOPMENT OF AIR CONDITIONING

Some of the principles of refrigeration as we know it today were known as long ago as ten thousand years before Christ (B.C.).

In Egypt a method was developed for removing the heat from the Pharaoh's palace. The walls of the palace were constructed of huge stone blocks weighing over a thousand tons, and every night three thousand slaves dismantled the walls and moved the heavy stones to the Sahara Desert. Because the temperature in the desert is cool during the night, the stones gave up the heat they had absorbed during the day. Before daybreak the slaves moved the stones back to the site and reassembled the walls.

It is believed that the Pharaoh enjoyed temperatures of about 80°F (26.6°C) inside the palace while the temperature outside soared to about 130°F (54.4°C). Three thousand men toiled all night to perform a task that modern refrigeration now easily handles. Although the work effort is less today, the same principle of refrigeration is applied to present systems as it was in the Pharaoh's time; that is, heat is removed from one space and transferred to another space.

First Air-Conditioned Dwelling

Shortly after the beginning of the twentieth century, Colonel T. C. Northcott, a heating and ventilating engineer, became the first man known to history to have a home with central heating and air conditioning. Northcott built his house on a hill above the famous Caverns of Luray. Because of his experience, he knew that air filtered through limestone was free of dust and pollen. This fact was important because Northcott and his family suffered from hay fever.

Northcott drilled a 5-foot (1.5-meter) shaft 105 feet behind his house through the ceiling of the cavern. He installed a 42-inch (1.07-meter) fan powered by a 5-horsepower (HP) electric motor in the shaft to pull 8,000 cubic feet (2,265 cubic meters) of air per minute (cfm) through the shaft. A shed over the shaft was connected to the house through a duct system. The duct was divided into two chambers, one above the other. The upper chamber carried air from the cavern and was heated by the sun. This chamber, supplemented by steam heat coils when necessary, provided air to warm the house on cool days. The lower chamber carried cool air from the cavern to cool the house on warm days.

The humidity (moisture content) of the air from the cavern was controlled in a chamber in which air from both ducts was mixed. Because warm air contains a greater amount of moisture than cool air, Northcott was able to control humidity to about 50 percent. Conditioned air from the mixing chamber of the air system was directed to all of the rooms in his house through a network of smaller ducts, Figure 1–2. During the winter season auxiliary heat was provided by steam coils located in the base of each of the branch ducts.

Each year more than 500,000 people visit the Caverns of Luray, where the temperature is a constant 54°F (12.22°C) to 56°F (13.33°C) at 87 percent relative humidity and the air is always free of dust and pollen.

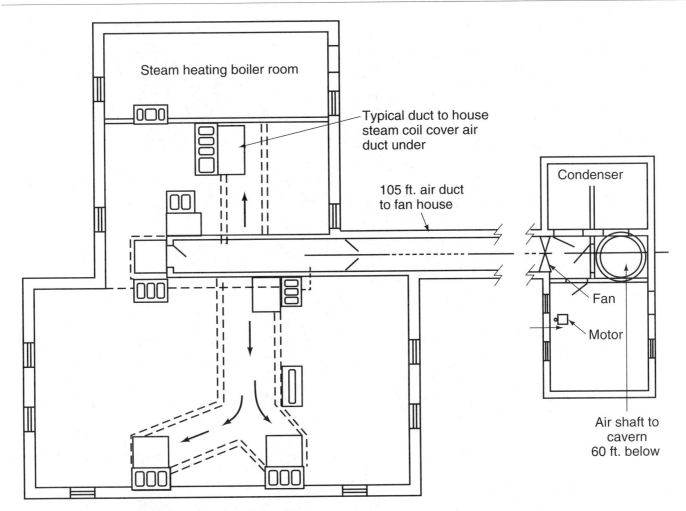

FIGURE 1–2 Duct plan in T. C. Northcott's basement.

Domestic Refrigeration

Domestic refrigeration systems first appeared in 1910, even though ice had been artificially made since about 1820. In 1896, the Sears, Roebuck and Company catalog offered several *refrigerators* for sale, Figure 1–3. Refrigeration was provided by 25 pounds of ice and was useful only for the short-term storage of foods. The selling price for this refrigerator was $5.65.

The first patented cooling system, which was invented in 1831 by Jacob Perkins, was installed in the British Parliament in 1837. It became known as the first building in the world to be cooled. In 1848 the Broadway Theater in New York City announced, "The public is respectfully informed that an Extensive Apparatus for the Perfect **Ventilation** of the Entire Building is now in operation." This machinery was patented by J. E. Coffee.

Touted as the "father of modern day air conditioning," Dr. John Gorrie invented an ice-making machine in 1851 to lower the temperature of his patients who were suffering from yellow fever. A repli-

ca of his ice machine, Figure 1–4, is on display in the Gorrie Museum in Apalachicola, Florida.

In 1868 William Davis, a Detroit fish dealer, patented his "ice box on wheels." He designed the first refrigerated railroad car the following year, but, G. F. Swift was given credit as the pioneer developer in 1870. At the same time G. H. Hammond invented a system whereby air was forced over ice then circulated through the car; he was also a pioneer of the system.

In 1902 Willis H. Carrier was issued a patent for a "scientific system to clean, circulate, and control air temperature and humidity in buildings." His 1906 patent was for "passing hot soggy air through a fine spray of water, condensing moisture on the droplets, leaving drier air behind."

The first manually operated refrigerator was produced by J. Larsen in 1913, and five years later the Kelvinator Company produced the first automatic refrigerator. The acceptance of this new refrigerator, however, was very slow. By 1920, only about 200 had been sold.

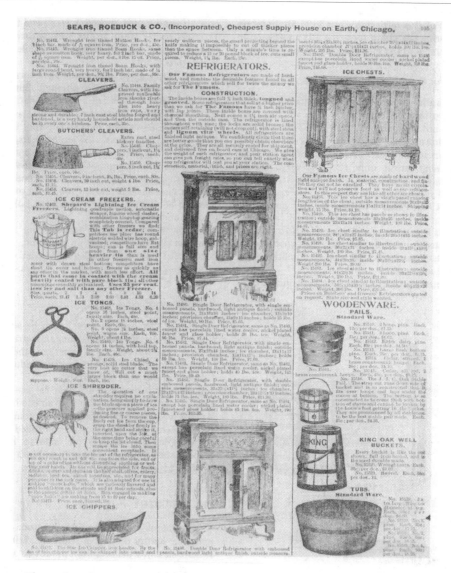

FIGURE 1–3 The early refrigerator was actually an icebox (1896). *(Courtesy of Sears, Roebuck and Co.)*

In 1926, the first hermetic (sealed) refrigerator was introduced by General Electric. The following year, Electrolux introduced an automatic absorption unit. In 1929 Frigidaire introduced the first room air conditioner. A 4-cubic-foot (ft.^{3}) refrigerator was introduced by Sears, Roebuck and Company in 1931. The refrigerator cabinet, Figure 1–5, and the refrigeration unit were shipped separately and required assembly.

In 1931 the Baltimore and Ohio (B&O) Railway introduced the first air-conditioned passenger train and, in 1936, United Airlines introduced the first air-conditioned airplane, all paving the way for mobile air conditioning.

Mobile Air Conditioning

Rugged construction, rigid suspension, and skinny tires produced an uncomfortable ride in the early automobile.

To add to this discomfort, the driver and passengers had to bundle up to keep warm in the winter season. In the summer months the only air conditioning was provided by the breeze produced while driving at a top speed of about 15 mph (24 km/h). "Fresh air" vents placed in the floor brought in more dirt and dust than cool air. In 1884, William Whiteley provided blocks of ice in an enclosure under horse-drawn carriages. "Cool" air was blown inside the carriage by a fan attached to an axle. A bucket of ice placed near the floor vent was the equivalent method of cooling a horseless carriage.

Advertised as an option in some cars in 1927, "air conditioning" meant that the vehicle could be supplied with a heater, a ventilation system, and a method of filtering the air. In 1938 Nash introduced "air conditioning, heating, and ventilation." In this system, fresh air was heated, filtered, and circulated around inside the vehicle by an electric fan.

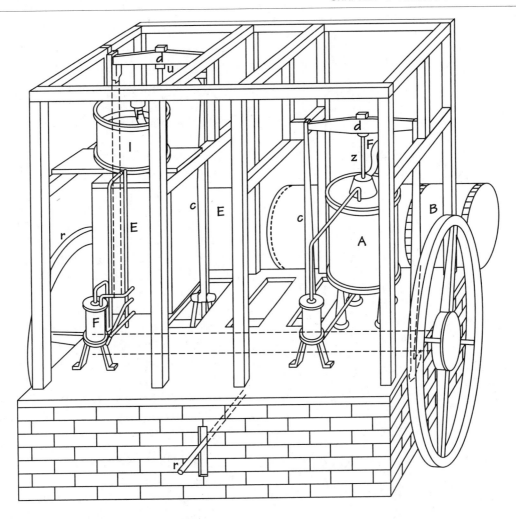

FIGURE 1–4 Gorrie's ice machine.

A few passenger buses that were air conditioned in 1938 used a converted commercial air-conditioning system with a belt-driven compressor. The first passenger car with a mechanical refrigeration system was manufactured by Packard in 1939 and was offered to the general public the following year, Figure 1–6. The evaporator of this system was a converted commercial refrigeration unit that took up the entire trunk space. The belt-driven compressor and condenser were located under the hood. The only control was a switch to turn the blower on and off. Packard advertised, "Forget the heat this summer in the only air conditioned car in the world."

By 1940 heaters and defrosters were standard equipment on many models. **Aftermarket** heaters were available for most any application. An evaporative cooling system that worked on the principle that air passing over water has a temperature reducing effect was also available. This system was called the "Weather Eye" by Nash and is still available for vans and recreational vehicles (RVs). It is most effective for use in dry areas.

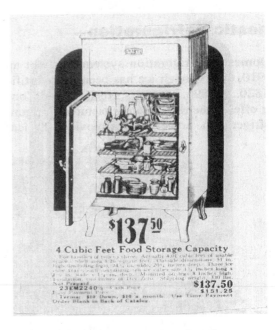

FIGURE 1–5 Early refrigerator (1931).
(Courtesy of Sears, Roebuck and Co.)

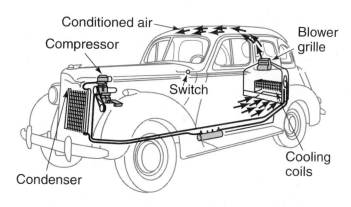

FIGURE 1–6 1939 Packard with air conditioner.

In 1941, 300 Cadillacs were manufactured with an air-conditioning system. Chrysler also produced some of their cars that year with an air-conditioning system. But neither the Cadillac nor the Chrysler systems, similar to the Packard system, had a compressor clutch. This meant that the compressor would run whenever the engine was running. To stop the compressor one had to remove the belt. It was then necessary to reinstall the belt when one wanted air conditioning.

Accurate records of sales were not kept in the early days of automotive air conditioning. However, it is known that before World War II between 3,000 and 4,000 units were installed in Packards. Defense priorities for materials and manufacturing prevented the improvement of automotive air conditioning until the early 1950s. At that time, the demand for air-conditioned vehicles began in the Southwest.

In 1948 the Automotive Refrigerated Air-Conditioning (ARA) Company was first to offer aftermarket automotive air-conditioning systems. By the mid-1950s there were more than fifteen companies offering aftermarket air-conditioning systems in kit form. One could be purchased virtually anywhere, such as Montgomery Wards, Sears, Roebuck and Company, automotive specialty stores, independent garages, and gasoline stations. The early aftermarket systems, which were installed in about a day, had truck-mounted evaporators with under-dash-mounted controls, eventually changing to under-dash-mounted evaporators and controls.

After World War II Cadillac offered a new "high-tech" feature—handy controls for the factory-installed air-conditioning system. Originally, the controls were located on the rear package shelf and all one had to do was climb into the backseat to turn the air-conditioning system on and off. In 1940 Nash offered the *Weather Eye*, a refrigerated air-conditioning system with all components mounted under the hood and cowl area.

In 1955 Chevrolet was the first to move everything under the hood and on the dash. That same year Nash was first to offer an electric clutch. Also, in 1955 Studebaker added a means of temperature control, and Pontiac introduced a split duct system that allowed outside air to pass over the evaporator, thereby providing better control of air quality and temperature. ARA introduced the first aftermarket air-conditioning systems designed specifically for pickup trucks.

The first of today's modern automotive air-conditioning systems was introduced by Cadillac in 1960. Their two-level system could cool the top level of the vehicle interior while heating the lower level, thus providing a method of controlling humidity.

The first fleet installation occurred in 1967 when all of the state police cars on Florida's Turnpike (now Ronald Reagan Turnpike) were air conditioned. Large firms reported increased productivity and sales after air-conditioning systems were installed in the cars of their salespeople. Truck lines realized larger profits because drivers who have air-conditioned cabs average more mileage than those who do not.

The Option Craze

In the late 1970s and early 1980s, people began moving to warmer states at about the same time that the "option craze" hit the automotive industry. The demand for cars with all of the "bells and whistles" meant extra profit for automobile dealers. Cars ordered for their floor plans were loaded with options. Faced with the prospect of several weeks' delivery, however, customers would decide on an "in stock" car for immediate delivery. This car generally included an air-conditioning system—one of the most expensive options. Soon car and light truck owners began to like this option, and as air-conditioning systems got better, customers insisted that their next car also be air conditioned.

Today's Vehicle

Today's vehicles have very efficient and reliable heating, ventilating, and air-conditioning systems. Modern automatic temperature control (**ATC**) air-conditioning systems are dependable, and computers help ensure that both the passenger and driver are comfortable. Individual controls ensure the comfort zone of each occupant. Future air-conditioning systems will feature more electronics and better compressor designs. New refrigerants will ensure a safe and healthful environment.

Mobile air conditioning is not only found in cars, trucks, and buses. Mobile air-conditioning application has been expanded for use in such farm equipment as tractors, harvesters, and thrashers. Additionally, mobile air-conditioning systems have been developed for use in other off-the-road equipment such as backhoes, bulldozers, and graders. Actually, mobile air conditioning may be found in almost any kind of domestic, farm, or commercial equipment that has an enclosed cab and requires an onboard operator.

We have come a long way since that 15 mph breeze or Whiteley's block of ice. With a simple push of a button conveniently located on the dash, one can now select warm or cool air.

Yesterday and Today

From this brief history it can be seen that although some forms of air-conditioning and refrigeration systems were used thousands of years ago, the refrigeration industry did not experience rapid growth until the late twentieth century. One must wonder what modern technology will bring in the twenty-first century.

Atomic submarines are able to remain submerged indefinitely due, in part, to air conditioning. Modern medicine and delicate machine components are developed and perfected in an atmosphere that is scientifically humidity and temperature controlled. Computer centers are able to function properly because they are kept within a specific range of temperature and humidity levels.

THE INDUSTRY

Today, mobile air conditioning is not a luxury—it has become a necessity. Millions enjoy the benefits it produces.

Business people are able to drive to appointments in comfort and arrive fresh and alert. People with allergies are able to travel without the fear of coming into contact with excessive dust and airborne pollen and pollution. Because of the extensive use of the automobile, air conditioning is playing an important role in promoting the comfort, health, and safety of travelers throughout the world. Over 90 percent of all vehicles manufactured in the United States today are equipped with an air-conditioning system.

The number of cars, trucks, and RVs equipped with air-conditioning systems has increased rapidly during the past 30 or so years. Although expected to level off in the United States, the numbers are rapidly growing in Europe and Asia.

It is easy to understand how air conditioning has become the industry's most sought-after product. In the South and Southwest, many specialty auto repair shops base their entire trade on selling, installing, and servicing vehicle air conditioners throughout the year. Many of these specialty shops offer cooling system service as well.

SOURCES OF HEAT AND COLD

Most of the conditions that the heating, ventilation, and air-conditioning (**HVAC**) system must regulate are generated by ambient air temperature conditions around the vehicle. There are some other environmental effects, however, that must be considered. Solar radiation, for example, heats a vehicle interior regardless of what the outside ambient air temperature may be. Other factors, Figure 1–7, include the heat generated by the engine, transmission, exhaust system, road heat, and occupant body heat.

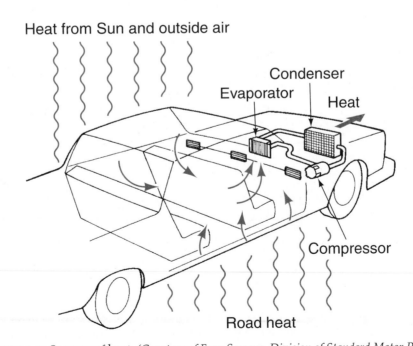

FIGURE 1–7 Sources of heat. *(Courtesy of Four Seasons, Division of Standard Motor Products)*

Solar Radiation

The interior temperature of a vehicle in direct sunlight will rapidly rise to a temperature that is higher than the ambient air temperature. The average temperature of a typical vehicle parked in direct sun with an ambient air temperature of 105°F (40.6°C) is about 205°F (96°C), Figure 1–8.

This heat is caused by solar radiation in the form of ultraviolet (UV) light, which easily penetrates the glass windows and is absorbed into the interior surfaces such as the dashboard and upholstery, thereby raising their temperatures.

These hot surfaces then radiate this heat energy into the vehicle interior as infrared (IR) radiation. Unlike ultraviolet (UV) radiation, infrared (IR) radiation cannot penetrate the glass to leave the vehicle interior and is therefore trapped inside.

This is the same principle used to warm a greenhouse, so this condition is referred to as the **greenhouse effect.** If the vehicle interior is ventilated or air conditioned, much of this hot air can be moved outside the vehicle to provide comfort for the occupants. The circulation of fresh or conditioned air over the interior surfaces helps to cool them.

Weather

Today's drivers expect their heating, ventilating and air-conditioning (HVAC) system to provide comfort during extremes of ambient air temperatures, which may vary from –60°F (–51.1°C) or below to 130°F (54.4°C) or above.

Heat radiating from a hot road surface into the vehicle interior must be considered. On a hot summer day an asphalt road surface temperature can reach 140°F (60°C) or more. Some of this road surface heat is absorbed through the floorboard of the vehicle, adding to the heat load that the air-conditioning system must remove to cool the vehicle.

Conversely, in winter months, ice and snow caked to the underside of a vehicle increase the heat required to warm the vehicle interior. The HVAC system also prevents interior and exterior condensation on the windows and helps remove freezing rain and snow.

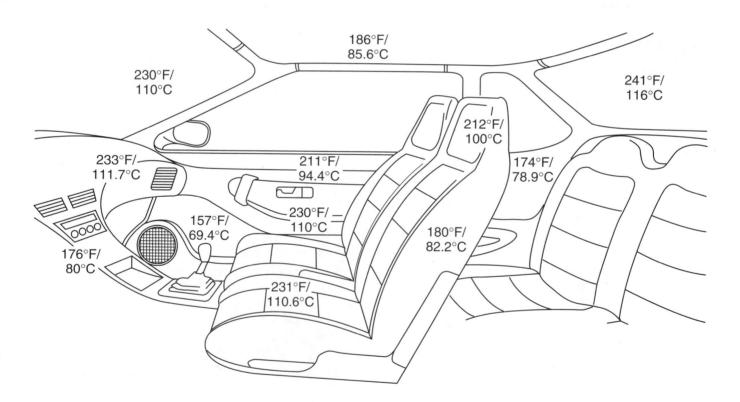

FIGURE 1–8 With all of the windows closed, the average temperature inside a typical car parked in direct sun at an ambient temperature of 105°F (40.6°C) was 205°F (96°C), almost two times that of the ambient air temperature. Note that temperatures exceeding the boiling point of water (H_2O) were recorded in several areas of the car. (*Courtesy Curtis Dwiggins*)

Engine Heat

A major source of heat generated in the vehicle interior comes from the engine. About two-thirds of all heat generated by the engine is given off into the ambient airstream outside the vehicle. Some of the heat from the engine, transmission, mufflers, and catalytic converter, however, warms the vehicle interior through the floorboard.

The Occupants

Another source of vehicle interior heat is the occupants—the driver and passengers. Body heat is constantly being transferred to the air from the surface of the skin as well as from exhaled air warmed in the lungs.

Humidity also rises inside a vehicle interior as the occupants exhale air moistened in their lungs. On rainy or cold days this moisture tends to condense as a mist on interior glass surfaces. The HVAC system removes excess humidity by replacing moisture-laden air with drier outside ambient air or by dehumidifying it.

Another function of an HVAC system is to replace or refresh the air inside a vehicle by changing the air once or twice each minute. Fresh air drawn in from out-side and generally heated or cooled to the selected temperature expels air that has been contaminated with odors such as tobacco smoke.

The ventilating system also improves heater and air-conditioner performance by maintaining air currents inside the vehicle interior to ensure that all areas receive heated, cooled, or fresh air, Figure 1–9.

THE PURPOSE OF AUTOMOTIVE HVAC SYSTEMS

There are three major functions of the heating, ventilating, and air-conditioning (HVAC) system in the modern vehicle:

❏ Regulates Temperature. The HVAC system warms or cools the vehicle passenger compartment air temperature as selected by adding or removing heat as required.

❏ Controls Humidity. The HVAC system prevents humidity from rising inside the passenger compartment, thereby preventing moisture from condensing on the windows. Dehumidification helps keep the air inside the vehicle drier and more comfortable for the occupants.

❏ Circulates the Air. The HVAC system circulates the air and replaces stale air while at the same time, maintaining a constant preselected interior temperature.

COST OF OPERATION

Many believe that the ever-increasing cost of fossil fuel will put an early end to the luxury of automotive air conditioning. The air-conditioning system does place an extra load on the engin, and because any engine load requires fuel it seems apparent that the use of an air conditioner will reduce gasoline mileage. This is true, but only for stop-and-go driving.

At highway speeds, air-conditioned vehicles with their windows closed and the air conditioning operating actually average 2 percent to 3 percent better gasoline mileage than do vehicles with their windows down. The aerodynamic design considerations of the vehicle body are based on having the windows closed. So it seems that reduced wind resistance, when the windows are closed, offsets the load demand of the air-conditioning system on the engine.

A few years ago, **EXXON** published in its *Happy Motoring News* (Vol. 19), "At 40 miles per hour (mph), or faster, you'll use more gasoline by driving with your windows open than you will by operating an air conditioner with all the windows closed. Open windows cause that much 'drag.'"

The number of annual new vehicle sales has decreased slightly in recent years. However, the number

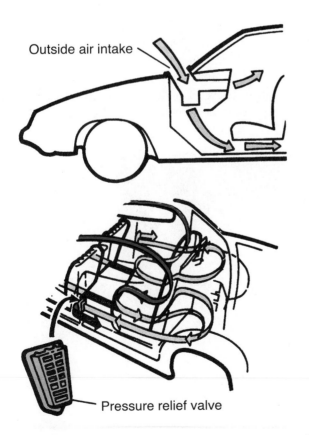

Outside air intake

Pressure relief valve

FIGURE 1–9 Flow-through ventilation system. *(Courtesy of General Motors Corporation—Service Operations)*

of new vehicles sold with air-conditioning systems has increased by about the same percentage. Owners are keeping their vehicles longer and in better repair, which means that aftermarket sales, service, and repairs on mobile air conditioners have increased as well.

THE AUTOMOBILE AND THE ENVIRONMENT

Environmental concerns have resulted in some very strict federal regulations relating to the mobile heating and cooling system industry. One concern of significant importance to the automotive air-conditioning technician is the depletion of the ozone (O_3) layer.

There is significant evidence that ozone (O_3) depletion is partially brought on by the release of certain man-made chemicals into the atmosphere. Unfortunately, it has been determined that the biggest offender is in a group of chemicals known as **CFCs.**

CFC identifies a family of "Group I" refrigerant chemicals known as chlorofluorocarbons. These chemicals contain a mixture of chlorine (Cl), fluorine (F), and carbon (C). One of the CFCs, dichlorodifluoromethane (Refrigerant-12), has always, with a few exceptions in the "early days," been used as the refrigerant for automotive air conditioning.

The biggest offender of CFCs is, by far, Refrigerant-12 (R-12 or CFC-12), Figure 1–10. Worldwide use of R-12 for automotive applications was on the order of 20 percent of production. In the United States, however, over 34 percent of R-12 consumption was for automotive use. Accordingly, the automotive air-conditioning industry was the first to be regulated.

More information is provided about the ozone and its destruction in Chapter 2 under the heading, "Ozone and the Environment."

A "NEW" REFRIGERANT

Virgin CFC-12 is no longer manufactured in the United States, and it is illegal to import CFC manufactured in any other country. The air-conditioning and refrigeration industry will have to rely on recovered and recycled CFC-12 or retrofit present air-conditioning systems in order to use a "new" refrigerant.

The new refrigerant to replace CFC-12, chosen by the automotive industry, is Refrigerant-134a (R-134a, or **HFC**-134a), Figure 1–11. It is from an environmentally friendly group of refrigerants known as HFCs. HFC identifies the hydrofluorocarbon group of refrigerants that contains hydrogen (H), fluorine (F), and carbon (C). The HFC group of refrigerants does not contain chlorine (Cl).

Several other alternative refrigerants are on the market, some suggested as a **"drop-in"** replacement for R-12. Although R-134a is not a drop-in substitute for R-12, it is the *only* refrigerant that is recommended by the automotive industry as a suitable replacement for CFC-12. There are certain system modifications that *must* be made, however, before replacing R-12 with R-134a.

Although there are refrigerant manufacturers who claim to have a direct replacement **(drop in)** refrigerant, all automobile manufacturers warn that these alternate refrigerants, though approved by the Environmental Protection Agency **(EPA),** are not approved for automotive use. Some refrigerants on the market are not

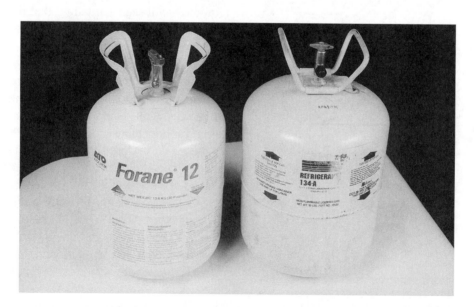

FIGURE 1–10 Refrigerant cylinders are designated for a particular type of refrigerant: 30 pounds (12.08 kilograms) cylinder CFC-12 (A) and 30 pounds (12.08 kilograms) cylinder HFC-134a (B).

FIGURE 1–11 Disposable cylinder. (From Whitman-Johnson, *Refrigeration and Air Conditioning Technology by Delmar Publishers*)

approved by the EPA and are considered hazardous; it is a violation of federal as well as state laws to use them in mobile applications.

Before using any other refrigerant, one must ensure that it is approved for use. Check with the vehicle manufacturer for approval because failure to do so may void certain vehicle warranties. More information is provided about alternate refrigerants and their use in Chapter 6.

TECHNICIAN CERTIFICATION

There are two types of **certification** available to the air-conditioning service technician. One type of certification is in the proper use and handling of refrigerant, which is required by the EPA for any technician who services mobile air-conditioning systems.

The other type of certification is by voluntary testing available through the National Institute of Automotive Service Excellence (ASE) so that technicians may demonstrate their to employers, peers, and customers their proficiency and exceptional knowledge in air-conditioning system troubleshooting and repair service.

EPA Certification

Section 609 of the **Clean Air Act (CAA)** Amendments of 1990 establishes standards specifically for the service of motor vehicle air conditioners **(MVACs).** Technicians who repair or service MVACs must be trained and certified by an EPA-approved program. These programs are specifically designed to cover MVAC recycling equipment in accordance with Society of Automobile Engineers (SAE) Standards and Section 609 regulatory requirements. After completing a required training program, MVAC technicians must pass a test to become certified, Figure 1–12. These tests are different from the Section 608 certification tests, Figure 1–13. Procedures involving MVACs that are not covered by Section 609, such as the disposal of MVACs, are covered by Section 608.

Under Section 608 the EPA has established four types of certification. Unlike Section 609, all training and review classes for Section 608 are voluntary. Passing the appropriate test is, however, mandatory. The four types of certification under Section 608 are:

FIGURE 1–12 Typical wallet card for 609 certification.

FIGURE 1–13 Typical wallet card for 608 certification.

❏   Type I   Small appliances
❏   Type II   High-pressure appliances, except small
             appliances and MVACs
❏   Type III   Low-pressure appliances
❏   Type IV   Universal—All appliances except
             MVACs

In addition, technicians who service or repair **MVAC-like appliances,** such as farm equipment and other off-road vehicles, may choose to be certified under either the Section 609 program or Section 608, Type II, program. Because of the similarities between MVAC and MVAC-like appliances, however, the EPA recommends that technicians servicing MVAC-like appliances consider certification under Section 609.

It should be noted that although buses using CFC-12 are MVACs, buses using HCFC-22 are neither MVACs nor MVAC-like appliances. They are considered high-pressure equipment and are covered under Type II of the Section 608 certification test.

ASE Certification

ASE's voluntary certification program is a vehicle for technicians to demonstrate their abilities to themselves as well as to their employers and their customers. By passing the ASE test one earns the most valuable credential available to the automotive technician. Having this credential that is recognized throughout the nation proves to yourself and others that you are among the elite in your profession.

There are currently more than 440,000 professionals certified as ASE technicians. They work in every segment of the automotive service industry, from car and truck dealerships to independent garages, fleets, service stations, and independent franchises.

Certified technicians promote customer trust and improve the image of the automotive repair industry. Trust and professionalism are the first steps to a better and more prosperous future. Certification is a valuable yardstick by which to measure the knowledge and skills of individual technicians as well as the commitment to quality of the repair facility employing ASE-certified technicians. ASE-certified technicians usually wear a blue and white ASE shoulder insignia, Figure 1–14. Employers often display their technicians' credentials, Figure 1–15, in the customer waiting area. Customers look for facilities that display the ASE Blue Seal of Excellence logo on outdoor signs, in the customer waiting area, in the Yellow Pages, and in other advertisements.

The Heating and Air Conditioning test is one of eight tests in ASE's Automotive Technician Certification series. You can become certified as an ASE Automobile Technician by passing one or more tests and having two years or more work experience in automotive repair.

The test questions are written by service industry experts familiar with all aspects of automotive heating and air-conditioning repair. They are entirely job-related and test the skills that you need to know to properly service today's domestic and import vehicles. The tests are administered by American College Testing (ACT) and stress real-world diagnostic and repair problems, not theory. All questions are pretested and quality checked on a national sample of technicians. They must meet ASE's standards of quality and accuracy to be included in the test.

To become familiar with ASE's test question format, many of the unit end questions in this textbook are written in ASE's style. The types of knowledge and skills needed to pass the test are:

❏   Basic Technical Knowledge. To test your knowledge of what is in a heating and air-conditioning system and how the system works, as well as the proper procedures and precautions to be followed in making repairs and adjustments.
❏   Correction or Repair Knowledge Skill. To test your ability and understanding to apply generally accepted repair procedures and precautions in assembly, disassembly, and reconditioning operations as well as your ability to use shop manuals and specialized tools.
❏   Testing and Diagnostic Knowledge and Skill. To test your ability to recognize repair problems and to use measurement and testing equipment to diagnose the condition as well as your ability to trace the effects of a condition to determine the cause.

FIGURE 1–14 ASE certification shoulder patches worn by (left) automotive technicians and (right) master automotive technicians.

FIGURE 1–15 A typical ASE certification.

TRADE ASSOCIATIONS

There are two trade associations dedicated exclusively to the mobile air-conditioning industry. They are the International Mobile Air Conditioning Association (IMACA) and the Mobile Air Conditioning Society (MACS). Both provide members with pertinent information about trends, problems, services, legislation, and specifications relating to air-conditioning system service.

IMACA

Founded in 1958 as the Automotive Air Conditioning Association (AACA) the name was changed in 1970 to the International Mobile Air Conditioning Association (IMACA) Figure 1-16. IMACA is a not-for-profit trade association dedicated to serving the needs and interests of the worldwide mobile air-conditioning industry.

With main offices in Fort Worth, Texas, IMACA's services and programs are made possible through the support of its members, including system manufacturers and suppliers, air-conditioning service facilities, and educators from over thirty countries.

MACS

With main offices in Lansdale, Pennsylvania, the Mobile Air Conditioning Society Worldwide (MACS) was founded in 1981. MACS, Figure 1–17, is a leading forum for a specialized segment of the automotive aftermarket whose purpose is to fill the industry needs for comprehensive technical information, training, and

communication. MACS is a full-service society that strives to provide its members with valuable programs to help them in their businesses. Governed by a volunteer board of directors elected by the membership,

IMACA
International Mobile Air Conditioning Association

A not-for profit trade association serving the Mobile Air Conditioning Industry and the Motoring Public since 1958

FIGURE 1–16 The IMACA logo.

FIGURE 1–17 MACS worldwide logo. (*Courtesy of Mobile Air Conditionaing Society Worldwide*)

MACS represents more than 1,600 members, including service shops, installers, distributors, component suppliers, and manufacturers in the United States, Canada, and countries around the world.

THE SERVICE TECHNICIAN

How does all of this affect the student of mobile air conditioning? As the popularity of air conditioning in vehicles increases, it is obvious that the need for installation, maintenance, and service technicians will also increase. Many shops that just a few years ago added air-conditioning service as a sideline now find it to be their primary business.

The air-conditioning technician must have a thorough working knowledge and understanding of the operation and function of the mechanical and electrical circuits and controls of the automotive air conditioner. A good knowledge of the equipment, special tools, techniques, and skills of the trade is also essential.

Federal and state laws mandate that all mobile air-conditioning systems be serviced in licensed shops by certified technicians using EPA-approved equipment. Class I refrigerants can no longer be purchased "over-the-counter" in "pound" cans. They cannot be purchased in any size container unless the purchaser is a certified technician and approved for such purchase. This essentially means that the "do-it-yourselfer" or "shade tree mechanic" can no longer service automotive air-conditioning systems.

Air conditioning has made it possible for the Space Age to become part of the twentieth century. What was fiction at the turn of the century is now commonplace. The service technician's contribution to the industry may help make today's fiction commonplace in the twenty-first century.

REVIEW

Select the correct answer from the choices given.

1. How did moving the stones of Pharaoh's palace into the desert help to keep the palace cool?
   a. The palace was given a chance to air out.
   b. The stones gave up heat during the day.
   c. The stones gave up heat during the night.
   d. The stones could be easily rotated for reassembly.

2. The first advertised air conditioning for a car was:
   a. in 1940 by Packard.
   b. in 1927 consisting of a heater, a ventilation system, and a filter.
   c. in 1941 by Chrysler.
   d. after the war in the 1950s.

3. What is perhaps the greatest technical accomplishment that was made possible, at least in part, by air conditioning?
   a. The space program
   b. Modern medicine
   c. Both A and B
   d. Neither A nor B

4. What percentage of the total domestic car production is expected to be equipped with an air-conditioning system this year?
   a. About 80 percent
   b. Between 70 and 80 percent
   c. 70 percent or less
   d. 90 percent or more

5. Generally, at a speed of 40 mph or faster, less fuel will be used with the air conditioner:
   a. off and the windows open.
   b. on and the windows closed.
   c. off and the windows closed.
   d. on and the windows open.

6. What under hood part(s) become(s) hot when the engine is running?
   a. Radiator
   b. Exhaust manifold
   c. Both A and B
   d. Neither A nor B

7. Technician A says one must be certified under Section 609 to service MVAC-like appliances. Technician B says one must be certified under Section 608 to service MVACs. Who is right?
   a. A only
   b. B only
   c. Both A and B
   d. Neither A nor B

8. The interior temperature of a vehicle parked in the sun at 105°F (40.6°C) will be:
   a. less than ambient temperature.
   b. the same as ambient temperature.
   c. either A or B.
   d. neither A nor B.

9. All of the following add to the heat load, EXCEPT:
   a. (UV) radiation.
   b. (IR) radiation.
   c. moisture.
   d. ventilation.

10. Technician A says that a MVAC system provides heated and/or cooled air. Technician B says that a HVAC system circulated fresh and/or conditioned air. Who is right?
    a. A only
    b. B only
    c. Both A and B
    d. Neither A nor B

11. What federal agency regulates the sale of refrigerant?
    a. EPA
    b. UL
    c. OSHA
    d. SAE

12. What is the refrigerant most recommended to replace CFC-12?
    a. HCFC-134a
    b. HFC-134a
    c. CFC-134a
    d. Any approved drop-in refrigerant

13. In what year did domestic refrigeration first appear?
    a. 1900
    b. 1905
    c. 1910
    d. 1915

14. What is the definition of the term "humidity"?
    a. Dampness found in the Caverns of Luray
    b. Moisture content of the air
    c. Both A and B
    d. Neither A nor B

15. One had to climb into the backseat to turn on the air conditioner in a:
    a. Packard.
    b. Cadillac.
    c. Nash.
    d. Studebaker.

16. One had to stop and remove a belt to turn off an air-conditioning system in:
    a. a Packard
    b. a Cadillac
    c. both A and B.
    d. neither A nor B

17. Cooling by evaporation was first offered by:
    a. Packard
    b. Cadillac
    c. Nash.
    d. Whiteley.

18. What methods of air conditioning are still available today?
    a. Mechanical
    b. Evaporative
    c. Both A and B
    d. Neither A nor B

19. Who advertised "Forget the heat this summer in the only air conditioned car in the world"?
    a. Chrysler
    b. Nash
    c. Cadillac
    d. Packard

20. What promoted the sales of automotive air-conditioning systems in the United States?
    a. The vehicle "options craze"
    b. Sales of "floor plan" vehicles
    c. Both A and B
    d. Neither A nor B

TERMS

Write a brief description of the following terms:

1. aftermarket
2. air conditioning
3. ATC
4. CAA
5. certification
6. CFC
7. Clean Air Act
8. drop in
9. environment
10. EPA
11. greenhouse effect
12. HFC
13. humidity
14. HVAC
15. MVAC
16. MVAC-like appliance
17. option
18. refrigeration
19. temperature
20. ventilation

CHAPTER 2

HEALTH AND SAFETY

OBJECTIVES

On completion and review of this chapter, you should be able to:

❑ Recognize many of the hazards associated with the automotive repair industry.
❑ Identify many of the hazardous conditions that may be found in an automotive repair facility.
❑ Explain the need for a health and safety program.
❑ Discuss the philosophy regarding health and safety.
❑ Explain the importance of the ozone layer.
❑ Discuss what is being done about the ozone depletion problem.
❑ Describe how ozone is created and destroyed.
❑ Compare and identify unsafe and safe tools.
❑ Understand the limitations by design of hand tools.

PERSONAL SAFETY

Technicians working in the automobile repair industry may be exposed to a wide variety of gases, dusts, vapors, mists, fumes, and noise, as well as ionizing or nonionizing **radiation.** An automotive air-conditioning technician may not work directly with these materials, but must be aware of all potential hazards that may exist in the facility. Some of the most common hazards include asbestos, **carbon monoxide,** caustics, solvents, paints, glues, radiation, **refrigerant,** and **oxygen** deficiency.

ASBESTOS

A serious hazard found in an automotive repair facility is the exposure to asbestos fibers such as used in brake linings and clutch plates until recent years. Exposure to asbestos can result in asbestosis and/or lung cancer. Less hazardous materials have replaced most asbestos applications but must still be considered a hazard. Always follow all applicable procedures associated with good hygiene

any time there is a possibility of airborne fibers. A high-efficiency particulate air (HEPA) filter, Figure 2–1, is used to remove asbestos fibers before they become airborne.

> CAUTION
> THOSE WHO SMOKE WHILE WORKING
> WITH ASBESTOS HAVE ALMOST 90 TIMES
> GREATER RISK OF CANCER THAN THOSE
> WHO DO NOT SMOKE.

CARBON MONOXIDE

Carbon monoxide (CO) is emitted from the tailpipe, Figure 2–2, whenever **propane-,** diesel-, or gasoline-powered internal combustion engines are in operation, as well as during some hot work operations, such as welding. A technician's exposure to carbon monoxide (CO) may be excessive when such operations are conducted in low-ceiling or confined areas. A technician should always work in a well-ventilated area to avoid exposure to excessive vapors and fumes.

17

FIGURE 2–1   A HEPA vacuum can remove asbestos without letting it escape into the shop environment. *(Courtesy of White/Pullman-Holt)*

FIGURE 2–2   Avoid hazardous exhaust fumes.

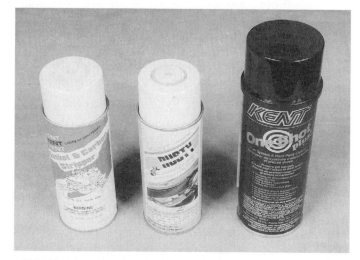

FIGURE 2–3   Many hazardous solvents may be used in the automotive shop.

Caustics

Many caustics, such as acids, solvents, paints, glues, and adhesives, Figure 2–3, are used in the automotive industry. Some of the common organic chemicals may result in dizziness, headaches, and sensations of drunkenness and may affect the eyes and respiratory tract. The use of some chemicals may result in various types of skin irritations and, in extreme cases, dermatitis. The availability and proper use of appropriate protective equipment, such as gloves, goggles or face shields, aprons, and respirators are essential.

Any hazardous materials striking the skin should be washed off immediately. An eyewash fountain, Figure 2–4, or safety shower, Figure 2–5, should be available for an emergency. Adequate ventilation in the work area is essential to avoid excessive exposure to fumes and vapors during operation in confined spaces.

Oxygen-Deficient Atmosphere

The repair of an automotive air-conditioning system in a confined space can be very dangerous. Not only may the technician be exposed to various toxic gases, but the **atmosphere** may also be deficient in oxygen (O), which would immediately be dangerous to life. Other vapors, although not harmful in themselves, displace oxygen (O), which is essential for life.

Radiation

Lasers used for some wheel alignment and other sighting procedures may produce intense non-ionizing radiation. Although to be avoided, this minor radiation is not generally considered harmful. Welding, however, produces ultraviolet **(UV)** light, which is harmful to the eyes and skin. If proper safeguards are not observed, both nonionizing and ionizing radiation can be very harmful.

REFRIGERANTS

The greatest problem that may occur during installation, servicing, and repair of an air-conditioning system is the

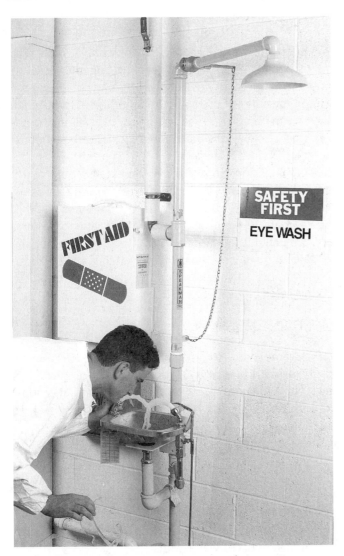

FIGURE 2–4 Typical emergency eyewash fountain.
(Courtesy of DuPont Automotive Finishes)

FIGURE 2–5 Showers should be available for personal safety. *(Courtesy of BET, Inc.)*

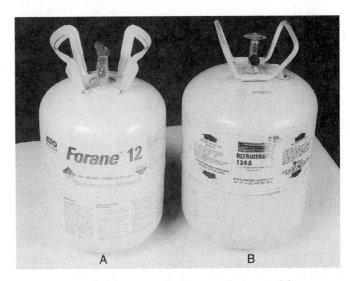

FIGURE 2–6 Refrigerant cylinders are designated for a particular type of refrigerant: 30 pounds (12.08 kilograms) cylinder CFC-12 (A) and 30 pounds (12.08 kilograms) cylinder HFC-134a (B).

unintentional release of refrigerant, Figure 2–6. Refrigerants may be considered in the following classes:

❏ Non**flammable** substances in which the toxicity is slight, such as some fluorinated hydrocarbons, for example, Refrigerant-12. Although considered fairly safe, this refrigerant may decompose into highly toxic gases, such as hydrochloric acid or **chlorine**, on exposure to hot surfaces or open flames.

❏ Toxic and corrosive refrigerants such as ammonia, which is often used in **absorption** refrigerators such as those found in recreational vehicles (RVs). Ammonia may be flammable in concentrations exceeding 3.5 percent by volume. Ammonia is also very irritating to the eyes, skin, and respiratory system.

❏ Highly flammable and/or explosive substances such as propane must be used with strict controls and safety equipment. Although propane is not approved for use as a refrigerant in mobile refrigeration, it is a fuel often found in mobile applications.

If a refrigerant escapes, action should be taken for its removal from the work area. Refrigerant is heavier than air so exhaust from the floor area should be provided. Exhaust from the ceiling should be provided for lighter than air gases.

The use and safe handling of refrigerants is covered in Chapter 6.

ANTIFREEZE

There are basically two types of antifreeze available for motor vehicle service: ethylene glycol (EG) based and propylene glycol (PG) based, Figure 2–7.

Ethylene glycol (EG) based antifreeze is a danger to animal life. Properly handled and installed, however, it presents little or no problems. It can be very dangerous if it is carelessly installed, improperly disposed of or leaks from a vehicle's cooling system.

EG-based antifreeze causes thousands of needless pet deaths in the United States each year. Animals are attracted to EG antifreeze because of its sweet taste. As little as 2 ounces can kill a dog and only one teaspoon is enough to poison a cat. This type of antifreeze is a danger to children as well. As little as 2 tablespoons can be fatal to a small child.

Propylene glycol (PG) based antifreeze is a much safer alternative. Unlike EG antifreeze, PG-based antifreeze is essentially nontoxic and, therefore, safer for animal life, children, and the environment. PG antifreeze is classified as "Generally Recognized as Safe" (GRAS) by the United States Food and Drug Administration (USFDA).

PG-based antifreeze coolant protects against freezing, overheating, and corrosion the same as conventional EG-based antifreeze coolants do. Although considered safe, the safety advantage will be lost if PG-based antifreeze coolants are mixed with EG-based antifreeze coolants.

Keep all antifreeze coolants, new or used, in a sealed container away from animals and out of the reach of children until safely disposed of. The safe handling of antifreeze is covered in Chapter 17.

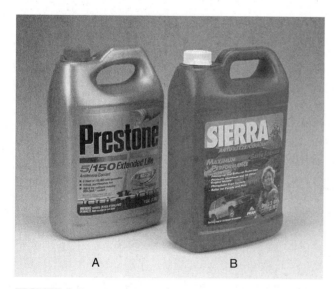

FIGURE 2–7 EG (A) and PG (B) antifreeze.

Welding

Fumes from welding, burning, and soldering actually contain traces of the metals being welded together, such as cadmium (Cd), zinc (Zn), lead (Pb), iron (Fe), or copper (Cu). Fumes may also include traces of the filler material, flux, and the coating on the welding rods. Welding operations may also generate other gases such as carbon monoxide (CO), arsine, and **ozone (O_3)**, at concentrations that may be hazardous to health. When extensive hot work operations, such as welding, particularly in confined areas, are performed, there could be an excessive fume exposure to these materials and adequate ventilation and/or respiratory protection may be required. Eye protection for the welder and for other technicians working in or in the vicinity of welding operations should be provided due to the production of ultraviolet (UV) light.

SAFETY IN THE SHOP

The automotive air-conditioning technician may be involved in all phases of automotive service, including electrical and mechanical repairs relating to air-conditioning malfunctions. The technician may therefore encounter some of the more common occupational safety and health problems as revealed during walk-around inspections of service repair facilities, such as:

❑ Poor housekeeping. Refuse and nonsalvageable materials not being removed; electrical cords and compressed gas lines scattered on floors; and oily, greasy spots or water spills

❑ Poor or nonexistent guard rails and/or toe-boards around open pit areas

❑ Use of unsafe equipment such as damaged creepers or improperly grounded electrical tools, Figure 2–8

❑ The unsafe stacking of stock and other material

❑ Failure to identify "safety" zones

❑ Pulleys, gears, and the "point of operation" of equipment guards missing

❑ Lack of adequate ventilation or acceptable respirator programs for operations in confined spaces

❑ Skin problems or dermatitis due to the handling of resins, cements, oils, and solvents without protection

❑ Electrical hazards, such as "U" ground missing from power tools, ungrounded extension cords, and frayed, damaged, or improper power cords

❑ Unsecured and improper storage of compressed gas cylinders

❑ Fire hazards due to improper storage and use of flammable and combustible materials and the presence of various ignition sources

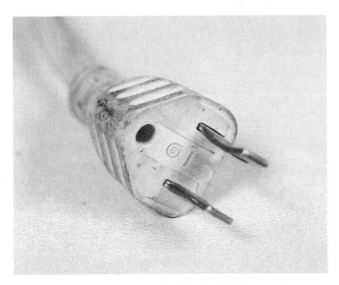

FIGURE 2–8 An unsafe male electrical plug.

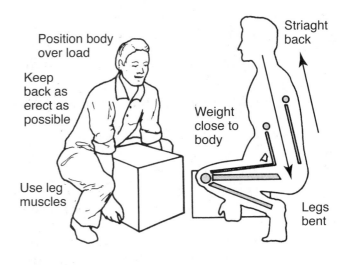

FIGURE 2–9 Lift with the legs, not the back.

❏ Improper lifting and material handling techniques, Figure 2–9

❏ Unsafe work practices, which could result in burns from hot work operations such as welding, burning, and soldering

HEALTH AND SAFETY PROGRAM

Hazardous conditions or practices not covered in the 1970 Occupational Safety and Health Act **(OSHA)** standards are covered under the general duty clause of the act, which states, "Each employer shall furnish to each employee a place of employment which is free from recognized hazards that are causing or are likely to cause death or serious physical harm."

A poster outlining employer safety responsibilities, such as that shown in Figure 2–10, must be posted in a common area where all employees will see it. It must be posted in other languages for all to read and understand, when applicable.

A health and safety program is an effective method to assist in providing for a safe working environment. The purpose of such a program is to recognize, evaluate, and control hazards and potential hazards in the workplace.

In a classroom learning situation the instructor may assign students to safety and health management responsibilities. Regular meetings and informal discussions should be held to discuss safety promotions and actual or potential hazards. The students assigned the responsibility for carrying out the program must be delegated the authority and have the instructor's support.

In the learning as well as work environment, persons may be exposed to excessive levels of a variety of

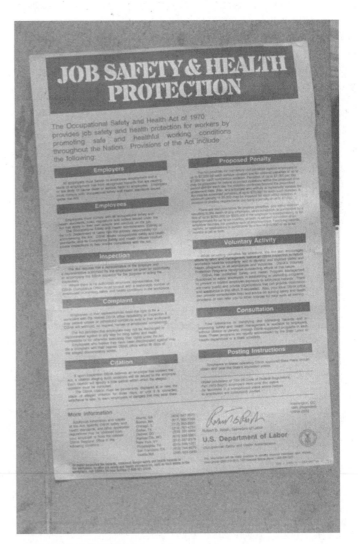

FIGURE 2–10 OSHA poster.

harmful materials. These include gases, dusts, mists, vapors, fumes, certain liquids and solids, and noise, as well as heat or cold.

Of the illnesses reported, respiratory problems due to dusts, fumes, and other toxic agents are the most prevalent. Often health hazards are not recognized because the materials used are identified only by trade names.

A further complication arises from the fact that materials tend to contain mixtures of substances, making identification still more difficult.

To begin identifying occupational health hazards, a materials analysis or product inventory is made and all hazardous substances are listed and evaluated. If the composition of a material cannot be determined, the information should be requested from the manufacturer or supplier, who must provide a material safety data sheet (MSDS), Figure 2–11, for the product. These sheets contain safety information about materials, such as toxicity levels, physical characteristics, protective equipment requirements, emergency procedures, and incompatibilities with other substances.

A process analysis should be performed noting all chemicals used and all products and by-products formed. When doing such an analysis, allied activities such as maintenance and service operations should be included. Examples to watch for are:

- Welding performed around chlorinated materials, such as Refrigerant-12, may cause the formation of toxic gases in addition to welding fumes.
- Exhaust gases vehicles with internal combustion engines contain carbon monoxide (CO) and should be included in the analysis.
- Sometimes poisonous gases, such as chlorine (Cl), are formed when certain cleaning solutions are mixed.

It should be noted that skin conditions such as chemical burns, skin rashes, and dermatitis constitute over half of all occupational health problems. The use of protective creams or lotions, proper personal protective clothing and other protective equipment, and good **personal hygiene** practices can often prevent these problems.

Technicians should wash their hands after every job and before eating. Technicians should not eat around toxic chemicals or in contaminated areas. If chemicals such as caustic epoxies or resins get on the skin, they should be washed off immediately. Clothing should be changed and washed daily if it becomes contaminated with toxic chemicals, dusts, fumes, or liquids.

SAFETY RULES FOR OPERATING POWER TOOLS

The following rules apply to those who use and operate power tools. To ensure safe operation the technicians must:

- Know the application, limitations, and potential hazards of the tool used.
- Select the proper tool for the job.
- Remove chuck keys and wrenches before turning on the power.
- Not use tools with frayed cords or loose or broken switches.
- Keep guards in place and in working order.
- Have electrical grounds in place.
- Maintain working areas free of clutter.
- Keep alert to potential hazards in the working environment, such as damp locations or the presence of highly combustible materials.
- Properly store combustibles.
- Dress properly to prevent loose clothing from getting caught in moving parts.
- Use safety glasses, dust or face shields, Figure 2–12,

MATERIAL SAFETY DATA SHEET

Section 1: Product & Company Identification

Product Name: Clean-R-Carb™ Carburetor Cleaner
Product Number (s): 05079, 05079 P, 05081, 05081-6, 05081C SK, 75081, 85081

Manufactured By: 674-4300	CRC Industries, Inc.	(215)
	885 Louis Drive, Warminster, PA 18974	
424-9300	24-Hour Emergency Information: CHEMTREC	(800)

Section 2: Composition/Information on Ingredients

Component OTHER LIMITS	%	CAS NUMBER	ACGIH TLV	OSHA PEL
Toluene NE	33-43	108-88-3	100 ppm	100 ppm
Methanol NE	26-36	67-56-1	200 ppm	200 ppm
Acetone NE	17-27	67-64-1	750 ppm	750 ppm
Carbon Dioxide NE	< 10	124-38-9	5000 ppm	10000 ppm

Section 3: Hazards Identification

Emergency Overview

Appearance & Odor: A clear, colorless liquid, aromatic odor.

Danger: Extremely Flammable. Vapor Harmful. May be fatal or cause blindness if swallowed. Eye & skin irritant. Contents Under Pressure.

Potential Health Effects:
Inhalation: Irritation, anesthetic or narcotic effects.
Eyes: Irritation
Skin: Irritation
Ingestion: NA

FIGURE 2–11 Typical MSDS. (*Courtesy of CRC Industries*)

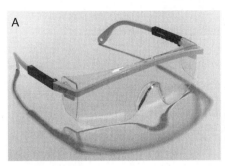
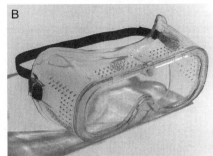
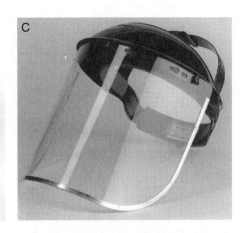

FIGURE 2–12 Eye protection. (A) Safety glasses; (B) Goggles; (C) Face shield. *(Courtesy of Goodson Shop Supplies)*

or other protective clothing and equipment when necessary.

❑ Not surprise or distract anyone using a power tool.
❑ Ensure that all appropriate guards are in place before operating equipment.

GOOD HOUSEKEEPING HELPS PREVENT FIRES

Maintaining a clean and orderly workplace reduces the danger of fires. Rubbish should be disposed of regularly. If it is necessary to store combustible waste materials, a covered metal receptacle is required, Figure 2–13.

Cleaning materials can create hazards. Combustible sweeping compounds such as oil-treated sawdust can be a fire hazard. Floor coatings containing low-flash-point solvents can be dangerous, especially near sources of ignition. All oily mops and rags must be stored in closed metal containers.

WALKING AND WORKING SURFACES

All areas, passageways, storerooms, and maintenance shops must be maintained clean, orderly, sanitary, and as much as possible in a dry condition. Spills should be cleaned up promptly. Floor areas must be kept clear of parts, tools, and other debris. Areas that are constantly wet should have nonslip surfaces where personnel normally walk or work.

Where mechanical handling equipment such as lift trucks are used, sufficient safe clearances must be provided for aisles at loading docks, through doorways, and wherever turns or passage must be made. Low obstructions that could create a hazard are not permitted in the aisles.

All permanent aisles must be easily recognizable. Usually aisles are identified by painting or taping lines on the floors, Figure 2–14.

OZONE AND THE ENVIRONMENT

Since the discovery of a hole in the ozone layer over Antarctica, there has been concern about the consequences for human health and for the environment. "Ozone depletion," together with "global warming" resulting from the "greenhouse effect," have attracted widespread media attention and well-founded concern around the world.

Oily
Rag
Safety
Container

Gasoline
Safety
Container

FIGURE 2–13 Keep combustibles in safety containers.

FIGURE 2–14 Aisles should be identified for safety.

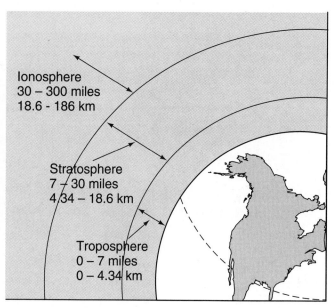

FIGURE 2–15 The atmosphere extends skyward for hundreds of miles.

What is Ozone?

Ozone is described as a gas having a penetrating refreshing odor while providing an exhilarating influence. Ozone is a molecular form of oxygen that has a different chemical property. It is then known as an "allotrope" of oxygen. In large concentrations ozone is considered a poisonous gas. The ozone layer, however, protects life from damaging ultraviolet radiation.

Contrary to the definition, to most of us ozone has a pungent odor, described by many as irritating. In high concentrations it has a pale blue color, in contrast with oxygen, which is colorless, tasteless, and has no odor. Ozone (O_3) is a form of oxygen (O); each molecule of ozone (O_3) contains three atoms of oxygen (O) instead of the normal two atoms of a molecule of oxygen (O_2).

THE ATMOSPHERE

The atmosphere extends skyward for hundreds of miles, Figure 2–15. The lowest part of the atmosphere, up to about 7 miles (11 kilometers), is known as the troposphere. Above that to an altitude of about 30 miles (48 kilometers) is the stratosphere. Above that is the ionosphere.

Ozone in the Atmosphere

Unlike other atmospheric gases, which are concentrated in the troposphere, about 90 percent of the ozone (O_3) occurs in the stratosphere, mainly between 9 and 22 miles (15 and 35 km). Even at its highest concentration ozone (O_3) does not exceed 10 ppm: only 1 ozone (O_3) molecule in every 100,000 molecules. For example, if all the ozone (O_3) in the atmosphere were concentrated at sea level it would form a layer less than 0.125-inch (3-mm) thick. There are about 3,000 million tons of ozone

(O_3) in the atmosphere, which is equivalent to about 1,600 pounds (726 kg) per person on earth. Compared with the total mass of the atmosphere, however, the amount of ozone (O_3) is negligible.

Ozone (O_3) is formed by the action of electrical discharges. More frequently it is formed by the action of ultraviolet radiation on oxygen (O) in the stratosphere. The oxygen (O) atoms in the oxygen (O) molecules split apart, and the separated atoms recombine with other oxygen (O) molecules to form the triatomic ozone (O_3).

Because sunlight is essential for the formation of stratospheric ozone (O_3), it is mainly formed over the equatorial region, where solar radiation is highest. From there it is distributed throughout the stratosphere by the slight global wind circulation. Stratospheric ozone (O_3) levels vary throughout the world, being lowest at the equator and highest toward the poles.

ABSORPTION OF ULTRAVIOLET RADIATION

Incoming radiation from the sun is of various wavelengths, ranging from ultraviolet (UV) through visible light to infrared. Ultraviolet radiation can be very damaging to living organisms, causing sunburn, skin cancer, damage to eyes including cataracts, and premature aging and wrinkling of the skin. UV radiation can also break the food chain by destroying minute organisms such as plankton in the ocean, thereby depriving certain species of their natural food. Plant life and crops can also be devastated by excessive UV radiation.

The damaging forms of UV radiation are absorbed by ozone (O_3) in the atmosphere and do not reach the earth, Figure 2–16. The ozone (O_3) layer acts as a giant sunscreen or umbrella enveloping the earth, protecting life from dangerous UV radiation. Depletion of ozone (O_3) allows more UV radiation to strike the earth and all living organisms.

Another consequence of the absorption of solar energy by ozone (O_3) is that the upper stratosphere is somewhat warmer than at lower altitudes and this helps to regulate the earth's temperature. Stratospheric ozone (O_3) absorbs about 3 percent of incoming solar radiation, thus serving as a heat sink. Loss of ozone (O_3) will decrease the temperature of the stratosphere, which will, in turn, affect the troposphere and, consequently, the weather and climate at the earth's surface.

THE OZONE HOLE

The term *ozone hole* refers to the loss of the blocking effect of ozone (O_3) to UV radiation. With the depletion of the ozone (O_3) barrier, a hole has been created, which allows a much greater amount of ultraviolet radiation to penetrate to the earth. The ozone (O_3) hole is akin to an umbrella with millions of tiny pinholes, each of which allows dangerous UV radiation to pass through to earth.

How Ozone Is Being Destroyed

Ozone (O_3) is both created and destroyed by the action of UV radiation on oxygen (O) molecules. Chlorine **(Cl)** is

the major gas causing the destruction of ozone (O_3); it starts chain reactions in which a single molecule of chlorine (Cl) can destroy a hundred thousand ozone (O_3) molecules over time, Figure 2–17. Such reactions can continue for many years, even a century or more, until the chlorine (Cl) drifts down into the troposphere or is bound into another compound.

The main sources of chlorine (Cl) are chlorofluorocarbons (CFCs), also referred to as Freons and halons. CFCs are artificially man-made chemicals first developed in 1928 and comprised of chlorine (Cl), fluorine (F), carbon (C), and (often) hydrogen (H).

CFCs are considered very stable chemicals and are nonflammable, nonirritating, nonexplosive, noncorrosive, odorless, and relatively low in toxicity. They vaporize at low temperatures, which makes them very suitable for use as a coolant in refrigerators and air con-

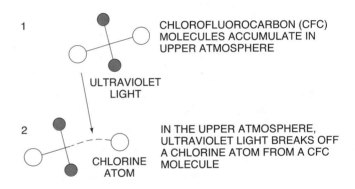

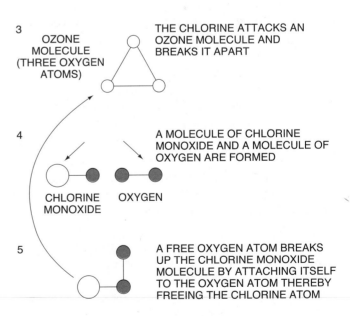

FIGURE 2–17 How CFCs destroy the ozone. (*Courtesy of Thomas Nelson Holdings*)

1 - CFCs released
2 - CFCs rise into ozone layer
3 - UV releases Cl from CFCs
4 - Cl destroys ozone
5 - Depleted ozone -> more UV
6 - More UV -> more skin cancer

FIGURE 2–16 Ozone absorbs UV radiation.

ditioners, as a solvent in cleaning electronics, in blowing bubbles in certain types of foam-blown plastics such as sponge plastic and food packaging, and in cleaning solvents for dry cleaning.

Halons are mainly used in fire extinguishers and are not discussed in this text.

Current worldwide use of CFCs is about 700,000 tons (635,460 metric tons) each year. The scheduled phaseout has caused drastic reductions, however. During the 1960s and 1970s, aerosol use was widespread due to the stable nature and nonflammability of CFCs. Production of CFCs peaked in 1974 and then dropped as their use for aerosols declined.

Although industrialized nations are the major consumers of CFCs, developing nations with larger populations such as China and India have enormous potential to require CFCs for refrigerators and other uses.

The United States' consumption of CFCs on a per capita basis is among the highest in the world, a reflection of our affluence and the popularity and use of air conditioners.

It is not the CFCs, as such, that cause the destruction of the ozone layer but rather the chlorine (Cl) released by the CFCs. The research of British scientists at Halley Bay during the 1980s, together with the international research program in late 1987 in which samples of stratospheric air were obtained by high altitude flights over Antarctica, proved the link between CFCs and ozone (O_3) destruction. A further factor contributing to the loss of ozone (O_3) is the polar stratospheric clouds that form during the Antarctic winter in the very cold stratospheric air. These comprise tiny particles of frozen water (H_2O) vaposr, which condense and form clouds in spring. The clouds act as reservoirs of frozen chlorine (Cl) during winter until thawed in spring, when the chlorine (Cl) is released and begins to react with the ozone (O_3) during the following five to six weeks until the vortex breaks up and the stratosphere becomes less stable. During the late 1980s, the clouds persisted until October.

A chlorine (Cl) atom reacts with an ozone (O_3) molecule, splitting it apart and attaching itself to one of the oxygen (O) atoms to form chlorine monoxide. A free oxygen (O) atom splits the chlorine monoxide molecule to reform a molecule of oxygen (O) and the chlorine atom is free to attack another ozone (O_3) molecule.

The concern is that the current hole and depletion have resulted from CFCs released about twenty years ago. Also, it is noted that CFCs and halons take from six to eight years to rise up through the atmosphere. The chlorine (Cl) used in swimming pools and the chlorine (Cl) bleach used in laundering is unstable and breaks down rapidly without rising into the atmosphere.

OZONE AND THE GREENHOUSE EFFECT

In the public's mind, the loss of ozone (O_3) and the greenhouse effect are similar. In fact they are totally different phenomena, although CFCs are a common agent in both. The greenhouse effect, or global warming, is the result of the release of increasing amounts of so-called "greenhouse gases" into the atmosphere, gases such as carbon dioxide **(CO_2)**, methane (CH_4), and CFCs. These gases trap some of the heat from the sun as it is radiated out from the earth, thus acting as a blanket, which retains the heat. Without the greenhouse effect, the earth would be about 60°F (33°C) colder, too cold to support life as we know it.

The year 1998 was significantly warmer than any year in over 100 years. As a matter of record, eleven of the thirteen warmest years ever recorded have occurred since 1983. There are many adverse effects of global warming, such as:

❑ Glaciers are being depleted around the world. Because polar regions have warmed more than the global average, ice coverage in athe Arctic Ocean has declined by about 40 percent in the last quarter of the twentieth century.

❑ Rising ocean temperatures are believed to be the cause of ever-increasing coral bleaching and disease and are adversly affecting all marine life.

❑ There has been an increase inrainfall since the early 1900s and thus flooding due to heavy rainfall events. More and frequent droughts exacerbate our water supplies and increase the possibility of wildfires.

❑ *El Niño* and *La Luna* events have been more frequent and intense since the late 1970s. Natural disasters increased tenfold in the last decade of the twentieth century.

There is a growing concern by the scientific community that global warming is occurring and that measures must be taken to reduce certian gases, such as carbon dioxide (CO_2), that are emitted into the atmosphere. Although there have been considerable efforts to reduce greenhouse emissions, it seems that voluntary measures are not enough. During the 1990s, for example, in the United States, the number one emitter of greenhouse gas—emissions—have increased by nearly 10 percent.

An effort toward increased recycling and the production of energy-efficient vehicles and buildings today can lead to the reduction of greenhouse gases and cleaner air tomorrow. Modern technology has advanced to the level that can result in lower emissions and have a direct positive impact, not only on our health and environment, but on our economy as well.

The Greenhouse Effect

The greenhouse effect is a natural process of warming, just as the ozone (O_3) layer is a natural function of the earth's atmosphere that protects life. Both have been affected by the release of pollutants caused by human activities; these pollutants have accelerated the greenhouse effect, thus resulting in increased warming and depletion of the ozone (O_3) layer, exposing life to damaging UV radiation.

THE CLEAN AIR ACT

The most significant legislation to affect the automotive air-conditioning industry in the United States is the Clean Air Act (CAA). The CAA was signed into law by President George Bush on November 15, 1990. Most of the rules and regulations of the CAA are a result of the recommendations made at the **Montreal Protocol.**

The Montreal Protocol, and later, the Copenhagen Amendments, deals with the environmental problems and issues created by certain refrigerants depleting the ozone (O_3) on an international level. The CAA deals with this problem on a national level. The Montreal Protocol is structured so that periodic meetings must take place in order to reassess the ozone (O_3) problem. As new facts about the impact of refrigerants are brought to fore, the protocol will be modified accordingly. The majority of protocol modifications will also result in the CAA being modified accordingly.

The CAA was somewhat more specific than the protocol in addressing the ozone (O_3) depletion problem. The Clean Air Act gave the EPA the authority to establish environmentally safe procedures in the use and reuse of refrigerants. In addition, the EPA has established standards for certification and service of refrigeration equipment and those who service the equipment. These standards were derived mainly from information that was furnished by private sector organizations.

STRATOSPHERIC OZONE PROTECTION -TITLE VI

A section in the CAA is called Title VI -Stratospheric Ozone Protection. Title VI establishes regulations for the production, use, and phaseout of CFCs, halons, and HCFCs. Other chemicals such as carbon tetrachloride (CCl_4), also covered by Title VI, are not covered in this text. Title VI breaks the substances to be regulated into two classes: Class I and Class II.

The chemical that we are primarily concerned with in the automotive industry is CFC-12, Figure 2–18, a Class I refrigerant. This refrigerant has been phased out and is no longer being manufactured in the United States. It is illegal to import this refrigerant from another country. The industry, then, must rely on reserves of virgin refrigerant, recovered and reclaimed refrigerant, or equipment conversion to an alternate refrigerant.

USE OF TOOLS

The proper use of tools is an important consideration for the technician. This notion may be divided into two major categories: the use of safe tools and the safe use of tools. The two go hand-in-hand, for without one there cannot be the other. An illustration of the tools discussed in this chapter is shown in Chapter 3. The formula for tool safety may be covered by three general rules: (1) use safe tools; (2) maintain tools in a safe condition; and (3) the right tool for the task.

Rule #1: Use Safe Tools

The safe use of tools and equipment is fundamental to any automotive technician's safety program. The first step in any tool safety program is to upgrade tools to maintain minimum safety standards. This means inspect all current tools on hand and replace any of them that are defective or do not meet minimum quality standards.

FIGURE 2–18 Typical "pound" can of CFC-12. (*Courtesy of BET, Inc.*)

Quality Standards for Tools.   Tools that are used to serve the automotive industry lead a rugged life. To perform safely on day-in-day-out applications, they must be designed and manufactured to rigid quality standards.

Some of the more important points for consideration when selecting tools include:

❑ Tools should be made of alloy steel. Finer grade alloys impart toughness to the metal used in the manufacture of tools.

❑ Tools should be tempered by heat treating. Tool strength and lasting quality are enhanced by precision heat treatment of the metal.

❑ Accuracy of machining. For a tool to fit the intended application accurately without slipping or binding, machining must be held to close tolerances, Figure 2–19.

❑ Safe by design. Firm, safe tool control with a minimum of effort should be provided by a lightweight, balanced design, including those that prevent slipping or accidental separation.

Identifying and discarding unsafe tools, Figure 2–20, are importants aspects of tool safety. In addition to those tools that are easily recognized as below standard, broken, or otherwise damaged, consider homemade and reworked tools.

❑ Homemade tools — Few repair facilities are equipped to work steel into tools suitable for high-leverage automotive repair applications. Homemade tools are often heavy and awkward to handle.

Lighter, stronger, and tougher tools for virtually any purpose are commercially available and should replace all homemade relics.

❑ Reworked tools — Grinding or otherwise reworking a tool to fit a particular application usually results in a tool that no longer measures up to safety requirements. Grinding a tool removes metal needed for strength. These tools should also be replaced.

Rule #2: Maintain Tools in a Safe Condition

When safe tools are on the job they must be kept in safe condition. This involves routine inspection on a regular basis and repairing or replacing those tools that are worn or no longer safe.

Screwdrivers.   Screwdrivers with worn, chipped, or broken tips are a potential menace to the technician who works with them. Such tools have little grip on the screw head, frequently jumping the slot and leaving the technician open to injury.

Some use a screwdriver as an all-purpose tool to take the place of punches, chisels, and prybars. The usual result is a damaged tool and possible injury to the user. Screwdrivers should be used only to turn screws.

It is very common practice for those who are not familiar with tools to try to turn a Phillips head screw with a standard tip screwdriver designed for use on slotted screw heads. The result is usually a tool slipping off the screw, a nicked screwdriver tip, and a hopelessly chewed-up fastener. Only a Phillips tip screwdriver should be used on a Phillips screw. Tools to turn "look-alike" screws are not interchangeable; one screwdriver will not seat properly in the other screw head, Figure 2–21.

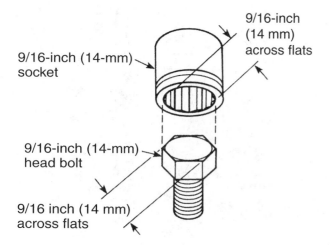

FIGURE 2–19 The size of the correct socket is the same as the size of the bolt head or nut.

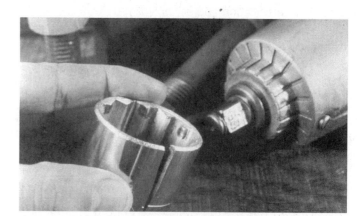

FIGURE 2–20 Broken tools should be replaced, such as this socket.

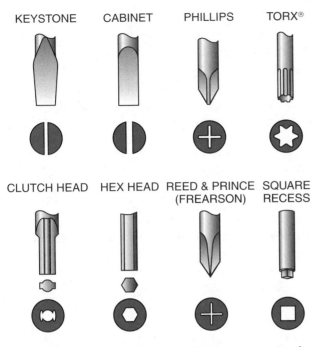

KEYSTONE   CABINET   PHILLIPS   TORX®

CLUTCH HEAD   HEX HEAD   REED & PRINCE   SQUARE
                          (FREARSON)   RECESS

FIGURE 2–21 Ensure that the screwdriver properly fits the fastener.

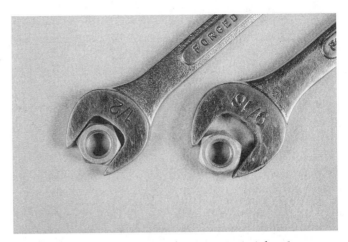

FIGURE 2–22 Oversize wrenches damage bolt heads.

Wrenches and Sockets. Worn-out or oversized wrenches and sockets that get only a partial "bite" on the corners of a nut, Figure 2–22, are most likely to slip on a heavy pull. A regular inspection of the toolbox for worn tools will prevent a lot of mishaps. Look for the following:

❑ Open-end wrenches with battered, spread-out jaw openings
❑ Sockets or "box sockets" whose walls have been battered and rounded by use
❑ Tools that have been abused, such as standard thin-wall (hand-use) sockets with lapped-over metal around square-drive opening, and wrenches or handles bearing hammer marks

Rule #3: The Right Tool for the Task

Safe tools, in safe condition are only half of the tool safety story. The other half rests with the technician who uses the tools. Every technician in the shop who works with tools should be familiar with the ideas listed here.

Hand tools are available in an endless assortment of types, styles, shapes, and sizes—perhaps over 5,000 in all. Each tool is designed to do a certain job quickly, safely, and easily. The safest tool is one that fits the job squarely and snugly. Misuse can lead to tool slipping or breaking, which can lead to injury. Every tool has a design safety limit. Exceeding the design limits can result in tool failure.

Wrenches and Socket Wrenches. Some bolt-turning tools have definite safety advantages over others. Here is a list of tools suitable for bolt turning in no order of preference.

The "safety limit" of a wrench or socket wrench is determined by the length of its handle. Use of a pipe extension or other "cheater" to move a tightly rusted nut can overload the tool past its safety limit.

When the tool being used will not turn the nut, a heavier-duty tool is required. Both open-end and **box socket** wrenches are available in a heavy-duty series that can be safely used with tubular handles from 15" to 36" in length, and can be substituted for a wrench that is too light for the job.

Wrenches. The adjustable wrench, often referred to by its trade name, *Crescent,* is recommended only for light-duty applications where time is an important factor and the proper tool is not readily available. Adjustable wrenches are prone to slip because of the difficulty encountered in setting the correct wrench size. They also have a tendency for the jaws to "work" as the wrench is being used. For these reasons, an adjustable wrench should not be considered as an all-purpose tool.

Sockets and box sockets should fit squarely. When these tools are "cocked," they are likely to break even under a moderate load. This is due to "binding," which concentrates the entire strain at one point, rather than spreading it evenly over the tool.

Selection of the correct wrench size is important. A wrench or socket one size too large will not grip the corners of the nut securely. The result can be a bad slip during a heavy pull. There is a correct wrench size available for virtually every nut or bolt made, whether English or metric. Wrench size is determined by measuring the nut or bolt head across the flats. For the most secure grip, open-end wrench jaws should contact the entire length of two flat surfaces of the nut or bolt head. When it is necessary to reach the fastening at extreme angles, there is a danger that the wrench will slide off. This can usually be avoided by the use of crowfoot, offset-head, or taper-head open-end wrenches.

Pliers. The use of pliers should be limited to gripping and cutting operations for which they were designed. Though often used for that purpose, pliers are not on the list of tools that are recommended to be used for bolt-turning work because their flexible jaws allow them to slip, and they leave tool marks on the surface being gripped. Also, slipping on the nut or bolt head often rounds the corners so badly that it becomes extremely difficult to service even with a proper wrench.

> WARNING
> THE INSIDE OF CUTTING JAWS SHOULD POINT AWAY FROM THE USER'S FACE TO PREVENT POSSIBLE INJURY FROM FLYING CUTTINGS.

Pullers. The puller is the only quick, easy, and safe tool for forcing a gear, wheel, pulley, or bearing off a shaft. Use of pry bars or chisels often causes the part to cock on the shaft, making removal even more difficult.

A puller provides a mechanical advantage that reduces the amount of force required. Furthermore, a puller is so designed that the force used is always under control.

Safety Accessories. Safety glasses or goggles, Figure 2–23, are essential eye protection when metal strikes metal,

or when grinding metal tools or parts on a power grinder. Safety glasses are recommended for everyone working in a shop. In many situations safety glasses, hard hats, and hearing protection are required for anyone entering the premises. Goggles may be used by those with prescription glasses, and a face shield may be used by anyone who cannot otherwise wear glasses or goggles.

PERSONAL SAFETY PRECAUTIONS

The following basic precautions apply to almost any type of automotive service. They are not, however, intended to supersede any instructions or precautions given by the manufacturer, supervisor, or instructor for any particular procedure.

- ❑ Wear safety glasses or goggles.
- ❑ Set the parking brake. Place the gear select in *park* (in *neutral* if a manual transmission).
- ❑ Ensure that the ignition switch is in the *off* position (unless otherwise required for the procedure).
- ❑ Operate the engine *only* in a well-ventilated area (when required by a procedure).
- ❑ Keep clear of all moving parts.
- ❑ Remove rings, watches, and loose-hanging jewelry.
- ❑ Avoid loose clothing.
- ❑ Use personal hearing protection in noisy areas, Figure 2–24.
- ❑ Tie long hair securely behind the head.
- ❑ Keep hands, clothing, tools, and test leads away from the radiator cooling fan because electric cool-

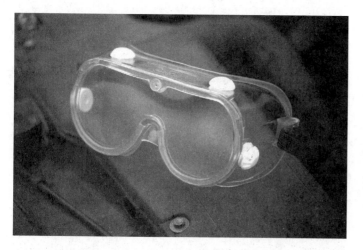

FIGURE 2–23 Use personal eye protection. *(Courtesy of Goodson Shop Supplies)*

FIGURE 2–24 Use personal hearing protection. *(Courtesy of Seibe North, Inc.)*

ing fans can start without warning even when the ignition switch is in the off position.

❑ Invest in an inexpensive personal eyewash system if working in an area heavily laden with airborne particles.

❑ Avoid contact with hot parts such as radiator, exhaust manifold, and high-side refrigeration lines.

❑ Follow the procedures outlined in manufacturers' service manuals when disconnecting the battery. (Sensitive computer circuits may be interrupted if specific procedures are not followed.)

❑ When in doubt, consult the service manual or ask your supervisor or instructor. *Do not take chances.*

It should be recognized that procedures that may be used by technicians performing automotive air-conditioning service may vary greatly. It is not possible to anticipate all the ways or conditions under which this service may be performed. Therefore, it is not possible to provide precautions for every conceivable hazard that may result. For example, an air bag may be inadvertently inflated while working on the air duct system under the dash if specific precautions are not taken by the technician. Specific manufacturers' specifications, references, and repair manuals must always be consulted before attempting any repair.

Also, one must be ever mindful of environmental concerns, such as the probation of the federal Clean Air Act on refrigerant venting, as follows:

Section 608 (c) PROHIBITIONS. — (1)

Effective July 1, 1992, it shall be unlawful for any person, in the course of maintaining, servicing, repairing, or disposing of any appliance or industrial process refrigeration, to knowingly vent or otherwise knowingly release or dispose of any class I or class II substance used as a refrigerant in such appliance (or industrial process refrigeration) in a manner which permits such substance to enter the environment.

Bracing Against a Backward Fall. Always pull on a wrench handle; never push on it. It is far easier to brace against a backward fall than against a sudden lunge forward should the tool slip or break. Therefore, when pulling on a wrench the technician should always brace against a backward fall. This is easily accomplished by placing one foot behind the other.

The danger of pulling an open-end wrench right off the nut or bolt can be minimized by using the proper size wrench and by making sure that the wrench is positioned so that the jaw opening faces in the direction of the pull.

CONCLUSION

Safety sense with tools pays off. This reminds the technician to think safety whenever applying a tool to the task. Some of the tips presented in this chapter may seem to be nothing more than common safety sense; they have been included only because technicians have been injured when they are overlooked.

Safety sense reminds the technician to protect against the possibility of something going wrong. Whenever tools are used, there is a risk of tools breaking or slipping. And there is also a risk that the part on which tools are used may break loose.

SUMMARY

Often, hand tool accidents are caused by poor housekeeping. Safety, as well as good workmanship, dictate that tools be properly stored and cleaned. Tools should always be kept in tote trays, boxes, or chests when not being used.

Tools with oily handles can be slippery and dangerous. Technicians should get in a habit of wiping off tools with a dry shop rag before putting them away. With tools, as with everything else, good housekeeping makes good safety sense.

❑ The many hazards associated with the automotive repair industry include exposure to asbestos, carbon monoxide, radiation, caustics, solvents, glues, and paints.

❑ Hazardous conditions, such as cluttered floors, missing guard rails, lack of safety zones, inadequate ventilation, and improper storage of combustible materials are often found in the repair facility.

❑ A health and safety program is an effective method to assist in providing for safe working conditions.

❑ Unsafe acts and conditions become apparent and may be corrected through the use of a health and safety program.

❑ Substandard tools, those that are not made of alloy steel, should be replaced with tools made of industry-standard high-alloy steel.

❑ Tools must never be "worked" beyond their design capabilities.

REVIEW

Select the correct answer from the choices given.

1. Ozone is being discussed. Technician A says ozone has a pale green color. Technician B says ozone is a form of oxygen. Who is right?
   a. A only
   b. B only
   c. Both A and B
   d. Neither A nor B

2. What harmful gas is produced by an internal combustion engine?
   a. Ozone (O_3)
   b. Carbon dioxide (CO_2)
   c. Argon
   d. Carbon monoxide (CO)

3. Technician A says ultraviolet radiation is absorbed by the ozone layer. Technician B says ultraviolet radiation may be hazardous to human health. Who is right?
   a. A only
   b. B only
   c. Both A and B
   d. Neither A nor B

4. Which of the following gases is responsible for the destruction of the ozone?
   a. Chlorine (Cl)
   b. Fluorine (F)
   c. Hydrogen (H)
   d. Argon (Ag)

5. Ionizing radiation may be caused by which of the following operations?
   a. Running an internal combustion engine
   b. Discharging an air conditioning system
   c. Using a laser sited thermometer
   d. Welding a trailer hitch to a vehicle frame

6. Technician A says a lack of ozone is responsible for the greenhouse effect. Technician B says a lack of ozone is responsible for excessive UV radiation. Who is right?
   a. A only
   b. B only
   c. Both A and B
   d. Neither A nor B

7. A refrigerant found in absorption refrigeration may be:
   a. toxic.
   b. corrosive.
   c. explosive.
   d. all of the above.

8. The Clean Air Act (CAA) was signed into law:
   a. January 1, 1990.
   b. November 15, 1990.
   c. January 1, 1995.
   d. November 15, 1995.

9. Title VI of the CAA establishes regulations for the phaseout of all of the following chemicals, EXCEPT:
   a. HCFC.
   b. CFC.
   c. halon.
   d. HFC.

10. Technician A says EG is a type of antifreeze. Technician B says BG is a type of antifreeze. Who is right?
    a. A only
    b. B only
    c. Both A and B
    d. Neither A nor B

11. Welding operations may produce all of the following gases, EXCEPT:
    a. carbon monoxide (CO).
    b. cadmium (Cd).
    c. ozone (O_3).
    d. carbon dioxide (CO_2).

12. Most of the job-related illnesses reported are due to:
    a. stress or are stress related.
    b. respiratory problems.
    c. infections.
    d. heat exhaustion.

13. Which poisonous gas may be formed when mixing certain cleaning solutions?
    a. Chlorine (Cl)
    b. Fluorine (F)
    c. Carbon monoxide (CO)
    d. Carbon dioxide (CO_2)

14. Technician A says tools should be made of carbon steel. Technician B says tools should be tempered by heat treating. Who is right?
    a. A only
    b. B only
    c. Both A and B
    d. Neither A nor B

15. Which of the following tools is recommended when a heavy pull is required?
    a. Box socket wrench
    b. Open-end wrench
    c. Vice grip pliers
    d. Slip-joint pliers

16. Technician A says a pipe may be used to gain leverage with an adjustable wrench. Technician B says a hammer may be used with a box socket to "break loose" a stubborn nut. Who is right?
    a. A only
    b. B only
    c. Both A and B
    d. Neither A nor B

17. Technician A says electric cooling fans may start without warning. Technician B says to ensure that the ignition switch is *off* to prevent the fans from starting. Who is right?
    a. A only
    b. B only
    c. Both A and B
    d. Neither A nor B

18. Which of the following components is *least* likely to be hot when the engine is running?
    a. Radiator
    b. Cooling system hoses
    c. Air-conditioning system hoses
    d. Exhaust manifold

19. Which of the following antifreeze types are the most toxic?
    a. Propylene Glycol (PG)
    b. Ethylene Glycol (EG)
    c. Both A and B
    d. Neither A nor B

20. All of the following statements about refrigerant are *false*, EXCEPT:
    a. Refrigerants are lighter than the ambient air.
    b. Refrigerant may be vented to the atmosphere.
    c. Refrigerants may be mixed in mobile air-conditioning systems to achieve maximum performance.
    d. Refrigerants that contain propane are not to be used in mobile air conditioning systems.

TERMS

Write a brief description of each of the following terms:

1. absorption
2. atmosphere
3. box socket
4. carbon monoxide
5. chlorine
6. Cl
7. CO_2
8. ethyl glycol
9. flammable
10. Montreal Protocol
11. OSHA
12. O_3
13. oxygen
14. ozone
15. personal hygiene
16. propane
17. radiation
18. refrigerant
19. technician
20. UV

CHAPTER 3

TOOLS AND MEASURING SYSTEM

OBJECTIVES

On completion and review of this chapter, you should be able to:

❑ Identify the various hand tools used by the technician.
❑ Identify the special tools used by the air-conditioning service technician.
❑ Compare the English system to the metric system of measure.
❑ Convert English to metric and metric to English values.

INTRODUCTION

This chapter begins with the systems of measurement used in repairing modern automobile air-conditioning systems. An automotive technician must learn and understand two systems of measurement: the British Imperial (English) system, referred to as the United States Customary System (USCS), and the metric system, known as SI **metrics.**

The British system, which uses fractions and decimals, has been the basic measuring system in England, the United States, and Canada until recent years. This system is based on inches, feet, and yards. In the twelfth century, one inch was decreed to be a length equal to three barley corns end to end, and one yard was the distance from King Henry's nose to the end of his thumb. The bases for this system of measurement, although somewhat ridiculous, have been used for many years.

THE METRIC SYSTEM

The international system (SI) of measurement is known as the metric system. The basis of the metric system is the meter, which is equivalent to 39.37 inch in the English system. In 1792 the French Academy of Sciences decided that the meter, also spelled metre, should be ten-millionth of a quarter of the earth's circumference in order to provide mankind with a univer-

sal measurement of length. The meter is easily broken down into smaller units, such as the centimeter (1/100 meter) and millimeter (1/1,000 meter).

The first system of metric units was called the CGS (centimeter-gram-second) system and was based on the centimeter (cm) as the unit of length, the gram (g) as the unit of mass, and the second (s) as the unit of time.

This system was used a little over 100 years until, at the turn of the twentieth century, a more practical set of metric units known as the MKS (meter-kilogram-second) system was developed. This system, based on the meter (m), kilogram (kg), and second (s), was used for about 50 years. In about 1950, the ampere (A) was added as the fourth unit, making it possible to link electrical units with mechanical units. The name was then changed to the MKSA system.

SYSTEM OF UNITS (SI METRICS)

Later, another system of metric units was established. It was based on the four MKSA units plus the kelvin (K) as the unit of temperature, the candela (cd) as the unit of luminous intensity, and the mole (mol) as the unit of substance. Previous systems had been named on the basis of the first letter of designated units, so this system might have been called the MKSAKM system. Instead, it was called the *Système International d'Unités* or, more commonly, the SI system of units (metrics).

Some liberty has been taken with the system standards, however. The SI metric standard for a unit of temperature is the kelvin (K). The Kelvin scale, used primarily for engineering, indicates that the freezing temperature of water (H_2O) at sea level is 273°K, and the boiling point is 373°K. The American Society of Heating, Refrigeration, and Air-Conditioning Engineers (**ASHRAE**) has adopted the **Celsius** (C) scale for practical technical application. In the Celsius scale, the freezing point of water (H_2O) at sea level is O°C, and the boiling point is 100°C.

The United States passed the Metric Conversion Act in 1975 in an attempt to get the American industry and the general public to use the metric system, as the rest of the world does. Although the general public has been slow to drop the customary measuring system of inches, gallons, and pounds, many industries, led by the automotive industry, are now predominantly metric.

Nearly all vehicles are now built to metric standards. Today's technicians must be able to measure and work with both systems of measurement.

Other American manufacturers are slowly changing their tooling to the metric system to be competitive with the rest of the world. Technicians working on domestic automobiles are seeing more and more metric parts and therefore need tools with both metric and English sizes.

DECIMAL SYSTEM

Metric units are based on the decimal system of measure. Fractions are not used in the metric system. Users need only think in terms of *ten*. Multiples or submultiples of any unit are related to the unit by powers of ten. Compared to working with fractions, calculations in metrics are made quite rapidly. The chance of error is also greatly reduced. These characteristics made the SI metric system simple to learn and use.

Metric Engine Size Measurement

The metric measurement of volume is the **liter,** which is slightly larger than an English quart. A liter is the quantity of liquid that will fill a cube measuring 1/10 of a meter (3.937 in.) on each edge.

NOTE: There are 61.02 cubic inches in one liter. One liter is equal to 1,000 cubic centimeters (cc); thus, a 5,000 cc engine is also called a 5-liter engine, or about 305 cubic inches in the English measure.

DIVISION AND POWER

In the metric system, four or more numbers are separated by a space (actually a half space) instead of a comma. The space may or may not be used with four-digit numbers; for example, 1,000 may be expressed as 1000 or 1 000. But the space is always used in numbers of more than four digits. For example, 10,000 becomes 10 000; 100,000 becomes 100 000; 1,000,000 becomes 1 000 000, and so on.

All units of measurement in the metric system are related to each other by a factor of 10. Every metric unit can be multiplied or divided by the factor of 10 to get larger units (multiples) or smaller units (submultiples). This makes the metric system much easier to use, with less chance of math errors than when using the English system of measure.

Another way of expressing metric numbers is by power. Power indicates how many times a prime number is multiplied by itself. For example, the number 100 may be expressed as 10^2 ($10 \times 10 = 100$); 1 000 may be expressed as 10^3 ($10 \times 10 \times 10 = 1000$); 10 000 expressed as 10^4 ($10 \times 10 \times 10 \times 10 = 10\ 000$), and so on. Either way of expression, by whole number or by power, is acceptable.

Powers of 10 are referred to as *multiples* and *submultiples*. In metrics, these powers have both a prefix and a symbol. Figure 3–1 includes the prefixes and symbols most commonly used.

SYMBOLS

It is important to use symbols correctly so as not to change their meaning. When capital letters are required,

Number	Power	Prefix	Symbol
1 000 000 000 000	10^{12}	tera	T
1 000 000 000	10^9	giga	G
1 000 000	10^6	mega	M
1 000	10^3	kilo	k
100	10^2	hecto	h
10	10^1	deka	da
0.1	10^{-1}	deci	d
0.01	10^{-2}	centi	c
0.001	10^{-3}	milli	m
0.000 001	10^{-6}	micro	μ
0.000 000 001	10^{-9}	nano	n
0.000 000 000 001	10^{-12}	pico	p

FIGURE 3–1 Powers, prefixes, and symbols commonly used in metrics.

Term	Symbol
Celsius	C
centimeter	cm
gram	g
kilogram	kg
kilometer	km
kilopascal	kPa
liter	L
meter	m
milliliter	mL
millimeter	mm

FIGURE 3–2 Common metric terms and symbols.

lowercase letters must not be used. For example, the capital letter *M* is the symbol for the prefix mega, whereas the lowercase letter *m* is the symbol for the prefix milli. There is a great difference between mg (milligrams) and Mg (megagrams). One mg is 1/1 000 of a gram, whereas 1 Mg is 1 million grams. The common metric terms and symbols used by the automotive air-conditioning technician are found in figure 3–2.

The metric term *Celsius* (C) is often confused with the term *centigrade* (C). Centigrade, often used to denote temperature, was changed in 1948 by international agreement to Celsius (C). In some parts of the world, such as France, centigrade is used to denote fractions of a right angle.

The lowercase letter *m* is used as the abbreviation symbol for the unit *meter* as well as for the prefix *milli*. As illustrated in figure 3–3, when used in combination, **mm** becomes *millimeter* (1/1 000 of a meter). It is therefore very important to understand and be able to use abbreviations and symbols in the metric system.

CONVERSIONS

Formulas have been developed for the rapid conversion of English-metric and metric-English units. Figure 3–4

gives the formulas most commonly used by the automotive air-conditioning technician.

Fraction to Decimal Conversions

The decimal system must be used to convert English values to metric values. Therefore, any English fractional value must first be converted to an English decimal. This is easily accomplished by dividing the numerator by the denominator. The *numerator* is the number above the line; the *denominator* is the number below the line.

Whole Numbers Remain Whole Numbers. The whole number is separated from the fraction by a decimal point. For example, 1 and 1/2 becomes 1.5. A decimal point is placed after the whole number and the fraction is converted to a decimal.

Those who use a calculator will find both the decimal and metric systems most welcome. Calculators may be used to convert fractions to decimals as well. Remember to hold decimal equivalents to three places. The fraction 5/64, for example, converts to 0.078125 on a calculator. The three-place decimal of 0.078 should be sufficient for most calculations.

If the decimal is of a fraction only and does not include a whole number, it is preceded by a zero (0). For example, the conversion for 1/4 is 0.25, not simply .25; the zero (0) is used to indicate that there is no whole number.

COMMON METRICS

The automotive air-conditioning technician uses only a few of the metric system conversions. These units include weight, measure, temperature, and **pressure.**

Weight

The metric unit of weight is the *gram,* which is the weight of the amount of water it takes to fill a cube that measures 1/100 of a meter (0.394") on each side. The kilogram, which

A	ampere	kPa	kiloPascal	mm	millimeter
°C	degrees Celsius	L	liter	oz.	ounce
°F	degrees Fahrenheit	lb.	pound	V	volt
ft.	foot	lb./in.²	psi	Pa	pascal
gal	gallon	m	meter	psi	pounds square inch
g	gram	mA	milliampere	psia	psi, absolute
in.	inch	mi.	mile	psig	psi, gauge
kg	kilogram	mL	milliliter	W	watt

FIGURE 3–3 Typical abbreviations.

METRIC TO ENGLISH

Multiply	By	To Get
Celsius (°C)	1.8 (+32)	Fahrenheit (°F)
gram (g)	0.035 3	ounce (oz.)
kilogram (kg)	2.205	pound (lb.)
kilometer (km)	0.621 4	mile (mi.)
kilopascal (kPa)	0.145	lb./in.^{2} (psi)
liter (L)	0.264 2	gallom (gal.)
meter (m)	3.281	foot (ft.)
milliliter (mL)	0.033 8	ounce (oz.)
millimeter (mm)	0.039 4	inch (in.)

ENGLISH TO METRIC

Fahrenheit (°F)	(−32) 0.556	Celsius (°C)
foot (ft.)	0.304 8	meter (m)
fluid ounce (fl. oz.)	29.57	milliliter (mL)
gallon (gal.)	3.785	liter (L)
inch (in.)	25.4	millimeter (mm)
mile (mi.)	1.609	kilometer (km)
ounce (oz.)	28.349 5	gram (g)
pound (lb.)	0.453 6	kilogram (kg)
lb./in.^{2} (psi)	6.895	kilopascal (kPa)

FIGURE 3–4 Metric to English and English to metric conversion factors.

equals about 2.2 pounds, is the most common use of the gram. All metric measurement units are relative. For example, 1,000 cc of water weighs 1 kilogram.

The customary ounce **(oz.)** is the gram (g) in the metric system. One ounce is equal to 28.349 grams. The standard "pound" can of refrigerant actually contains 12 or 14 ounces (340.2 or 396.9 grams) of refrigerant. Due to the metric system, this packaging may change to contain 350 or 400 grams, which is a little more than 12 (12.4 oz.) or 14 ounces (14.108 oz.), respectively.

The customary pound **(lb.)** becomes the kilogram (kg) in the metric system. One pound is equal to 0.454 kilogram. The standard 15-pound cylinder of refrigerant, containing 6.804 kilograms, may be repackaged to contain, 7 kilograms (15.4 lb.).

Again, to convert ounces to grams, multiply by 28.350; to convert grams to ounces, multiply by 0.035. To convert pounds to kilograms, multiply by 0.454; to convert kilograms to pounds, multiply by 2.205. One fluid ounce (fl. oz.) is equal to 29.573 milliliters (mL).

Measure

The customary inch (in.) becomes the millimeter (mm) in the metric system. One inch is equal to 25.4 millimeters. The customary foot (ft.) becomes the meter (m). One foot is equal to 0.305 meter.

To convert inches to millimeters, multiply by 25.4; to convert millimeters to inches, multiply by 0.039. To convert feet to meters, multiply by 0.305; to convert meters to feet, multiply by 3.281.

Temperature

On the English Fahrenheit scale, water freezes at 32°F and boils at 212°F. In the metric system, temperature is often given as degrees centigrade. The correct term, however, is expressed as degrees Celsius. The temperature at which water freezes is 0°C. Water boils at 100°C.

In the English system, the normal body temperature is 98.6°F. The metric equivalent is 37°C. At sea level, the freezing point of the water (H_2O) is 32°F or 0°C, and the boiling point is 212°F or 100°C.

To convert Fahrenheit (F) to Celsius (C), first subtract 32 and then multiply by 0.555. To convert Celsius (C) to Fahrenheit (F), first multiply by 1.8 and then add 32.

Pressure

The metric system equivalent to pounds per square inch (psi) is expressed as **kiloPascals,** but is often expressed as kilograms per square centimeter (kg/cm^2). Both measurements are used in measuring atmospheric pressure at sea level.

- ❑ Atmospheric pressure is one BAR in the metric system (1 kilogram per square centimeter).
- ❑ Atmospheric pressure in the English system is called one atmosphere (14.696 pounds per square inch, **gauge**).

The customary pounds per square inch (psi), pounds per square inch gauge **(psig),** and pounds per square inch absolute (psia) become kiloPascal (kPa) in the metric system. One psi is equal to 6.895 kPa. One kPa equals 0.145 psi.

Early confusion when using the metric system led many to take the stand that the conversion from pounds per square inch (psi or lb.-in.^{2}) was to kilograms per square centimeter (kg/cm^2). To be correct, kg/cm^2 is a metric term that is generally accepted for force, not pressure. To avoid confusion, the American Society of Refrigeration and Air Conditioning Engineers (ASHRAE) endorsed the

American Standard Metric Practice Standards (IEEE 268-1979 and ASTM 380-79), as supplied by the U.S. Department of Commerce, National Bureau of Standards. The following are excerpts from those standards:

3.4.6 Pressure and Vacuum.

Gauge pressure is absolute pressure minus ambient pressure (usually atmospheric pressure). Both gauge pressure and absolute pressure are properly expressed in Pascal, using SI prefixes as appropriate. Absolute pressure is never negative. Gauge pressure is positive if above ambient pressure and negative if below. Pressure below ambient is often called **vacuum;** whenever the term *vacuum* is applied to a numerical measure it should be made clear whether negative gauge pressure or absolute pressure is meant. See 3.5.5 for methods of designating gauge pressure and absolute pressure.

3.5.5. Attachments to Unit Symbols.

No attempt should be made to construct SI equivalents of the abbreviations "psia" and "psig," so often used to distinguish between absolute and gauge pressure. If the context leaves any doubt as to which is meant, the word pressure must be qualified appropriately. For example: at a gauge pressure of 13 kPa or, at an absolute pressure of 13 kPa.

Where space is limited, such as on gauges, nameplates, graph labels, and in table headings, the use of a modifier in parentheses, such as kPa (gauge) or kPa (absolute), is permitted.

TORQUE

Torque measurements are expressed in pounds-foot (lb.-ft. or lb.-in.) in the English system. The metric equivalent is expressed in Newton-meters (N•m). A newton is approximately 1/10 of a kilogram. To convert lb.-ft. to N•m multiply by 0.735, and to convert N•m to lb.-ft. multiply by 1.365. Torque conversion formulas are given in Figure 3–5.

METRIC REFERENCES IN THIS TEXT

So that there is no doubt, unless otherwise noted, all references to the metric system of pressure throughout this text will be for gauge pressure based at sea level ambient.

When applicable and practical, the metric equivalent of all English terms will be given in parentheses. For example, "The boiling point of water (H_2O) at atmospheric pressure is 212°F (100°C)."

No parenthetical notation will be made for those English units that do not convert to a standard metric unit. For example, "Remove the six 1/4-inch capscrews holding the seal plate." Figure 3–6 compares English and metric fasteners for #8 (English) through M27 (metric). Note that there is no metric fastener that will replace an English fastener.

> CAUTION
>
> DO NOT MIX ENGLISH AND METRIC FASTENERS. ALTHOUGH SOME MAY APPEAR TO BE THE SAME, THERE ARE SLIGHT DIFFERENCES. FOR EXAMPLE, AN ENGLISH 5/16-24 MAY LOOK EXACTLY LIKE A METRIC M8-25. AS A MATTER OF FACT, THERE IS VERY LITTLE DIFFERENCE; ABOUT ONE THREAD PER INCH (0.04 THREAD PER MM) AND 0.003 INCH (0.076 MM) IN DIAMETER, AS NOTED IN FIGURE 3–6. THEY ARE NOT, HOWEVER, INTENDED TO BE INTERCHANGEABLE.
>
> IF THERE ARE ANY DOUBTS, USE A THREAD PITCH GAUGE, SUCH AS THAT SHOWN IN FIGURE 3–7, TO DETERMINE THE NUMBER OF THREADS PER INCH (TPI).

TOOLS

Servicing the air-conditioning system on the modern automobile requires the use of various tools. Many of these tools are common hand and power tools used everyday by a technician. Other tools are very specialized and are only used for specific repairs on specific systems and vehicles. This chapter covers some of the more commonly used hand, power, and specialty tools that every technician must be familiar with.

TO CONVERT FROM	TO	MULTIPLY BY
Newton-meters (N•m)	pound-inch (lb.-in.)	8.851 (8.8512)
Newton-meters (N•m)	pound-foot (lb.-ft.)	0.738 (0.73758)
pound-inch (lb.-in.)	Newton-meters (N•m)	0.113 (0.11298)
pound-foot (lb.-ft.)	Newton-meters (N•m)	1.356 (1.3558)

FIGURE 3–5 Torque conversion formulas.

English Series				Metric Series			
Size	Diameter		Threads Per Inch	Size	Diameter		Threads Per Inch (prox)
	in	mm			in	mm	
#8	0.164	4.165	32 or 36				
#10	0.190	4.636	24 or 32				
1/4	0.250	6.350	20 or 28	M6.3	0.248	6.299	25
				M7	0.275	6.985	25
5/16	0.312	7.924	18 or 24	M8	0.315	8.001	20 or 25
3/8	0.375	9.525	16 or 24				
				M10	0.393	9.982	17 or 20
7/16	0.437	11.099	14 or 20				
				M12	0.472	11.988	14.5 or 20
1/2	0.500	12.700	13 or 20				
9/16	0.562	14.274	12 or 18	M14	0.551	13.995	12.5 or 17
5/8	0.625	15.875	11 or 18				
				M16	0.630	16.002	12.5 or 17
				M18	0.700	17.780	10 or 17
3/4	0.750	19.050	10 or 16				
				M20	0.787	19.989	10 or 17
				M22	0.866	21.996	10 or 17
7/8	0.875	24.765	9 or 14				
				M24	0.945	24.003	8.5 or 12.5
1	1.000	25.400	8 or 14				
				M27	1.063	27.000	8.5 or 12.5

FIGURE 3–6 Comparison of English and metric fasteners.

Tools used by the automotive air-conditioning technician are generally those that may be found in any shop. Other tools required for specific operations are found in shops specializing in particular services, such as compressor rebuilding. The following general description covers most of the common tools as well as many of the special tools.

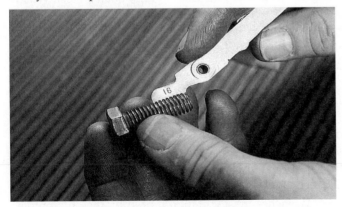

FIGURE 3–7 Using a thread pitch gauge.

SCREWDRIVER

A screwdriver has a metal shank with a handle on one end and a special shape on the other end that may be used to remove or replace screws and screw-like fasteners or to hold screws while nuts are placed and tightened on them. Two of the most common types of screwdriver used are shown in Figure 3–8.

The end of a screwdriver blade is shaped to fit a slot or recess in the head of a screw or a bolt. The shaped end of a screwdriver is placed in the recess in the fastener while the handle is turned.

Different types of fasteners are used to hold components together, characterized by the shape of the recess in the head, as shown in Figure 3–9. Two of the most common types are those that have a slot and those that have a cross-shaped recess.

Screwdrivers are available in various sizes. The technician should have a selection of sizes in each of the popular types.

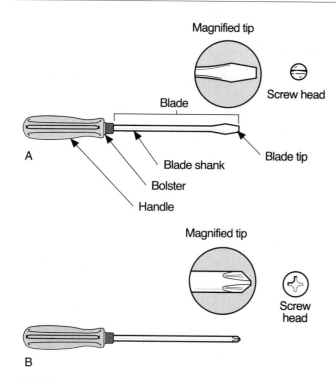

FIGURE 3–8 Standard flat tip (A) and Phillips head (B) screwdrivers.

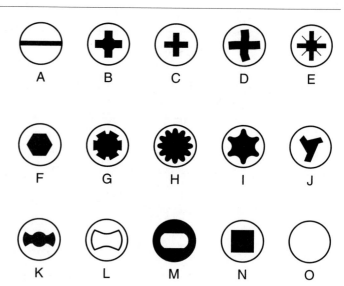

FIGURE 3–9 Typical fastener head types. (A) Slotted, (B) Phillips*, (C) Reed and Prince, also called Frearson, (D) Torq-Set*, (E) Pozidriv*, (F) Hex Cap, (G) Multi Spline, (H) Triple Square, (I) Torx*, (J) Tri-Wing, (K) Type A Clutch, (L) Type G Clutch, (M) Slab Head, (N) Scrulox*, and (O) Carrage *(Courtesy of BET, Inc.)* *indicates tradenames

NUT DRIVER

A set of nut drivers is recommended for an automotive air-conditioning service technician. A nut driver looks like a socket wrench on the end of a screwdriver shank. The tool is used like a screwdriver on nuts, bolts, sheet metal screws, and other such fasteners. Nut drivers are available in sizes ranging from 3/16 in. to 1/2 in. or 5 mm to 13 mm.

WRENCH

A wrench is used to hold a bolt while a nut is placed on it or used to loosen or tighten a nut or a bolt. A wrench has a head that fits on or around a nut or bolt head and has a handle that permits the leverage necessary to turn the nut or bolt.

There are several types of wrenches, each designed and used for specific applications, as follows:

Adjustable Wrench

Several sizes of adjustable wrench, Figure 3–10, are available. The jaws in the head are movable so an adjustment can be made to fit nuts or bolt heads within a certain range of sizes. Common sizes of 6", 8", and 10" identify their handle length, not their jaw opening

size. Because of their movable head, they are not considered as durable as a wrench with a fixed head. Adjustable wrenches are not the wrench of choice for the service technician. They are most handy, however, when the correct wrench is not available.

Socket Wrench

A socket wrench is, perhaps, the most useful type of wrench. The socket, separate from the handle, can easily be attached securely to it. A racheting handle, Figure 3–11, is used with a number of different-sized sockets. The sockets may also be used on different types of handles for different applications.

The sides of the socket (that fit over a nut or bolt head) are forged of one piece of high quality steel. When the tool is placed on a nut or bolt head it is completely enclosed by the walls of the socket. It is a good wrench to use when a nut needs to be very tight. Socket wrench sets are available for all sizes of nuts and bolts.

The socket is available in 8 points for square-head fasteners and 6 or 12 points for hex head fasteners. Handles are available in 1/4" drive for light duty, 3/8" drive for medium duty, and 1/2" for heavy-duty service as well as flexible, reacceding, rigid, or a combination. Socket wrenches are available for virtually any size fastener.

Wrenches most commonly used by the automotive air-conditioning technician include the open- and box-end, combination, flex head, and flare nut.

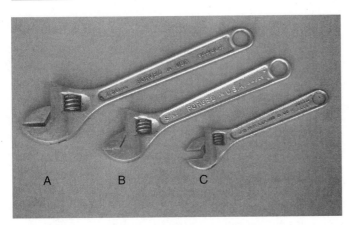

FIGURE 3–10 Adjustable wrench (A) 10", (B) 8", (C) 6".

Open-End Wrench

An open-end wrench is the most common type of wrench. The head of the wrench is open so it can be placed over a nut or bolt head. Both sides of the head are parallel so they fit snugly on the flat sides of a nut or bolt head.

Open-end wrenches are commonly available in a wide range of English and metric sizes from 1/4 in. (6 mm) to over 1-1/4 in. (32 mm). An open-end wrench may be straight or offset. An open-end wrench is also available in the "crowfoot" design for use in tight areas.

An open-end wrench is not as durable as a box-end wrench but in many applications it is much easier to use. The open-end wrench is used for general holding and tightening but is not recommended for heavy-duty service.

Box-End Wrench

A box-end wrench is similar to open-end wrenches except that the head is completely enclosed. For use the wrench is placed over the top of the nut or bolt head. Racheting box wrenches, Figure 3–12, are available and may be used in tight spots. Box-end wrenches are available in the same sizes as open-end wrenches. A box-end wrench may be straight or offset.

The closed construction of a box-end wrench, available in 6 or 12 points, Figure 3–13, provides the extra strength required for heavy-duty holding or tightening.

Combination Wrench

The combination wrench, as its name implies, has a box wrench on one end and an open wrench on the other end. Both open and box are of the same size. Generally, English sizes range from 1/4 to 1-1/2 inches. Metric sizes range from 7 mm to 30 mm. Other types of combination

FIGURE 3–11 An assortment of ratchets. *(Courtesy of Snap-on Tools Company)*

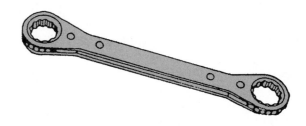

FIGURE 3–12 A ratcheting box wrench. *(Courtesy of Snap-on Tools Company)*

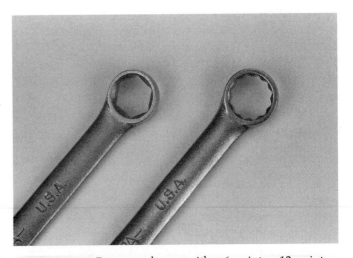

FIGURE 3–13 Box wrenches are either 6-point or 12-point. *(Courtesy of Snap-on Tools Company)*

wrench include those with an open wrench on one end and a swivel socket on the other end. These are generally available from 3/8 to 3/4 inch.

Flare-Nut Wrench

A flare-nut wrench combines the features of an open-end wrench and a box-end wrench and is used on flare nuts on refrigerant hoses and lines. The open end of the wrench head is placed around the refrigerant line and the head is then slipped over the flare nut. The box shape of the head allows it to fit snugly over the nut.

Flare-nut wrenches are generally available in 1/4 in. through 13/16 in. English and 6 mm through 21 mm in metric sizes. Flare-nut wrenches are also available in the "crowfoot" design for use in tight areas.

Torque Wrench

Torque wrenches, Figure 3–14, are used with an appropriate socket or crowfoot wrench to tension a bolt, nut, or other fastening device to manufacturer specifications. There are basically three types: the least expensive beam type, the versatile click type, and the most accurate dial type. Torque wrenches are calibrated in English inch-pounds (in.-lb.) or foot-pounds (ft.-lb.) or in metric Newton-meters (N•m).

Hex-key Wrench

A **hex-key wrench,** Figure 3–15, often referred to by the trade name "Allen wrench," is used to remove and replace screws that have a recess hexagon such as a setscrew, which is used to secure parts to each other. Hex wrenches are commonly available in English sizes ranging from 5/64 in. through 3/8 in. and in metric sizes

from 1.5 mm through 10 mm. The hex recess is often mistaken for a Torx® recess, Figure 3–16.

PLIERS

Pliers are a hand-held tool that have jaws on one end that can be opened and closed by manipulating the handle. Pliers are used to hold nuts or bolts, to twist wires together, and to hold parts when mechanical work is being done. Pliers are available in many different types, each designed for a specific use.

It is not recommended that pliers be used as a substitute for the proper wrench. The use of pliers on some fasteners distort the head so much that it is no longer possible to use the proper tool. The most popular sizes, 6 in. and 8 in. designate their length, not their jaw opening capacity. Some of the more common types of pliers are:

Diagonals

The diagonal cutting plier, Figure 3–17, often referred to as "dykes" are used for wire cutting or stripping insulation from wires.

FIGURE 3–15 Top: A handy tool containing many different Allen wrenches. Bottom: Tee-handle Allen wrenches designed for better gripping and easier torque application. *(Courtesy of Snap-on Tools Company)*

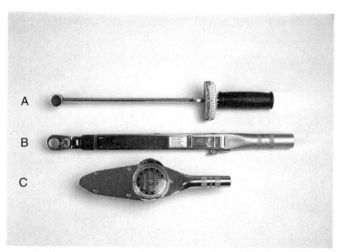

FIGURE 3–14 Torque wrenches: (A) A beam scale torque wrench, (B) A "click" wrench, (C) A dial indicator torque wrench.

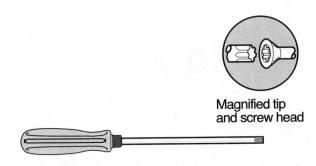

FIGURE 3–16 A Torx screwdriver. *(Courtesy of Stanley Works)*

FIGURE 3–17 Various diagonal-cutting pliers. *(Courtesy of Snap-on Tools Company)*

Needle Nose

There are two types of needle nose pliers: curved and straight. Curved needle nose pliers are used for holding parts or fasteners in close areas where access would be difficult with straight nose pliers. Straight needle nose pliers, Figure 3–18, are used for reaching into close areas to hold parts.

Wire Cutter

Wire cutters, similar to "diagonals" are often called "nippers." They are used for cutting wires close to a fastener or for other wire-cutting operations.

Water Pump

Water pumps, Figure 3–19, are offset adjustable joint pliers often referred to as "groove joint," "tongue and groove," or "slip joint" pliers used for holding parts and for tightening fasteners in areas where other types of pliers do not easily fit. Water pump pliers are available in a wide range of sizes from 6 in. to 15 in., which identify their length, not their jaw opening capacity.

Slip Joint

The straight slip joint pliers, Figure 3–20, often called a "combination pliers" is perhaps the most common and often used pliers and is available in several different sizes for most applications.

Slim-Nose

The jaws of slim nose pliers are extended and tapered, making them useful for reaching into close areas or for working on small parts. They often have a wire cutter on one side of the jaws for electrical service work.

Vise Grip

Vise grip pliers, Figure 3–21, often called "lock joint" or "locking pliers" are used to grip and hold. The opening

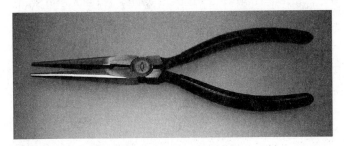

FIGURE 3–18 Typical needle nose pliers.

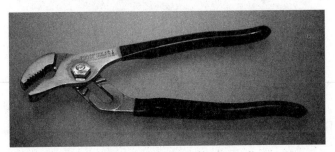

FIGURE 3–19 Typical adjustable pliers.

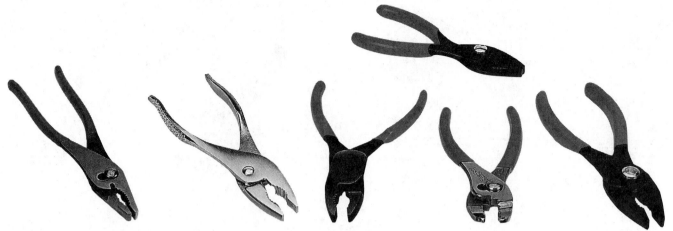

FIGURE 3–20 Various combination pliers. (*Courtesy of Snap-on Tools Company*)

of the jaws can be adjusted and a cam leverage of one handle allows the user to lock the pliers closed. These pliers are very useful when it is necessary to hold an object while it is being worked on. Some have special jaws or provisions for gripping a particular shape, such as an engine flywheel. Typically, vise grip pliers are available in straight or curved jaws from 5 in. to 10 in.

Snap Ring Pliers

Snap ring pliers, Figure 3–22, are not really pliers at all, although they are manipulated in the same scissor-like action as a pliers. Snap ring pliers are used to remove and replace snap rings. There are two types: internal and external. An internal snap ring pliers is used to remove or replace a snap ring inside a cavity. An external snap ring pliers, as its name implies, is used to remove or replace an external snap ring, such as on a shaft.

HAMMER

There are a number of types of hammers used by the service technician. The most common are ball peen, dead blow, and soft faced, more often referred to as a "mallet."

Ball Peen

Ball peen hammers, Figure 3–23, are used for general automotive service. They are available in 8-, 12-, 16-, 24-, and 32-ounce sizes.

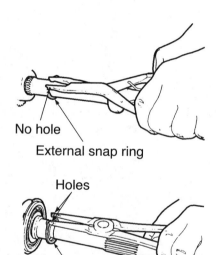

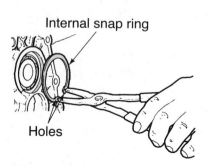

No hole
External snap ring

Holes
External snap ring

Internal snap ring
Holes

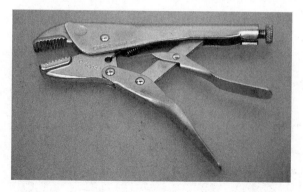

FIGURE 3–21 Typical locking or vise-grip pliers.

FIGURE 3–22 Types of snap ring pliers. (*Courtesy of Ford Motor Company*)

FIGURE 3–23 Ball peen hammer.

Dead Blow

A dead blow hammer head is filled with lead shot to provide a heavy striking force without rebound. It is used to add extra force to a hammer blow and where there is a danger of sparks or of marring a surface. Dead blow hammers are generally available with 1-, 1.5-, and 2-pound heads.

Soft-Faced

A soft-faced hammer or mallet, Figure 3–24, has ahead of rubber, plastic, or a combination of both to provide nonmarring blows to the work surfaces. This type of hammer provides a somewhat softer blow than the ball peen hammer.

FIGURE 3–24 Various soft-faced hammers. *(Courtesy of Snap-on Tools Company)*

PULLERS

Two and three jaw pullers, Figure 3–25, are available for service procedures, such as removing a bearing from a shaft. There are general purpose pullers available that may be used for specific procedures, such as removing a compressor rotor. There are also dedicated pullers that are used for particular procedures, such as for removing an air conditioner compressor shaft seal, Figure 3–26.

SAWS

Those who install aftermarket air-conditioning systems may find a hacksaw, Figure 3–27, a handy tool for removing unwanted metal protrusions. A hole saw,

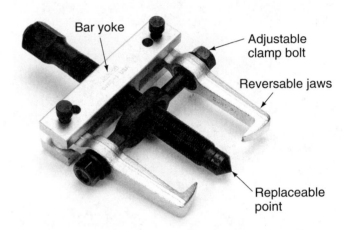

FIGURE 3–25 Typical puller.

FIGURE 3–26 Dedicated puller, such as for removing a compressor shaft seal.

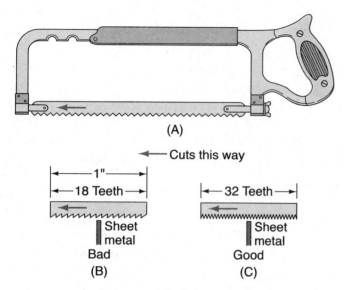

FIGURE 3–27 Hacksaws: (A) The blade is installed with the teeth facing forward. (B) A hacksaw blade that is too coarse will not work well with sheet metal. (C) A fine-toothed blade is used with thin metal.

FIGURE 3–28 A typical hole saw.

Figure 3–28, is a necessity for drilling holes in the fire wall and radiator shroud to route the refrigerant hoses. Generally, saw blades for automotive service should have 32 teeth per inch (tpi).

DRILLS

Corded or cordless drill motors are often used by the technician for drilling holes to mount components and accessories or to pass hoses through. Drill motors are classified by their maximum chuck capacity. The most popular are 1/4 inch and 3/8 inch.

Drill Bits

Drill bits are available in English sizes from 1/16 inch through 1/2 inch in 1/64-inch increments. They are also available in comparable metric sizes from 1.5 mm through 13 mm.

MANIFOLD AND GAUGE SET

The **manifold** and gauge set, Figure 3–29, generally consists of a manifold with two hand valves and two gauges, a compound gauge and a pressure gauge, and three hoses. There are several types of manifold and gauge sets. Regardless of the type, they all serve the same purpose.

Note that the manifold has fittings for the connection of three hoses. The hose on the left, below the compound gauge, is the low-side hose; on the right, below the pressure gauge, is the high-side hose. The center hose is used for system service, such as for evacuating and charging (see Chapter 8). Some manifolds may be equipped with two center hoses: a small hose, generally 1/4 in. or 6 mm for refrigerant **recovery** and charging; and a large hose, generally 5/16 in. or 8 mm, for evacuation.

Regardless of the type selected, a separate and complete manifold and gauge set with appropriate service hoses that have unique fittings are required for each type of refrigerant that is to be handled in the service facility. This means that a minimum of three sets are generally required: one set for CFC-12 refrigerant, one set for HFC-134a refrigerant, and a set for contaminated refrigerant.

Low-Side Gauge

The low-side gauge, also referred to as the compound gauge, indicates either a vacuum or pressure. Generally, this gauge will be calibrated from 30 in. Hg (0 kPa absolute) vacuum to 250 psig (1,724 kPa) pressure. Actually, the pressure calibration is to 150 psig (1,035 kPa) with retard to 250 psig (1,724 kPa) maximum. That means that pressures to 150 psig (1,035 kPa) may be read with reasonable accuracy, while pressures to 250 psig (1,724 kPa) may be applied without damage to the gauge movement. The low-side gauge is found at the left of the manifold.

High-Side Gauge

The high-side gauge is usually calibrated from 0 psig (0 kPa) to 500 psig (3,448 kPa). Insomuch as the high side of the system will never go into a vacuum, pressures below 0 psig (0 kPa) are not indicated on the high-side gauge. The high-side gauge is also often referred to as the pressure gauge. The high-side gauge is found at the right of the manifold.

The Manifold

There are "circuits" in the manifold that connect the hose ports to the gauge ports. When both hand valves are closed, both the low- and high-side hose ports are connected only to the low- and high-side gauges. If the hoses

FIGURE 3–29 Typical manifold and gauge set. *(Courtesy of Uniweld Products)*

were connected from the manifold to an air-conditioning system, the low-side gauge will indicate the pressure on the low side of the system and the high-side gauge will indicate the pressure on the high-side of the system.

More detailed information is given on the manifold and gauge set in Chapter 7.

Hoses

Specific requirements for service hoses used in automotive air-conditioning service are given in the Society of Automotive Engineers (SAE) standards J2196 and J2197 and are covered in more detail in Chapter 7.

CAN TAP

The **can tap**, Figure 3–30, is used to dispense refrigerant from a "pound" can. Pound disposable cans actually contain 12 or 14 oz. (340 or 397 g) of refrigerant and are available in either screw top or flat top. Some can taps are designed to fit either type.

SAFETY GLASSES

There are several types of safety glasses, (Figure 3–31) available. A safety shield-type goggle may be used with or without eyeglasses. It is important to note that glasses or goggles selected should be a type that is approved for working with liquids or gases, meeting ANSI Z87.1-1989 standard.

SPECIAL TOOLS

Basically, three special tools expand service and repair capabilities considerably: a thermometer, a leak detector, and a vacuum pump.

Thermometer

A glass-type or a dial-type thermometer may be used. The glass type is usually less expensive, but it is more easily broken. Regardless of the type it is suggested that the temperature range be from 0°F to 220°F (–18°C to 104°C). For more accurate and reliable service, an electronic digital thermometer, Figure 3–32, is recommended.

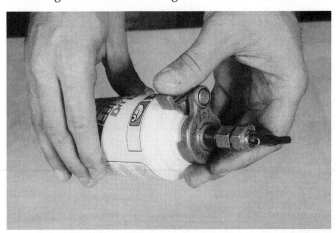

FIGURE 3–30 A typical can tap valve.

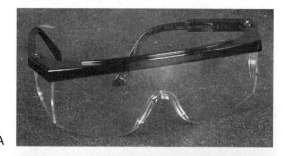

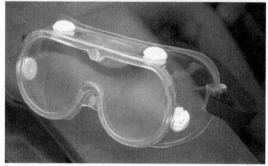

FIGURE 3–31 Safety glasses (A) and goggles (B). *(Courtesy of Goodson Shop Supplies)*

FIGURE 3–32 Typical thermometers: (A) digital, (B) analog, and (C) electronic.

Leak Detector

In many cases, leaks can be detected by the use of a soap solution. A good dish washing liquid mixed with an equal amount of clean water and applied with a small brush will indicate a leak by bubbling. A commercially available product, such as Leak Finder® can also be used.

Electronic Leak Detector. Electronic leak detectors, Figure 3–33, called **halogen leak detectors,** although considerably expensive, are desirable because they offer great sensitivity and can pinpoint a leak as slight as 0.5 oz. (14 g) per year. It should be noted that halogen leak detectors that can be used to test either CFC-12 or HFC-

134a refrigerants are available. When refrigerant vapor enters a halogen leak detector's search probe, the device emits an audible and/or visual signal.

CAUTION

A HALOGEN LEAK DETECTOR MUST NOT BE USED IN A SPACE WHERE EXPLOSIVES SUCH AS GASES, DUST, OR VAPOR ARE PRESENT. USE ONLY IN A WELL-VENTI-LATED AREA. BY-PRODUCTS OF DECOMPOSING CFC AND HCFC REFRIGERANTS, SUCH AS HYDROCHLORIC (HCl)AND HYDROFLUORIC ACID (H_2F_2), ARE A HEALTH HAZARD.

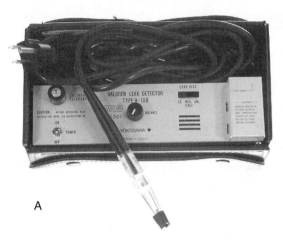

A

FIGURE 3–33 Electronic leak detectors. (A *Photo by Bill Johnson;* B *Courtesy of Robinair Division, SPX Corporation)*

Fluorescent Leak Detector. Fluorescent leak detectors, Figure 3–34, are becoming more and more popular. A fluorescent dye is injected into the system, where it remains without affecting cooling performance. When a leak is suspected, an ultraviolet lamp will quickly and efficiently pinpoint the problem area.

Vacuum Pump

It is necessary to remove as much moisture and air from the system as possible before charging it with refrigerant. This is accomplished with the use of a vacuum pump, Figure 3–35. The vacuum pump is usually provided by the service facility. Some refrigerant recovery equipment has a vacuum pump incorporated in the equipment. High-volume service facilities, however, cannot generally tie up recovery equipment to be used as a vacuum pump for the time required to adequately evacuate an automotive air-conditioning system.

REFRIGERANT IDENTIFIER

To determine what type of refrigerant is in a system, a refrigerant identifier, Figure 3–36, should be used before servicing the refrigeration system of any vehicle. The refrigerant identifier is used to identify the purity and quality of a gas sample taken directly from a refrigeration system or a refrigerant storage container. The identifier will display the following:

D **R-12:** If the refrigerant is CFC-12 and its purity is better than 98 percent by weight.

U **R-134a:** If the refrigerant is HFC-134a and its purity is 98 percent or better by weight.

LI **FAIL:** If neither CFC-12 or HFC-134a have been identified or if it is not at least 98 percent pure.

D **HC:** If the gas sample contains hydrocarbon, a flammable material. A horn will also sound.

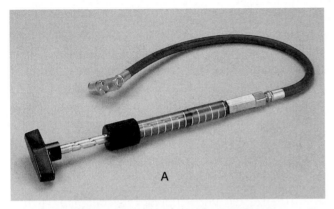
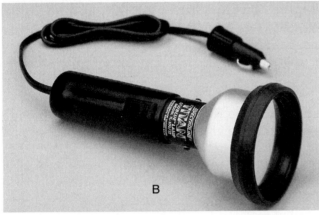

FIGURE 3–34 Fluorescent refrigerant leak detection system using additive (A) with high-intensity ultraviolet lamp (B). *(Courtesy of Spectronics Corporation, Westbury, NY)*

FIGURE 3–35 A typical high vacuum pump. *(Courtesy of Robinair Division, SPX Corporation)*

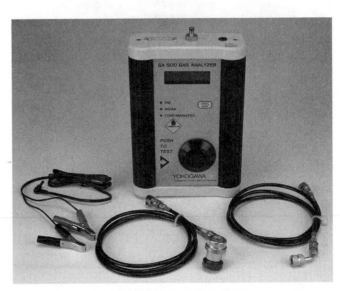

FIGURE 3–36 A typical refrigerant identifier.

OTHER SPECIAL TOOLS

Other special tools available to the service technician are generally supplied by the service facility. These tools include a refrigerant recovery and recycling system, antifreeze recovery and recycle system, an electronic thermometer, and an electronic scale. Also, special testers are available for automatic temperature control (ATC) testing, and special tools are available for servicing and repairing compressors.

The following is a brief description of these tools. They are covered in more detail in the appropriate chapters of this manual.

Refrigerant Recovery and Recycle System

The service center must have a recovery, recycle, and recharge machine, Figure 3–37, for each type of refrigerant to be serviced. Some refrigerant recover, recycle, and recharge machines, however, may be used for both CFC-12 and HFC-134a. A single-pass system with an onboard microprocessor controls the evacuation time as well as the amount of refrigerant charged into the system. The mixing of CFC-12 and HFC-134a refrigerants is prevented by the use of a sliding lockout panel, allowing only one set of manifold hoses to be connected at any time. Also, the fittings on the hoses prevent them from being connected to the wrong port. Each type of refrigerant has a separate dedicated set of hoses and recovery tank. A self-clearing loop removes residual refrigerant from the machine before connecting the other set of hoses and recovery tank.

Other desirable features of a recover, recycle, recharge system include an automatic air purge, a high-performance vacuum pump, and an automatic shut-off when the tank is full.

Antifreeze Recovery and Recycle System

An antifreeze recovery and recycle machine, such as Prestone's ProClean Plus™ Recycler, Figure 3–38, is a self-contained system that drains, fills, flushes, and pressure tests the cooling system. It can also be used to recycle coolant. A typical cooling system drain, recycle, and refill takes about 20 minutes.

Some recovery systems add an additive during the recycle phase to bond heavy metals, such as lead (pb) and other contaminants. This renders them into non-leachable solids that are not hazardous as defined by the EPA. Additives separate the contaminants so they can be easily removed. Also, inhibitors may be added to protect against corrosion and acid formation.

Other desired onboard functions of the machine include standard coolant exchange, flushing procedures,

FIGURE 3–37 A typical refrigerant recovery unit. *(Courtesy of Robinair Division, SPX Corporation)*

FIGURE 3–38 A typical antifreeze recovery/recycle machine. *(Courtesy of RobinairDivision, SPX Corporation)*

pressure testing for leaks, and vacuum fill for adding coolant to an empty system.

Electronic Thermometer

A single- or two-probe, hand-held electronic thermometer, Figure 3–39, is commonly used in mobile air-conditioning system diagnosis and service. The two-probe model is used to quickly and accurately measure

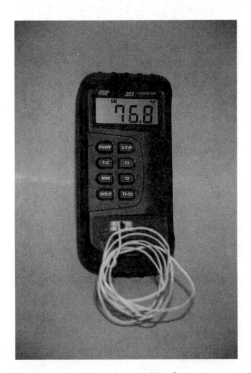

FIGURE 3–39 Digital-type electronic thermometer. *(Photo by Bill Johnson)*

FIGURE 3–40 Electronic charging scale.

superheat as required for critically charging some HFC-134a air-conditioning systems.

The range for most battery-powered digital electronic thermometers is generally –50°F to 2000°F (–46°C to 1093°C) and have an accuracy greater than ±0.3 percent.

Electronic Scale

An electronic scale, Figure 3–40, may be used for CFC-12 or HFC-134a refrigerants to deliver an accurate charge by weight, manually or automatically. Automatic charging is generally accomplished by programming the amount of refrigerant to be charged into the onboard solid-state microprocessor. The charge is stopped and an audible tone signals that the programmed weight has been dispersed. A liquid crystal display is used to keep track of the refrigerant dispersed. Some models have a switchable pounds/kilograms readout with a resolution of 0.05 lb. (0.02 kg). The scale platform should handle up to a 50 lb. (23 kg) bulk tank of refrigerant and be equipped with control panel fittings and two hoses to accommodate both CFC-12 and HFC-134a refrigerants.

Automatic Temperature Control (ATC) Testers (Scan Tool)

There are many different types of testers available. The scan tool is very popular and is used to enhance troubleshooting efforts to quickly get to the root of a problem. Scan tools are available in a wide variety of brands, prices, and capabilities. One such tool is shown in Figure 3–41. One good feature is that not only can a scan tool be used to retrieve trouble codes, some allow the technician to monitor and view sensor and computer information. This fea-

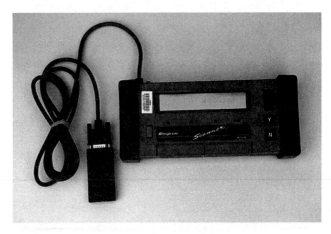

FIGURE 3–41 A typical automatic temperature control (ATC) tester.

ture, known as serial data or the data stream, helps to pinpoint a heating, ventilation, or air-conditioning (HVAC) problem. A scan tool can sometimes even take the role of a manifold and gauge set by obtaining system pressure readings through transducers in refrigerant lines.

Compressor Tools

There are several special tools required for compressor clutch and shaft seal service such as those shown in Figure 3–42. Clutch plate tools are used to remove the clutch plate to gain access to the shaft seal. They are also used for reinstalling the clutch plate after service. These tools should be compact in design for working in close quarters so that it may often be possible to service the compressor clutch and shaft seal without having to remove the compressor from the vehicle.

Basically, a shaft seal service kit includes an adjustable spanner wrench, clutch plate remover/installer, snap-ring pliers, ceramic seal remover/installer, seal seat remover/installer, shaft seal protector, seal assembly remover/installer, thin wall socket, O-ring remover, and O-ring installer. For clutch pulley and bearing service, service tools include a pulley puller, pulley installer, bearing remover/installer, and rotor and bearing installer.

Special compressor service tools are designed to

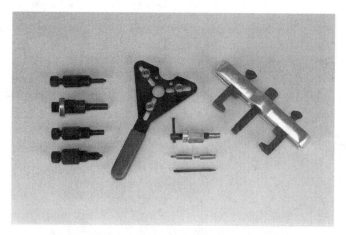

FIGURE 3–42 Typical compressor service tools.

fit a particular application. Although some are interchangeable, most are not.

SUMMARY

It is very important to always refer to the equipment manufacturer's instructions for specific and proper tool usage, and to specific local regulations relating to refrigerant and antifreeze handling and disposal requirements.

REVIEW

Select the correct answer from the choices given.

1. All of the following are a metric term used in the MVAC industry, EXCEPT:
   a. centigrade.
   b. meter.
   c. kiloPascal.
   d. liter.

2. Technician A says the metric system is only used in certain parts of the world. Technician B says the metric system is based on powers of 10. Who is right?
   a. A only
   b. B only
   c. Both A and B
   d. Neither A nor B

3. Convert the following fractions to decimals:
   a. 1/10 _____
   b. 5/32 _____
   c. 1/25 _____
   d. 3/8 _____

4. Convert the following temperature scales:
   a. 42°F _____
   b. 200°F _____
   c. 42°C _____
   d. 100°C _____

5. Convert the following measure:
   a. 10 ft. _____
   b. 10 mi. _____
   c. 254 mm _____
   d. 254 cm _____

6. Convert the following weights:
   a. 5 lb. _____
   b. 27 oz _____
   c. 250 g _____
   d. 25 kg _____

7. Convert the following liquid measure:
   a. 10 L _____
   b. 200 mL _____
   c. 40 gal _____
   d. 20 oz _____

8. Convert the following pressure values:
   a. 10 kPa _____
   b. 25 psig _____
   c. 10 psig _____
   d. 26 kPa _____

9. What is sea level atmospheric pressure?
   a. 14.696 kPa
   b. 14.696 psi
   c. 100 psia
   d. 100 kPa

10. Technician A says a manifold and gauge set is an essential tool for air-conditioning system service. Technician B says a manifold and gauge set is required to remove or add refrigerant to an air-conditioning system. Who is right?
    a. A only
    b. B only
    c. Both A and B
    d. Neither A nor B

11. Which of the following wrenches is *most likely* to cause bolt head damage?
    a. Box-end wrench
    b. Open-end wrench
    c. Combination wrench
    d. Adjustable wrench

12. Technician A says a hex wrench is often called a Crescent wrench. Technician B says a hex recess looks a lot like a Torx® recess. Who is right?
    a. A only
    b. B only
    c. Both A and B
    d. Neither A nor B

13. Water pump pliers are often called either of the following, EXCEPT:
    a. slip joint.
    b. tongue and groove.
    c. groove joint.
    d. offset joint.

14. What standards must safety glasses meet?
    a. ANSI
    b. OSHA
    c. EPA
    d. SAE

15. Technician A says a vacuum pump is used to remove moisture. Technician B says a vacuum pump is used to remove air. Who is right?
    a A only
    b. B only
    c. Both A and B
    d. Neither A nor B

16. Technician A says a refrigerant identifier is used to determine the amount of impurities there are in a sample of refrigerant. Technician B says a refrigerant identifier is used to determine the percentage of cross contamination in a sample of refrigerant. Who is right?
    a. A only
    b. B only
    c. Both A and B
    d. Neither A nor B

17. Technician A says that one electronic scale may be used to meter both CFC-12 and HFC-134a refrigerants into an air-conditioning system. Technician B says that one vacuum pump may be used to evacuate either a CFC-12 or an HFC-134a air-conditioning system. Who is right?
    a. A only
    b. B only
    c. Both A and B
    d. Neither A nor B

18. All of the following may be considered a compressor repair tool, EXCEPT:
    a. can tap.
    b. O-ring installer.
    c. snap ring pliers.
    d. spanner wrench.

19. A nut driver resembles a _____.
    a. box wrench
    b. open-end wrench
    c. hex wrench
    d. socket wrench

20. A combination wrench _____.
    a. has a box wrench on one end and an open-end wrench on the other end
    b. has an open-end wrench on one end and a socket wrench on the other end
    c. both A and B
    d. neither A nor B

TERMS

Provide a brief description of the following terms:

1. ASHRAE
2. can tap
3. Celsius
4. flare-nut wrench
5. gauge
6. halogen leak detector
7. hex key wrench
8. kiloPascal
9. lb.
10. liter
11. manifold
12. metrics
13. mm
14. oz.
15. pressure
16. psig
17. recovery
18. snap ring pliers
19. torque
20. vacuum

CHAPTER 4

COMFORT

OBJECTIVES

On completion and review of this chapter, you should be able to:

❏ Discuss the fundamentals of heat transfer.
❏ Explain the terms *sensible heat* and *latent heat.*
❏ Describe how heat flows.
❏ Understand the effects of air movement.
❏ Discuss the conditions that affect human comfort.
❏ Understand the effects of humidity.
❏ Discuss odor problems and how to eliminate them.

HUMAN COMFORT CONTROL

A human being is sensitive to impurities such as dust, smoke, and pollen that cause irritation to the nose, lungs, and eyes, thus the need for clean air. A human also requires fresh air to renew its oxygen supply as well as to dilute undesirable odors. Five properties of the air must be considered for treatment to provide a comfortable and healthful environment:

❏ Its temperature; by cooling or heating.
❏ Its moisture content; by humidifying or dehumidifying.
❏ Its movement; by circulation.
❏ Its cleanliness; by filtration.
❏ Its ventilation; by introducing fresh outside air to replace stale air.

It is important to first examine how the human body adjusts to the temperature changes of its surroundings because air conditioning is primarily concerned with providing a comfortable environment for humans.

The inside temperature of the average human body, better known as "subsurface" or "deep tissue" temperature is approximately 98.6°F (37°C). The temperature at the skin surface, however, is about 70°F (21.1°C). If the temperature of the surrounding air or any substance immediately adjacent to the skin should rise above the skin surface temperature, we feel "hot"; if below skin surface temperature we feel **"cold."**

Whenever the nerve endings at the skin surface send a signal to the brain that the surrounding temperature is decreasing, the brain sends a signal to the pores of the skin, causing them to close up tightly to prevent releasing body moisture. If the temperature becomes lower, humans must don additional clothing to build a cushion of warm air next to the skin to confine body heat.

Whenever the nerve endings at the skin surface send a signal to the brain that the surrounding temperature is rising, the brain sends a signal to the pores in the skin, causing them to open, releasing body fats in the form of a liquid, known as **perspiration.** Most of this liquid is evaporated when it comes in contact with the ambient air. This **evaporation** is accompanied by the cooling of the skin surface, which, in turn, releases its heat by **convection** to the surrounding air. Because of this convection, a breeze will lower the skin surface temperature by carrying away heat faster as it speeds up the evaporation.

The body's interior heat now flows to the cooler skin by conduction to cool the entire body. A balance is thus maintained so that the deep tissue heat of a healthy body never deviates more than plus or minus (±) one degree **Fahrenheit** (1°F) from its normal temperature, regardless of the outside ambient temperature.

COMFORT ZONE

Although the deep tissue body heat remains nearly constant, this does not mean that one is always comfortable. The comfort zone that feels comfortable for most of us covers a range of a few degrees below and above the skin surface temperature of about 70°F (21.1°C). Outside this zone most of us will feel too hot or too cold. This comfort zone, however, varies somewhat from person to person. The **relative humidity** (RH) as well as the temperature and movement of the surrounding air are important factors for producing comfort or discomfort.

There is no one specific set of temperature and humidity conditions, or comfort zone, that is comfortable for everyone. People react differently to different conditions. The American Society of Heating, Refrigerating, and Air-Conditioning Engineers (ASHRAE) has conducted extensive research for many years in an attempt to determine the range of combined temperature, humidity, and air movement that provides the most comfort for the most people. Known as the comfort zone, each combination of

temperature, humidity, and air velocity is called an **effective temperature** (ET). For example, for a given air velocity there are several different combinations of dry-bulb temperatures and relative humidity conditions that provide the same feeling of comfort to over 90 percent of those involved in the survey.

A comfort zone, Figure 4–1, was developed from this study. From the shaded area it can easily be determined that the higher the humidity, the lower the dry-bulb temperature can be.

There are three main conditions that affect personal body comfort as illustrated in Figure 4–2. They are *air movement, temperature,* and *humidity.*

Air Movement

Air movement, or velocity, is an important factor in considering human comfort conditions. The comfort zone just discussed is based on an air velocity of 15 to 25 ft./min. The effective temperature drops sharply as air velocity is increased, a desirable trait for summer air conditioning. Conditioned air is usually introduced at about 12°F to 15°F (11.1°C to 9.4°C) below ambient conditions. If the velocity should approach 100 ft./min., uncomfortable cold drafts would be experienced.

Warm-air heating is more subject to drafts, particularly when the blower first comes on. The human

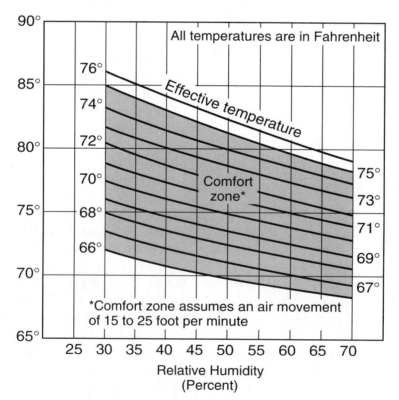

FIGURE 4–1 Typical comfort zone for the average person.

Personal Comfort

Relative Humidity

Air Motion

Temperature

FIGURE 4–2 Three conditions that affect personal comfort.

body, it seems, reacts more quickly to warm air currents than to cold air currents.

It is recommended that both heating and cooling air velocity be 70 ft./min. maximum. On the other hand, too little air velocity should also be avoided, because people tend to feel closed in.

Airflow considerations are not generally a problem for the automotive air-conditioning technician. The air-conditioning system in today's modern vehicles have three-, four-, five-, or infinite-speed blower motors so the operator can regulate the air velocity. Some systems have dual controls—one for the driver and one for the front seat passenger. Others have dual controls but one is for front seat occupants and the other is for rear seat occupants.

Another factor that affects the ability of the body to give off heat is the movement of air around the body. The following processes occur as the air movement increases:

❑ The evaporation process of removing body heat speeds up because moisture in the air near the body is carried away at a faster rate.
❑ The convection process increases because the layer of warm air surrounding the body is carried away more rapidly.
❑ The radiation process increases because the heat on the surrounding surfaces is removed at a faster rate.

As a result, heat radiates from the body at a faster rate. As the air movement decreases, the process of evaporation, convection, and radiation decreases. A transfer of heat, however, cannot take place unless there is a difference in temperature between the two objects in which the transfer is to take place. The greater the difference in temperature, often referred to as *delta T* (Δ_T), the greater the heat transfer.

Wind Chill Factor. The windchill factor was developed in 1941 to measure relative personal discomfort due to the combination of the temperature and wind speed. A windchill factor chart, shown in Figure 4–3, is based on physiological studies of the rate of heat loss for various combinations of ambient temperature and

THE EFFECT OF WIND CHILL

AIR TEMP (°F)	WIND SPEED (MPH)							
	5	10	15	20	25	30	35	40*
40	37	28	23	19	16	13	12	11
30	27	16	9	4	1	–2	–4	–5
20	16	3	–5	–10	–15	–18	–20	–21
10	6	–9	–18	–24	–29	–33	–35	–37
0	–5	–22	–31	–39	–44	–49	–52	–53
–10	–15	–34	–45	–53	–59	–64	–67	–69
–20	–26	–46	–58	–67	–74	–79	–82	–84
–30	–38	–58	–72	–81	–88	–93	–97	–100
–40	–47	–71	–85	–95	–103	–109	–113	–115

*Winds above 40 mph (64 km/h) have little additional effect on windchill factor.

FIGURE 4–3 Windchill factor.

wind speed. The windchill factor is based on the actual air temperature when the wind speed is 4 mph (6.4 **km/h**) or less. At higher wind speeds, the windchill temperature is lower than the air temperature; the windchill factor measures the increased cold stress and discomfort associated with the wind.

The air temperature is not lowered by the windchill factor. Regardless of how strong the wind, the air temperature remains constant. The windchill factor is a measure of how rapidly heat is being removed from the human body. If, for example, the air temperature is 40°F (4.4°C) and the wind speed is 20 mph (32 km/h), it feels the same as 19°F (–7°C) would feel with no wind blowing.

A windchill factor near or below 0°F (–17.8°C) is an indication that there is a risk of frostbite or other injury to exposed human flesh. At temperatures between 10°F and 15°F (–12.2°C and –9.4°C) there is little danger; between –30°F and –70°F (–35.6°C and –57.6°C) there is danger that human flesh may freeze within one minute of exposure. Below -75°F (–59.4°C) there is great danger that human flesh may freeze within 30 seconds of exposure.

The actual effects of **windchill,** however, depend on several factors, such as the amount of clothing worn, health, age, gender, and body weight of the individual.

Temperature

Cool air increases the rate of convection; warm air slows it down. Cool air lowers the temperature of the surrounding surfaces. Therefore, the rate of radiation increases. Because warm air raises the surrounding surface temperature, the radiation rate decreases. In general, cool air increases the rate of evaporation and warm air slows it down. The evaporation rate also depends on the amount of moisture that is already in the air (humidity) and the amount of air movement (wind).

Humidity

Water is required inside the body for survival. When one is deprived of water, one becomes **dehydrated.** The body must also be surrounded by water in the form of vapor, commonly referred to as humidity. Because of evaporation, the atmosphere always contains some moisture in the form of vapor. The weight of the water contained in a volume of air is the absolute humidity and is usually expressed as a percentage of the vapor by weight.

Generally, the air does not have sufficient water vapor to produce saturation. The ratio of water vapor that exists in the air and the amount needed to produce saturation is called relative humidity (RH).

Perhaps you have heard the expression, "It's not the temperature; it's the humidity." One can be uncomfortable due to humidity regardless of the temperature. The moisture content of air, when low, is most often sensed by a feeling of dryness or a muggy, sticky feeling, when high.

Humidity refers to moisture that has been evaporated into the ambient air as an invisible vapor. A wet-bulb thermometer is used to determine humidity. A wet-bulb thermometer, Figure 4–4A, is an ordinary thermometer, Figure 4–4B, with a wick placed over its bulb. By wetting the wick and passing air over it, moisture will be evaporated until it is in balance with the moisture content of the air passing over it. At that point no more evaporation takes place and the thermometer will indicate the **wet-bulb temperature,** which is always lower than the dry-bulb temperature.

Another instrument used for determining humidity is the sling psychrometer, Figure 4–5, which takes dry-bulb and wet-bulb temperatures simultaneously. Two

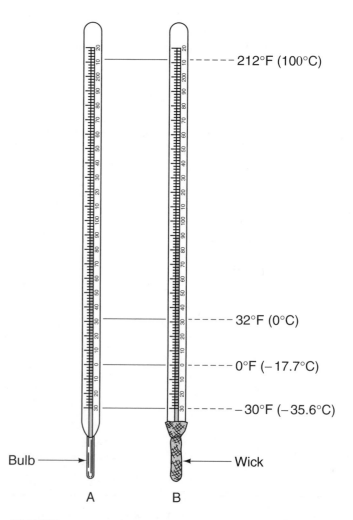

FIGURE 4–4 Typical dry bulb (A) and wet bulb (B) thermometers.

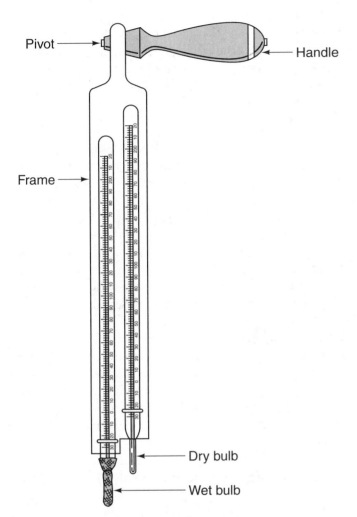

Pivot

Handle

Frame

Dry bulb

Wet bulb

FIGURE 4–5 Typical sling psychrometer.

matching thermometers that have been scaled and calibrated the same are mounted on a common frame and attached to a handle so the technician can rotate (sling) the instrument.

After two or three rotations per second for about a minute, the dry-bulb and wet-bulb thermometer readings are compared to a chart to determine the actual amount of moisture in the air as compared to the maximum amount that air could hold at that dry-bulb temperature.

If both thermometer readings were the same the relative humidity would be 100 percent. The difference between the two readings is known as the wet-bulb depression. For example, as may be noted in Figure 4–6, if the dry bulb were 72°F and the wet bulb were 61°F, a wet-bulb depression of 11°F is indicated. The relative humidity, according to the chart, will be 53 percent.

Controlling relative humidity is important so the air will be dry enough in the summer time to absorb body perspiration for comfort. In the winter time the air should be humid enough so the skin, nose, and throat do not have a sensation of dryness. Also, high humidity is a contributing factor in causing mold, **mildew,** and rust problems.

Relative humidity, then, defines the amount of moisture in the ambient air relative to any given temperature. For example, an RH of 60 at 90°F means that the ambient air contains 60 percent of all of the moisture that it can hold at that temperature. The RH range is generally 20 to 90 percent.

A low moisture content in the ambient air makes us feel uncomfortable. If moisture is too low, the evaporation process is accelerated and we feel cold. If, however,

DRY-BULB TEMP (°F)	RELATIVE HUMIDITY PERCENT									
65	90	79	70	61	53	43	35	27	20	12
70	91	81	72	64	55	47	39	31	28	19
75	91	82	74	66	58	51	43	38	32	25
80	91	83	75	68	61	54	47	41	35	29
85	92	84	76	70	63	56	50	44	38	32
90	92	85	78	71	65	58	52	48	42	37
95	93	86	79	72	66	60	54	50	45	40
100	93	86	80	73	68	62	56	51	47	42
105	93	87	80	74	69	63	58	53	48	44
Depression ☞	2	4	6	8	10	12	14	16	18	20

To determine relative humidity (RH) by Depression, subtract wet-bulb from dry-bulb readings.

FIGURE 4–6 Relative humidity based on wet-bulb and dry-bulb temperature readings.

moisture is too high, the evaporation of perspiration is slowed down because the air near the skin is so heavily laden with moisture that it cannot absorb more. This slows down the cooling process, and we perspire heavily even without exertion.

As discussed earlier, the moisture content of air is measured in terms of relative humidity (RH). The term *50 percent relative humidity,* for example, means that the air contains one-half the amount of moisture that it is capable of holding at a given temperature.

A low relative humidity permits heat to be taken away from the body by evaporation. Because low humidity means the air is relatively dry, it can readily absorb moisture. Conversely, a high relative humidity has the opposite effect. The evaporation process slows down in humid conditions; thus, the speed at which heat can be removed by evaporation decreases.

The comfort zone of the human body is between 30 and 70 percent relative humidity (RH) and a temperature of 65°F to 85°F (18.3 to 29.4°C). A discomfort index (DI) has been developed, which reduces the temperature and humidity index to one value. The formula for finding the DI in any given space is:

$$DI = 0.4 \, (DB + WB) + 15$$

For example, assume a dry-bulb temperature of 78°F and a wet-bulb temperature of 68°F. The formula becomes:

$$0.4 \times (78 + 68) + 15 \text{ or } 0.4 \times 146 + 15 = 73.4.$$

In this formula **DB** is the dry-bulb temperature and WB is the wet-bulb temperature, both in the English Fahrenheit scale. Nine out of ten, or 90 percent will feel comfortable at a DI of 70 or less. About half, or 50 percent, will feel uncomfortable at a DI of 75, but everyone will be uncomfortable at a DI of 79. Serious discomfort that affects one's ability to work occurs above a DI of 85.

The discomfort index (DI) is generally not a concern of the mobile air-conditioning system technician. Determining relative humidity (RH), however, may be required for performance troubleshooting an air-conditioning system. To determine RH, subtract the WB reading from the DB reading and find the RH in percent in Figure 4–6.

For health reasons it is recommended that the relative humidity (RH) for an occupied space be maintained between 40 percent and 60 percent. Studies of indoor air quality (IAQ) have shown that bacteria, fungi, and viruses become more active below 40 percent RH and above 60 percent RH. It is not uncommon to hear a complaint of a "musty mildew odor" coming from the air-conditioning system. This odor is caused by a mildew-type fungus that forms in the evaporator case during the off period of the air-conditioning system. The procedures for cleaning the evaporator core to correct this problem follow.

Odor Problems

Under some climatic and operating conditions, a musty odor is detected from the air-conditioning system due to mold growth on the evaporator core. This is generally a temporary problem and, as climate conditions change, will disappear on its own. Sometimes this problem can be eliminated by spraying an air freshener, deodorizer, or antibacterial product into the air-conditioning system fresh-air or recirculate-air intake. If the odor persists, however, it will be necessary to clean the evaporator core with an appropriate cleaner. A typical procedure for this service follows:

Deodorizing Procedure. Odors emitted from the air-conditioning system, primarily at startup, in hot humid climates may be the result of debris in the heater or evaporator case. More often, however, it is due to a growth of mold on the evaporator core. To correct this condition it is possible to clean the core without removing it from the vehicle.

Always follow the manufacturer's specific procedures for cleaning an evaporator core. The following procedure is given for cleaning the core in certain General Motors vehicles and is to be considered typical.

For this procedure a cleaning gun, **disinfectant** kit, hose, pedestal fan, rubber gloves, safety goggles, selected hand tools, 4-quart pan, and 2 quarts of clean water are required.

> CAUTION
> PERFORM THIS PROCEDURE ONLY:

- ❑ on a cold vehicle to prevent the disinfectant from coming in contact with hot engine components.
- ❑ while wearing personal protection, such as rubber gloves and goggles. If disinfectant gets into the eyes, hold the eyelids open and flush with a gentle stream of water for 15 to 20 minutes. Seek medical attention if irritation persists.
- ❑ in a well-ventilated area. Use a pedestal fan to provide cross ventilation.

Procedure

1. Don protective clothing, to include rubber gloves and safety goggles.

2. Mix disinfectant solution following applicable instructions.
3. Place a 4-quart pan under the drain tube under the vehicle. If necessary, install an additional hose onto the drain hose to ensure that all fluids will be recovered.

   NOTE: Ensure that the drain outlet is not restricted.

4. Remove the blower resistor leaving the electrical connector attached.

   NOTE: Ensure that the metal coils of the blower resistor do not become grounded to any metal surface.

5. Visually check the heater or evaporator case for debris and remove any found through the blower resistor opening.

   Note: If debris cannot be removed in this manner the core must be removed from the vehicle for cleaning.

6. Open all vehicle windows and doors.
7. Move the ignition switch to the ON position but do not start the engine.
8. Set the air-conditioning system mode selector to VENT, the temperature to full Cold, and the blower to LOW speed.
9. Position a pedestal fan to provide cross ventilation during the cleaning procedure. Turn the fan on to HIGH speed.
10. Insert the nozzle of the spray gun through the blower resistor opening.
11. While ensuring adequate coverage of the corners and edges, spray directly toward the evaporator face to completely saturate the core.
12. Turn off the air-conditioning system controls as well as the ignition switch.
13. Allow the core to soak for 5 to 10 minutes.
14. Check under the vehicle to ensure proper drainage. If necessary, carefully unclog any drain obstructions.
15. Repeat steps 7 and 8.
16. Using the spray gun, rinse the evaporator core with sufficient clean water to remove all disinfectant residue.
17. Repeat step 12.
18. Reinstall the blower resistor.
19. Dispose of the disinfectant and rinse water runoff collected in the drain pan in an approved manner.

HEAT

Heat is a form of **kinetic energy.** All things in nature, referred to as matter, contain heat, although some matter may contain more heat than others. Generally, heat may be measured on a thermometer. Heat content is often expressed in British thermal units (Btu). A **calorie,** generally associated with food, is actually a measure of heat.

The term *heat* is actually defined in many ways. The definition best suited for air-conditioning service describes heat as a form of energy that can be transferred from one place to another; that is, removed from an area where it is not wanted and placed in an area where it is not objectionable.

How the Body Produces Heat

All food that is taken into the body contains heat in the form of calories. The large or great calorie is used to express the heat value of food. The calorie, a metric term, is the amount of heat that is required to raise 1 kilogram (kg) of water (H_2O) 1°C. There are 252 calories in 1 British thermal unit (Btu). One Btu is equal to the amount of heat that is needed to raise the temperature of 1 pound (0.45 kg) of water (H_2O) 1°F (0.56°C). In the metric system, 1 calorie of heat is required to raise the temperature of 1 gram (g) or 0.035 ounce (oz) of water (H_2O) 1°C (1.8°F).

For example, when 1 pound (0.4536 kg) of water (H_2O) is heated from 75°F to 76°F (23.9°C to 29.4°C), Figure 4–7, 1 Btu of heat energy is absorbed into the water (H_2O). Therefore, 10 pounds (4.536 kg) of water (H_2O), Figure 4–8, would require 10 Btu of heat energy for an increase of 1°F (0.56°C). Consider another example: If 10 pounds (4.536 kg) of water (H_2O) were to be heated from 75°F to 85°F (23.9°C to 29.4°C), how much heat energy would be required? If 1 Btu is required to raise 1 pound (0.4536 kg) 1°F (0.56°C), then 10 Btu would be required to raise each pound (0.4536 kg) 10°F (5.56°C).

$$85°F - 75°F = 10°F$$

Thus, 10 pounds (4.536 kg) Figure 4–9, would require 100 Btu:

$$10 \times 10 = 100$$

As calories are taken into the body, they are converted into energy and stored for future use. This conversion process generates heat. All body movements use up the stored energy and in doing so, add to the heat generated by the conversion process.

The body consistently produces more heat than it requires. Therefore, for body comfort, all of the excess heat produced must be given off by the body.

The Body Rejects Heat

The constant removal of body heat takes place through three natural processes occurring at the same time. These three natural processes are known as convection, radiation, and evaporation.

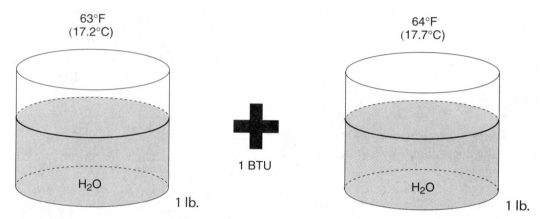

FIGURE 4–7   One Btu raises 1 lb. (0.436 kg) of water (H_2O) 1°F (0.56°C).

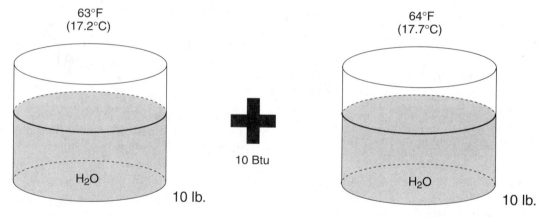

FIGURE 4–8   10 Btu are required to raise ten pound of water one degree Fahrenheit.

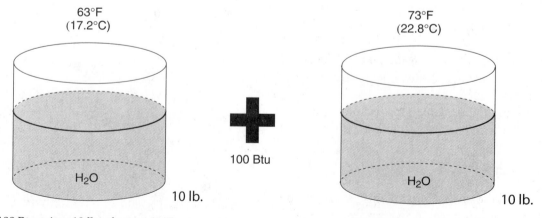

FIGURE 4–9   100 Btu raises 10 lb. of water 10°F.

Convection.

The convection process, Figure 4–10, of removing heat is based on two phenomena:

❏ Heat flows from a hot surface to a surface containing less heat. For example, heat flows from the body to the air surrounding the body when the air temperature is lower than the skin temperature.

❏ Heat rises. This is evident by watching the smoke from a chimney or the steam from boiling water.

When these two phenomena are applied to the body process of removing heat, the following occurs:

❏ The body gives off heat to the surrounding air, which has a lower temperature (less heat).

❏ The surrounding air becomes warmer and moves upward.

❏ As the warmer air moves upward, air containing less heat takes its place.

There are two types of convection: forced and natural. The following is a brief description of each.

Forced Convection.

A vehicle cooling system is a good example of heat transfer by forced convection. The coolant, generally a mixture of water (H_2O) and antifreeze, is circulated through the system by means of a water pump, usually driven by the engine. The coolant removes the heat created by the engine and carries it from the passages of the engine block to the tubes of the radiator. The heat is then dissipated into the air passing through the radiator.

Natural Convection.

Natural convection, based on the premise that hot air rises, is evidenced by the contents of a kettle of water (H_2O) boiling away atop a stove. Water (H_2O), in the form of steam vapor, from the kettle is much warmer than the surrounding air and it therefore rises.

Radiation.

Radiation is the process that moves heat from a heat source to an object by means of heat rays, as shown Figure 4–11. This principle is based on the phenomenon that heat moves from a hot surface to a surface containing less heat. Radiation takes place independently of convection. It does not require air movement to complete the heat transfer. Although this process is not affected by air temperature, it is affected by the temperature of the surrounding surfaces.

Radiation means that heat is being transmitted through a medium although the medium does not become hot. An example of this phenomenon is the manner in which one acquires a sunburn. The heat from the sun is transmitted to the skin through the air although the air is not heated.

It may be interesting to note that heat transferred in this manner diminishes by the square of the distance from its source. For example, if the distance from the heat source is doubled, then the heat intensity will be reduced to one-fourth.

Evaporation.

Evaporation is the process by which moisture becomes a vapor, as illustrated in Figure 4–12. As moisture vaporizes from a warm surface, it removes heat and thus lowers the temperature of the surface. This process takes place constantly on the surface of the body. Moisture is given off through the pores of the skin. As the moisture evaporates, it removes heat from the body.

Perspiration, which appear as drops of moisture on the body, indicates that the body is producing more heat than is being removed by convection, radiation, and normal evaporation.

Conduction.

Conduction means that heat is being transferred through a solid. A good example is food that is frying in a skillet. The heat from the stove burner is conducted through the skillet to the food. Food boiling

Convection

FIGURE 4–10 Body heat is removed by convection.

Radiation

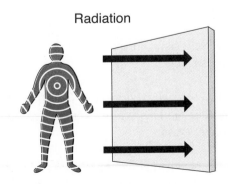

FIGURE 4–11 Body heat is removed by radiation.

FIGURE 4–12 Body heat is removed by evaporation.

in a pot is heated from the burner, through the pot, then through the water.

As a general rule, any material that is a poor conductor of electricity is also a poor conductor of heat. Copper (Cu), an excellent conductor of electricity, is also an excellent conductor of heat. Wood, on the other hand, is a poor conductor of both heat and electricity.

If a length of copper (Cu) rod and a length of wooden dowel were placed in a sunny area, the copper (Cu) rod would conduct much more heat than the wood dowel. The copper (Cu) rod will conduct heat and become too hot to hold long before the wooden dowel.

Sensible Heat

Sensible heat is a term used for any heat that can be felt and that can be measured on a thermometer. A good example of sensible heat is the heat content of the surrounding air. This air is called ambient air, and the temperature of this air is called the ambient temperature. When the temperature drops a few degrees, one feels cool. A few degrees' increase in temperature causes one to feel warm. In other words, we "sense" the changes in temperature, hence the term sensible heat.

What feels warm to one person may feel cool to another person. The sensation of warmth or coldness when one comes in contact with an object is somewhat relative. For example, when two objects are placed together they are said to be in thermal contact. If they are of a different temperature, the object with the higher temperature is cooled while the object with the lower temperature is warmed. At some point in time they will both be the same temperature and no more change will take place. When thermal changes between two objects stop, we say that they are in thermal equilibrium. To our senses, then, both objects would feel the same.

Whenever we say that something is cool or that something is warm we are speaking in relative terms. An English thinker, John Locke, developed an experiment over 300 years ago that is still used to demonstrate the term sensible heat. For the experiment you will need two pans large enough for one hand and one pan large enough for both hands. Two cups (0.94 liter) of hot water and a tray of ice cubes are also needed. Proceed as follows:

1. Place the pans on a level surface with the large pan in the middle.
2. Fill each pan with sufficient tap water to cover the hand(s). Allow sufficient time for the water to adjust to the ambient room temperature.
3. Put 2 cups hot water in the pan on the right.

CAUTION
ENSURE THAT THE WATER IS NOT SO HOT AS TO CAUSE PERSONAL INJURY.

4. Put the tray of ice cubes in the pan on the left.
5. Place both hands in the center pan for 15 to 20 seconds. What do you feel?

   NOTE: You should experience a feeling of neither warmth nor cold because the water in this pan is in thermal equilibrium with the room ambient temperature.

6. Place the left hand in the pan with the ice cubes. What do you feel?

   NOTE: You should experience a sensation of cold because this water is below room ambient temperature.

7. Place the right hand in the pan with warm water. What do you feel?

   NOTE: You should experience a sensation of warmth because this water is above room ambient temperature.

8. Remove your left hand from the left pan and place it in the center pan for a few seconds. What do you feel?

NOTE: The surface temperature of your left hand was lowered, so the water in this pan should now feel warm.

9. Remove your right hand from the right pan and place it in the center pan for a few seconds. What do you feel?

NOTE: The surface temperature of your right hand was raised, so the water should now feel cool.

10. Conclusion: In relative terms, the left hand felt warm when placed in the center pan because the temperature of the ambient water was higher than the iced water.

The right hand felt cool because the temperature of the ambient water was lower than the warmed water. It should now be obvious that the feeling of coldness or warmth is relative in relation to the temperature of an object or to that of the ambient environment.

Latent Heat

The definition of **latent heat** is "hidden heat," the heat that is required to cause a change of state of matter, as illustrated in Figure 4–13. Unlike sensible heat, latent heat cannot be measured on a thermometer and it cannot be felt. The British thermal unit (Btu) is also used as the standard measure for latent heat.

As illustrated previously, a change of state occurs when a solid changes to a liquid or a liquid changes to a gas or vice versa. Water (H_2O) at atmospheric pressure (14.696 psig) between 32°F (0°C) and 212°F (100°C) is called *subcooled liquid*. Water (H_2O) at 212°F (100°C) is called *saturated liquid*; that is, at 212°F (100°C) it contains all of the heat it can hold and still remain a liquid. Any additional heat will cause it to vaporize.

To change the state of 1 pound (0.453 6 kg) of water (H_2O) at 212°F (100°C) to 1 pound (0.453 6 kg) of steam at 212°F (100°C) requires an amount of heat equal to 970 Btu (244.44 kg-cal). This heat is called the *latent heat of vaporization*. Remember that this latent heat cannot be measured on a thermometer and does not cause a change in the temperature.

In addition, steam at 212°F (100°C) gives up 970 Btu (244.44 kg-cal) of heat per pound (0.453 6 kg) as it condenses into liquid at 212°F (100°C). The heat released in this process is called the *latent heat of condensation*.

The additional removal of heat at the rate of 1 Btu (0.252 kg-cal) per pound (0.453 6 kg) lowers the temperature of the water (H_2O). This temperature decrease can be felt as well as measured on a thermometer until it reaches 32°F (0°C).

At 32°F (0°C), all of the heat that can be removed without causing a change of state has been removed. The heat that must be removed so that 1 pound (0.453 6 kg) of liquid water (H_2O) at 32°F (0°C) can be changed to 1 pound (0.453 6 kg) of ice is 144 Btu (35.288 kg-cal). This value of heat energy is called the *latent heat of fusion*.

This principle governing the addition and removal of latent heat energy is the basis for refrigeration and air-conditioning theory. A refrigerant is selected for its ability to absorb and to give up large quantities of latent heat rapidly. Figure 4–14 illustrates the relative values of the latent heat of fusion and the latent heat of vaporization.

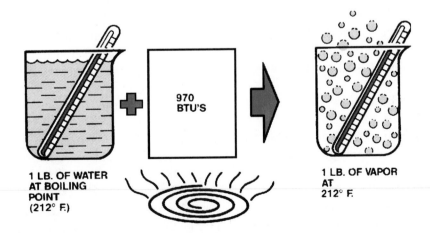

1 LB. OF WATER AT BOILING POINT (212° F.)

970 BTU'S

1 LB. OF VAPOR AT 212° F.

FIGURE 4–13 It takes 970 Btu of latent heat to make 1 pound of boiling water turn into steam without raising its temperature. *(Courtesy of Ford Motor Company)*

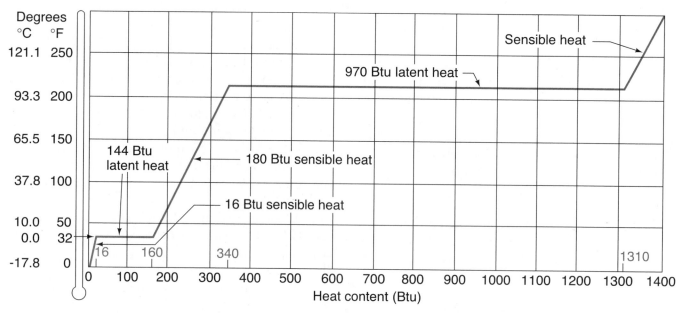

FIGURE 4–14 Latent and sensible heat values for water (H_2O).

Specific Heat

Every element or compound has its own heat characteristics. Every substance has a different capacity for accepting and rejecting heat energy. The capacity to accept (absorb) or reject (expel) heat energy is known as the *specific* or *thermal heat* of a substance. **Specific heat** is defined as the amount of heat that must be absorbed by a material if it is to undergo a temperature change of 1°F (0.56°C).

An experiment can be performed with three small balls, each made of a different substance, such as copper (Cu), steel, and glass. Heat the three balls in a container of hot oil until all three reach the same temperature. Using pliers, carefully place each of the three balls on a slab of paraffin. Observe what happens; each ball will sink to a different depth in the paraffin. The depth to which each ball sinks depends not so much on its weight but on the amount of heat that is emitted. This experiment illustrates that different materials, although at the same temperature, will absorb and emit different amounts of heat.

A scale is used to show the relationship of the abilities of various substances to absorb or emit heat. Water (H_2O) is the standard to which all other substances are compared. The value of water (H_2O) is given as 1, 1.00, or 1.000. When compared to water (H_2O), most other substances will require less heat per unit of weight to cause an increase in their temperature. Two exceptions to this rule are ammonia (NH_3), with a specific heat of 1.10, and hydrogen (H), with a specific heat of 3.410.

For example, the specific heat of glass is 0.194, so it requires less than 1/5 (19.4 percent) the number of Btu to raise its temperature as is required for an equal amount of water (H_2O). The specific heat of copper (Cu) is 0.093. This means that just under 1/11 (10.75 percent) the value in Btu is required to raise the temperature of copper (Cu) as required for an equal amount of water (H_2O).

When a comparison is made between glass or copper (Cu) and water (H_2O), it can be seen that only 0.194 and 0.093 times as much heat energy is required to change the temperature of these materials 1°F (0.56°C). This is true because materials vary in their ability to absorb, emit, and exchange heat. Thus, if equal amounts of copper (Cu), steel, glass, or any other substance are heated through equal changes of temperature, then each material will absorb a different amount of heat. Some commonly used specific heat values are listed in Figure 4–15.

Because it is known that the specific heat of water (H_2O) is 1.000 and that 1 Btu (0.252 kg-cal) is required to raise its temperature 1°F (0.56°C) per pound (0.453 6 kg), there is a simple way of determining how many degrees (°F or °C) per unit of weight other materials can be raised.

$$\frac{1.000}{\text{Specific Gravity}} \div \frac{\text{Weight}}{\text{Heat Energy}} = \frac{\text{Temperature}}{\text{Change}}$$

For example, the specific heat of aluminum (Al) is 0.230. By applying the previous formula, it is found that 1 Btu (0.252 kg-cal) raises the temperature of 1 pound (0.452 kg) of aluminum (Al) 4.35°F (2.42°C).

Air	.0.240	Nitrogen	.0.240
Alcohol	.0.600	Oxygen	.0.220
Aluminum	.0.230	Rubber	.0.481
Brass	.0.086	Silver	.0.055
Carbon dioxide	.0.200	Steel	.0.118
Carbon tetrachloride	.0.200	Tin	.0.045
Gasoline	.0.700	Water, fresh	.1.000
Lead	.0.031	Water, sea	.0.940

FIGURE 4–15 Specific heat values of selected solids, liquids, and gases.

English

$$\frac{1.00}{0.230} \times \frac{1\ \text{lb.}}{1\ \text{Btu}} = 4.347\,83 \div 1 = 4.34783\ (4.35°F)$$

Metric

$$\frac{1.00}{0.230} \times \frac{0.453\,6\ \text{kg}}{0.252\ \text{kg-cal}} = 4.347\,83 \div 1.8 = 2.41545\ (2.42°C)$$

Values for other material can be determined in the same manner. Thus, for lead (Pb), which has a specific heat of 0.031, 1 Btu (0.252 kg-cal) raises the temperature of 1 pound (0.453 kg) of lead (Pb) 32.25°F (17.92°C):

English

$$\frac{1.00}{0.031} \times \frac{1\ \text{lb.}}{1\ \text{Btu}} = 32.2581 \div 1 = 32.2581\ (32.26°F)$$

Metric

$$\frac{1.00}{0.031} \times \frac{0.4536\ \text{kg}}{0.252\ \text{kg-cal}} = 32.2581 \div 1.8 = 17.9211\ (17.92°C)$$

A certain type of specific heat is called a **heat load.** When dealing with automobiles, the heat load is an important factor in the efficiency of an air-conditioning system. Things that must be considered in determining the heat load of a vehicle include the color of the vehicle, amount of glass area, amount and type of insulation, and number of occupants.

When determining the refrigeration requirements for a particular application, the specific heat of all materials involved must be considered in the heat load calculations. For example, in a refrigerated truck body, the specific heat of the product being cooled is an important factor, as is the type and amount of insulation and the type of material in the body. How often the doors are opened and closed as well as the length of time the product is to be refrigerated are also important factors.

The interior temperature of a vehicle reaches very high levels when unattended. The illustration of Figure 4–16 depicts some of the temperature extremes that a vehicle air-conditioning system must overcome in order to provide comfort for the driver and passengers.

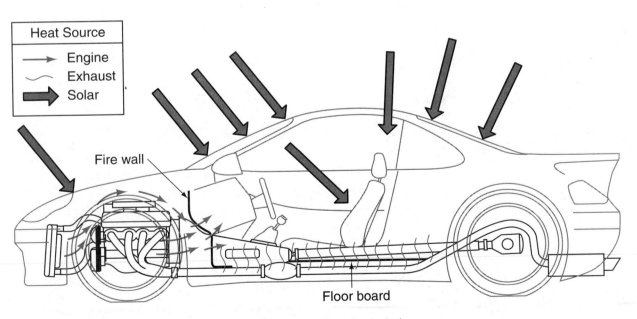

FIGURE 4–16 Some temperature extremes. *(Courtesy of Curtis Dwiggins)*

MEASUREMENT OF HEAT

As discussed earlier, heat energy is measured in terms of the calorie. There are large and small calories. The *gram* or *small calorie* is the smallest measure of heat energy. This discussion of measurements, however, is concerned with the *kilogram* or *large calorie*. One kilogram-calorie (kg-cal) is required to raise the temperature of 1 gram (g) of water (H_2O) from 14.5°C to 15.5°C.

In refrigeration and air-conditioning service, heat energy is expressed in British thermal units (Btu). There are 252 calories (0.252 kg-cal) in 1 Btu. It has been demonstrated that to raise the temperature of 1 pound (0.45 kg) of water (H_2O) 1°F (0.56°C), 1 Btu (0.252 kg-cal) is required.

Water (H_2O), at sea level atmospheric pressure is liquid between the temperatures of 32°F (0°C) and 212°F (100°C). This 180°F (100°C) temperature range is called the *subcooled liquid range* for water (H_2O), Figure 4–17. For each Btu (0.252 kg-cal) of heat added to 1 pound (0.453 kg) of water (H_2O) in this range, the temperature of the water (H_2O) is increased by 1°F (0.56°C). Thus, if 180 Btu (45.36 kg-cal) of heat energy are added to 1 pound (0.453 kg) of water (H_2O) at 32°F (0°C) its temperature is increased to 212°F (100°C).

To obtain the final value of the water (H_2O) temperature, divide the Btu (or kg-cal) being added by the number of pounds (or kilograms) of the water (H_2O). This value is then added to the original temperature (°F or °C) to obtain the new temperature:

$$\text{Btu} \div \text{pounds (lb.)} = H \text{ (heat)}$$
$$H + \text{original temperature} = \text{new temperature}$$

Thus, by adding 180 Btu (45.36 kg-cal) to (0.453 kg) of water (H_2O) at 32°F (0°C), the result is:

English

$$180 \text{ Btu} \div 1 \text{ lb.} = 180°F$$
$$180°F + 32°F = 212°F$$

Metric

$$45.36 \text{ kg-cal} \div 0.4536 \text{ kg} = 100°C$$
$$100°C + 0°C = 100°C$$

Now, if 180 Btu (45.36 kg-cal) of heat energy are added to 10 pounds (4.536 kg) of water (H_2O) at 32°F (0°C), the temperature is raised to only 50°F (10°C), as follows:

English

$$180 \text{ Btu} \div 10 \text{ lb.} = 18°F$$
$$18°F + 32°F = 50°F$$

Metric

$$45.36 \text{ kg-cal} \div 4.536 \text{ kg} = 10°C$$
$$10°C + 0°C = 10°C$$

Although the same amount of heat energy is added to both water (H_2O) samples and the temperature of both samples is increased, there is ten times as much water (H_2O) in the second sample. Therefore, the temperature of this sample is increased only one-tenth as much as the original sample. To raise the temperature of 10 pounds (4.536 kg) of water (H_2O) from 32°F (0°C) to 212°F (100°C), the addition of 1,800 Btu (453.6 kg-cal) of heat energy is required.

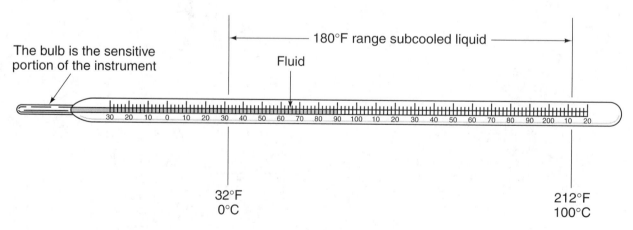

FIGURE 4–17   The subcooled liquid range for water is 32°F (0°C) to 212°F (100°C).

English

$$1800 \text{ Btu} \div 10 \text{ lb.} = 180°F$$
$$180°F + 32°F = 212°F$$

Metric

$$453.6 \text{ kg-cal} \div 4.536 \text{ kg} = 100°C$$
$$100°C + 0°C = 100°C$$

COLD: THE ABSENCE OF HEAT

What is meant by the word *cold*? Cold, by definition, is the absence of heat. If cold is to be understood, then the learner must first understand that heat is energy. It is present in all things. Although heat cannot be contained, it can be controlled. The molecular structure of all things, with few exceptions, is changed into one of three forms by heat.

Heat is molecular movement. As previously discussed, for example, water (H_2O) is liquid between 32°F (0°C) and 212°F (100°C). If heat is added at 212°F (100°C), then its molecular movement is increased. As a result, water (H_2O) vaporizes and turns to steam (vapor). When heat is removed at 32°F (0°C), its molecular movement is decreased. The water (H_2O) then solidifies and turns to ice (solid).

All matter generates heat called *specific heat*. The body generates heat that must be overcome if one is to feel cool. Food stored in a refrigerator generates heat that must be overcome if the food is to be kept at a safe temperature as required for short- or long-term storage. Any matter that is to be cooled must first have its specific heat removed or overcome.

The answer to the question "What is cold?" now appears to be that cold is the absence of heat. If this is true, then at what point is all the heat removed from matter? Ice, at 32°F (0°C), is said to be cold. But solid carbon dioxide (CO_2), commonly known as *dry ice*, is even colder at its normal temperature of –109.03°F (–165.8°C). Dry ice (CO_2) is so cold that, if touched, one would have the sensation of being burned. However, dry ice (CO_2) cannot be called cold because it still contains a large amount of heat as measured in Btu.

Absolute Cold

Absolute cold, then, is the absence of all heat. Complete absence of heat does not occur until the temperature of –459.67°F (–273.16°C) is reached. All temperatures above this value contain heat. For example, any substance at a temperature of –459°F still contains 0.67°F of heat; –273°C still contains 0.16°C of heat.

Absolute cold, like other absolutes, has not yet been achieved and may never be achieved, although scientists have made several attempts. A Dutch physicist (the first to come close), Wander de Haas, working at the University of Leiden (Holland), achieved a temperature of 0.0044°C above absolute zero. In 1957, Dr. Arthur Spohr, working at the U.S. Naval Research Laboratory, achieved a temperature of less than one-millionth of a degree Kelvin (K) above absolute zero (0.000 001°K). The Kelvin scale, used in physics, uses 0°K for absolute cold, 273°K as the freezing point for water, and 373 K as its boiling point.

To summarize, the term cold refers to the absence of all heat energy. According to current scientific theory, absolute zero is the point at which all molecular movement stops. Because it is the molecular movement that causes heat energy, it follows that if there is movement there is heat.

THE THERMOMETER

Long ago, it was recognized that it was desirable to have a device to measure the temperature of matter to determine the heat content of matter. Such a device became known as a *thermometer*.

In about 1585, Galileo Galilei constructed a crude water (H_2O) thermometer. Although this early device was very inaccurate, the principles stated by Galileo in its construction helped other scientists to design more accurate instruments.

More than a century later, in 1714, Gabriel D. Fahrenheit constructed a thermometer using a column of mercury (Hg). From the time of Galileo and until Fahrenheit's experiments, temperature measuring devices used tubes of alcohol or other substances as indicators.

Fahrenheit realized that even though many thermometers had been made, none of these devices had been constructed to a standard scale. Recognizing the need for a standard scale, Fahrenheit decided that a zero (0) should be placed on the tube to indicate the absence of heat. He then reasoned that all values above zero would be relative and would contain so many units of heat.

Fahrenheit then decided that it would be necessary to take his tube of mercury (Hg) for calibration to the coldest location that could be found in the world. After talking with several sailors, he decided that a particular seaport in Iceland was the ideal place to calibrate his thermometer; it was the "coldest place in the world," he was assured.

Once in Iceland, he waited until he was told by the local inhabitants that "This is the coldest day we have seen." He then made a mark on his glass tube at the mercury (Hg) level to indicate zero, the absence of heat he assumed. He then waited and, as it became warmer, he noted that as the ice melted his mercury (Hg) column had expanded by 32/1000 of its original volume. This experiment and measurement, freezing and thawing, was repeated several times. Each time, the same contraction and expansion of the mercury (Hg) column occurred. Fahrenheit, therefore, designated this point as 32.

He also noted that normal body temperature was 98.6/1000 of the mercury (Hg) column and that when water (H_2O) boiled, the column of mercury (Hg) expanded to 212/1000 of its original volume.

Fahrenheit's thermometer was accepted as the standard and was the most widely used device for determining the temperature for many years. There are, however, three other scales in use today.

While Fahrenheit was working on his thermometer, Anders Celsius, a Swedish astronomer, proposed the centigrade thermometer. On this thermometer he designated the temperature of melting ice as 0°C and the boiling point of water as 100°C. This scale, now known as the Celsius scale, is used in the metric system of measurement. Celsius is sometimes referred to as "centigrade." In some parts of the world, however, this term is also used in the metric system of measure to determine the angle of an arc.

In 1848, at the age of 24, W.T. Kelvin (Lord Kelvin) proposed the absolute scale of temperature that bears his name. The Kelvin scale is used in scientific work. A comparison of the various temperature scales is given in Figure 4–18.

The Rankine thermometer is named after its inventor W.J.M. Rankine, a Scottish engineer. On the Rankine scale, the freezing point of water is 492°R and its boiling point is 672°R.

Rankine temperature equivalents are obtained by adding 460° to the Fahrenheit temperature; the Fahrenheit equivalent is obtained by subtracting 460° from the Rankine temperature.

Several types of thermometers pictured in Figure 4–19, are available to the service technician. The most popular style for general service work is the stem/dial thermometer. It has an all-metal stem and an easy-to-read dial.

In general, the accuracy of this type of thermometer is ±1 percent throughout the entire indicated temperature scale. Several ranges are available: 0°F to 220°F and –40°F to +160°F. The latter range is the most popular for use with air-conditioning systems. Both ranges can be used on either the high side of a refrigeration system, as in Figure 4–20, or on the low side of the system, as in Figure 4–21.

Thermometers are also available with the metric scale or with both English and metric scales. A popular scale range for a metric thermometer is –10°C to +110°C. This scale is approximately equal to the English scale of 14°F to 230°F.

The electronic digital thermometer is becoming very popular. This type of thermometer is available in several styles. A two-probe model, as shown in Figure 4–22, may be used to determine temperature difference (Δ_t), which is ideal for the measurement of superheat. The measurement of superheat is required for air-conditioning system diagnosis in many late-model vehicles that use HFC-134a as a refrigerant. The use of thermometers is covered in chapter 6.

Temperature Conversion

Most of the servicing work on air conditioners and refrigeration systems has dealt with temperature values measured on the Fahrenheit (F) scale. However, as conversion to the metric system proceeds, the learner will

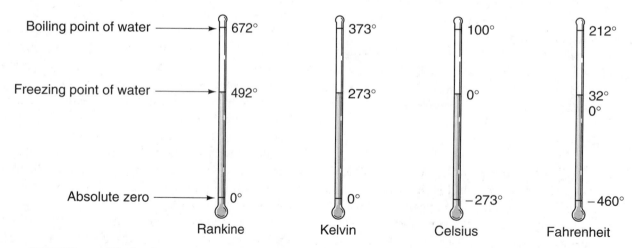

FIGURE 4–18 Kinds of temperature scales.

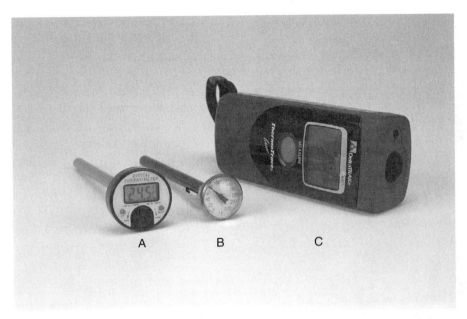

FIGURE 4–19 Typical thermometers: digital (A), analog (B), and electronic (C).

FIGURE 4–20 Thermometer used to check condenser temperature.

FIGURE 4–21 Measure discharge air at left register.

also need to be prepared to work with the Celsius (C) temperature scale. On many occasions, the technician will be required to make conversions between Fahrenheit (F) and Celsius (C) scales. These conver-sions are quite simple. For example, to change a Celsius reading to a Fahrenheit reading, multiply the Celsius reading by 1.8, then add 32.

Assume that it is necessary to convert a tempera-ture of 115°C to the Fahrenheit (F) equivalent. First, mul-tiply the value by 1.8: 115°C × 1.8 = 207°C. Then add 32. 207 + 32 = 239°F. Thus, 115°C = 239°F.

As another example, consider the boiling point of water (H_2O) at 100°C, or 212°F. Given the value of 100°C, it can be proven that the Fahrenheit equivalent is, in fact, 212°F.

$$100°C \times 1.8 = 180$$
$$180 + 32 = 212 \ (212°F)$$

To change a Fahrenheit (F) reading to the Celsius (C) equivalent, first subtract 32 and then multiply by 0.555. By applying these steps, the Celsius equivalent of a temperature of 221°F can be found.

$$221 - 32° = 189$$
$$189 \times 0.555 = 104.89 \ (105°C)$$

This formula can be proven, as in the previous case, by converting the boiling point of water (H_2O) on the Fahrenheit (F) scale, 212°F, to its Celsius (C) equivalent.

$$212 - 32° = 180$$
$$180 \times 0.555 = 99.9 \ (100°C)$$

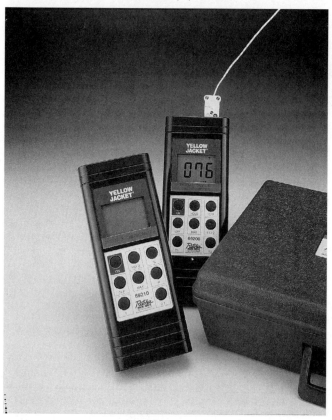

FIGURE 4–22 A typical electronic thermometer. *(Courtesy of Ritchie Engineering)*

REVIEW

Select the correct answers from the choices given.

1. An air conditioning system does all of the follow-ing, EXCEPT:
   a. heat or cool.
   b. humidify or dehumidify.
   c. clean and recirculate the air.
   d. prevent mold and destroy odors.

2. Technician A says a dry-bulb thermometer is used to measure the ambient temperature. Technician B says a Rankine scale thermometer is used to mea-sure relative humidity. Who is right?

   a. A only
   b. B only
   c. Both A and B
   d. Neither A nor B

3. The sensation of feeling hot or cold is affected by all of the following, EXCEPT:
   a. latent heat.
   b. sensible heat.
   c. insulation or clothing.
   d. season of the year.

4. Technician A says that the terms *conduction* and *convection* mean the same thing. Technician B says that the same is true for the terms *evaporation* and *radiation*. Who is right?
   a. A only
   b. B only
   c. Both A and B
   d. Neither A nor B

5. Theoretically, if two 1-pound containers of water, one at 50°F and the other at 100°F were mixed equally, the results would be:
   a. 2 pounds of water at 75°F.
   b. 2 pounds of water at 100°F.
   c. 2 pounds of water at 50°F.
   d. neither of the above.
      Explain. _____

6. Which of the following will contain the greatest amount of heat per cubic foot (ft.³)?
   a. Lead
   b. Copper
   c. Tin
   d. Silver

7. The latent heat required for the vaporization of 1 pound of water:
   a. is 970 Btu.
   b. is 144 Btu.
   c. is 180 Btu.
   d. depends on the water temperature.

8. The specific heat of water is:
   a. 0.100.
   b. 1.000.
   c. 0.093.
   d. 0.930.

9. Technician A says the metric thermometer scale is known as centigrade. Technician B says either Celsius or centigrade scales are used in the metric system. Who is right?
   a. A only
   b. B only
   c. Both A and B
   d. Neither A nor B

10. Technician A says to convert an English to a metric temperature value, one must first subtract 32, then multiply by 0.555. Technician B says to convert a metric to an English temperature value, one must multiply by 1.8, then add 32. Who is right?
    a. A only
    b. B only
    c. Both A and B
    d. Neither A nor B

11. The subsurface or deep surface temperature of the human body is:
    a. 98.6 °F (37 °C).
    b. 96.8 °F (36 °C).
    c. 70 °F (21.1 °C).
    d. none of the above.

12. Natural convection takes place about the human body:
    a. because warm air rises.
    b. when one perspires.
    c. if the surrounding air is cooler.
    d. due to all of the above.

13. All of the following statements are correct, EXCEPT:
    a. the evaporation rate does not depend on relative humidity.
    b. the evaporation rate depends on the air movement.
    c. cool air increases the rate of evaporation.
    d. cool air increases the rate of convection.

14. Technician A says a low relative humidity causes discomfort because one may feel too cold. Technician B says a high relative humidity causes discomfort because one may feel too warm. Who is right?
    a. A only
    b. B only
    c. Both A and B
    d. Neither A nor B

15. To prevent mold and mildew, the relative humidity should be maintained:
    a. between 40 and 60 percent.
    b. below 30 percent.
    c. between 30 and 50 percent.
    d. above 60 percent.

16. How many Btu are required to raise the temperature of 5 pounds (2.27 kg) of water from 50°F (10°C) to 75°F (23.9°C)?
    a. 50
    b. 100
    c. 75
    d. 125

17. As moisture vaporizes from a warm surface it:
    a. removes heat from the surface.
    b. lowers the temperature of the surface.
    c. both A and B.
    d. neither A nor B.

18. The comfort zone depends on an air velocity:
    a. between 15 and 25 mph.
    b. between 15 and 25 ft./min.
    c. below 16 mph.
    d. above 25 ft/min.

19. Technician A says the conditioned air should be about 12°F (6.7°C) to 15°F (8.3°C) below the ambient air temperature. Technician B says the velocity of the conditioned air should be about 100 ft./min. for the best personal comfort conditions. Who is right?

a. A only
b. B only
c. Both A and B
d. Neither A nor B

20. The blower motor and its purpose are being discussed. Technician A says that air velocity in a mobile air-conditioning system is controlled by varying the blower motor speed. Technician B says that the temperature of a mobile air-conditioning system is controlled by varying the blower motor speed. Who is right?
    a. A only
    b. B only
    c. Both A and B
    d. Neither A nor B

TERMS

Write a brief description of the following terms:

1. calorie
2. cold
3. convection
4. DB
5. dehydrated
6. disinfectant
7. effective temperature
8. evaporation
9. Fahrenheit
10. heat load
11. kinetic energy
12. km/h
13. latent heat
14. mildew
15. perspiration
16. relative humidity
17. sensible heat
18. specific heat
19. wet-bulb temperature
20. windchill

CHAPTER 5

PRESSURE AND TEMPERATURE

OBJECTIVES

On completion and review of this chapter, you should be able to:

❏ Understand the effects of heat energy as it relates to pressure.
❏ Understand the effects of pressure as it relates to heat energy.
❏ Discuss the structure of matter and the movement of **molecules.**
❏ Understand the function of the basic air-conditioning system.
❏ Understand the purpose and function of the engine cooling system.

INTRODUCTION

To develop a good understanding of the function of an air-conditioning system, the technicians must understand the effects of **heat** energy and pressure, as they relate to temperature within the system. The topics discussed in this chapter then include **matter,** heat, pressure, and the principles of refrigeration—the basic physical laws for developing an understanding of an air conditioning system.

MATTER

Matter is defined as anything that occupies space and has weight. All things are composed of matter and are found in one of three forms: solid, liquid, or gas, as illustrated in Figure 5–1. As an example, consider one of the more common substances, water (H_2O), which, in its natural form, is liquid. If enough of its natural heat is removed, it then turns to ice, which is a solid. If sufficient heat is added and its temperature is raised enough, it boils and vaporizes to gas, commonly referred to as steam.

Although various processes can be applied to matter to cause it to change from one state to another, most objects and things are usually thought of in their natural state. For example, water (H_2O), which is generally thought of as a liquid, has the ability to flow but cannot take a shape of its own. Therefore, it assumes the shape of the container in which it is placed. In a container, water (H_2O) exerts an outward and downward force. The greatest force is toward the bottom, and it lessens toward the top of the container.

In another example, water (H_2O) in the **vapor** state, referred to as steam, dissipates into the surrounding ambient air if it is not contained. When placed in a sealed or enclosed container, steam exerts pressure in all directions with equal force.

In yet another example, when water (H_2O) is in the solid state (ice), it holds the same shape and size as when

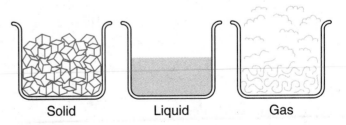

FIGURE 5–1 Three basic states of matter.

it was in the liquid state. As a solid, ice exerts force in a downward direction only.

THE ATOM

Each of the **elements** consists of millions, perhaps billions, of tiny particles called atoms. An **atom** is so small that it can only be seen with the most powerful of microscopes. Scientists have developed a means whereby they can measure and weigh it, and they have learned many things about its nature.

Without getting into great detail, an atom is defined as the smallest particle of which an element is composed that still retains the characteristics of that element. For example, an atom of copper (Cu) is still copper (Cu), and it is different from an atom of aluminum (Al). For our study purpose, consider the atom as indivisible and unchangeable; that is, it cannot be divided by ordinary means. Whenever an atom is divided, physically and chemically, it retains the characteristics of that element. Atoms of all of the elements are different. Iron (Fe) is composed of iron atoms; lead **(Pb)** of lead atoms; tin (Sn) of tin atoms, and so on.

An atom, illustrated in Figure 5–2, is composed of still smaller particles called *protons, neutrons,* and *electrons.* The proton has a positive (+) charge and the electron has a negative (−) charge. The neutron, as its name implies, is neutral and has neither a positive (+) nor a negative (−) charge.

Scientists have been able to split some kinds of elements, such as uranium (U). The automotive technician, however, should focus on the following facts regarding atoms:

❑   The atom is the very smallest possible particle of matter.
❑   All of the elements are composed of atoms.
❑   The atoms of different elements are different.

CHEMICAL COMPOUNDS

A molecule may consist of two or more atoms of different elements. In such an instance, the material becomes entirely different and usually does not resemble either of the elements that it is composed of. For example, if the molecule consists of one atom of iron (Fe) and one atom of oxygen (O), it becomes iron oxide (FeO), which is quite different from either iron (Fe) or oxygen (O).

How different the **compound** material itself can be from the elements of which it is composed is illustrated by water (H_2O). The molecule of water, in any form, consists of two atoms of the element hydrogen (H), which is a very light, highly flammable gas, and one atom of the element oxygen (O), which is also a gas that aids combustion. The combination of these two elements, both a gas in their natural state, produces a liquid, water (H_2O), which is unlike either hydrogen (H) or oxygen (O).

The molecule is often quite complex and may have several different kinds of elements in it. Refrigerant-12 (CFC-12), which is normally a colorless gas, is a good example of this. The CFC-12 molecule consists of one atom of **carbon** (C), normally a black solid; one atom of hydrogen (H), a colorless gas; two atoms of chlorine (Cl), normally a yellow-green gas; and two atoms of fluorine (F), normally a pale-yellow gas. The molecular structure of CFC-12 is CCl_2F_2 as shown in Figure 5–3. HFC-134a, an ozone-friendly refrigerant developed to replace CFC-12 in vehicles, consists of two carbon (C) atoms, four fluorine (F) atoms, and two hydrogen (H) atoms. The molecular structure of HFC-134a is $CH_2F\text{-}CF_3$ as shown in Figure 5–4.

Everything in nature is known as matter and is made up of one or more of the 106 known basic ele-

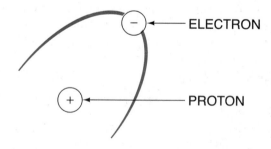

FIGURE 5–2 A simple hydrogen (H) atom is composed of one proton and one electron. (*Adapted from Whitman and Johnson, Refrigeration and Air Conditioning Technology by Delmar Publishers*)

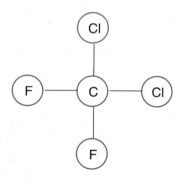

FIGURE 5–3 Composition of Refrigerant-12.

ments. Some of the more common and better known of these elements are:

Carbon

Carbon (C) is a nonmetallic element found in many inorganic compounds and in all organic compounds. Carbon is present in many chemical compounds and gases such as Refrigerant-12, a chlorofluorocarbon (CFC) refrigerant, and Refrigerant-134a, a hydrofluorocarbon (HFC) refrigerant.

At **atmospheric pressure** and temperatures, carbon normally exists as a solid. Nearly pure carbon (C) in crystalline form is known as a diamond.

Chlorine

Chlorine (CI) is a heavy greenish-yellow gas used for the purification of water (H_2O) and the manufacture of CFC and HCFC refrigerants. Chlorine, it has been determined, causes problems with the ozone layer of the atmosphere.

Aluminum

Aluminum (Al) is a lightweight ductile metal that does not readily tarnish or corrode. Aluminum and aluminum alloys have widespread use in the manufacture of automotive components and parts.

Lead

Lead (Pb) is a heavy soft blue-gray metal that was once used extensively as a filler material for soldering. It has been determined that lead is a health hazard and its use is limited and, in many cases, prohibited. Lead (Pb) is currently used in the construction of the popular lead-acid storage battery.

Nitrogen

The ordinary atmospheric air that we breathe is about 78 percent nitrogen (N). Nitrogen normally exists as a gas and is a very important element in all plant life. Nitrogen is an essential compound of all living things.

Oxygen

Oxygen (O) makes up 21 percent of the atmospheric air we breathe. Oxygen (O) is absolutely essential to all animal life. It is a very active element and combines readily with most of the other elements to form oxides or more complex chemicals. Oxygen normally exists as a gas.

Hydrogen

Hydrogen (H), is an important element found in oil, fuel, and acid as well as many other compounds. Hydrogen (H) has no odor or taste and it is colorless. It is normally a gas and is the lightest of the elements. Hydrogen rarely exists alone in nature; it is most commonly known as an element in the mixture with oxygen (O) to form water (H_2O).

Other Elements

The remaining one percent of the air we breathe in the atmosphere consists of argon (Ar), neon (Ne), krypton (Kr), helium (He), and xenon (Xe). There are also other trace nonelement gases in the atmosphere, such as **carbon dioxide** (CO_2) and ozone (O_3).

THE STRUCTURE OF MATTER

The structure of matter is shown in Figure 5–5. All matter, regardless of its state, is composed of small parts (particles) called *molecules*. Each molecule of matter is actually the smallest particle of a material that retains all the properties of the original material. For example, if a grain of salt (sodium chloride) is divided in two, each subsequent particle is divided again, and the process is continued until the grain is divided as finely as possible. Then the smallest stable

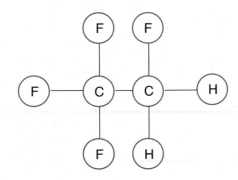

FIGURE 5–4 Composition of Refrigerant-134a.

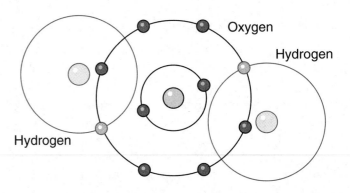

FIGURE 5–5 The structure of matter. Two hydrogen (H) atoms combined with one oxygen (O) atom results in one molecule of water (H_2O).

particle having all the properties of salt (sodium chloride) is a molecule of salt (sodium chloride). The word *stable* means that a molecule is satisfied to remain as it is.

Although the molecule may seem to be the smallest possible division, each molecule is made up of even smaller particles of matter known as *atoms*. The atoms within a molecule are not always stable. Atoms tend to join with atoms of other substances to form new and different molecules and substances.

The speed freedom (or position) and the number of molecules determine the state of the material, the temperature of the material, and the effect the material has on other parts or mechanisms of which it may be a part.

ARRANGEMENT AND MOVEMENT OF MOLECULES

Any material consists of millions and millions of molecules that are all alike. Different materials have different molecules. The characteristics and properties of different materials depend on the nature and arrangement of the molecules. In turn, the behavior of each molecule largely depends on the material (substance) of which the molecule is composed.

Regardless of the state of a material, the molecules within the material are continuously moving. This movement is called *kinetic energy* because it is an energy of motion. The addition of heat energy to a material increases the kinetic energy of the molecules of the material. In solids, the motion of the molecules is in the form of *vibration;* that is, the particles never move far from a fixed position. The addition of heat energy affects the motion of the molecules of water in each of its three states.

When heat energy is added to or removed from a material, a change in the state of the material occurs. Heat then is the factor that governs the movement of the molecules making up the substance. Removing heat causes the molecular action to slow down: gases become liquid and liquids solidify. Adding heat causes the molecular action to speed up: solids become liquids and liquids vaporize.

PRESSURE

To understand the basics of air conditioning one must understand pressure and the effects of pressure within a closed system. *Pressure*, as applied to air conditioning, is defined as a force exerted per unit of surface area. This is generally expressed in English terms as pounds per square inch (psi) and in metric terms as kiloPascal (kPa). The best example of the action of pressure and what it means is illustrated by the envelope of air (gas) around the planet Earth, known as the atmosphere. This gas consists primari-

ly of nitrogen (N), 78 percent by volume, and oxygen (O), 21 percent by volume. The remaining 1 percent consists of several other gases. The atmosphere extends nearly 600 miles (965.6 km) above the earth and is held in place by Earth's gravitational pull.

Atmospheric Pressure

The 600-mile (965.6 km) belt of gas surrounding the earth exerts a pressure measured in pounds per square inch (psi) in the English system or kiloPascals (kPa) in the metric system. For a 1 square-inch (1 in.^{2} or 6.45 cm^2) area, the pressure of the gas column is 14.696 **psia** (101.328 kPa absolute), Figure 5–6. This value is generally rounded off to 14.7 psia (101.3 kPa absolute) and is known as atmospheric pressure. This pressure, measured with an instrument called a *barometer,* is known as *barometric pressure,* which is important in making weather predictions. The barometric pressure is generally given as a part of the weather report along with the temperature and other atmospheric conditions.

PRESSURE MEASUREMENT

Most service manuals provided by vehicle manufacturers generally refer to the pressures of an air-conditioning

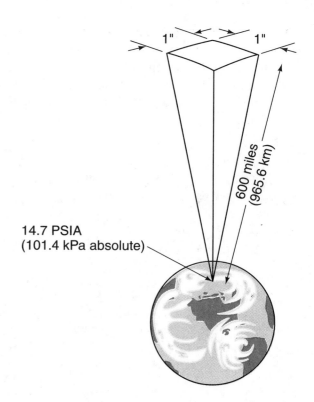

FIGURE 5–6 Atmospheric pressure.

system as pounds per square inch (psi) or pounds per square inch, gauge (psig). In some cases this pressure may also be referred to as pounds per square inch absolute (psia). The general metric reference to pressure is kiloPascal (kPa) or kiloPascal absolute (kPa absolute). It should be noted that there is a considerable difference in the actual meanings of the three English **abbreviations,** psi, psig, and psia.

English Abbreviations

The abbreviation for pounds per square inch is psi. This most commonly used term refers to the amount of pressure impressed and does not consider or compensate for variables, such as atmospheric pressure. In general, the phrase "the low-side gauge should read 30 psi" really means 30 psig.

Psig, the abbreviation for pounds per square inch gauge, is the amount of pressure in pounds per square inch that is impressed on a gauge that has been adjusted or calibrated to compensate for atmospheric pressure. Zero pressure on a gauge calibrated for psig compensates for the normal atmospheric pressure at sea level of 14.696 psi, rounded off to 14.7 psi.

Pounds per square inch absolute, abbreviated psia, is a term that refers to the amount of pressure measured from absolute zero. This value equals the gauge pressure plus 14.7 psi. Thus, a gauge calibrated in psia reads 14.7 psi without being connected to a pressure source such as an air-conditioning system.

A comparison of a psig and psia gauge is given in Figure 5–7. Note that the pointer is in the same place on both gauges. The difference is the gauge scale only. As an example, assume that two gauges are connected to the same test port of an air conditioner. One gauge is cali-brated in psig and reads 28 psi of pressure (zero reference). The other gauge is calibrated in psia and reads approximately 42.7 psia.

$$28 \text{ psi} + 14.7 = 42.7 \text{ psia}$$

Metric Abbreviations

Atmospheric and absolute metric gauges are calibrated in the same manner as an English calibrated gauge. Each psi in the English system of measurement is equal to 6.895 kPa in the metric system of measurement.

In the metric system, the sea level atmospheric pressure is 101.329 kPa (14.696 × 6.895 = 101.329) and is shown as 0 kPa on the atmospheric scale. This number is generally rounded off to 101 kPa. As in the previous English example, 28 psig would equal 193 kPa (28 × 6.895 = 193.06 kPa) on a metric atmospheric scale and 294 kPa (28 + 14.7 × 6.895 = 294.417) on a metric absolute scale.

TEMPERATURE AND PRESSURE

Any pressure above atmospheric is referred to as *gauge pressure*. Any pressure below atmospheric is referred to as a *vacuum*, Figure 5–8. Zero gauge pressure remains zero regardless of the altitude. On the English scale of measure, pressures above atmospheric pressure are recorded as pounds per square inch gauge or psig. Pressures below atmospheric pressure are recorded as inches of mercury (in. Hg). The abbreviation for inch is "in." and "Hg" is the chemical symbol for mercury.

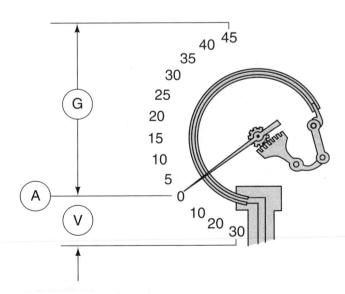

FIGURE 5–8 Atmospheric pressure (A), vacuum pressure (V), and gauge pressure (G).

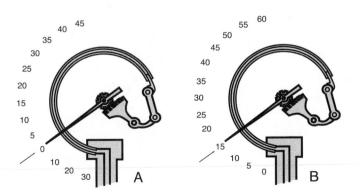

FIGURE 5–7 A comparison of a low side gauge scaled in (A) pounds per square inch gauge (psig) and (B) pounds per square inch absolute (psia).

On the metric scale of measure, pressures above atmospheric pressure are recorded as kiloPascals gauge (kPa gauge). Pressures below atmospheric pressure are recorded as kiloPascals absolute (kPa absolute). There is no metric equivalent for negative pressure in the metric system.

An example of a metric atmospheric gauge as compared to a metric absolute gauge is shown in Figure 5–9. Note that its movement is identical to the English gauge that was shown in Figure 5–7. The only difference is the scale printed on the face of the dial.

Effects of Pressure

At sea level, where the atmospheric pressure is equal to 14.7 psi or 101.3 kPa, the **boiling point** of water (H_2O) is 212°F (100°C). In areas higher than sea level, the atmospheric pressure is lower and so is the boiling point of water (H_2O). The point at which water (H_2O) boils decreases at the rate of 1.1°F (0.61°C) per 1,000 feet (304.8 meters) of altitude. To find the boiling point at any given altitude, multiply the altitude, in thousands of feet or in equivalent meters, by 1.1 for Fahrenheit or by 0.61 for Celsius. The result of this computation is then subtracted from 212°F or 100°C to obtain the new boiling point.

For example, at an altitude of 12,000 feet (3,657.6 m), water boils at about 198.8°F (92.7°C).

English

$$12,000 \text{ ft.} \div 1,000 = 12$$
$$12 \times 1.10 = 13.20$$
$$212°F - 13.2°F = 198.8°F$$

Metric

$$3657.6 \text{ m} \div 304.8 = 12$$
$$12 \times 0.61°C = 7.32°C$$
$$100°C - 7.32°C = 92.7°C$$

In Denver, Colorado, for example, water (H_2O) boils at a lower temperature than it does at sea level. At an elevation of 8,900 feet (2712.72 m), as in Figure 5–10, the boiling point can be determined as follows:

English

$$8,900 \text{ ft.} \div 1,000 = 8.9$$
$$8.9 \times 1.1°F = 9.8°F$$
$$212°F - 9.8°F = 202.2°F$$

Metric

$$2712.72 \text{ m} \div 304.8 = 8.9$$
$$8.9 \times 0.61°C = 5.4°C$$
$$100°C --5.4°C = 94.6°C$$

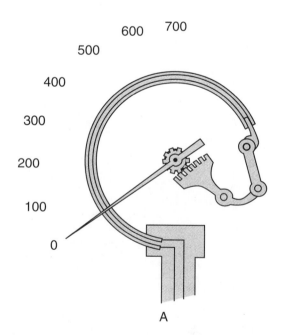

A

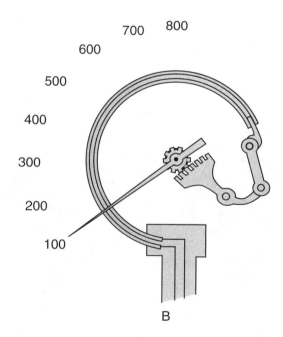

B

FIGURE 5–9 Low side scale for (A) kiloPascal atmospheric and (B) kiloPascal absolute.

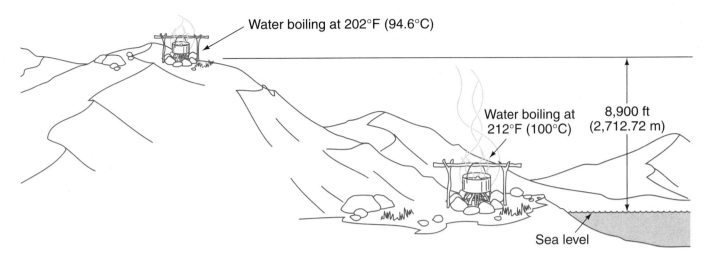

FIGURE 5–10 The boiling point of water at an elevation of 8,900 feet.

Thus, the boiling point of water (H_2O) at this elevation is about 202°F (94.6°C). It should be noted that the small error in the comparisons of temperatures (converting Fahrenheit to Celsius or Celsius to Fahrenheit) is due to the conversion factors of 1.1°F (0.61°C) and 1,000 feet (304.8 meters). If these factors were extended to four or five decimal places, the error would be greatly reduced.

It should also be noted that water (H_2O) boils when it contains all of the heat it can for a given condition. It is said to be *saturated*. Thus, when water (H_2O) boils at a lower temperature, it contains less heat; when it boils at a high temperature, it contains more heat.

If the boiling point of water (H_2O) is affected by a pressure drop, it is reasonable to assume that a pressure increase also affects its boiling point.

The principle is often put to use when food is prepared in a pressure cooker, as in Figure 5–11. The boiling point is increased because the pressure is increased. As the water (H_2O) changes from a liquid to steam, a pressure is created because the vapor cannot escape from the sealed vessel. As a result, the vapor is *superheated*. The food then cooks much faster because it is exposed to a greater temperature and pressure.

A vehicle's cooling system is another good example where the fluid temperature is increased by increasing the system pressure. Some vehicle manufacturers have increased the normal working pressure of a cooling system to 16 psi (110.3 kPa) or more. With each psi (6.895 kPa) of pressure increase, the boiling point of the water (or **coolant**) is increased about 2.53°F (1.4°C).

A cooling system having a pressure cap rated at only 10 psi (69 kPa) causes an increase in the boiling point of the coolant by about 25°F (14°C). The new boiling point of the coolant with the pressure cap in place was found in the following manner:

English

Multiply the rating of the cap, in psi, by 2.53, then add 212°F, as follows:

$$10 \text{ psi} \times 2.53°F = 25.3°F$$
$$25.3°F + 212°F = 237.3 \ (237°F)$$

Metric

Multiply the rating of the cap, in kPa, by 0.20, then add 100°C, as follows:

$$69 \text{ kPa} \times 0.20°C = 13.8°C$$
$$13.8°C + 100°C = 113.8 \ (114°C)$$

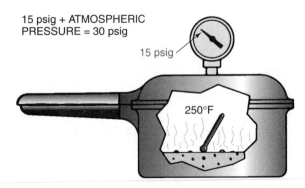

FIGURE 5–11 The water in the pressure cooker boils at 250°F. As heat is added, the water boils to make vapor. The vapor cannot escape, and the vapor pressure rises to 15 psig. The water boils at 250°F because the pressure is 15 psig.

The new boiling point of the coolant in a system having any rated pressure cap can be found in the same manner.

Also assume that a vehicle is known to hold the temperature of a coolant to 260°F (126.6°C). The required rating of the pressure cap may be determined by subtracting the normal boiling point of water from the desired temperature. This result is then divided by 2.53°F to find the psi rating in the English system, or divided by 0.2°C to find the kPa rating in the metric system. In this example a 19 psi (131 kPa) pressure cap would be required, as follows:

English

$$260°F - 212°F = 48°F$$
$$48°F \div 2.53°F = 18.9 \ (19 \ psi)$$

Metric

$$126.6°C - 100°C = 26.6°C$$
$$26.6°C \div .20°C = 133 \ kPa$$

If water (H_2O) boils at a higher temperature when pressure is applied and at a lower temperature when the pressure is reduced, it is obvious that the temperature can be controlled if the pressure is controlled. That is the basic **theory** of physics that determines and controls the temperature conditions of many air-conditioning systems.

The refrigerant in an air-conditioning system has similar boiling temperature and pressure relationships as described for water (H_2O). Refrigerants are covered in greater detail in Chapter 6. The following is a brief description of how heat is moved from the driver and passenger compartment to the outside of the vehicle.

AIR CONDITIONER CIRCUIT

Unwanted heat is picked up inside a vehicle's driver and passenger compartment from the air passing through the coils and fins of an **evaporator.** It is the heat that is picked up by the liquid refrigerant in the evaporator that causes it to change to a vapor. The heat-laden refrigerant vapor is then pumped by the air-conditioning system **compressor** into the **condenser,** usually located in front of the **radiator.** In the condenser, the refrigerant's heat is given up to the less-hot air passing across the coils and fins of the condenser. In giving up its heat the refrigerant changes back to a liquid.

The transfer of heat in an air-conditioning system is accomplished by two pressure and two temperature zones: a low-pressure zone of 21 to 35 psig (145 to 241 kPa) with a corresponding low temperature of 21 to 38°F (–6.1 to 3.3°C), and a high-pressure zone of 180 to 220 psig (1,241 to 1,517 kPa) with a corresponding high temperature of 88 to 100°F (31.1 to 37.7°C). In an air-conditioning system, as well as in most any enclosed system, when there is a pressure change, there will be a corresponding temperature change. During this pressure-temperature change with refrigerants R-12 and R-134a, there is also a change of state, Figure 5–12. In the low-pressure side, the change is from a liquid to a vapor; in the high-pressure side, the change is from a vapor to a liquid.

It is important to understand the pressure-temperature relationship of these two refrigerants when using the manifold and gauge set as a diagnostic tool. The manifold and gauge set, used as a diagnostic tool to determine system performance and problems that relate to normal and abnormal gauge pressures, is covered in Chapters 7 and 17.

Following the basic air-conditioning system of Figure 5–13, the pressure and temperature of the refrigerant are increased by the compressor, which pulls low-pressure refrigerant vapor from the evaporator and pumps it to the condenser at a high pressure and temperature. The pressure and temperature of the liquid refrigerant are decreased by the metering device (expansion valve or orifice tube) at the inlet of the evaporator.

There are five major components in an air-conditioning system:

- ❏ Compressor
- ❏ Condenser
- ❏ Evaporator
- ❏ Metering device
- ❏ Accumulator or receiver

The air-conditioning system and its components are covered in greater detail in Chapter 9.

THE COOLING SYSTEM

A vehicle's cooling system, Figure 5–14, if in good condition, must be capable of handling any heat load that may be encountered in day-to-day operation. If a vehicle is not equipped with a factory-installed air-conditioning system, its cooling system is most likely not designed to handle the additional heat load. Under normal circumstances, however, the addition of an aftermarket air-conditioning system will cause no problems with the automotive cooling system. An aftermarket installation kit often contains a smaller water pump pulley to provide increased coolant flow and a fan system to provide additional airflow.

If the vehicle is a few years old, the installation of an aftermarket air-conditioning system may cause an engine overheating problem. On the other hand, after a few years or so, unless the owner has performed periodic scheduled maintenance, cooling system problems generally develop just as do other problems.

An understanding of the vehicle's cooling system is helpful in diagnosing and correcting cooling system problems. The purpose of a vehicle's cooling system is to

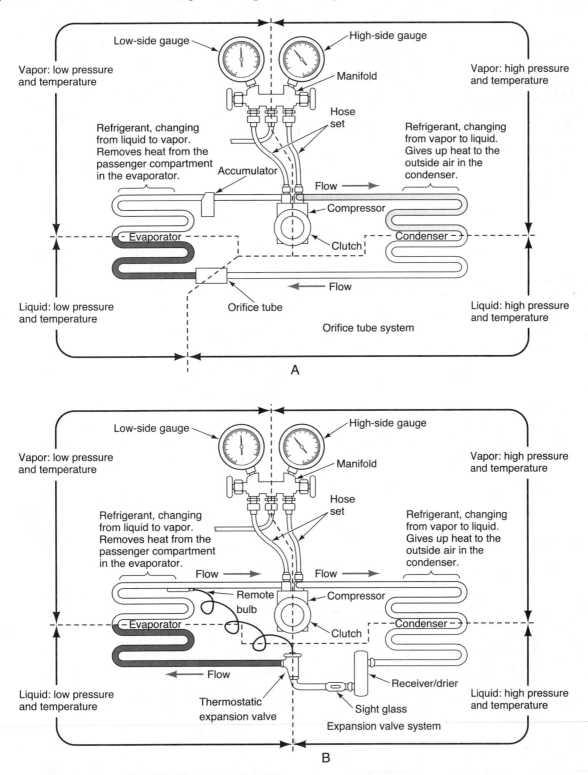

FIGURE 5–12 Typical automotive air-conditioning system: (A) orifice tube; (B) expansion valve.

carry heat that is generated by engine combustion, away from the engine, Figure 5–15. The heat generated in the combustion chambers creates internal engine exhaust valve temperatures that exceed 4,500°F (2,482°C). For the removal of most of this heat it is not hard to realize that the cooling system has a difficult task.

The "refrigerant" in a cooling system is generally water (H_2O) and an antifreeze solution. For maximum protection and performance, the recommended mix is 50 percent of each. The purpose of coolant in a cooling system is much the same as refrigerant in an air-conditioning system. While the cooling system is under pressure, picking up and dissipating heat, the coolant does not undergo a change of state as does refrigerant in an air-conditioning system.

About one-third of the heat generated by the vehicle's engine is given up in the radiator. This is accomplished by two heat transfer processes known as radiation and conduction. Radiation and conduction were covered in Chapter 4. The tubes and fins of a radiator provide an overall surface area of about 28 to 35 ft.^{2} (2.6 to 3.2 m^2). Their physical size, however, does not reveal that they have that much cooling area.

Most of the engine's heat is absorbed and dissipated by the cylinder walls, heads, and pistons. The job of the cooling system is to remove the remaining undesired heat. The cooling system's design function is to remove about 35 percent of the total heat produced by engine combustion.

Heat, as previously discussed, is always ready to flow from a hot to a less-hot object or area that contains

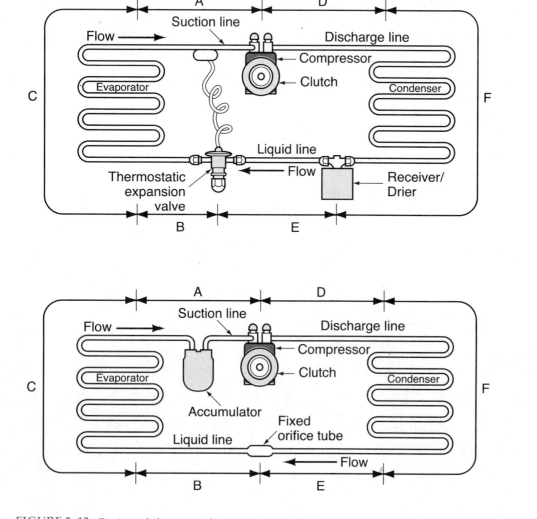

Legend:
Low Pressure
A - Vapor
B - Liquid
C - Liquid/Vapor

High Pressure
D - Vapor
E - Liquid
F - Vapor/Liquid

FIGURE 5–13   Basic mobile air-conditioning systems.

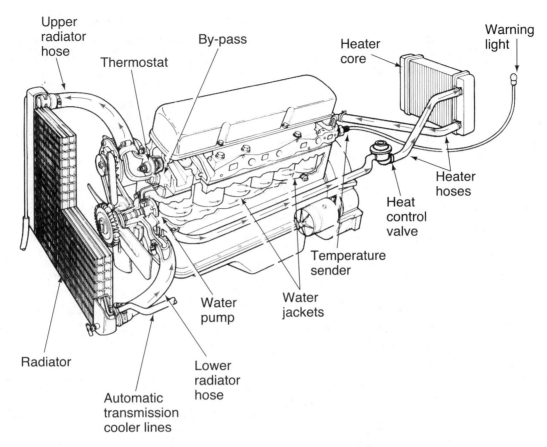

Upper
radiator
hose

Thermostat

By-pass

Heater
core

Warning
light

Heater
hoses

Heat
control
valve

Temperature
sender

Water
jackets

Water
pump

Lower
radiator
hose

Automatic
transmission
cooler lines

Radiator

FIGURE 5–14 Typical engine cooling system.

less heat. The heat of combustion is picked up in the engine block by the less-hot coolant. The now heat-laden coolant is given off to the less-hot air passing over the fins and coils of the radiator. The air movement across the radiator is created in two ways: (1) by the engine- or electrically driven fan, known as forced air, and (2) by the

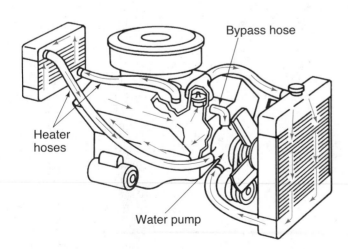

Bypass hose

Heater
hoses

Water pump

FIGURE 5–15 Heat is transferred by conduction.

forward motion of the vehicle, known as ram air.

The high-limit properties of engine lubricating oil necessitate proper heat removal to prevent destroying its formulated lubricating characteristics. Removing too much heat, on the other hand, lowers the thermal efficiency of the engine. To prevent the removal of too much heat, a condition known as overcooling, a thermostat is used in the engine outlet water passage. The thermostat, a temperature-sensitive device, controls the flow of coolant from the engine into the radiator.

To accomplish this, there are six major components of the cooling system in addition to the necessary hoses. The six major components are:

- ❏ Water pump
- ❏ Cooling fan
- ❏ Radiator
- ❏ Pressure cap
- ❏ Thermostat
- ❏ Heater core

The cooling system and each of its components are covered individually and in more detail in Chapter 16.

HEATING AND COOLING SYSTEMS

All of the components of the cooling system work together as a team. The cooling system transfers some of the heat of combustion in the engine to the radiator to be dissipated in the ambient air. Generally, this requires only an annual inspection following a periodic preventative maintenance (PM) schedule. If it becomes continuously necessary to add coolant, however, a leak or an overheating condition of the cooling system is indicated. An overheating engine may also indicate a malfunction of a cooling system component, such as a defective water pump, thermostat, or pressure cap.

Problems related to an overheating engine may sometimes be caused by an air-conditioning problem. Conversely, an air-conditioning problem may be caused by an overheating engine.

During the air-conditioning cycle, both sensible and latent heat are transferred from the air inside the vehicle to the ambient air outside the vehicle. This is accomplished by changing the high-pressure liquid refrigerant to a low-pressure liquid, then to a low-pressure vapor, then to a high-pressure vapor, then back to a high-pressure liquid to repeat the process.

The air-conditioning system and engine cooling system are covered in greater detail later in this text.

SUMMARY

- ❏ Everything in nature is made up of matter.
- ❏ Matter is in three forms: solid, liquid, or vapor (gas).
- ❏ An atom is made up of protons, neutrons, and electrons.
- ❏ A molecule consists of two or more atoms.

Each time the pressure is changed in an air-conditioning system the refrigerant must either pick up or reject heat while changing state, bringing about heat transfer. The end result is that the air inside the car is lower and more comfortable.

REVIEW

Select the correct answer from the choices given.

1. Technician A says a molecule consists of two or more different elements. Technician B says refrigerant used in vehicles consists of but one element. Who is right?
    a. A only
    b. B only
    c. Both A and B
    d. Neither A nor B

2. The most abundant gas in the atmosphere is:
    a. oxygen (O).
    b. nitrogen (N).
    c. hydrogen (H).
    d. chlorine (Cl).

3. All of the following are part of an atom, EXCEPT:
    a. proton.
    b. neutron.
    c. electron.
    d. micotron.

4. Another term for atmospheric air is:
    a. ambient air.
    b. surrounding air.
    c. either A or B.
    d. neither A nor B.

5. The chemical symbol for the element, carbon, is:
    a. Cl.
    b. Ca.
    c. Cb.
    d. C.

6. Which of the following is considered a nonelement gas?
    a. Ozone (O_3)
    b. Xenon (Xe)
    c. Krypton (Kr)
    d. Helium (He)

7. Technician A says matter is anything that occupies space. Technician B says matter is anything that has weight. Who is right?

a. A only
b. B only
✓ c. Both A and B
d. Neither A nor B

8. Sea level atmospheric pressure is being discussed. Technician A says it is about 14.7 psig. Technician B says it is about 101 kPa absolute. Who is right?
a. A only
b. B only
c. Both A and B
d. Neither A nor B

9. Refrigerant evaporates in the condenser at:
a. a low pressure.
b. a high pressure.
c. Both A and B.
d. Neither A nor B.

10. All of the following are major components of the air-conditioning system, EXCEPT:
a. condenser.
b. pressure cap.
c. evaporator.
d. receiver.

11. Technician A says that a cooling system problem may affect the air-conditioning system's performance. Technician B says that an air-conditioning problem may affect the cooling system's performance. Who is right?
a. A only
b. B only
c. Both A and B
d. Neither A nor B

12. The cooling system removes about ____ percent of the engine's heat.
a. 25
b. 35
c. 50
d. 75

13. A fluid (liquid) in a container:
a. assumes the shape of the container.
b. exerts pressure in all directions.
c. both A and B.
d. neither A nor B.

14. The smallest matter is called:
a. an atom.
b. a particle.
c. a molecule.
d. material.

15. Molecules in motion are/is known as:
a. vibrations.
b. activity.
c. change of state.
d. kinetic energy.

16. Technician A says that atmospheric pressure may be measured in pounds per square inch (psi). Technician B says that the atmospheric pressure may be measured with a barometer. Who is right?
a. A only
b. B only
c. Both A and B
d. Neither A nor B

17. When referring to pressure reading on a gauge, which of the following terms is *least* descriptive?
a. psi
b. psig
c. psia
d. kPa

18. On the English scale, pressures below atmospheric pressure, are properly referred to as:
a. in. hg.
b. psia.
c. both A and B.
d. neither A nor B.

19. A pressure of 101 kPa absolute is equal to:
a. 0 psia.
b. 14.7 psig.
c. both A and B.
d. neither A nor B.

20. All of the following are parts of the engine's cooling system, EXCEPT:
a. thermostat.
b. compressor.
c. pressure cap.
d. heater core.

TERMS

Write a brief description of the following terms:

1. abbreviation
2. atmospheric pressure
3. atom
4. boiling point
5. carbon
6. carbon dioxide
7. compound
8. compressor
9. condenser
10. coolant
11. element
12. evaporator
13. heat
14. matter
15. molecule
16. Pb
17. psia
18. radiator
19. theory
20. vapor

CHAPTER 6

REFRIGERANTS AND LUBRICANTS

OBJECTIVES

On completion and review of this chapter, you should be able to:

❏ Understand the composition of refrigerants and lubricants used in automotive air-conditioning systems.
❏ Discuss the importance of understanding the temperature and pressure relationships of refrigerants CFC-12 and HFC-134a.
❏ Understand the safety hazards associated with the handling of refrigerants and lubricants.
❏ Understand and discuss the importance of holding a **purity test** for refrigerants.

INTRODUCTION

The term *refrigerant* to the air-conditioning technician refers to the fluid used in an air-conditioning system to produce cool air by removing some of its heat. For many years, from the early development of automotive air-conditioning systems through the mid 1990s Refrigerant-12 (R-12) was used as the only refrigerant. R-12 had the highest human safety factor of any available refrigerant at that time and was capable of withstanding high pressures and temperatures without deteriorating or **decomposing.**

Nature has not provided a perfect refrigerant. Thus it was necessary to devise a compound suitable for automotive use. A fluorinated hydrocarbon known as carbon tetrachloride was selected. Carbon tetrachloride met the requirements most closely, requiring only a few minor changes.

Carbon tetrachloride consists of one atom of carbon (C) and four atoms of chlorine (Cl). The chemical symbol for this compound is CCl_4. To change carbon tetrachloride (CCl_4) into a suitable refrigerant, two of the chlorine (Cl) atoms are removed and replaced by two atoms of fluorine (F). The new compound, known as dichlorodifluo-

romethane, is Refrigerant-12, more commonly known as R-12. It had many applications in various types of domestic and commercial refrigeration and air-conditioning systems but was perhaps best known as the refrigerant used in automotive air-conditioning systems. The chemical symbol for Refrigerant-12 is CCl_2F_2. This means that one molecule of this refrigerant contains one atom of carbon (C), two atoms of chlorine (Cl), and two atoms of fluorine (F).

R-12 was considered ideal for automotive use because of its relatively low operating pressures, as compared to other available refrigerants. Its stability at high and low operating temperatures was also desirable. R-12 does not react with most metals such as iron (Fe), aluminum (Al), copper (Cu), or steel. However, liquid R-12 may cause discoloration of chrome (Cr) and stainless steel (SS) if large quantities are allowed to strike these surfaces.

R-12 is soluble in mineral oil and does not react with rubber. Some **synthetic** rubber compositions, however, may deteriorate if used as refrigerant hose. Synthetic rubber hose such as Buna "N", designated for refrigeration service, is used for R-12 service. More recently, **barrier hoses** are standard in automotive air-conditioning systems. A **barrier hose** is one that has a non-pervious lining of nylon or other such material.

R-12 is odorless in concentrates of 20 percent or less. In greater **concentrations,** it can be detected by the faint odor of its original compound, carbon tetrachloride (CCl_4). R-12 does not affect the taste, odor, or color of water (H_2O) or food. It was believed that it was not harmful to animal or plant life. Recent discoveries, however, dispel this belief. Unfortunately, it has been determined that R-12, a chlorofluorocarbon (CFC) is, by far, the leading single cause of ozone depletion and has contributed to the greenhouse effect problem. The United States and twenty-two other countries signed an agreement in 1987 known as the *Montreal Protocol*. These nations agreed that the manufacture and production of all chlorofluorocarbon (CFC) refrigerants would be phased out in a timely manner. The mobile air-conditioning industry was among the first to be regulated because it was the greatest offender. It had been determined that 30 percent of all R-12, also referred to as CFC-12, released into the atmosphere was from mobile air-conditioning systems.

THE BEGINNING OF THE END

After November 15, 1992, it became unlawful to sell or distribute CFC-12 in **containers** of less than 20 lb. (9 kg) to the general public. Proper certification is now required for the purchase of refrigerant in any quantity to ensure that those dispensing refrigerants are knowledgeable in their profession and may be held accountable for their actions. Some states also require special licensing of the service facility in addition to the federal certification requirements. Also, to legally service automotive air-conditioning systems the service facility must have proper, adequate, and EPA-approved refrigerant service equipment for each type of refrigerant it wishes to service.

The production of CFC-12 ended in the United States on December 31, 1995. Importing virgin CFC-12 into the United States from other countries is illegal. The only exceptions are for limited quantities that are used for medical purposes, such as metered-dose inhalers. It is, however, legal to import recovered CFC-12 under very close federal government scrutiny.

The industry must rely heavily on surplus and recycled CFC-12 or equipment conversion to another type of refrigerant such as HFC-134a.

A NEW BEGINNING

Fortunately, by the early 1990s, several **alternate refrigerants** had been developed. One of them, tetrafluoroethane,

was selected by the industry to take the place of CFC-12 for motor vehicle service. This new refrigerant, referred to as **R-134a,** has many of the same properties of CFC-12 but poses no threat to the ozone. It does not contain ozone-depleting chlorine. Its chemical formula is $C_2F_2H_4$.

Unfortunately, R-134a, a hydrofluorocarbon (HFC) refrigerant also identified as HFC-134a, is not compatible in a CFC-12 system. Specific guidelines must be followed in order to convert a CFC-12 system to an HFC-134a system. The guidelines for this procedure are found in Chapter 19.

It is important to remember that two refrigerants may not be mixed; CFC-12 or any other refrigerant, must not be mixed in with an HFC-134a system. Conversely, HFC-134a or any other refrigerant must not be mixed in with a CFC-12 system. Serious damage to the air-conditioning system could result if refrigerants are mixed. Also, some mixtures may create an explosive condition, causing a personal safety hazard.

To prevent the accidental mixing of refrigerants, the Society of Automotive Engineers **(SAE)** has developed guidelines for service hoses and service ports for all alternate refrigerants. These guidelines are known as ASE standards J-2196 and J-1639. Their purpose is to ensure that the service ports of the two systems may only be attached to the couplings on respective refrigerant cylinders and recovery/recycle/recharge equipment, as shown in Figure 6–1.

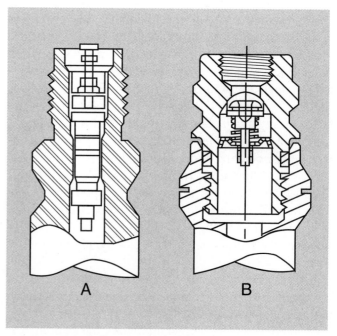

FIGURE 6–1 A comparison of (A) a CFC-12 and (B) an HFC-134a access fitting. *(Courtesy of BET, Inc.)*

It cannot be over-emphasized that CFC-12, a chloro-fluorocarbon refrigerant; HCF-22, a hydrochlorofluorocarbon; and HFC-134a, a hydrofluorocarbon refrigerant, are not compatible with each other. They must not be mixed under any circumstances, or in any other manner substituted one for the other. To do so even in small quantities may cause serious damage to the air-conditioning system. Mixing refrigerants will result in exceptionally high pressures that may cause damage to system components such as the evaporator and hoses. An improper refrigerant may cause damage to the system due to the incompatibility of the lubricant and desiccant. The appropriate equipment, such as manifold and gauge set, recovery system, and charging station must be used for each refrigerant.

There are many CFC-12 to HFC-134a refrigerant **retrofit** kits available for converting an automotive air-conditioning system. Unfortunately, not all of them contain adequate information and components for a successful conversion. Some early CFC-12 refrigerant systems use non-barrier-type hoses as well as O-rings not compatible with HFC-134a refrigerant. Also, the desiccant in some early systems is not compatible with the new refrigerant. Finally, the lubricant must be changed. The mineral oil used in CFC-12 systems is not compatible with HFC-134a. Before retrofitting an automotive air-conditioning system, always refer to the manufacturer's recommendations and instructions for parts requirements and procedures.

TEMPERATURE-PRESSURE RELATIONSHIP OF CFC-12

One of the characteristics of CFC-12 that made it a desirable refrigerant is the fact that the English temperature and system pressure values in the 20°F/psig to 80°F/psig range are very close. Figure 6–2 shows that there is only a slight variation between the temperature and pressure values of CFC-12 refrigerant in this range.

Temp. °F	Press. psig	Temp. °F	Press. psig	Temp. °F	Press. psig	Temp. °F	Press. psig	Temp. °F	Press. psig
0	9.1	35	32.5	60	57.7	85	91.7	110	136.0
2	10.1	36	33.4	61	58.9	86	93.2	111	138.0
4	11.2	37	34.3	62	60.0	87	94.8	112	140.1
6	12.3	38	35.1	63	61.3	88	96.4	113	142.1
8	13.4	39	36.0	64	62.5	89	98.0	114	144.2
10	14.6	40	36.9	65	63.7	90	99.6	115	146.3
12	15.8	41	37.9	66	64.9	91	101.3	116	148.4
14	17.1	42	38.8	67	66.2	92	103.0	117	151.2
16	18.3	43	39.7	68	67.5	93	104.6	118	152.7
18	19.7	44	40.7	69	68.8	94	106.3	119	154.9
20	21.0	45	41.7	70	70.1	95	108.1	120	157.1
21	21.7	46	42.6	71	71.4	96	109.8	121	159.3
22	22.4	47	43.6	72	72.8	97	111.5	122	161.5
23	23.1	48	44.6	73	74.2	98	113.3	123	163.8
24	23.8	49	45.6	74	75.5	99	115.1	124	166.1
25	24.6	50	46.6	75	76.9	100	116.9	125	168.4
26	25.3	51	47.8	76	78.3	101	118.8	126	170.7
27	26.1	52	48.7	77	79.2	102	120.6	127	173.1
28	26.8	53	49.8	78	81.1	103	122.4	128	175.4
29	27.6	54	50.9	79	82.5	104	124.3	129	177.8
30	28.4	55	52.0	80	84.0	105	126.2	130	182.2
31	29.2	56	53.1	81	85.5	106	128.1	131	182.6
32	30.0	57	55.4	82	87.0	107	130.0	132	185.1
33	30.9	58	56.6	83	88.5	108	132.1	133	187.6
34	31.7	59	57.1	84	90.1	109	135.1	134	190.1

FIGURE 6–2 English temperature-pressure chart for CFC-12.

The metric equivalent of this temperature-pressure chart is given in Figure 6–3. These variations in temperature and pressure can be detected by sensitive thermometers and **pressure gauges.** In this range, an assumption may be made that for each pound of pressure recorded, the temperature is about the same. For example, Figure 6–2 indicates that for a CFC-12 pressure of 23.1 psig, the corresponding temperature is 23°F. This value is the temperature of the evaporating refrigerant inside the evaporator. It is not the temperature of the outside surface of the **evaporator coil** or of the air passing through its fins.

The objective in air conditioning is to allow the fins and tubes of the evaporator to reach its coldest point without icing. Ice forms at 32°F (0°C), so the fins and cooling **coils** of the evaporator must not be allowed to reach a colder temperature. Because of the temperature rise through the walls of the cooling fins and coils, the temperature of the refrigerant may be several degrees cooler than that of the air passing through the evaporator. For example, a pressure gauge reading of 28 psig (193.06 kPa) in a CFC-12 system means that the evaporating temperature of the refrigerant inside the evaporator tubing is about 30°F (–1.1°C). Because of the temperature rise through the fins and coils, the air passing over the coil is about 34°F or 35°F (1.1°C or 1.7°C).

TEMPERATURE-PRESSURE RELATIONSHIP OF HFC-134a

Like CFC-12, HFC-134a also has its own unique English temperature and pressure relationship, as shown in Figure 6–4. The metric equivalent for HFC-134a is given in Figure 6–5. Note that the English evaporating temperature and pressure values of HFC-134a are reasonably close in the 10°F to 40°F (–12.2°C to +4.4°C) range. In fact, the temperature and pressure value is nearly the same at 15°F (–9.4°C).

A pressure gauge reading of 26 psig (179 kPa)

EVAPORATOR TEMPERATURE °C	EVAPORATOR PRESSURE GAUGE READING KILOPASCAL		AMBIENT TEMPERATURE °C	HIGH PRESSURE GAUGE READING KILOPASCAL (GAUGE)
	(GAUGE)	(ABSOLUTE)		
–16	73.4	174.7	16	737.7
–15	81.0	182.3	17	759.8
–14	87.8	189.1	18	784.6
–13	94.8	196.1	19	810.2
–12	100.6	201.9	20	841.2
–11	108.9	210.2	21	868.7
–10	117.9	219.2	22	901.8
–9	124.5	225.8	23	932.2
–8	133.9	235.2	24	970.8
–7	140.3	241.6	25	1 020.5
–6	149.6	250.9	26	1 075.6
–5	159.2	260.5	27	1 111.5
–4	167.4	268.7	28	1 143.2
–3	183.2	284.6	29	1 174.9
–2	186.9	288.3	30	1 206.6
–1	195.8	297.2	31	1 241.1
0	206.8	308.1	32	1 267.3
1	218.5	319.8	33	1 294.8
2	227.8	329.1	34	1 319.7
3	238.7	340.0	35	1 344.5
4	249.4	350.7	36	1 413.5
5	261.3	362.6	37	1 468.6
6	273.7	375.0	38	1 527.9
7	287.5	388.8	39	1 577.5
8	296.6	397.9	40	1 627.2
9	303.3	404.6	42	1 737.5
10	321.5	422.8	45	1 854.7

FIGURE 6–3 Metric temperature-pressure chart for CFC-12.

Temperature °F	Pressure psig	Temperature °F	Pressure psig
--5	4.1	39.0	34.1
0	6.5	40.0	35.0
5.0	9.1	45.0	40.0
10.0	12.0	50.0	45.4
15.0	15.1	55.0	51.2
20.0	18.4	60.0	57.4
21.0	19.1	65.0	64.0
22.0	19.9	70.0	71.1
23.0	20.6	75.0	78.6
24.0	21.4	80.0	86.7
25.0	22.1	85.0	95.2
26.0	22.9	90.0	104.3
27.0	23.7	95.0	113.9
28.0	24.5	100.0	124.1
29.0	25.3	105.0	134.9
30.0	26.2	110.0	146.3
31.0	27.0	115.0	158.4
32.0	27.8	120.0	171.1
33.0	28.7	125.0	184.5
34.0	29.5	130.0	198.7
35.0	30.4	135.0	213.5
36.0	31.3	140.0	229.2
37.0	32.2	145.0	245.6
38.0	33.2	150.0	262.8

FIGURE 6–4 English temperature-pressure chart for HFC-134a.

Temperature °C	Pressure kPa	Temperature °C	Pressure kPa
–15.0	63	5.0	247
–12.5	83	7.5	280
–10.0	103	10.0	313
–7.5	122	12.5	345
–5.0	142	15.0	381
–4.5	147	17.5	422
–4.0	152	20.0	465
–3.5	157	22.5	510
–3.0	162	25.0	560
–2.5	167	27.5	616
–2.0	172	30.0	670
–1.5	177	32.5	726
–1.0	182	35.0	785
–0.5	187	37.5	849
0.0	192	40.0	916
0.5	198	42.5	990
1.0	203	45.0	1066
1.5	209	47.5	1146
2.0	214	50.0	1230
2.5	220	52.5	1315
3.0	225	55.0	1385
3.5	231	57.5	1480
4.0	236	60.0	1580
4.5	242	65.0	1795

FIGURE 6–5 Metric temperature-pressure chart for HFC-134a.

means that the evaporating temperature of HFC-12 refrigerant in the evaporator is about 30°F (1.1°C). This may be favorably compared with the temperature and pressure relationship of CFC-12.

HANDLING REFRIGERANT

All refrigerants should be properly stored, handled, and used because liquid refrigerant can cause blindness if splashed into the eyes. Also, frostbite may result if liquid refrigerant comes into contact with the skin.

A refrigerant container should never be exposed to heat above 125°F (51.7°C). This means that they should not be allowed to come into contact with an open flame, any type of heating device, or stored in direct sunlight. As a result of excessive heat, the increase in refrigerant pressure inside the container, known as *hydrostatic pressure,* can become great enough to cause the container to explode.

If refrigerant is allowed to come into contact with

an open flame or heated metal, a poisonous gas is created. Anyone breathing this gas may become ill.

When charging an air-conditioning system with liquid refrigerant, first ensure that the compressor is not running. Carefully invert a **disposable-**type refrigerant cylinder to dispense liquid refrigerant. Reusable and recovery-type cylinders are not to be **inverted;** use the liquid valve for dispensing liquid refrigerant. Again, ensure that the compressor is not running.

CAUTION
REFRIGERANT IS NOT A TOY. REFRIGERANT SHOULD BE HANDLED ONLY UNDER THE DIRECTION OF A TRAINED REFRIGERATION SERVICE TECHNICIAN.

TRADE NAME

The term *Freon* is often used to identify any refrigerant. The term, however, properly refers to a group of refrigerants manufactured by E.I. duPont de Nemours and Company. Freon® and Freon-12® are registered trademarks of E.I. duPont. These terms should only be used when referring to refrigerant that is manufactured or packaged by this company or by processing plants licensed by duPont. A new term, **Suva**® is used by duPont to identify an ozone-friendly group of refrigerants that includes HFC-134a. This refrigerant, as well as others, are produced by several manufacturers and are packaged under various trade names.

PACKAGING

Refrigerants are generally available in sizes from "pound" cans to 1-ton (907 kg) cylinders. Although often referred to as a "pound can," a can of CFC-12 contains only 14 ounces (414 mL) and an HFC-134a can contains just 12 ounces (355 mL). Figure 6–6 shows various sizes of refrigerant containers. The most popular sizes are 15 pound (6.8 kg) and 25 pound (11.3 kg) cylinders.

RESTRICTIONS

On November 15, 1992, it became unlawful to sell or distribute CFC-12 to the general public. Licensing by an accredited agency approved by the Environmental Protection Agency (EPA) is required for the purchase of refrigerant in any quantity. In many states, the model and serial number of the recovery equipment must be on file with the selling vendor in order to purchase refrigerant. This is to ensure that those purchasing refrigerant are knowledgeable in their profession and may be held accountable for their actions.

IDENTIFICATION

The industry standard color code for a CFC-12 container is white. Light blue identifies an HFC-134a container, which must not be confused with the dark blue container of R-114. If the type of refrigerant is not known, or if there is any doubt as to its purity, a purity test should be conducted.

PURITY TEST

A **refrigerant identifier,** such as the Sentinel® by Robinair, Figure 6–7, quickly and simply and safely identifies the purity and type of refrigerant in a vehicle air-conditioning system or tank. A display shows if the refrigerant is at least 98 percent CFC-12 or HFC-134a. If purity or type is not identified, the display will indicate UNKNOWN.

With the transition to CFC-free air-conditioning systems, the likelihood of **cross mixing** refrigerants is a growing concern. Different refrigerants, as well as their lubricants, are not compatible and should not be mixed. It is, however, possible for the wrong refrigerant to be mistakenly charged into an air-conditioning system or for refrigerants to be mixed in the same recovery tank. Also, because recovery/recycling equipment is generally designed for a particular refrigerant, inadvertent mixing can cause damage to the equipment.

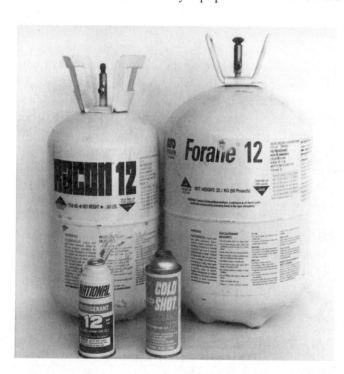

FIGURE 6–6 Various size refrigerant containers: 12 ounces (340 g) to 50 pounds (22 kg).

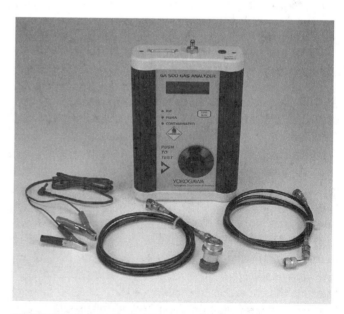

FIGURE 6–7 A typical refrigerant identifier.

A refrigerant identifier tester is used to determine if the refrigerant in a system or tank is at least 98 percent pure. The tester is far superior to pressure-temperature comparisons because at certain temperatures the pressures of CFC-12 and HFC-134a are too similar to differentiate with a standard gauge. This is easily noted in the chart of Figure 6–8. For example, at 90°F both 95 percent CFC-12 and 95 percent HFC-134a have about the same pressure, 111 and 112 psig, respectively. Considering this chart is accurate to ±2 percent, there is really no way of determining which type of refrigerant is in the air-conditioning system or tank. Also, because other substitute refrigerants and blends may have been introduced into the automotive air-conditioning system, they can contaminate a system or tank and may not be detected by the pressure-temperature method. A refrigerant identifier would call our example UNKNOWN.

Use of a refrigerant identifier, often called a purity tester, should be the first step in servicing an automotive air-conditioning system so that one does not have to be concerned about customer dissatisfaction or damage to the vehicle.

Testing refrigerant protects refrigerant supplies and recovery/recycling equipment. At today's prices preventing just one tank of refrigerant from contamination can save several hundred dollars plus the high cost of disposing of the contaminated refrigerant.

Always follow manufacturers' instructions when using any type of test equipment. The following is typical when using the Sentinel Refrigerant Identification Instrument:

1. Turn on the MAIN POWER switch; the unit automatically clears the last refrigerant sample and is made ready for a new sample.
2. When READY appears on the display, connect a service hose from the tester to the vehicle air-conditioning system or tank of refrigerant being tested.
3. The tester automatically pulls in a sample and begins processing it; TESTING shows on the display.
4. In about one minute the display will show R-12, R-134a, or UNKNOWN. If UNKNOWN is displayed, the refrigerant is a mixture or is some other type of refrigerant. In either case, it should not be added to previously recovered refrigerant. It also should not be recycled or reused.
5. Turn off the MAIN POWER switch and disconnect the service hose.

If no other method of refrigerant identification is available and there is any doubt as to the condition of the refrigerant in an air-conditioning system, the following purity test may be used. It should be noted, however, that for a pressure-temperature test to be valid there must be some liquid refrigerant in the system.

The purity test should be performed on the refrigerant in the vehicle air-conditioning system before performing service. This is especially important if the refrigerant is suspected of being contaminated. If a tester is not available, the procedure for holding a purity test is as follows. For this test to be valid, there must be some liquid refrigerant in the air conditioning system or tank. If the

AMB TEMP		R-12/R-134a PERCENT BY WEIGHT										
°F	°C	100/0	98/2	95/5	90/10	75/25	50/50	25/75	10/90	5/95	2/98	0/100
65	18.3	64	67	71	74	83	84	78	73	70	67	64
70	21.1	70	74	79	82	90	92	87	81	77	74	71
75	23.9	77	81	85	91	99	101	96	89	85	83	79
80	26.7	84	88	93	99	107	110	105	98	95	92	87
85	29.4	92	96	101	108	116	120	114	106	103	100	95
90	32.2	100	105	111	116	125	130	125	116	112	109	104
95	35.0	108	114	119	126	135	140	135	126	122	119	114
100	37.8	117	123	127	135	145	151	145	136	133	130	124
105	40.6	127	132	138	146	158	164	159	149	144	141	135
110	43.3	136	142	147	156	170	176	173	164	157	152	146
115	46.1	147	152	159	166	183	192	184	175	168	163	158
120	48.9	158	164	170	177	195	205	196	187	181	176	171

FIGURE 6–8 CFC-12/HFC-134a Cross Contamination Chart. All pressures are given in psig. For kPa, multiply psig by 6.895. For example, 100 percent R-12 at 95°F (35°C) is 108 psig or 744.7 kPa.

refrigerant has leaked to the point that only vapor remains, the pressure will be below that specified at any given temperature. Proceed as follows:

1. Park the vehicle or place the tank inside the shop in an area that is draft free and the ambient temperature is not expected to go below 70°F (21°C).
2. If a vehicle, raise the hood.
3. Determine the type of refrigerant that should be in the system or tank: CFC-12 or HFC-134a.
4. Attach a 0–150 psig (0–1000 kPa) gauge of known accuracy, appropriate for the refrigerant type (Figure 6–9).
5. Place a thermometer of known accuracy, Figure 6–10, in the immediate area of the vehicle or tank to measure the ambient temperature.
6. Note and record the pressure reading indicated on the gauge first thing the following morning.
7. Compare the gauge reading with the appropriate table:
   ❏ Figure 6–11: (A) English (B) metric for CFC-12 refrigerant
   ❏ Figure 6–12: (A) English (B) metric for HFC-134a refrigerant

Allow for reasonable inaccuracies in the gauge, the thermometer, and the reader. The pressure should reasonably match that expected for any given ambient temperature if the refrigerant is pure.

Other factors may give an inaccurate reading. For example, if there is air in the refrigerant an accurate reading may not be obtained. If, however, there is any doubt as to the purity of the refrigerant, it should be treated as if it were contaminated.

COMPATIBILITY

It cannot be overemphasized that CFC-12, a chlorofluorocarbon refrigerant, and HFC-134a, a hydroflorocarbon refrigerant, are not compatible. They must, under no circumstances, be mixed or in any other manner substituted one for the other. To do so even in small quantities can cause serious damage to the air-conditioning system. Dedicated service tools, such as manifold and gauge set, recovery system, and charging station must be used for each refrigerant.

ALTERNATIVE REFRIGERANTS

Several alternative refrigerants have been developed and are available as "replacements" for CFC-12 in an automotive air-conditioning system. These refrigerants, such as R-176™, GHG-Refrigerant-12™, and OZ-12™, although approved by the EPA for use in automotive air-conditioning systems, are not accepted by the automotive industry. The Mobile Air-Conditioning Society (MACS), International Mobile Air Conditioning Association (IMACA), and others have advised that some available refrigerants contain flammable components, such as isobutane, butane, and propane. Flammable refrigerants are not approved for mobile air-conditioning use by the EPA. Also, most states have passed laws that prohibit the use of flammable refrigerants in mobile air-conditioning systems. As MACS reports, although specific wording may differ, most state statutes require that "Mobile air-conditioning equipment must be manufactured, installed and maintained with regard for the safety of the occupants of the vehicle and the public, and may NOT use or contain a flammable refrigerant or a refrigerant that is toxic."

FIGURE 6–9 A pressure test gauge. *(Courtesy of BET, Inc.)*

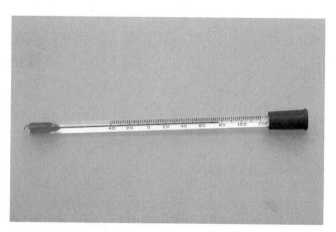

FIGURE 6–10 A spirit thermometer used to measure ambient temperature. *(Courtesy of BET, Inc.)*

Temperature Fahrenheit	Pressure psig	kPa	Temperature Fahrenheit	Pressure psig	kPa
70	80	551	86	103	710
71	82	565	87	105	724
72	83	572	88	107	738
73	84	579	89	108	745
74	86	593	90	110	758
75	87	600	91	111	765
76	88	607	92	113	779
77	90	621	93	115	793
78	92	634	94	116	800
79	94	648	95	118	814
80	96	662	96	120	827
81	98	676	97	122	841
82	99	683	98	124	855
83	100	690	99	125	862
84	101	696	100	127	876
85	102	703	101	129	889

A

Temperature Celsius	Pressure psig	kPa	Temperature Celsius	Pressure psig	kPa
21.1	551	80	30.0	710	103
21.7	565	82	30.5	724	105
22.2	572	83	31.1	738	107
22.8	579	84	31.7	745	108
23.3	593	86	32.2	758	110
23.9	600	87	32.8	765	111
24.4	607	88	33.3	779	113
25.0	621	90	33.9	793	115
25.6	634	92	34.4	800	116
26.1	648	94	35.0	814	118
26.7	662	96	35.6	827	120
27.2	676	98	36.1	841	122
27.8	683	99	36.7	855	124
28.3	690	100	37.2	862	125
28.9	696	101	37.8	876	127
29.4	703	102	38.3	889	129

B

FIGURE 6–11 Temperature-pressure table for CFC-12 refrigerant in English (A) and in metric (B).

Temperature Fahrenheit	Pressure PSIG	kPa	Temperature Fahrenheit	Pressure PSIG	kPa
70	76	524	86	102	703
71	77	531	87	103	710
72	79	545	88	105	724
73	80	551	89	107	738
74	82	565	90	109	752
75	83	572	91	111	765
76	85	586	92	113	779
77	86	593	93	115	793
78	88	607	94	117	807
79	90	621	95	118	814
80	91	627	96	120	827
81	93	641	97	122	841
82	95	655	98	125	862
83	96	662	99	127	876
84	98	676	100	129	889
85	100	690	101	131	903

A

Temperature Celsius	Pressure PSIG	kPa	Temperature Celsius	Pressure PSIG	kPa
21.1	524	76	30.0	703	102
21.7	531	77	30.5	710	103
22.2	545	79	31.1	724	105
22.8	551	80	31.7	738	107
23.3	565	82	32.2	752	109
23.9	572	83	32.8	765	111
24.4	586	85	33.3	779	113
25.0	593	86	33.9	793	115
25.6	607	88	34.4	807	117
26.1	621	90	35.0	814	118
26.7	627	91	35.6	827	120
27.2	641	93	36.1	841	122
27.8	655	95	36.7	862	125
28.3	662	96	37.2	876	127
28.9	676	98	37.8	889	129
29.4	690	100	38.3	903	131

B

FIGURE 6–12   Temperature-pressure table for HFC-134a refrigerant in English (A) and in metric (B).

SUBSTITUTE REFRIGERANTS

The main component of five of the ten substitute refrigerants acceptable for automotive use by the EPA, Significant New Alternatives Policy **(SNAP),** is HCFC-22. The use conditions for the refrigerants R-406/GHG/McCOOL™, GHG-X4/autofrost/Chil-it™, Hot Shot/Kar Kool™, GHG-HP™, and GHG-X5™, in addition to unique fittings, labels, and compressor shut-off switch, require barrier hoses. The other five refrigerants, HFC-134a, FRIGC FR-12™, Free Zone RB-276™, Ikon-12™, and Freeze-12™, approved by the EPA, have the same use conditions except for the requirement for barrier hoses.

Currently, with the exception of HFC-134a, no vehicle manufacturer has approved the use of any of these refrigerants in any of their air-conditioning systems as a substitute or replacement for CFC-12.

There are three refrigerants at the present time that are not acceptable for use by the EPA due to their flammability. These refrigerants are OZ-12, HC-12A©, and Duracool-12. Also, refrigerant R-176 is not acceptable because it contains CFC-12, and R-405A is unacceptable because of extremely high global warming potential and high stratospheric lifetime.

It must be noted that the EPA approves refrigerant for use in certain applications, such as motor vehicle air conditioners (MVAC). This approval, however, does not mean that the EPA recommends or otherwise endorses any particular refrigerant for any particular use. They do recognize that HFC-134a is currently the accepted refrigerant for vehicle use by the industry.

Dedicated service and storage equipment are required by the EPA for each type of refrigerant used in a service facility. For the average facility, that means two systems: one for CFC-12 and one for HFC-134a. If one decides to service vehicles using, for example, Freeze-12™, a third set of service and storage equipment must be purchased for use. This is required although Freeze-12™ contains 80 percent HFC-134a. The other 20 percent, HCFC-142b, would contaminate the HFC-134a equipment.

For the latest update and information on refrigerant approval or any other stratospheric ozone issue, one may contact the EPA. Contact information, toll-free numbers, fax numbers, and Web site addresses are found in the Appendix.

REPLACEMENT REFRIGERANTS

With the exception of HFC-134a, all approved alternate refrigerants are blends; that is, they contain two or more refrigerants. In addition to HFC-134a, the following alternate refrigerants are available. One must be cautioned that not all are approved by the EPA for use in motor vehicle air conditioners (MVAC) and MVAC-like appliances. Some are considered dangerous and very heavy penalties are imposed on those who use them. The EPA makes no exceptions; the rules are simple: Use it — get caught — pay the penalty. There are no excuses. There are one or more questions in the certification exam that attest the technician "knows the law." If there are any doubts, call the EPA hotline and ask. Their toll-free number is included in the Appendix.

Freeze-12

Freeze-12™, a blend of 80 percent HFC-134a and 20 percent HCFC-142b, is acceptable for automotive use subject to fittings, labeling, and compressor shutoff switch. It is not a drop-in replacement for CFC-12 or HFC-134a. The high-side service port shall be 3/8-24 right-hand thread and the low-side service port shall be 7/16-20 right-hand thread. The label background color shall be yellow.

Free Zone/RB-276

Free Zone/RB-276, (Figure 6–15) a blend of 79 percent HFC-134a, 19 percent HCFC-142b, and 2 percent lubricant is acceptable for automotive use subject to fittings, labeling, and compressor shutoff switch. It is not a drop-in replacement for CFC-12 or HFC-134a. The high-side service port shall be 1/2-13 right-hand thread and the low-side service port shall be 9/16-18 right-hand thread. The label background color shall be light green.

Hot Shot/Kar Kool

Hot Shot™, a blend of 50 percent HCFC-22, 39 percent HCFC-124, 9.5 percent HCFC-142b, and 1.5 percent R-600a is acceptable for automotive use subject to fittings, labeling, barrier hoses, and compressor shut-off switch. It is not a drop-in replacement for CFC-12 or HFC-134a. Although this refrigerant contains hydrocarbons (R-600a, isobutane) it is not flammable as blended. The high-side service port shall be 5/8-18 left-hand thread and the low-side service port shall be 5/8-18 right-hand thread. The label background color shall be medium blue.

GHG-HP

GHG-HP™ is a blend of 65 percent HCFC-22, 31 percent HCFC-142b, and 4 percent R-600a. It is acceptable for automotive use subject to fittings, labeling, barrier hose, and compressor shutoff switch conditions. Although it contains hydrocarbons (R-600a, isobutane) it is not considered flammable as blended. It is not a drop-in replacement for CFC-12 or HFC-134a. The fitting size and label background color requirements were not yet determined as of this writing. Call EPA for this information.

GXG-X4/Autofrost/Chill-It

GXG-X4/Autofrost/Chill-it[TM], a blend of 51 percent HCFC-22, 28.5 percent HCFC-124, 16.5 percent HCFC-142b, and 4 percent R-600a is acceptable for automotive use subject to fittings, labeling, barrier hoses, and compressor shutoff switch. Although it contains hydrocarbon (R-600a, isobutane) it is not flammable as blended. It is not a drop-in replacement for CFC-12 or HFC-134a. The high-side service port shall be 0.305-32 right-hand thread and the low-side service port shall be 0.368-26 right-hand thread. The label background color shall be red.

GXG-X5

GXG-X5[TM], a blend of 41 percent HCFC-22, 15 percent HCFC-142b, 40 percent HFC-227ea, and 4 percent R-600a is acceptable for automotive use subject to fittings, labeling, barrier hose, and compressor shutoff switch conditions. This refrigerant contains 4 percent isobutane, a hydrocarbon, but is not considered flammable as blended. It is not a drop-in replacement for CFC-12 or HFC-134a. The high-side service port shall be 1/2-20 left-hand thread and the low-side service port shall be 9/16-18 left-hand thread. The label background color shall be orange.

R-406A/GHG

R-406A/GHG[TM] is a blend of 55 percent HCFC-22, 41 percent HCFC-142b, and 4 percent R-600a. It is acceptable for automotive use subject to fittings, labeling, barrier hose, and compressor shutoff switch conditions. It is not considered flammable as blended although it contains isobutane (R-600a), a hydrocarbon. It is not a drop-in replacement for CFC-12 or HFC-134a. The high-side service port shall be 0.305-32 left-hand thread and the low-side service port shall be 0.368-26 left-hand thread. The label background color shall be black.

Ikon-12

Ikon-12[TM] was approved for automotive air-conditioning system use in mid-1996. The manufacturer, Ikon Corporation, claims the composition of this refrigerant as confidential business information. Subject to use requirements are unique fittings, distinctive label color, and high-side pressure switch. It is not yet known if barrier hoses are required. Fitting size requirements and label color are not developed at the time of this writing. Call EPA's hotline or the manufacturer for more information.

Frigc FR-12

Frigc FR-12[TM] is a blend of 39 percent HCFC-124, 59 percent HFC-134a, and 2 percent R-600. It is acceptable for automotive use subject to fittings, labeling, and compressor shutoff switch. It is not considered flammable as blended though it contains a hydrocarbon, butane (R-600). It is not a drop-in replacement for CFC-12 or HFC-134a. The high-side and low-side service ports shall be quick-disconnect but different from the R-134a service ports. The label background color shall be gray.

OZ-12

OZ-12®, a hydrocarbon Blend A is not SNAP approved by the EPA, who claims that it contains a flammable blend of hydrocarbons and that insufficient data were submitted to demonstrate safety.

R-176

This refrigerant contains CFC-12, HCFC-22, and HCFC-142b. It is not SNAP approved by the EPA, who claims that it is not appropriate to use a CFC-12 blend as a CFC-12 substitute.

HC-12a

HC-12a®, a hydrocarbon Blend B is not SNAP approved by the EPA, who claims that it contains a flammable blend of hydrocarbons and that insufficient data were submitted to demonstrate safety.

Duracool 12a

This refrigerant is not SNAP approved by the EPA. It is identical to HC-12a® in composition but is made by a different manufacturer.

R-405A

This refrigerant is not SNAP approved by the EPA because it contains perfluorocarbon, which has extremely high global warming potential and lifetime.

MT-31

MT-31 is a blend proposed as a CFC-12 substitute that is not approved by the EPA for use in any application because of the toxicity of one of its components.

SPECIAL SAFETY PRECAUTIONS

It is important for technicians to be aware of the hazards involved in the handling and use of any refrigerant. Safety procedures must be observed (Figure 6–13). Recall that most refrigerants are:

- ❏ Odorless
- ❏ Undetectable in small quantities
- ❏ Colorless
- ❏ Nonstaining

Refrigerant is dangerous because of the damage it can cause if it strikes the human eye or comes into contact with the skin. Suitable eye protection must be worn to protect the eyes in case of splashing. If refrigerant does enter the eye, the eye can freeze, resulting in blindness. The following procedure is suggested if refrigerant enters the eye(s):

1. Do not rub the eye.
2. Splash large quantities of cool water (H_2O) into the eye to increase the temperature.
3. Tape a sterile eye patch in place over the eye to prevent dirt from entering the eye.
4. Go immediately to a doctor or hospital for professional care.
5. DO NOT ATTEMPT SELF-TREATMENT.

If liquid refrigerant strikes the skin, frostbite can occur. The same procedure outlined for emergency eye care can be used to combat the effects of skin contact. Refrigerant in air is harmless unless it is released in a confined space. Under this condition, refrigerant displaces oxygen in the air and may cause drowsiness, unconsciousness, or even death. However, the automobile owner and the service technician need not be concerned about the safety of the automotive air-conditioning system under normal conditions. The small capacity of the system compared to the large area of the car interior minimizes the concentration of any contamination.

Refrigerant must not be allowed to come into contact with an open flame or a very hot metal. Many texts, including previous editions of *Automotive Air-Conditioning*, state that fluorocarbon refrigerants such as CFC-12 produce phosgene gas when exposed to hot metal or an open flame. The proper name for phosgene gas is carbonyl chloride ($COCl_2$). In 1933, shortly after the development of CFC-12, tests made by Underwriters' Laboratory, Inc. (UL) indicated that CFC-12 produced this highly toxic gas during decomposition. In recent years, however, using advanced technology equipment, the UL has determined

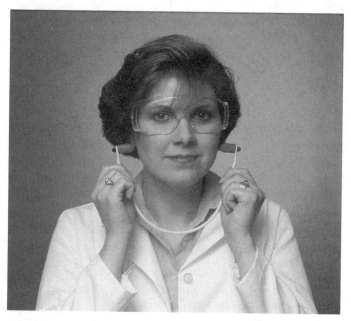

FIGURE 6–13 Use personal sight and hearing protection. *(Courtesy of Wilson Safety Company)*

that phosgene gas is not produced in this manner. Decomposition, however, results in the formation of carbonyl fluoride (COF_2) and carbonyl chlorofluoride ($COClF$) with small amounts of free chlorine (Cl_2).

Although 20 to 50 times less toxic than phosgene, the decomposed gases of CFC-12 must be avoided. At high concentrations, lack of oxygen (O), which causes asphyxiation, is the real hazard. A primary rule, then, is to *avoid breathing these or any other fumes.* The human body requires oxygen (O) in the quantity found in noncontaminated air. Diluting air with any foreign gas can reduce the available oxygen (O) to a level that may be harmful or, in some cases, fatal.

The following rules must always be observed when handling refrigerants:

1. Above 130°F (54.44°C), liquid refrigerant completely fills a container, and hydrostatic pressure builds up rapidly with each degree of temperature rise. To provide for some margin of safety, never heat a refrigerant cylinder above 125°F (51.7°C) or allow it to reach this temperature.

NOTE: It is the practice of some service technicians to place containers of refrigerant into a pan of warm water as an aid in speeding up the charging process. This practice is not recommended and is especially discouraged for the inexperienced. Even the "pros" are sometimes injured with this practice.

2. Never apply direct flame to a refrigerant cylinder or container.

3. Never place an electrical resistance heater near or in direct contact with a container of refrigerant.

4. Do not abuse a refrigerant cylinder or container.
   a. To avoid damage, use an approved valve wrench for opening and closing the valves.
   b. Secure all cylinders in an upright position when storing and withdrawing refrigerant.

5. Do not handle refrigerant without suitable eye protection.

6. Do not discharge (vent) refrigerant into the atmosphere.

7. Remove refrigerant from a system using approved recovery equipment only.

8. Use only Department of Transportation (DOT)-approved refrigerant recovery cylinders, Figure 6–14.

9. Do not fill recovery cylinders beyond 80 percent of their capacity at 70°F (21°C).

10. Do not mix refrigerants. Refrigerants are not compatible. Any mixture of refrigerant increases the cost of recycling for reuse or of disposal.

11. Do not mix refrigeration oil. Mineral oil and some of the new synthetic lubricants, for example, polyalkalyne glycol (PAG), are *not* compatible.

12. For MVAC systems, do not introduce anything but pure approved refrigerant and refrigeration lubricants into the system.

Under controlled conditions, recent tests, have shown that HFC-134a can be combustible under certain conditions. This occurs at above atmospheric pressure with an air concentration of 60 percent or more by volume. Although an ignition source, not generally found in a MVAC system, is required for combustion, the mixture is potentially dangerous and must be avoided.

Refrigerant Used for Leak Testing

Most leaks in a MVAC system can be located with a system pressure of 50 psig (434.75 kPa). This pressure is easily obtainable using only refrigerant as a source. For a difficult to locate leak, however, it is often desirable to increase the system pressure to about 100 psig (689.5 kPa). This is easily accomplished using a pressure-regulated dry nitrogen charge to increase the pressure of the refrigerant. There are some technicians, on the other hand, who prefer to use only refrigerant-22 (HCFC-22), Figure 6–15, to pressurize the system without the use of nitrogen. Unlike federal requirements and regulations for refrigerants CFC-12 and HFC-134a, there is no law that requires dedicated equipment be used for HCFC-22. The same fittings and equipment that are used to service and leak test CFC-12 air-conditioning systems can be used to service and leak test HCFC-22. It should be noted, however, that several of the EPA-approved blend refrigerants that contain HCFC-22 do require dedicated fittings and equipment for service.

FIGURE 6–14 Color-coded refrigerant recovery cylinders. The bodies of the cylinders are gray; the tops are yellow. *(Courtesy of White Industries)*

FIGURE 6–15 A small container of HCFC-22. *(Courtesy of BET, Inc.)*

Before introducing HCFC-22 into the system for leak testing, all residual refrigerant must first be removed. This is not a simple task, because refrigerant has a tendency to migrate into the lubricant. Also, after leak testing it is important that all HCFC-22 be removed from the system before introducing any other refrigerant. Again, due to migration, this is not an easy task.

On the bottom line, the use of HCFC-22 or any other "foreign" refrigerant for leak testing is discouraged. Residual traces left in the air-conditioning system may affect system performance or cause premature failure.

CAUTION

UNDER NO CIRCUMSTANCES SHOULD ANY EQUIPMENT OR SYSTEM BE PRESSURE TESTED USING COMPRESSED AIR (SHOP AIR). AIR CONTAINS 21 PERCENT OXYGEN (O), WHICH CREATES A HAZARDOUS ENVIRONMENT WITHIN THE AIR-CONDITIONING SYSTEM.

REFRIGERANT CYLINDERS

All refrigerant cylinders are manufactured to specifications established by the U.S. Department of Transportation (DOT). The DOT has regulatory authority over all hazardous material that may be commercially transported. The disposable cylinders, such as those used for CFC-12 and HFC-134a, are manufactured to DOT specifications 39. They are often referred to as "DOT 39 cylinders" in trade jargon. DOT 39 cylinders are not intended to be reused for any reason. Transporting refilled DOT 39 cylinders is illegal and subject to severe penalties and possible imprisonment. The use of refilled DOT 39 cylinders for any reason is a violation of the Occupational Safety and Health Administration (OSHA) workplace regulations as well as state laws. An empty DOT 39 cylinder should be ruptured, as demonstrated in Figure 6–16, to ensure that it cannot be reused. Although kits are available to convert an empty cylinder to an air tank, the practice is discouraged. Air contains moisture which will eventually cause rust from inside the tank and could create a personal hazard for the user.

Refillable cylinders are available that are approved by DOT to transport recovered and used refrigerant. They are available in sizes ranging from 15 pounds (6.8 kg) to 1,000 pounds (453.6 kg). Refillable cylinders are fitted with a "Y" configured combination valve, shown in Figure 6–17, with Compressed Gas Association's (CGA) #165 fittings. CGA's #165 fittings are identical to the SAE's 1/4-inch flare fitting. To avoid the problems of mixing refrigerants, HFC-134a cylinders are equipped with CGA's #167 fittings, which are the same as SAE's M14 × 1.5 fittings.

A   B

FIGURE 6–16   (A) Rupture disc for disposable cylinder; (B) punch hole before discarding to prevent cylinder reuse. *(Courtesy of BET, Inc.)*

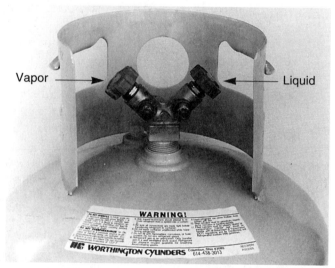

FIGURE 6–17 The "Y" configured combination valve of a reusable refrigerant recovery cylinder. *(Courtesy of BET, Inc.)*

When dispensing refrigerant, the blue valve is for liquid and the red valve is for vapor. To remove any doubt, however, these valves are marked LIQUID and VAPOR. The color code for recovery cylinders for any used nonrecycled refrigerant is a gray body with a yellow top. The contents are identified by using a label such as the one shown in Figure 6–18.

Refillable cylinders *must* be inspected every five years. Refrigerant cylinders must not be filled to greater than 80 percent of their capacity. Vapor pressure must not exceed 318 psig (2193 kPa) at a temperature of 130°F (54.4°C). These conditions are based on R-502, a refrigerant with the greatest vapor pressure. It is not likely that either CFC-12 or HFC-134a will approach this pressure under the same conditions.

REFRIGERATION LUBRICANT

The moving parts of a compressor assembly must be lubricated to prevent damage during operation. A lubri-

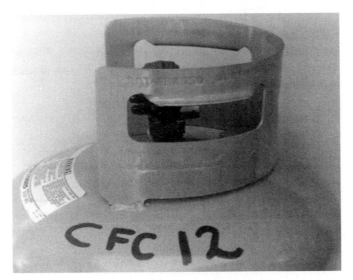

FIGURE 6–18 Identifying the contents of a refrigerant recovery cylinder helps to avoid cross-contamination. *(Courtesy of BET, Inc.)*

cant especially formulated for use with refrigerants, is used on these moving parts and on the seals and gaskets as well. In addition, a small amount of lubricant is mixed with the refrigerant, which circulates through the system. This refrigerant and lubricant combination helps maintain the thermostatic expansion valve and other internal moving parts in proper operating condition.

The lubricant that must be used in an automobile air-conditioning system is a nonfoaming, sulfur-free grade specifically formulated for use in certain types of air-conditioning systems. This special lubricant is known as *refrigeration oil*, depicted in Figure 6–19, and is available in several grades and types. The grade and type is determined by the compressor manufacturer as well as the type of refrigerant in the system. To replace lubricant that may be lost due to a refrigerant leak, *pressurized oil* is available in disposable cans. The container shown in Figure 6–20 contains 2 fluid ounces (59 mL) of oil and a like measure of refrigerant. The refrigerant provides the necessary pressure required to force the oil into the system. A word of caution: Ensure that the lubricant and refrigerant are compatible with that contained in the air-conditioning system.

Mineral oil that is used with CFC-12 in air-conditioning systems is clear to light yellow in color. Some

FIGURE 6–19 Refrigeration oil in 1 gallon (3.785 L) container. *(Courtesy of BET, Inc.)*

FIGURE 6–20 Refrigerant oil charge for adding oil to system. *(Courtesy of BET, Inc.)*

synthetic lubricants used with HFC-134a in an air-conditioning system may be blue or an other color. An impurity in any lubricant can cause a color change ranging from brown to black. Mineral oil is practically odorless; a strong odor indicates that the oil is impure. Some pure synthetic lubricants, on the other hand, have a pungent odor. In either case, to ensure optimum system protection and performance, impure oil must be removed and replaced with clean, fresh oil. The receiver-drier or suction-accumulator should also be replaced and a good pump down (system evacuation) performed before the air-conditioning system is recharged with refrigerant.

These procedures are given in Chapter 8.

LUBRICANTS FOR HFC REFRIGERANTS

Some of the alternative automotive refrigerants are compatible with conventional mineral oil or alkyl benzene lubricants. However, R-134a, a hydrofluorocarbon (HFC) refrigerant, is not miscible with conventional mineral oil or alkyl benzene lubricants. Lack of miscibility can lead to system operational problems where the lubricant travels with the refrigerant and is present in the condenser and evaporator. When the two fluids are miscible, the lubricant is carried back to the compressor. When not miscible, the lubricant can accumulate in the condenser and evaporator.

Lubricant that has accumulated in the condenser can reduce heat transfer and restrict the flow of liquid refrigerant. This condition can cause vapor pockets to form in the liquid stream as it flows through the metering device into the evaporator. Lubricant that has accu-

mulated in the evaporator will reduce heat transfer and restrict refrigerant vapor flow. Poor lubrication return to the compressor often leads to excessive wear of the compressor due to lubricant starvation.

Synthetic Lubricants

Two different types of synthetic lubricant, polyalkylene glycols (PAG) and neopentyl polyol esters (POE), Figure 6–21, are available for use with HFC refrigerants. Although new to MVAC systems since 1992, PAG is not a new lubricant. It has been used for years in compressors for natural gas production and for compressors handling other difficult gases.

Polyalkylene Glycol

Polyalkylene glycols (PAGs) are extremely hygroscopic and can absorb several thousand parts per million (ppm) of water (H_2O) when exposed to moist air; thirty times more than mineral oil, which generally contains less than 100 ppm of water (H_2O). High water (H_2O) content causes corrosion and copperplating problems in some refrigeration systems when PAG lubricants are used. PAG is also sensitive to chlorine-containing contaminants such as residual R-12 remaining in a system that has been converted.

PAG, however, was selected in 1992 as the lubricant of choice for automotive air-conditioning systems in which HFC-134a was used as a refrigerant. This selection was based on previous success and due to deadline time constraints to phase out CFC-12. It is the opinion of many in the industy that there will be an eventual shift to poly-

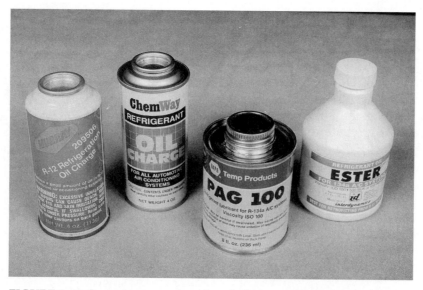

FIGURE 6–21 Several types and grades of oil are used in automotive air-conditioning systems.

ol ester (POE) lubricants for automotive application. For the time, however, the lubricant of choice for use with most HFC-134a automotive air-conditioning systems is PAG. PAG is available in low- and high-viscosity grades. Always follow the manufacturer's specifications when adding or changing compressor lubricant.

Polyol Ester

Neopentyl polyol esters (POEs) are a group of organic esters that have an application in systems that have a wide operating temperature range where good lubricating characteristics are desired. These lubricants are also desired because of their good miscibility with HFCs.

Various POE compositions enhance miscibility with HFCs and their resistance to copperplating in refrigeration systems. They also have excellent thermal stability, low volatility, low deposit-forming tendencies, high flash points and high auto-ignition temperatures.

On the downside, POEs are hygroscopic, susceptible to hydrolysis, and incompatible with certain elastomers. These esters can absorb several thousand parts per million of water (H_2O), however, they are less absorbent than most PAGs. Therefore, like PAGs, POEs must be handled so excess moisture does not enter the refrigeration system. POE should only be used in automotive air-conditioning systems that specifically require this type of lubricant.

Safety

Personal protection equipment, such as rubber- or PVC-coated gloves or barrier creams and safety goggles should be worn when handling lubricants. Prolonged skin contact or any eye contact can cause irritations and discomfort, such as stinging and burning. One should avoid breathing the vapors produced by these lubricants, and they should only be used in a well-ventilated area. Keep them in tightly sealed containers to prevent moisture contamination due to humidity and to ensure that their vapors do not escape.

THE CLASSIFICATION OF REFRIGERATION LUBRICANT

The classification of refrigeration lubricant is based on three factors: *viscosity*, **compatibility** *with refrigerant*, and *pour point*.

Viscosity. The viscosity rating for a fluid is based on the time in seconds that is required for a measured quantity of the fluid to pass through a calibrated orifice when the temperature of the fluid is 100°F (37.8°C). The resistance to flow of any fluid is judged by its viscosity rating; the thicker the fluid, the higher the viscosity number.

The *Saybolt Universal Viscosity* (SUV) is defined as the time in seconds that is required for 60 cubic centimeters (3.67 in.^{3}) of lubricant at 100°F (37.8°C) to flow through a standard Saybolt orifice.

Compatibility. Refrigeration lubricant must be compatible with the refrigerant that it is to be used with. It should be noted that refrigeration lubricant used in a CFC-12 system is a mineral oil designated "YN-9." This lubricant cannot be used in an HFC-134a system. Conversely, a synthetic PAG lubricant, designated "YN-12," for an HFC-134a air-conditioning system with a reciprocating compressor should not be used in a CFC-12 air conditioning system. A second PAG lubricant is used in a rotary compressor. Actually, there may be three or more synthetic lubricants, each formulated for a particular application. Incompatible lubricant mixtures may cause serious damage to the air-conditioning system compressor.

To be compatible with the refrigerant used in the system, refrigeration lubricant must be capable of remaining a lubricant when mixed with the refrigerant. In other words, the lubricant must not change or become separated by chemical interaction with the refrigerant.

The compatibility of a refrigeration lubricant with a refrigerant is determined by a test called *floc test F.* This test is performed by placing a mixture containing 90 percent lubricant and 10 percent refrigerant in a sealed glass tube. The mixture is then slowly cooled until a waxy substance appears. The temperature at which the substance forms is recorded as the *floc point.*

Pour Point. The temperature at which a lubricant will just flow is its *pour point.* This temperature is recorded in degrees Fahrenheit. The pour point is a standard of the American Society for Testing Materials (ASTM).

The pour point temperature for oil and lubricant used in a high-temperature refrigeration such as a motor vehicle air-conditioning (MVAC) system is between –40°F (–40°C) and –10°F (–23.3°C), depending on its grade and type.

SERVICING TIPS

If practical, the lubricant should be checked each time the air-conditioning system is "opened" for repair or service. If there is evidence of a large lubricant leak—due to a ruptured hose, for example—the lubricant level in the system must be checked. Always check the manufacturer's recommendations before adding lubricant to the air-

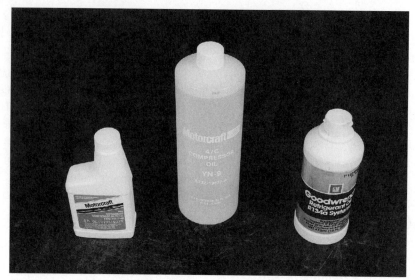

FIGURE 6–22 Refrigeration oil is packaged in several container sizes.

conditioning system. The procedures for checking and adding lubricant to compressors and air-conditioning systems can be found in Chapter 10.

When the lubricant is not being dispensed, the container must remain capped. Always be sure that the cap is in place and tightly secured. Refrigeration lubricant is very hygroscopic; that is, it readily absorbs moisture. This is especially true for HFC-134a-compatible lubricants. It should be noted that polyalkylene glycol lubricant is ten times more hygroscopic than mineral-based lubricants. It is also more expensive. For these reasons, it is suggested that refrigeration lubricant be purchased in small containers, Figure 6–22. Most lubricant is packaged in 1 pint (0.473 L), 1 quart (0.946 L), 1 gallon (3.785 L), 5 gallon (18.925 L), and larger containers.

SUMMARY

The properties of a good refrigeration lubricant are low wax content, good thermal and chemical stability,
low viscosity, and a low pour point. The following simple rules should be observed when handling refrigeration lubricants:

DO

- ❏ Use only approved refrigeration lubricant.
- ❏ Be sure the cap is tight on the container when not in use.
- ❏ Replace lubricant if there is any doubt of its condition.
- ❏ Avoid contaminating the lubricant.
- ❏ Dispose of used lubricant in a proper manner.

DO NOT

- ❏ Transfer lubricant from one container to another.
- ❏ Return used lubricant to the container.
- ❏ Leave the lubricant container uncapped.
- ❏ Use a grade or type of lubricant other than that recommended for the air conditioner being serviced.

REVIEW

Select the correct answer from the choices given.

1. Technician A says that lubricant used in a CFC air-conditioning system has a mineral oil base. Technician B says that although mineral oil is generally blue, it turns from brown to black if contaminated. Who is right?
   a. A only
   b. B only
   c. Both A and B
   d. Neither A nor B

2. Technician A says that unused refrigeration oil should be returned to its container as soon as possible to prevent contamination. Technician B says that the ambient air is readily absorbed into the lubricant, forming a contaminant. Who is right?
   a. A only
   b. B only
   c. Both A and B
   d. Neither A nor B

3. Which refrigerant was used in motor vehicle air-conditioning systems until the mid-1990s ?
   a. HFC-134a
   b. CFC-12
   c. HCFC-22
   d. HC-12a

4. All of the following material are basic elements, EXCEPT:
   a. steel.
   b. iron.
   c. copper.
   d. aluminum.

5. Unlike CFC-12, HFC-134a does not contain:
   a. fluorine.
   b. hydrogen.
   c. carbon.
   d. chlorine.

6. Technician A says the use of CFC-12 was a major cause of ozone depletion. Technician B says the use of CFC-12 contributed to the greenhouse effect. Who is right?
   a. A only
   b. B only
   c. Both A and B
   d. Neither A nor B

7. Technician A says that the production of CFC-12 ended in the United States on December 31, 1996. Technician B says the United States continued to import CFC-12 until December 31, 1998. Who is right?
   a. A only
   b. B only
   c. Both A and B
   d. Neither A nor B

8. When retrofitting an air conditioning system, the following must always be changed, EXCEPT:
   a. service fittings.
   b. metering device.
   c. lubricant.
   d. nonbarrier hoses.

9. Technician A says the term *Freon*® is a trade name for a brand of refrigerant. Technician B says the term *Suva*®, also a trade name, identifies a competitor's brand of refrigerant. Who is right?
   a. A only
   b. B only

c. Both A and B
d. Neither A nor B

10. The "pound" cans of refrigerant are being discussed. Technician A says an HFC-134a can generally contains only 15 oz. (444 mL). Technician B says a CFC-12 can generally contains 17 oz (503 mL). Who is right?
    a. A only
    b. B only
    c. Both A and B
    d. Neiter A nor B

11. The color code for an HFC-134a refrigerant disposable cylinder is:
    a. light blue.
    b. dark blue.
    c. white.
    d. green.

12. A refrigerant purity tester is used to determine if the refrigerant is at least ____ pure.
    a. 92
    b. 94
    c. 96
    d. 98

13. All of the following are lubricants used in a MVAC system, EXCEPT:
    a. POE.
    b. PAG.
    c. mineral oil.
    d. castor oil.

14. Lubricant accumulation in the evaporator is being discussed. Technician A says this problem will restrict refrigerant flow. Technician B says this problem will reduce heat transfer. Who is right?
    a. A only
    b. B only
    c. Both A and B
    d. Neither A nor B

15. Which of the following components are *not* approved for use in an ozone-friendly refrigerant?
    a. R-12
    b. R-406
    c. R-22
    d. R-142b

16. Technician A says a refrigerant cylinder should never be heated above 150°F (65.6°C). Technician B says never use a direct flame or resistance heater to warm a refrigerant cylinder. Who is right?
    a. A only
    b. B only
    c. Both A and B
    d. Neither A nor B

17. Technician A says a high concentration of air in an HFC-134a system may be combustible. Technician B says even a low concentration of hydrocarbon in a CFC-12 system may be combustible. Who is right?
    a. A only
    b. B only
    c. Both A and B
    d. Neither A nor B

18. The ability and resistance of a fluid to flow is known as its:
    a. viscosity.
    b. pour point.
    c. both A and B.
    d. neither A nor B.

19. When should the lubricant in an air-conditioning system be checked?
    a. Annually, as part of a preventative maintenance schedule
    b. Any time that the system is opened for service or repair
    c. Both A and B
    d. Neither A nor B

20. Which of the following is a drop-in replacement for CFC-12?
    a. HFC-134a
    b. HCFC-22
    c. Both A and B
    d. Neither A nor B

TERMS

Write a brief description of the following terms:

1. alternate refrigerant
2. barrier hose
3. coil
4. compatibility
5. concentration
6.
7.

11. inverted
12. pressure gauge
13. purity test
14. refrigerant identifier
15. R-134a
16. retrofit
17. SAE
18. SNAP
19. Suva®
20. synthetic

CHAPTER 7

SPECIAL SERVICE TOOLS

OBJECTIVES

On completion and review of this chapter, you should be able to:

❏ Identify the various special tools used in automotive air-conditioning service.
❏ Discuss the purpose and use of special service tools.
❏ Understand the need for special service tools.
❏ Identify required and alternative services and the special tools required.

INTRODUCTION

The air-conditioning service technician must be familiar with and know how to properly use **specialized tools** and equipment relating to automotive air-conditioning service. Very basic automotive service tools, such as the various types of pliers, screwdrivers, and wrenches, were covered in Chapter 3. This chapter contains a description of the many specialized tools that are required to properly repair various components of the mobile air-conditioning system. Other special tools are discussed, where applicable, throughout this text.

MANIFOLD AND GAUGE SET

A basic tool for the air-conditioning service technician is the manifold and gauge set, shown in Figure 7–1. Because system pressures accurately indicate air-conditioning system performance, a means must be provided to take the pressure measurements of an air-conditioning system while it is operational. The manifold and gauge set is essential in making these measurements. The servicing of most vehicular air conditioners requires the use of a two-gauge manifold set. Some early systems required a three-gauge set such as that shown in Figure 7–2.

For a two-gauge set, the gauge on the left is used on the low (suction) side of the system. The other gauge, on the right, is used on the high (discharge) side of the system. They are connected to the respective low- and high-side **fittings** of an air-conditioning system. Those that require a third gauge have a second low-side fitting. There is no significant difference between gauges that are designated for CFC-12 refrigerant when compared with those designated for HFC-134a refrigerant. To prevent mixing these refrigerants, however, federal regulations require that two manifold and gauge sets be used for motor vehicle air-conditioning (MVAC) system service. It is interesting to note, however, that this regulation does not apply to air-conditioning technicians who do not service MVAC systems.

ELECTRONIC MANIFOLD

An electronic battery-powered manifold and gauge set does not require that hoses be connected to the air-conditioning system for evaluation and testing. This prevents the possibility of cross contamination, as well as venting, of refrigerant. This device displays instant digital readout information of both low and high sides of the air-conditioning system relative to system conditions on a split screen. This information includes:

- ❏ Low- and high-side pressure (**psia** or kPa)
- ❏ Low- and high-side and external temperature (°F or °C for CFC-12 or HFC-134a)
- ❏ **Superheat** and subcooling (minimum/maximum/average)
- ❏ Vacuum (in. hg. or micron)
- ❏ Low manifold batteries (4 AA cells)

As with a **standard** manifold and gauge set, there is no need to have a different set for each type of refrigerant. Because there are no hoses to entrap refrigerant, the electronic manifold may be used for both CFC-12 and HFC-134a service.

For service procedures the leads have a built-in fitting for a standard service hose. Although the same electronic manifold may be used for both refrigerants, dedicated service hoses are required to prevent cross-contamination.

Manifold

The gauges are connected into the air-conditioning system through a manifold and high-pressure hoses. The manifold contains provisions for fittings to which gauges and hoses can be connected. In addition, two hand shutoff valves are provided on the manifold to control the flow of refrigerant through the manifold. Some manifolds, like the one in Figure 7–3, have additional hand shutoff valves to allow the technician to recover, evacuate, and charge the system without disconnecting a hose.

The gauges are attached to the manifold by 1/8-inch **NPT** (pipe) connections. Hoses are connected to the CFC-12 refrigerant manifold with 1/8-inch NPT × 7/16-20 (flare) half unions.

HFC-134a hoses are attached to the manifold with 1/8-inch NPT × 1/2-16 acme half unions. Some manifolds may have this fitting as part of their casting, eliminating the need for a half union. A description of the service hoses used with the manifold is described later in this chapter.

The low-side hose fitting is directly below the low-

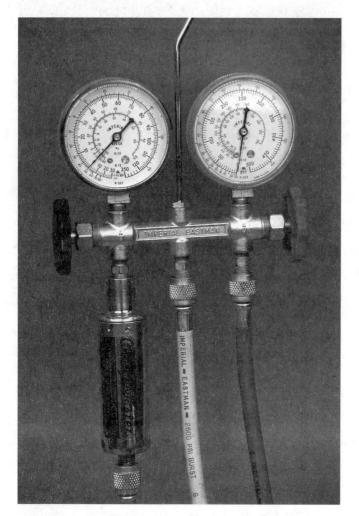

FIGURE 7–1 Manifold and gauge set, side wheel. *(Photo by Bill Johnson)*

FIGURE 7–2 Manifold and three–gauge set, front wheel. *(Courtesy of BET, Inc.)*

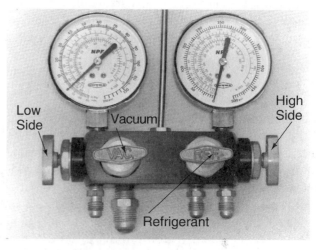

FIGURE 7–3 A manifold that allows refrigerant recovery, evacuation, and recharging without disconnecting a hose. *(Courtesy of BET, Inc.)*

side gauge and the high-side hose fitting is directly below the high-side gauge. The center hose fitting of the manifold is used for recovery, charging, evacuation, or any other service that may be required.

The hand shutoff valves regulate the flow of refrigerant through the manifold. The manifold is closed when the hand valve is turned all the way to the right in a clockwise **(cw)** direction. The air-conditioning system pressure, however, is exerted on its respective gauge. Figure 7–4 shows both manifold hand valves in the closed position. Note that the air-conditioning system

pressure can still be recorded on each respective gauge.

The hand shutoff valve is opened by turning its handle to the left or counterclockwise (ccw). When the hand valve is open, the manifold circuit is opened to the center hose port of the manifold set. If there is nothing connected to the center manifold port, refrigerant will escape. This condition, therefore, is used only when refrigerant is intended to be allowed to enter or be removed from the system.

When the low-side manifold hand valve is opened, as in Figure 7–5, the passage is complete between the low-side port and the center port only. The low-side gauge will indicate the low-side pressure only. The high side remains closed and the high-side gauge indicates only the high-side pressure.

Similarly, when only the high-side hand valve is opened, as in Figure 7–6, the passage is complete between the high-side port and the center port and the high-side gauge indicates only high-side pressure. As shown, the low-side and high-side gauges indicate only the pressure in their respective sides.

If both hand valves are opened as in Figure 7–7, however, both the low-side and high-side ports are open to the center port. The pressures indicated on the gauges are not accurate when both hand valves are opened. Some of the high-side pressure feeds through the manifold to the low-side gauge, with the result that the high-side pressure indication is decreased and the low-side pressure indication is increased.

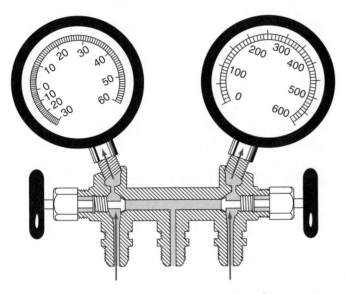

FIGURE 7–4 Fluid passages are open only to the respective gauges when both hand valves are closed.

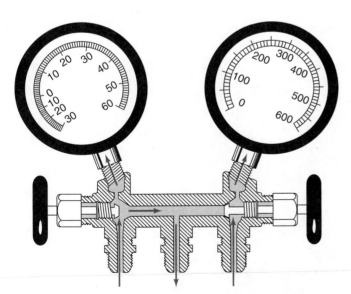

FIGURE 7–5 When low-side hand valve is open the fluid passage is to the low-side gauge and between the low-side service hose and the center service hose. The high-side fluid passage is open to the high-side gauge only.

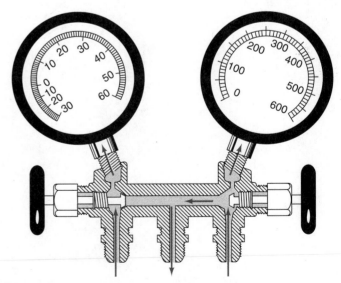

FIGURE 7–6 When the high-side hand valve is open the fluid passage is to the high-side gauge and between the high-side service hose and the center hose. The low-side fluid passage is open to the low-side gauge only.

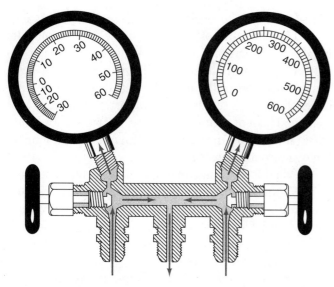

FIGURE 7–8 Front–wheel type manifold and gauge set. (*Courtesy of BET, Inc.*)

FIGURE 7–7 When both hand valves are open all fluid pressures are equal in the manifold and neither gauge indicates accurate pressure. This *is not* a desired condition for system testing. It is, however, a condition that may be used for system evacuation procedures.

The manifold and gauge set is used to perform nearly all of the air-conditioning system mechanical tests and diagnostic procedures. The manifolds are available in a front-wheel type, as shown in Figure 7–8, a side wheel type, as shown in Figure 7–9; an offset wheel type, as shown in Figure 7–10; and a multi-port sight glass type, as shown in Figure 7–11.

Low-Side Gauge

The **low-side** gauge, as its name implies, is used to measure the pressure present in the low side of the air-conditioning system. There is no particular difference in the construction of an English and metric low-side gauge except for the dial **calibration.**

English. The English scaled gauge used on the low side of the system is called a *compound gauge*. This gauge is designed to give both vacuum and pressure indications. It is connected through the manifold and the high-pressure hose to the low side of the air-conditioning system.

FIGURE 7–9 Make sure that manifold hand shutoff valves are closed. (*Courtesy of Uniweld Products Inc.*)

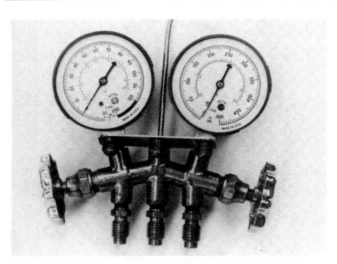

FIGURE 7–10 An offset wheel manifold. (*Courtesy of BET, Inc.*)

FIGURE7–11 A mulitport HFC–134a manifold and gauge set with sight glass.

The vacuum **scale** of a compound gauge is generally calibrated to show pressures from 30 inches of mercury (in. Hg) to 0 inches of mercury (in. Hg). The pressure scale is calibrated to indicate pressures from 0 pounds per square inch gauge (psig) to 120 pounds per square inch gauge (psig). The compound gauge is constructed to prevent damage to the gauge if the pressure reaches a value as high as 250 to 350 psig. The gauge described in this paragraph is designated in the following manner: *30"-0-120 psi, with retard to 250 psi.*

Pressures above 80 psig are rarely experienced in the low side of the system. However, such pressures may result if the manifold hoses are crossed so that the manifold gauges are connected backward to the air-conditioning system. (Even experienced service technicians can make this type of error.)

Metric. The metric gauge scale to be used on the low side of the system seems, at this time, to be rather confusing. Some sources are using millimeters of mercury (mm Hg) for below atmospheric pressure, and kilograms per square centimeter (kg/cm^2) are being used for above atmospheric pressure. But, as noted in Chapter 3, these scales are incorrect. Other sources are using negative kiloPascal (–kPa) values (3.38 kPa = 1 in. Hg) for below atmospheric pressure and positive kiloPascal (+kPa) values (6.895 kPa = 1 psig) for above atmospheric pressure. According to IEEE Standard 268-1979, 3.4.6, "**Absolute** pressure is never negative."

Therefore, it seems likely that the correct English/metric low-side gauge scale will be one of those shown in Figure 7–12. Figure 7–13 has a standard in. Hg/psig English scale and a kPa absolute metric scale.

FIGURE 7–12 Typical R–134a low-side gauge.

Figure 7–14 gives both metric and English scales in absolute values.

It seems reasonable that the scale of Figure 7–13 will be used, particularly if a kPa (gauge) scale is used as a high-side gauge. If, on the other hand, the high-side gauge is scaled kPa (absolute), the low-side gauge scale of Figure 7–14 will be used.

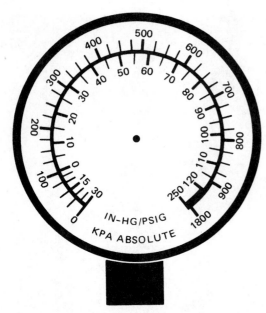

FIGURE 7–13 Composed gauge showing psi (gauge) and kPa (absolute) pressure scales for comparison.

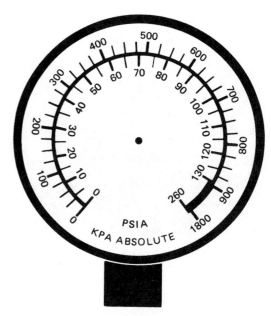

FIGURE 7–14 Compound gauge scaled for absolute pressure.

High-Side Gauge

The high-side gauge as its name implies, is used to measure the pressure present in the high side of an air-conditioning system. There is no difference in the construction of an English and metric high-side gauge except for the dial calibration.

English. The high-side gauge, shown in Figure 7–15, indicates the pressure in the high side of the system. Pressures in the high side of an air-conditioning system, under normal conditions, seldom exceed 250 psig. However, as a **safety factor,** particularly with an HFC-134a system, it is recommended that the minimum scale indication of the high-side gauge be 500 psig. This is a popular scale for the high-side gauge. The high-side gauge is not calibrated for below atmospheric pressure and is so designed that it will not be damaged when the air-conditioning system is pulled into a vacuum.

Metric. High-side metric pressure gauges are being scaled by some at kilograms per square centimeter (kg/cm^2). As noted in Chapter 3, this scale is not correct according to SI metrics. The correct conversion is to kiloPascals (kPa), whereby 1 psi equals 6.895 kPa. Either of the high-side metric gauge scales is correct; Figure 7–16 is scaled kPa (gauge) and uses atmospheric pressure as a zero

FIGURE 7–15 Typical R–134a high-side gauge.

reference. The pressure gauge in Figure 7–17 shows psia and kPa absolute pressure scales for comparison. The low end of this gauge is at atmospheric pressure, or 14.696 psia (rounded off to 15 psia) English or 101.328 kPa absolute (rounded off to 100 kPa absolute) metric.

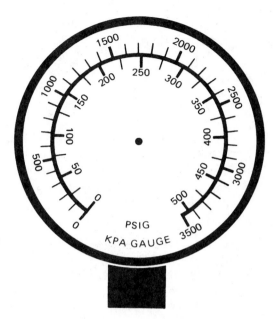

FIGURE 7–16 Pressure gauge showing psi (gauge) and kPa (gauge) pressure scales for comparison.

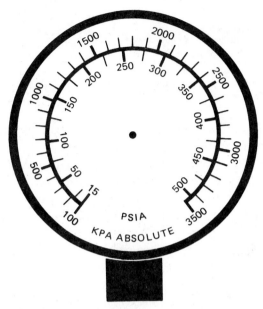

FIGURE 7–17 Pressure gauge showing psi (absolute) and kPa (absolute) pressure scales for comparison.

High-side pressure in a mobile air-conditioning system does not go below atmospheric, so the scale of Figure 7–16 seems to be the logical choice. As for the low-side metric gauge scale, only time will tell. In time, the "standard" will become whatever is accepted by the automotive industry.

Gauge Calibration

Many gauges are provided with calibration adjustment screws. Others may be adjusted by removing and replacing the pointer. A quality gauge is reasonably accurate to about 2 percent of the total scale reading when it is calibrated. The pointer, or needle, should rest on zero when there is no applied pressure.

To calibrate a gauge, it is necessary to remove the retaining ring (bezel) and glass or plastic cover. A small screwdriver is then used to turn the adjusting screw in either direction until the pointer is lined up with the zero mark, as illustrated in Figure 7–18. The adjusting screw must not be forced; to do so can damage the gauge.

Many gauges have inner scales that indicate the temperature-pressure relationship of three popular types of refrigerants. The gauge used for CFC-12 shows the corresponding evaporating temperatures of R-12, R-22, and R-502. The HFC-134a gauge has an inner scale indicating the temperature-pressure relationship of R-134a refrigerant. The temperature and pressure relationships of CFC-12 and HFC-134a refrigerants are outlined in detail in Chapter 5.

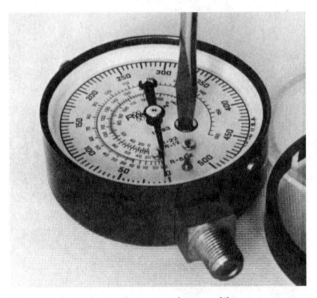

FIGURE 7–18 Screwdriver used to recalibrate gauge to zero. *(Courtesy of BET, Inc.)*

HOSES

The **charging hose** shown in Figure 7–19 is constructed to withstand working pressures in excess of 500 psi (3,448 kPa). Some charging hoses may have a burst-pressure rating of up to 2,000 psi (13,790 kPa).

Hoses designated for CFC-12 service are available in white, yellow, red, and blue. A standard color code can be helpful when connecting the hoses. Blue is used on the low side, red is used on the high side, and white or yellow is used for the center port. Color coding of the hoses lessens

FIGURE 7–19 Typical charging hose. Note shutoff valve. *(Courtesy of Uniweld Products, Inc.)*

the chance of accidentally reversing the manifold connections to the air-conditioning system.

Color coding for hoses designated for HFC-134a service is similar; blue with black stripe or black with blue stripe for the low side, red with black stripe or black with red stripe for the high side, and yellow or green with black stripe or black with green or yellow stripe for the service hose.

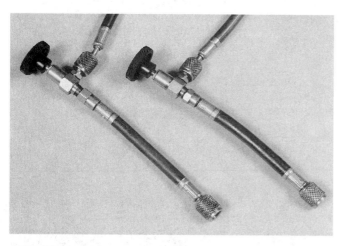

FIGURE 7–20 There must be a shutoff valve within 12 inches (30.5 cm) of the service hose end.

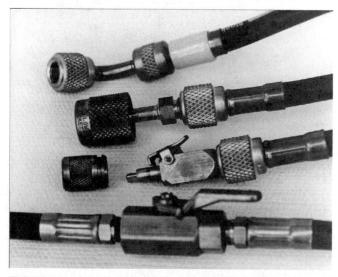

FIGURE 7–21 Typical hose shutoff valves to prevent refrigerant emissions during service procedures. *(Courtesy of BET, Inc.)*

Hoses are usually purchased in standard lengths. Any length, however, may be obtained by special order. Standard lengths are 24 inches (610 mm), 36 inches (914 mm), 48 inches (1.22 m), 60 inches (1.52 m), 72 inches (1.83 m), 84 inches (2.13 m), 8 feet (2.44 m), and 12 feet (3.66 m). The most popular hose length is 36 inches (914 mm). This length is sufficient for most service needs.

Regardless of the hose length, there must be a shutoff valve located within 12 inches (30.5 cm) of the service end, Figure 7–20. This valve is to be closed at all times when the hose is not in use. A variety of valves, illustrated in Figure 7–21, is used to help to reduce the amount of refrigerant that may be vented to the atmosphere during any service procedure.

The fitting at the end of a standard CFC-12 charging hose is designed to fit the **flare** fittings of the manifold set and air-conditioning system access ports. The hose fittings are equipped with replaceable nylon, neoprene, or rubber gasket inserts. These gaskets are always a potential source of leaks during recovery, evacuation, and charging procedures and should be replaced periodically. Hoses are available with a built-in pin on one end, as in Figure 7–22, for use on **Schrader**-type access ports. The hose end without the pin attaches to the manifold set. The hose end with the pin attaches to the service port of the system. If the hose is not equipped with a pin, a Schrader-type adapter is available. If the pin (or adapter) is not used, gauge pressures cannot be determined and

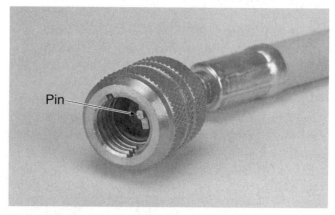

Pin

FIGURE 7–22 A pin in the manifold hose fitting is used to offseat the Schrader of the system service valve

system servicing is not possible on systems equipped with Schrader-type valves.

Hoses designated for HFC-134a service are not **interchangeable** with CFC-12 service hoses. Air-conditioning system service ports and manifold fittings are different for the two systems. Only the proper hose may be connected to its respective fitting. This helps prevent the intermixing of the two refrigerants.

The Third Gauge

Some older air-conditioning systems that require a third gauge may still be in service. This additional gauge is used on systems having some type of low-side pressure control for the evaporator. Two low-side gauges are used, one on each side of the pressure control, to determine pressure drop across the control device.

LEAK DETECTORS

In many instances, the leak in an air-conditioning system can be either a **cold leak** or a **hot leak**. A cold leak occurs when the system is not at its operating temperature and pressure, such as when a vehicle is parked overnight. A hot leak occurs at periods of high pressure within the system, such as when a vehicle is moving slowly in heavy traffic.

The methods of detecting leaks in an air-conditioning system range from a soap solution to an expensive self-contained electronic instrument. Until recently, the most popular leak detector was the halide gas torch. Its popularity was a result of its low cost and ease of handling. It is now seldom used, however, primarily because it is not reliable for detecting HFC-134a refrigerant. Also,

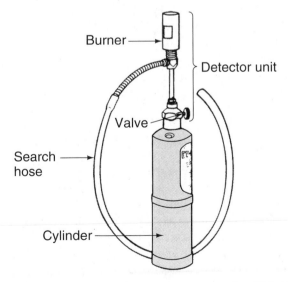

Burner

Detector unit

Valve

Search hose

Cylinder

FIGURE 7–23 A typical halide leak detector for CFC–12.

potential health hazards are associated with using an open flame with refrigerants.

Halide Leak Detector. The halide leak detector, Figure 7–23, was once a very popular tool for leak testing. It is still used by some technicians for detecting CFC-12 leaks. It can detect a quantity equal to 1 pound (0.4536 kg) over 10 years. Air is drawn through the search hose into the burner and the area of the copper reactor plate. When the gas and air mixture is ignited, the flow is regulated until the flame burns about 1/4 inch above the opening in the reactor plate, which is heated to a red-hot temperature. When the search hose comes into contact with leaking refrigerant, refrigerant is drawn in, displacing the oxygen to the reactor plate. As a result, the flame turns violet. In some cases, if the leak is severe enough to displace atmospheric oxygen, the flame is extinguished.

> CAUTION
> THE USE OF A HALIDE LEAK DETECTOR IS DISCOURAGED. IT MUST BE USED IN A WELL-VENTILATED AREA AND NEVER IN AN AREA WHERE EXPLOSIVES, SUCH AS GASES, DUST, OR VAPORS, ARE PRESENT. FUMES FROM A HALIDE LEAK DETECTOR MAY BE TOXIC.

Soap Solution

A soap solution is an efficient method of locating small leaks. Because leaks often occur in areas of limited access, an electronic leak detector cannot be used easily to locate them. If a commercial product like the one shown in Figure 7–24 is not available, mix 1/2 cup of soap powder with water to form a thick solution. The solution should be just light enough to make suds with a small brush. When the solution is applied to the area of a suspected leak, as in Figure 7–25, bubbles should reveal the leak.

Dye Leak Detection

A dye solution may be introduced into the air-conditioning system to locate a leak. Dye is available in either yellow or red form. It is formulated to be safe for an air-conditioning system and will not affect the operation of the system. After injection, the vehicle is driven a few days and the leak area can be easily detected by the visible dye trace at the point of the leak.

It is important to note that many packaged leak detector dye solutions are only compatible with certain

FIGURE 7–24 Typical liquid leak detector.
(Courtesy of BET, Inc.)

FIGURE 7–25 Bubbles reveal point of leak when using a soap solution.
(Courtesy of BET, Inc.)

systems. For example, a dye solution intended for a CFC-12 system is not recommended for use in an HFC-134a system. Conversely, HFC-134a dye should not be introduced into a CFC-12 system. When retrofitting a system that contained dye, it is important that all residual dye be flushed out of the system.

Before using a dye solution be sure to determine that it is approved for such use and will not affect air-conditioning system performance or any vehicle warranty conditions. This assurance should be made by the product manufacturer in writing.

At least one manufacturer, E.I. duPont de Nemours and Company, produces refrigerants with a red dye called Dytel®. Dye is charged into the air-conditioning system in much the same manner as refrigerant. Typical procedures for adding dye to an air-conditioning system follow.

Adding Dye or Trace Solution

NOTE: Be certain that the dye trace solution that is to be injected into the air-conditioning system is compatible with the refrigerant, CFC-12 or HFC-134a.

Procedure.
1. Connect the manifold and gauge set to the system as outlined in Chapter 19. If the system is charged

with refrigerant, recover it from the system as outlined in Chapter 19.
2. Prepare to inject dye into the system.
   a. Install a dye injector in the low-side service hose if using a trace similar to that shown in Figure 7–26.
   b. Install a can tap on center hhose if using packaged dye solution similar to that shown in Figure 7–27.
3. As appropriate:
   a. Fill the dye injector and connect the center manifold hose to an appropriate refrigerant source.
   b. Install a can tap on packaged dye solution.
4. Depending on the method used:
   a. "Tap" the packaged dye can and open the valve.
   b. Open the refrigerant source valve.
5. Start the vehicle engine and operate it at fast idle speed.
6. Set the air-conditioning system controls for maximum cooling.
7. Open the low-side manifold hand valve slightly and allow the trace dye to slowly enter the air-conditioning system.
8. Add additional refrigerant sufficient to charge the air-conditioning system to at least half capacity.
9. Allow the air conditioner to operate for 15–20 minutes to allow time for the dye to circulate throughout the air-conditioning system.

FIGURE 7–26 Typical dye solution that is compatible with Refrigerant–12. *(Courtesy of BET, Inc.)*

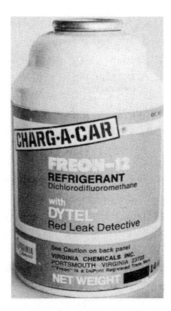

FIGURE 7–27 Typical "pound" can of Refrigerant–12 with a red dye additive. "Freon" and "Dytel" are trademarks of E.J. DuPont. *(Courtesy of BET, Inc.)*

10. Shut off the air conditioner and stop the vehicle engine.
11. Observe the hoses, fittings, and other components for signs of the dye solution, indicating a leak.
12. If no signs of a leak are noted, arrange to have the vehicle available the following day for further diagnosis and repairs.
13. If one or more leaks are detected repairs are necessary.
14. After locating the leak(s), recover the test refrigerant from the air-conditioning system as outlined in Chapter 19.
15. Repair the leak as required.
16. Check the compressor lubricant as outlined in Chapter 11.
17. Add lubricant, if required, and recheck for leaks following steps 5, 6, and steps 8 through 11.
18. If no leaks are found, recover the test refrigerant, and evacuate and charge the air-conditioning system as outlined in Chapter 19.
19. Remove the manifold and gauge set.
20. Replace the protective caps.

NOTE: The dye solution will remain in the air-conditioning system lubricant without causing harm to the system or affecting its cooling performance.

Fluorescent Leak Detectors

There are several manufacturers of **fluorescent** leak detectors. A typical model is shown in Figure 7–28. A metered

FIGURE 7–28 Fluorescent leak detector. *(Courtesy of Tracer Products)*

amount of ultraviolet-sensitive dye is introduced into the system. The air conditioner is operated for a few minutes to allow time for the dye to circulate. An ultraviolet lamp is then used to pinpoint the leak. Although it is rather expensive, the ultraviolet method of leak detection is the most effective for pinpointing small leaks.

Some vehicle manufacturers add fluorescent dye, called "scanner solution," to factory-installed air conditioners. Ford Motor Company started this practice in 1996. Many technicians add scanner solution to systems being serviced for future troubleshooting. The important thing to note, as with any additive, is to make sure that the proper scanner solution is used to ensure system compatibility. More important than the type of refrigerant is the type of lubricant in the system. The scanner solution has either mineral, alkyl benzine, PAG, or polyol ester base stock to match the system lubricant.

Using an improper scanner solution can contaminate an otherwise healthy system. Just 0.3 ounce (0.89 milliliter) is sufficient to treat a system with a refrigerant capacity of up to 2.9 pounds (1.21 liters). The average system, however, requires 0.5 ounce (147.86 milliliter) of scanner solution. This is sufficient for a system with a capacity of up to 4.9 pounds (2.33 liters) of refrigerant.

Because the scanner solution is soluble in, and separates out, from the lubricant, the refrigerant is recyclable and is accepted by most manufacturers and reclaimers. If retrofitting a system be sure that all trace of the scanner solution is removed with the lubricant before introducing a new refrigerant.

Electronic Leak Detectors

An electronic leak detector is perhaps the most sensitive of all leak detection devices. The initial purchase price of an electronic leak detector is soon offset by time saved due to its effectiveness in detecting leaks.

Electronic leak detectors are also known as halogen leak detectors. Such a device can detect a CFC-12 rate of loss as low as 1/2 ounce (14.1 g) per year. This value corresponds to one hundred parts of refrigerant to one million parts of air (100 ppm).

Two examples of the electronic leak detector are the corded and cordless types, both shown in Figure 7–29. The corded leak detector operates on 120 volts (V), 60 hertz (Hz). The cordless leak detector is portable and operates from a rechargeable battery. Both units are simple to operate and easy to maintain.

Some halogen leak detectors, such as that manufactured by Thermal Industries, shown in Figure 7–30, are automatically adjusted by calibrating themselves "while in use" to ignore ambient concentrations of refrigerant gas and pinpoint leaks much more easily. Also, this leak detector may be used to detect CFCs, HCFCs, or HFCs (R-12, R-22, or R-134a).

Ultrasonic Leak Detector

An **ultrasonic** leak detector is an ultrasensitive device that "hears" the sound of a leak. Finding leaks in an air-conditioning system using this device is fast and accurate because it eliminates false triggering. There is generally no need to recalibrate the device to compensate for background noise.

An ultrasonic leak detector, Figure 7–31, should be

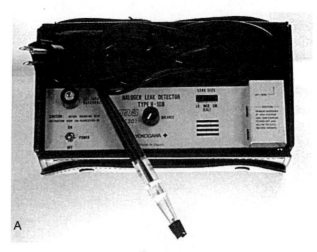

FIGURE 7–29 Two types of electronic leak detector: (A) corded and (B) cordless. *([A] Courtesy of Bill Johnson; [B] Courtesy of Robinair, SPX Corporation)*

considered as a complement to other types of refrigerant leak detectors. It is effective only on systems that are pressurized or under vacuum, because it senses the frequency of escaping gases. An ultraviolet or electronic leak detector is recommended for an empty system or for one having a large leak.

An ultrasonic leak detector can also be used to find other leaks, such as in the vacuum system. It can also be used in a preventative maintenance program to detect sounds that indicate that a component, such as a valve, solenoid, or bearing, is starting to fail.

To use the ultrasonic leak detector, put on the headset, turn on the unit, adjust the sensitivity level, and place the sensor in the vicinity of the suspected leak area. The amplified sound of the leak, heard in the headset, will get louder as the sensor is positioned nearer to the leak. The unit automatically suppresses background noise, wind, stray gases, or other contaminants so not to trigger a false alarm.

Other Types. Many other types, makes, and models of leak detectors are available. Space does not permit the description of each type of leak detector in this text. At a recent International Air Conditioning, Heating, and Refrigeration Exposition there were fifty-two suppliers of leak detectors, each with several different models to choose from. Technicians should contact local refrigeration suppliers for additional information so that they can compare the different makes and models before purchase.

VACUUM PUMP

The vacuum pump, Figure 7–32, is used to remove air

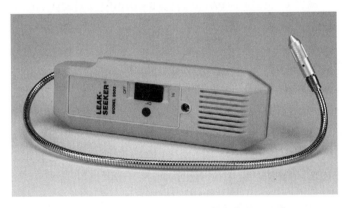

FIGURE 7–30 A portable electronic leak detector that may be used for CFC–12 or HFC–134a refrigerants.

and moisture from an air-conditioning system. Although some technicians may argue that a vacuum pump will not remove moisture from an air-conditioning system, the manufacturers of quality vacuum pumps advise that if three conditions are met, excess moisture will be removed. Those three conditions are:

- ❏ The right size pump
- ❏ The right vacuum pump oil
- ❏ Proper maintenance

The Right Size Pump

A vacuum pump is rated by how much air it can move. The more air it moves, the faster it can reach an acceptable vacuum. The minimum size pump recommended for mobile vehicle air-conditioning (MVAC) service is 1.0 cubic foot per minute (cfm). The following may be used as a guide:

SYSTEM	SIZE	PUMP (cfm)
Passenger cars and light trucks	Up to 10 tons*	1.2
Panel trucks and recreational vehicles	Up to 30 tons*	4.0
Tractor trailers and buses	Up to 50 tons*	6.0

*1 ton = 12,000 Btu/hr

FIGURE 7–31 An ultrasonic vacuum leak detector.

FIGURE 7–32 A lightweight single–stage high–vacuum pump ideal for air-conditioning service. *(Courtesy of BET, Inc.)*

The Right Oil

Only use oil that is especially formulated for vacuum pump service, which is generally available in pint, quart, and gallon containers. Vacuum pump oil must be thermally stable in that it must resist breaking down due to high heat over an extended period of time. It must also be formulated to have a low moisture content. Moisture degrades the oil's purity, thinning it, and reduces the vacuum pump's ability to reach the desired vacuum.

Proper Maintenance

The performance of a vacuum pump relies on the purity and quality of the oil. Proper maintenance, to ensure that the vacuum pump operates at peak efficiency, means that the oil must be changed regularly. Moisture that is removed from an air-conditioning system mixes with the oil in the vacuum pump. This moisture thins the oil, thereby reducing the pump's efficiency. For maximum efficiency, change the oil:

❏ after a system known to have a high moisture content has been evacuated.
❏ after a system known to contain contaminants has been evacuated.
❏ if the oil looks cloudy or milky.
❏ if the pump will not pull a vacuum to factory specifications.
❏ after each 10 hours of operation.

The proper use and maintenance of vacuum pumps are covered in Chapter 8.

FLUSH KIT

Some technicians recommend that an air-conditioning system be flushed, but some claim that more harm may be done by flushing than by not flushing. Those against flushing claim that the flush agent "washes" the bearings and all internal parts of residual lubricant. It also leaves traces of the flush agent, which degrades the performance of the air-conditioning system somewhat.

Some of the reasons given in favor of flushing include: as a matter of routine maintenance, whenever repairs are necessary, and before retrofitting procedures.

Routine Maintenance. An older air-conditioning system should be flushed to remove oil, sludge, and particles that can clog the screen in an expansion valve or orifice tube as well as the receiver or accumulator.

Repairs. To ensure that the air-conditioning system will be most efficient after a repair, some technicians recommend flushing when making repairs or replacing components. Failed components may contain metal shavings, sludge, and other debris that can damage the system or degrade its performance.

Before Retrofitting. Some retrofit procedures require the complete removal of existing lubricant in a CFC-12 air-conditioning system before changing fittings and recharging with HFC-134a. Flushing is the most thorough method of removing the lubricant and also picks up any residual CFC-12 lingering in the system. Flushing combined with a correct charge of new lubricant and pulling a deep vacuum is the best assurance that the system is properly prepared for the new refrigerant.

There are many different makes and models of flush kits available. Some are self-contained and use a universal flush solvent compatible with CFC and HFC refrigerants. Others are an accessory to the recovery/recycle equipment, Figure 7–33, and use liquid refrigerant from the recovery tank for flushing. Regardless of the type used, follow the procedures included with the flush kit for proper and safe service.

RECOVERY SYSTEMS

There are many manufacturers of refrigerant recovery systems and each offers several models. Actually, refrigerant can be successfully recovered from a motor vehicle

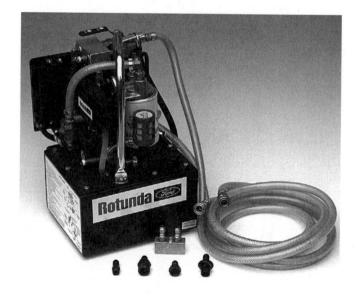

FIGURE 7–33 Typical flush system.

air-conditioning (MVAC) system by immersing the recovery cylinder in an ice pack, as shown in Figure 7–34. However, this procedure is very time consuming and not practical for the average automotive service technician. Moreover, if a recovery machine is not available, the service technician may be charged with "intent to vent." When considering recovery equipment, there are three terms to remember: *recover, reclaim,* and *recycle.*

To recover refrigerant is to remove it, in any condition, and store it in an external container without necessarily processing it further. Under certain conditions, this refrigerant is returned to the system from which it was removed. It may also be sold to a reclamation center where it is processed to new product specifications.

To reclaim refrigerant is to remove it from a system and reprocess it to new product specifications (ARI 700-88

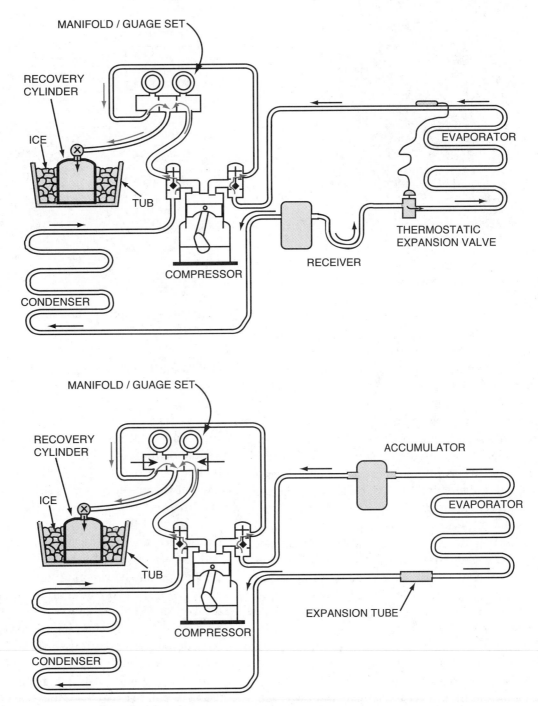

FIGURE 7–34 Removing refrigerant from the (A) expansion valve or (B) orifice tube system by placing the recovery cylinder in ice. *(Courtesy of Whitman–Johnson, Refrigeration and Air Conditioning Technology, by Delmar, a division of Thomson Learning)*

FIGURE 7–35 A typical recovery unit. *(Courtesy of Robinair Division, SPX Corporation)*

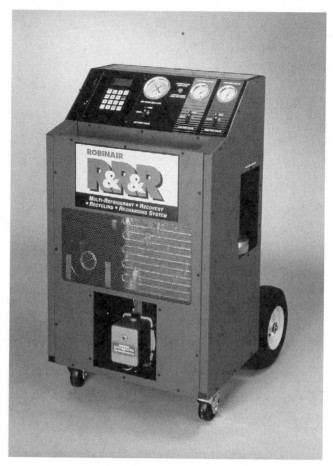

FIGURE 7–36 Refrigerant recovery recycling station. *(Courtesy of Robinair Division, SPX Corporation)*

standards). Analytical testing is required. This is an off-site procedure by a laboratory equipped to make such tests.

To recycle refrigerant is to remove refrigerant from a system and reduce contaminants by oil separation and filter drying to remove moisture, acid, and particulates. This is an on-site procedure and analytical testing is not required.

The make and model of the equipment selected should be based on the needs of the service facility. For example, if the air-conditioning service is "occasional," then a manual recovery unit like the one shown in Figure 7–35 may suffice. This unit is inexpensive and can recover about 1 pound (0.4536 kg) per minute. It cannot be used for recycling, however. An electric recovery unit, as shown in Figure 7–36, frees up the technician for other service while the system is being evacuated. This unit costs about twice as much as the manual unit and removes about 0.78 pound (0.354 kg) per minute. This unit, however, cannot be used for recycling.

Most recovery systems are designed to be used either for CFC and HCFC or HFC recovery. Some, having an oil-less compressor, may be used for CFCs, HCFCs, or HFCs. The automotive service technician, however, is only concerned with two refrigerants: R-12 (a CFC) and

R-134a (an HFC). It is suggested, however, that the service facility have three recovery systems: one for CFC-12, as shown in Figure 7–37, one for R-134a, as shown in Figure 7–38, and one for contaminated refrigerants. The Environmental Protection Agency (EPA) regulations require dedicated equipment for each type of refrigerant. Although recovery equipment is an expensive investment at the onset, considering the high cost of refrigerants, there can be an early payback.

Some of these systems, manufactured for automotive service, are used for recovery, evacuation, recycling, and recharging. This eliminates the need for a separate vacuum pump and charging station.

The make and model of the equipment selected for recovery and recycling of refrigerants should be based on the needs of the service facility. For example, if the air-conditioning service is occasional, a simple recovery unit (Figure 7–39) should suffice. This inexpensive unit can recover about 0.5 lb. (0.227 kg) per minute. A separate recovery unit frees up the service technician for other service work while a system is being evacuated.

FIGURE 7–37 Typical CFC–12 recovery system (*Courtesy of BET, Inc.*)

FIGURE 7–38 Typical HFC–134a recovery system (*Courtesy of BET, Inc.*)

FIGURE 7–39 A mechanical refrigerant pump used to recover refrigerant. (*From Whitman–Johnson, Refrigeration and Air Conditioning Technology, by Delmar, a division of Thomson Learning*)

Many recovery systems are designed to be used for CFC and HCFC recovery only. Some recovery systems may be used for CFC and HFC refrigerant. These systems have special provisions to prevent mixing refrigerants. The automotive service technician is primarily concerned with two refrigerants: R-12, a CFC, and R-134a, an HFC. For full service, if a CFC/HFC combination system is not available, the service facility must have two recovery systems, one dedicated for R-12 service (see Figure 7–37) and another dedicated for R-134a service (see Figure 7–38). This is an expensive investment at the onset but, considering the high cost of refrigerants, an early payback may be realized.

Many systems manufactured for automotive air-conditioning system service are used for recovery, evacuation, recycling, and recharging. Some recover-recycle-recharge systems, Figure 7–36, are microprocessor controlled and automatically control the functions of the equipment. This eliminates the need for personal attention and the requirement for a separate vacuum pump and charging station. Some recycling machines have automatic air purge provisions that detect and vent air from the storage cylinder.

A single-pass recycling machine cleans and filters refrigerant as it is being recovered. A multipass recycling machine recovers the refrigerant in one operation and recycles it through multiple filters, driers, and separators in another operation.

It is important that the manufacturer's maintenance and operational instructions be followed for optimum equipment performance and service. Improper startup procedures, for example, may induce unwanted air into the system.

SCALES

Electronic scales, although more expensive than other types, are often preferred over dial-type or beam-type scales. Electronic scales, often called "charging meters," are more convenient for measuring a proper charge of refrigerant into an air-conditioning system. These scales may be adjusted to zero with a full cylinder of refrigerant. As refrigerant is added to the system, the scale will read a positive value. For example, if 24 ounces (0.7 liter) of refrigerant is required, the refrigerant cylinder is placed on the scale platform and the scale is set to 0. As the refrigerant leaves the cylinder, the scale counts upward, generally in quarter ounce (7.4 mL) increments. When 24 ounces have been dispensed, the refrigerant flow is stopped. This timesaving feature avoids the sometimes erroneous calculations that result from using a mechanical scale. Figure 7–40 shows an electronic scale that may be programmed for the desired refrigerant charge. A solid-state microprocessor controls a solenoid that starts the charging process and stops it when the selected programmed weight has been dispensed.

FIGURE 7–40 Typical electronic charging meter (scale).

REVIEW

Select the correct answer from the choices given.

1. Technician A says that federal regulations require that different manifold and gauge sets be used for CFC-12 and HFC-134a refrigerants. Technician B says that federal regulations require that different hose sets be used for CFC-12 and HFC-134a refrigerants. Who is right?
   a. A only
   b. B only
   c. Both A and B
   d. Neither A nor B

2. Technician A says the electronic manifold and gauge set may be used for CFC-12 or HFC-134a refrigerant. Technician B says that the same hose set may be used with an electronic manifold for either type of refrigerant. Who is right?
   a. A only
   b. B only
   c. Both A and B
   d. Neither A nor B

3. When the low-side manifold hand valve is closed, the circuit is between:
   a. the low-side hose and service hose only.
   b. the low-side hose, low-side gauge, and center service hose.

   c. the low-side gauge and center service hose only.
   d. the low-side hose and low-side gauge only.

4. Technician A says that a hot leak most likely occurs when a vehicle is being driven in heavy traffic. Technician B says a cold leak most likely occurs when a vehicle is not being driven. Who is right?
   a. A only
   b. B only
   c. Both A and B
   d. Neither A nor B

5. The low-side gauge is also called a:
   a. compound gauge.
   b. high-pressure gauge.
   c. low-pressure gauge.
   d. dual-pressure gauge.

6. Technician A says that English pressures below atmospheric may be given as psig (absolute). Technician B says that metric pressures below atmospheric may be given as kPa (absolute). Who is right?
   a. A only
   b. B only
   c. Both A and B
   d. Neither A nor B

F=ma kg×m· kg m²/s

7. The following are popular methods of leak detection, EXCEPT:
   a. halogen.
   b. halide.
   c. ultrasonic.
   d. fluorescent.

8. A high-side gauge is usually scaled to:
   a. 200 psig or 1,400 kPa.
   b. 300 psig or 2,000 kPa.
   c. 400 psig or 2,800 kPa.
   d. 500 psig or 3,500 kPa.

9. All of the following *may* be used to indicate pressure, EXCEPT:
   a. N•m.
   b. psig.
   c. kg/cm².
   d. kPa.

10. The most effective method of detecting small leaks is with the use of _____ leak detector.
    a. a fluorescent
    b. an electronic
    c. a halide
    d. a halogen

11. Sea level atmospheric pressure is:
    a. 14.696 psig.
    b. 101.328 kPa absolute.
    c. either A or B.
    d. neither A nor B.

12. The *minimum* burst pressure rating for a service hose should be:
    a. 500 psig.
    b. 750 psig.
    c. 1,000 psig.
    d. 2,000 psig.

13. An ultrasonic leak detector detects a refrigerant leak by:
    a. sight.
    b. sound.
    c. both A and B.
    d. neither A nor B.

14. A vacuum pump is used to remove _____ from an air conditioning system.
    a. air
    b. moisture

c. Both A and B
d. Neither A nor B

15. A 4.0 cfm vacuum pump is recommended for use on air-conditioning systems up to:
    a. 50 tons.
    b. 30 tons.
    c. 10 tons.
    d. 5 tons.

16. Technician A says that vacuum pump oil must be changed if it looks cloudy or milky. Technician B says that vacuum pump oil must be replaced after 25 hours of operation. Who is right?
    a. A only
    b. B only
    c. Both A and B
    d. Neither A nor B

17. Technician A says that flushing an air-conditioning system removes the residual lubricant from all internal parts. Technician B says that residual flushing agents degrade the performance of the air-conditioning system. Who is right?
    a. A only
    b. B only
    c. Both A and B
    d. Neither A nor B

18. The three "R"s of motor vehicle air-conditioning service are:
    a. "Recover, Recycle, and Recharge."
    b. "Recover, Reclaim, and Recharge."
    c. "Recover, Restore, and Recycle."
    d. "Recover, Reclaim, and Recycle."

19. Technician A says a single-pass recovery unit cleans and filters refrigerant as it is being recovered. Technician B says a multipass recovery unit may be used to recycle recovered refrigerant. Who is right?
    a. A only
    b. B only
    c. Both A and B
    d. Neither A nor B

20. A *charging meter* is a term often used for:
    a. a charging cylinder.
    b. microprocessors.
    c. a manifold and gauge set.
    d. electronic scales.

TERMS

Write a brief description of the following terms:

1. absolute
2. calibration
3. charging hose
4. cold leak
5. cw
6. fittings
7. flare
8. fluorescent
9. hot leak
10. interchangeable

11. low side
12. NPT
13. psi
14. safety factor
15. scale
16. Schrader
17. specialized tools
18. standard
19. superheat
20. ultrasonic

CHAPTER 8

MOISTURE AND MOISTURE REMOVAL

OBJECTIVES

On completion and review of this chapter, you should be able to:

❏ Understand and discuss the importance of evacuating an air-conditioning system.

❏ Understand the effects of ambient temperature on moisture removal.

❏ Discuss the alternate methods of moisture removal.

❏ Understand and discuss the effects of moisture in an air-conditioning system.

❏ Understand the importance of proper vacuum pump maintenance.

INTRODUCTION

For all practical purposes, refrigerant can be considered to be **moisture** free. The moisture content of new refrigerant should not exceed 10 parts per million (**ppm**). If new refrigerant and refrigeration lubricant are used in an air-conditioning system, then any moisture found in the system must have come from an outside source, such as a break in a line or from improperly fastened hoses or fittings on the installation.

Whenever a component is removed from the air-conditioning system for repair or replacement, air is most likely to be introduced into the system. As a result, there is always a danger of moisture entering the unit because air contains moisture, known as *relative humidity* (**RH**). Refrigerant and refrigeration lubricant, particularly R-134a and its lubricant, **absorb** moisture readily when exposed to ambient air. To keep the system as moisture free as possible, all air-conditioning systems use a receiver-drier, Figure 8–1, or an **accumulator-drier**, Figure 8–2. This component contains a bag of drying agent, known as **desiccant**. A desiccant is a chemical drying agent that is able to absorb and hold a small quantity of moisture. Any moisture intro-

duced into the system in excess of the amount that the desiccant can handle is free in the system. Even one drop of free moisture, or water (H_2O), cannot be controlled and can cause erratic or degraded cooling. Moisture in the system may also cause irreparable damage to the internal parts of the air-conditioning system.

Moisture in concentrations greater than 20 parts per million (ppm) may cause serious damage. To illustrate

FIGURE 8–1 A typical receiver–drier.

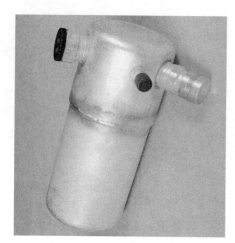

FIGURE 8–2 A typical accumulator.

how small an amount 20 ppm is, one small drop of water in an air-conditioning system having a refrigerant capacity of 2 pounds (0.91 kg) amounts to 60 ppm, or six times the moisture content that is permitted.

Refrigerant reacts chemically with water (H_2O) to form **hydrochloric acid** (HCl). The heat generated in the air-conditioning system speeds up the acid-forming process. The greater the concentration of moisture in the system, the more concentrated is the corrosive acid formed. The hydrochloric acid (HCl) corrodes all of the metallic parts of the system, particularly those made of steel. Iron (Fe), copper (Cu), and aluminum (Al) parts are damaged by the acid as well. The corrosive process creates oxides that are released into the refrigerant as particles of metal that form a sludge. Further damage is caused when oxides plug the fine-mesh screens in the metering device, compressor inlet, and the receiver-drier or accumulator-drier.

Some "remedies" for a CFC-12 system suggest the addition of alcohol, or methanol, to prevent freeze-up. The claim is that system freeze-up can be avoided by adding 0.07 fluid ounce (2.0 mL) of alcohol per pound (0.45 kg) of refrigerant. However, the addition of alcohol to the system can cause even greater damage because the desiccant in the drier seeks out alcohol even more than moisture. Thus, the desiccant releases all of its moisture to the system, which can now cause more severe damage to the system components. Once a system is saturated with moisture, irreparable damage is done to its interior unless it is removed in short order. If the moisture condition is neglected long enough, pinholes caused by corrosion appear in the evaporator and condenser coils and in any metal tubing used in the system. Any parts so affected by moisture must be replaced to restore air-conditioning system integrity.

Additives, under various trade names, claiming to increase cooling effect and/or "stop leaks" are available. As a rule of thumb, however, additives are not recommended for an air-conditioning system. It has long been known in the motor vehicle industry that there is no *magic bullet* for repairing an air-conditioning system.

MOISTURE

Whenever there is evidence of moisture in a system, a thorough system clean-out is recommended. This may mean removing individual components, such as the evaporator and condenser, so they can be thoroughly flushed to remove all residue. A thorough clean-out should be followed by the installation of a new receiver-drier or accumulator-drier. The air-conditioning system should then be pumped down using a vacuum pump.

The air-conditioning technician can generally prevent the introduction of unwanted moisture and dirt into a system by following these few simple rules:

- ❑ When servicing the air-conditioning system, always uncap and install the receiver-drier or accumulator-drier last.
- ❑ When servicing the air-conditioning system, always cap the open ends of hoses and fittings immediately after they are disconnected.
- ❑ Never work around water (H_2O), in the rain, or in very humid locations.
- ❑ Do not allow new refrigerant or refrigeration lubricant to become contaminated.
- ❑ Always keep the refrigeration lubricant container capped when not being used.
- ❑ Develop clean habits; do not allow dirt to enter the system.
- ❑ Keep all service tools free of grease and dirt.
- ❑ Never fill an air-conditioning system with refrigerant without first ensuring that as much air and moisture as possible have been removed.

MOISTURE REMOVAL

As previously discussed, many problems can arise due to excessive moisture in an air-conditioning system. After any repairs have been completed, pump down or **evacuate** the air-conditioning system to remove any moisture. This chapter discusses how an air-conditioning system is pumped down and explains how a vacuum is used to remove moisture. A vacuum is any pressure below atmospheric pressure. Recall, in Chapter 5, that a pressure below zero (0) pounds gauge pressure (0 psig or 0 kPa gauge) is referred to in terms of inches of mercury (in. **Hg**), English, or kiloPascals absolute (kPa absolute), metric.

The removal of moisture from an air-conditioning system without the proper tools can cause serious problems for the service technician. A vacuum pump for adequate moisture removal is a must for air-conditioning service. Although there are other methods that may be used, the vacuum pump is the most efficient means of moisture removal. Figure 8–3 and Figure 8–4 show typical vacuum pumps.

Moisture is removed from an air-conditioning system by creating a vacuum within the system. In a vacuum, the boiling point of water (H_2O) is reduced in proportion to the absolute pressure. Any moisture in the system actually boils and the pumping action of the vacuum pump then pulls the moisture, in the form of a vapor, from the system. As its pressure is increased on the discharge side of the pump, the vapor again liquefies. This process usually occurs inside the pump and the water (H_2O) mixes with the vacuum pump oil.

Although it is possible to use the system's air-conditioning compressor to evacuate the system, this procedure is not recommended. A minimum of 30 minutes is required at a compressor speed of about 1,500 revolutions per minute (**rpm** or r/min) just to remove the air. A much longer period would be required to remove the moisture. The compressor is lubricated by refrigeration lubricant in its sump. An oil pump usually circulates the lubricant, some of which is picked up in the refrigerant vapor. Because the compressor may run dry of oil when operated as a vacuum pump, the compressor can become seriously damaged. In addition, the vacuum pressures are exposed to atmospheric pressure above the discharge valve plate and most vapor moisture will again liquefy in the compressor discharge cavity (inside the compressor).

Thus, if the automotive compressor is used to evacuate the system, the moisture-laden vapor is pulled out of the system and most of it is deposited inside the compressor. Little to nothing is gained in this procedure because the moisture is still inside the system.

An introduction of moisture removal was given in Chapter 5 with information relating to the way a vacuum pump accomplishes moisture removal. Chapter 5 also covers temperature-pressure relationships and the boiling point of water (H_2O) at a lower temperature at higher altitudes. But remember, at higher altitudes, the atmospheric pressure has a lower value than at sea level. A vacuum pump can simulate conditions at a higher altitude by mechanical means. A good vacuum pump is capable of evacuating a system to a pressure of 29.76 in. Hg (0.81 kPa absolute) or better. At this pressure, water (H_2O) boils at 40°F (4.44°C). In other words, if the ambient temperature is 40°F (4.44°C) or higher, the moisture will boil out of the system, assuming the vacuum pump is properly sized and in good working order.

Recall that at 0 in. Hg (101.3 kPa absolute) at sea level, water (H_2O) boils at 212°F (100°C). To find the boiling point of water (H_2O) in a vacuum (absolute pressure), use Figure 8–5 (English) or Figure 8–6 (metric). Note that the boiling point is lowered only 112°F (62.2°C) to 100°F (37.7°C) as the pressure is decreased from 0 in. Hg (101.3 kPa absolute) at sea level to 28 in. Hg (0.98 kPa absolute). However, the boiling point drops by 120°F (66.6°C) as the pressure decreases from 28 in. Hg (0.98 kPa absolute) to 29.91 in. Hg (0.30 kPa absolute).

The degree of vacuum achieved and the amount of time the system is subjected to a vacuum determine the amount of moisture removed from the system. The recom-

FIGURE 8–3 Standard vacuum pump. (*Courtesy of BET, Inc.*)

FIGURE 8–4 A typical high–vacuum pump. (*Courtesy of Robinair, SPX Corporation*)

System Vacuum Inches Mercury	Temperature °F Boiling Point
24.04	140
25.39	130
26.45	120
27.32	110
27.99	100
28.50	90
28.89	80
29.18	70
29.40	60
29.66	50
29.71	40
29.76	30
29.82	20
29.86	10
29.87	5
29.88	0
29.90	−10
29.91	−20

FIGURE 8–5 Boiling point of water under a vacuum (English system).

System Vacuum kilopascals absolute	Temperature °C Boiling Point
19.66	60.0
15.61	54.4
12.02	48.8
9.07	43.3
6.80	37.7
5.08	32.2
3.75	26.6
2.77	21.1
2.03	15.5
1.15	10.0
0.98	4.4
0.81	−1.1
0.60	−6.7
0.47	−12.2
0.44	−15.0
0.40	−17.8
0.33	−23.0
0.30	−28.8

FIGURE 8–6 Boiling point of water under a vacuum (metric system).

mended minimum pumping time is 30 minutes. If time allows, however, a 4-hour **pump down** achieves much better results. The vacuum pump manufacturer's specifications and recommendations for changing oil on a regular basis to ensure maximum **efficiency** should be followed.

The removal of moisture from a system can be compared to the boiling away (vaporization) of water (H_2O) in a saucepan. It is not enough to cause the water (H_2O) to boil; time must be allowed for it to boil away.

MOISTURE REMOVAL AT HIGHER ALTITUDES

The information given for moisture removal by a vacuum pump is true for normal atmospheric pressures at sea level, 14.7 psig (101.3 kPa absolute). It also holds true for higher pressures at higher altitudes providing that the pressure is reduced, thereby reducing the boiling point to a point below the ambient temperature.

As indicated in Chapter 5, moisture (H_2O) boils at a lower temperature at higher altitudes. However, it must be pointed out that a vacuum pump's efficiency is reduced at higher altitudes. For example, the altitude of Denver, Colorado, is 5,280 feet (1,609 m) above sea level. Water (H_2O) boils at 206.2°F (96.78°C) at this altitude. The maximum efficiency of a vacuum pump is also reduced. At this altitude a vacuum pump that can pump

29.92 in. Hg (0.27 kPa absolute) at sea level can only pump 25.44 in. Hg (15.44 kPa absolute). Note in Figure 8–5 and Figure 8–6 that water (H_2O) boils at about 130°F (54.4°C) at this pressure.

The English formula for determining vacuum pump efficiency at a given atmospheric pressure is given in Figure 8–7 and the metric formula is given in Figure 8–8.

Assume that a vacuum pump has a rated efficiency of 29.92 in. Hg at sea level (0.27 kPa absolute) and the atmospheric pressure at Denver is 12.5 psia (86.18 kPa absolute). To determine the actual efficiency at this location, use the formula given in Figure 8–9.

Assuming the same conditions previously mentioned, the formula is applied in the following manner:

$$\frac{AP_L}{AP_S} \times RPE = APE$$

Where: AP_L = Atmospheric pressure in your location
AP_S = Atmospheric pressure at sea level
PRE = Pump rated efficiency
APE = Actual pump efficiency

FIGURE 8–7 Altitude adjustment for vacuum pump efficiency (English values).

Atmospheric pressure at sea level	−	Atmospheric pressure in your location	+	Original efficiency	=	Actual efficiency

FIGURE 8–8 Formula for determining vacuum pump efficiency (metric values).

$$\frac{12.5}{14.7} \times 29.92 = 25.44 \text{ in. Hg}$$

FIGURE 8–9 Assume that a vacuum pump has a rated efficiency of 29.92 in. Hg at sea level (0.27 kPa absolute) and the atmospheric pressure in Denver is 12.5 psia (86.18 kPa absolute).

$$101.32 - 86.18 + 0.27 = 15.41 \text{ kPa absolute}$$

In this example, the ambient temperature must be raised above 130°F (54.44°C) if the vacuum pump is to be efficient for moisture removal. To increase the ambient temperature under the hood, the automobile engine can be operated with the air conditioner turned off. The compressor, condenser, and some of the hoses may be heated sufficiently; however, some other parts, such as the evaporator and the receiver-drier or accumulator-drier, will not be greatly affected.

EVACUATING THE SYSTEM

The air-conditioning system must be evacuated whenever the system is serviced or repaired. The evacuation process rids the air-conditioning system of any air and moisture that was allowed to enter during the service procedure. At or near sea level, a good vacuum pump is one that can achieve a value of 29.7 in. Hg (1.0 kPa absolute) or better. For each 1,000 feet (305 m) of elevation, the reading is about 1 in. Hg (3.4 kPa absolute) higher. For example, at 5,000 feet (1,524 m) the vacuum reading is about 24 in. Hg (20.3 kPa absolute).

As the internal pressure inside the air-conditioning system is lowered, the boiling temperature of any moisture that is present in the system will also be lowered. The moisture (water vapor) can then be pulled out of the air-conditioning system with the same device that lowered its pressure, the vacuum pump. The table in Figure 8–10 illustrates the effectiveness of moisture removal for a given vacuum.

Some refrigerant recovery units may also be used to evacuate and recharge an air-conditioning system.

System Vacuum, in. Hg	Temperature, °F
27.99	100
28.89	80
29.40	60
29.71	40
29.82	20
29.88	0

FIGURE 8–10 Boiling point of water in a vacuum (English).

Whenever using a recovery unit for this purpose, follow the manufacturer's procedures to achieve the best results. This service procedure is provided for the use of an independent vacuum pump or charging station.

Tools. •Hand tools, as required •Fender covers •Manifold and gauge set with service hoses •Vacuum pump or charging station

Procedure

1. Connect the manifold and gauge set to the system as shown in Figure 8–11.

   NOTE: Make sure that the high- and low-side manifold hand valves are in the closed (clockwise) position.

2. Remove the protective caps from the inlet and exhaust of the vacuum pump or charging station.

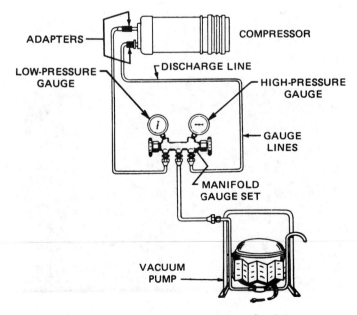

FIGURE 8–11 Connections for evacuation of system.

NOTE: Make sure the port cap is removed from the exhaust port to avoid damage to the vacuum pump.

3. Connect the center service hose of the manifold to the inlet of the vacuum pump or charging station.
4. Start the vacuum pump.
5. Open the low-side manifold hand valve (counterclockwise) and observe the **compound gauge** needle.

NOTE: The needle should be pulled down to indicate a slight vacuum within 15–20 seconds, Figure 8–12.

6. After about 5 minutes observe the gauges.

NOTE: The compound gauge should indicate 20 in. Hg (33.8 kPa absolute) or less, Figure 8–13, and the high-side gauge needle should be slightly below the zero index of the gauge, Figure 8–14.

FIGURE 8–12 Open the low–side manifold hand valve and observe the low–side gauge needle. The needle should be immediately pulled down to indicate a slight vacuum.

7. If the high-side needle did not drop below 20 in. Hg (unless restricted by a stop), a system blockage is indicated.

NOTE: If the system is blocked, discontinue the evacuation. Repair or remove the obstruction. If the system is clear, then continue the evacuation as follows.

8. Operate the pump for 15 minutes and observe the gauges.

NOTE: The system should be at a vacuum of 24-26 in Hg (20.3-13.5 kPa absolute) minimum if there is no **leak.**

9. If the system is not down to 24–26 in. Hg, (20.3-13.5 kPa absolute), close the low-side hand valve and observe the compound gauge.
10. If the compound gauge needle rises, indicating a loss of vacuum, there is a leak that must be repaired before the evacuation is continued.

NOTE: Leak check the system as outlined in Chapter 7.

11. If no leak is indicated, open the high-side manifold hand valve and continue the pump down for 30 minutes, or longer, if time permits.
12. After pump down, close the high- and low-side manifold hand valves.
13. Close all valves, shut off the vacuum pump, disconnect the manifold hose, and replace the protective caps. Note the compound gauge reading; it should be 29 in. Hg (3.4 kPa absolute), or below, Figure 8–15.
14. Observe the compound gauge after 5 minutes.

FIGURE 8–13 The compound gauge should indicate 20 in. Hg (33.8 kPa absolute) or below.

FIGURE 8–14 The high–side gauge should drop below zero.

FIGURE 8–15 Observe the compound (low–side) gauge. If the needle rises, indicating a loss of vacuum, there is a leak that must be repaired before the evacuation is continued.

NOTE: The compound gauge needle should not rise at a rate faster than 1 in. Hg (3.4 kPa absolute) in 5 minutes.

15. If the system fails to meet this requirement, then it either has a leak or was not pumped down long enough.

NOTE: Proceed with step 16 if a leak is suspected. Proceed with step 18 if there are no irregularities.

16. A partial charge must be installed and the system must again be leak checked as outlined in Chapter 7.
17. After the leak is detected and repaired, the refrigerant must be recovered and the air-conditioning system completely evacuated.
18. If the system holds the vacuum as specified, continue with the charging procedure (or other procedures as required).

OTHER METHODS

Other methods used for moisture removal included the *sweep* and *triple evacuation* methods. The sweep method is no longer used because of the ban on venting refrigerant. With this method, one simply introduced refrigerant into the high side and vented it to the atmosphere from the low side. The idea was that the refrigerant would displace or sweep out the impurities in the air-conditioning system. Not only is this method now illegal, it would also be very expensive due to the high cost of refrigerant. The triple evacuation method, however, should be sufficient to reduce the moisture content to a safe level if the system is otherwise sound and a new receiver-drier or accumulator-drier has been installed.

TRIPLE EVACUATION METHOD

The triple evacuation method procedures assume that the system is sound after the refrigerant has been removed and after repairs, if any, have been made. Procedures for connecting the manifold and gauge set into the system and refrigerant recovery are covered in Chapter 7. The use and operation of the vacuum pump is covered in this chapter.

Procedure

1. Connect a manifold and gauge set to the system, to the vacuum pump, and to a regulated nitrogen supply. Ensure that all hoses and connections are tight and secure.

   NOTE: Regulate the nitrogen pressure to 3–5 psig (20.7–34.5 kPa).

2. Start the vacuum pump and pump a vacuum to its highest efficiency for 15–30 minutes.
3. Stop the vacuum pump and break the vacuum by adding dry nitrogen. Pressurize the air-conditioning system to 1–2 psig (6.8–13.7 kPa).
4. Wait 30 minutes to allow sufficient time for the nitrogen to "stratify" the system.
5. Start the vacuum pump and pump a vacuum to the highest efficiency for 15–30 minutes.
6. Repeat step 3.
7. Repeat steps 4. and 5 Allow the pump to run for 30 minutes or longer.
8. The system is now ready for charging. Follow procedures as outlined in Chapter 19.

RECEIVER AND ACCUMULATOR

The *receiver* and *accumulator* are storage tanks for excess refrigerant in an air-conditioning system. They also hold the desiccant for adsorbing excess moisture from the refrigerant. The receiver or receiver-drier is found in the liquid line of a thermostatic expansion valve (TXV) system and the accumulator or accumulator-drier is found in the suction line of fixed orifice tube (FOT) air-conditioning systems.

THE RECEIVER-DRIER

The receiver-drier, most often referred to as a drier, is an important part of a motor vehicle air-conditioning (MVAC) system that uses a thermostatic expansion valve (TXV) as a metering device. The heat load on the evaporator is constantly variable due to temperature and humidity variations within the vehicle. Also, some air-conditioning systems may lose refrigerant due to small leaks. Thus, a storage tank is necessary to hold extra refrigerant until it is needed by the evaporator. The receiver portion of the drier provides this storage area. Because of its function in storing liquid refrigerant, the use of a receiver generally means that it is not necessary to measure precisely the charge of refrigerant into the system. A refrigerant charge of plus or minus a few ounces (mL) is of no consequence.

Several ounces (mL) over or under the recommended refrigerant charge generally makes little difference in overall air-conditioning system performance. In early automotive air-conditioning systems, a separate tank was often used as a receiver and a separate dehydrator or drier for the desiccant was found in the liquid line. The drier, which is usually an in-line type, contains a **filter** and a desiccant or drying material. A **sight glass,** at the outlet of the drier, gives the service technician a means of observing refrigerant flow in the system. The sight glass, however, is seldom found in an air-conditioning system that uses HFC-134a as a refrigerant. Figure 8–16 illustrates the typical location of the components of the device, such as the sight glass, pickup tube, and desiccant.

The Desiccant

A desiccant is a solid or granular substance that is used to remove moisture from a gas, liquid, or solid. The desiccant commonly used in a CFC-12 system drier is usually **silica gel;** molecular sieve; or Mobil-Gel;® commonly classified as "XH-5." This desiccant may not be used in an HFC-134a system, however. Desiccants classified as XH-7 or XH-9 are used in HFC-134a applications. These desiccants, incidently, may also be used in R-12 systems.

The desiccant may be placed between two screens (which also act as **strainers**) within the receiver, shown in Figure 8–17, or it may be placed in a metal mesh bag and suspended from a metal spring. In some cases, the bag of desiccant is simply placed in the tank and is not held in place. It is not uncommon to shake a receiver-drier or accumulator-drier tank and hear the desiccant move. This sound does not necessarily mean that the component is damaged.

The capacity of the desiccant for adsorbing moisture depends on the volume, type of desiccant used, and the temperature. For example, 5 cubic inches (81.94 cm^3) of silica gel can adsorb and hold about 100 drops of water (H$_2$O) at 150°F (65.56°C).

Additives claiming to prevent excessive system moisture from freezing are available in "one-shot" package form. This additive is nothing more than some type of alco-

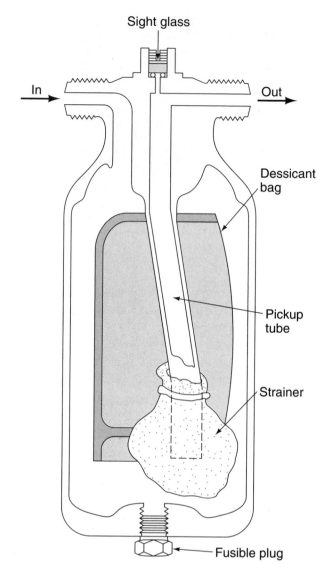

FIGURE 8–16 Cutaway view of receiver-drier showing details.

hol. As discussed earlier, some desiccants have a greater affinity for alcohol than for water and release moisture to the system in order to absorb the alcohol. Other desiccants are rapidly deteriorated by alcohol. Although this additive probably does prevent system freeze-up, it is not recommended due to its adverse affect.

The Filter

Most receiver-driers and accumulator-driers contain filters through which the refrigerant must pass before it leaves the tank. The filtering material prevents desiccant dust and other solids from being carried with the refrigerant into the air-conditioning system. Some driers have one filter on each side of the desiccant. The refrigerant must pass through both filters and the desiccant before

The Pickup Tube

The pickup tube, as a component of the receiver-drier, ensures that 100 percent liquid refrigerant is fed to the thermostatic expansion valve. Because the refrigerant entering the receiver can be a mixture of gas and liquid, the tank also acts as a separator. The liquid refrigerant drops to the bottom of the tank and the gaseous part of the refrigerant remains at the top. The pickup tube extends to the bottom of the tank, thus ensuring that a constant supply of gas-free liquid is delivered to the thermostatic expansion valve.

As a component of the accumulator-drier, the pickup tube ensures that 100 percent vapor is fed to the compressor. Because the refrigerant entering the accumulator can be a mixture of gas and liquid, the tank also acts as a separator. The liquid refrigerant drops to the bottom of the tank and the gaseous part of the refrigerant remains at the top. The U-shaped pickup tube ensures that a constant supply of liquid-free gas is delivered to the compressor. A metered orifice in the U of the tube allows a small amount of lubricant to pass back to the compressor.

The Strainer

The strainer is made of fine wire mesh and is placed in the tank to aid in removing impurities (in particle form) as the refrigerant passes through the receiver-drier or accumulator-drier. Some tanks have two strainers, one on each side of the desiccant. These strainers also serve to hold the desiccant in place (in a manner similar to that of filters). Although some driers may not have a filter, all driers should have one or more strainers. Refrigerant must pass through either filter(s) or strainer(s) before leaving the tank.

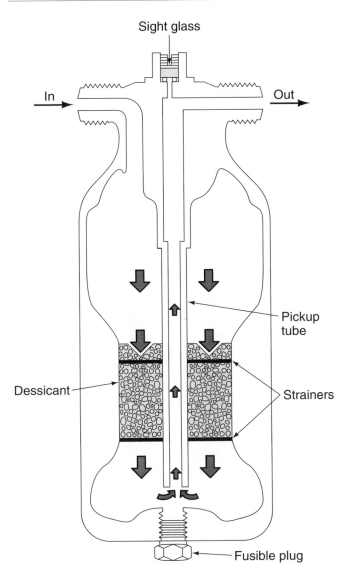

FIGURE 8–17 Cutaway of receiver–drier showing construction details.

leaving the tank. Driers without a filter rely on the strainer to catch all foreign particles that otherwise would pass into the receiver-drier or accumulator-drier and then into the system proper. It is recommended that a service filter be installed in the system before compressor replacement if failure is due to internal causes. This will assist the main filter in the receiver or accumulator to clean the refrigerant and remove any **debris** or contamination caused by compressor failure.

Usually service filters are installed in the liquid line between the condenser and metering device and in the suction line between the evaporator outlet and compressor inlet. There are variations to the installation of the service filters depending on system type. An instruction sheet is generally included with the service filter for a particular application.

The Sight Glass

The sight glass is used to visually observe the flow of refrigerant in the system and to determine if the system is undercharged. The sight glass may be located in either the liquid or outlet side of the receiver-drier or at any point in the liquid line. From these locations, the service technician can readily observe the state of the refrigerant within the system. When the system is operating properly, a steady stream of liquid free of bubbles can be observed in the glass. The presence of bubbles or foam often indicates a system malfunction or a loss of refrigerant.

Using the sight glass to determine refrigerant charge is only valid for an R-12 system if the ambient temperature is above 70°F (21.1°C). It is normal for continuous bubbles to appear in the sight glass on a cool day. If the sight glass is generally clear and the air conditioner is operating satisfactorily, occasional bubbles do not indicate a shortage of refrigerant. This condition may occur when the heat load changes and/or the compressor cycles OFF and ON.

It should be noted, however, that not all CFC-12 systems and very few HFC-134a systems have a sight glass. In such cases, system conditions must be diagnosed with the manifold and gauge set or by superheat study.

Location

The location of the receiver-drier has a direct bearing on its ability to absorb and hold moisture. As indicated previously, 5 cubic inches (81.94 cm³) of silica gel desiccant can hold 100 drops of moisture at 150°F (65.56°C). As the temperature increases, the ability of a desiccant to hold moisture decreases. Thus, the ability of a desiccant to hold moisture is directly proportional to the surrounding (ambient) temperature.

Some technicians may recommend the addition of a small quantity of alcohol to the air-conditioning system. However, the presence of alcohol in the system also decreases the drier's capacity to hold moisture. Although the alcohol prevents any moisture in the system from freezing, the presence of alcohol is detrimental in the long run. It is extremely important to remove the moisture because it reacts with refrigerant to form hydrochloric (HCl) acid. This acid attacks all metal parts, so the thermostatic expansion valve, compressor valve plates, and service valves will be damaged if moisture is allowed to remain in the system.

When the air-conditioning system is operating in the late evening and early morning hours when the outside temperatures are lower, the desiccant holds the moisture and prevents it from circulating in the air-conditioning system. Temperature increases during the day also cause the temperature of the desiccant to increase. When the desiccant temperature reaches its saturation point, some of its moisture will be released into the air-conditioning system.

As little as one droplet of moisture can collect inside the metering device and change to ice in the metering device orifice. This ice then blocks the flow of refrigerant into the evaporator and the cooling action stops.

The evidence of moisture in an air conditioning system is not always easy to detect when the vehicle is in the shop for repairs because it takes some time for the droplets to form and turn to ice. Diagnosis of this condition is made easier if the customer has the following complaint:

"The air conditioner works fine for about 15 minutes or so, but then it just quits. It even puts out hot air. I can turn it off for a few minutes, then turn it on, and it works fine for another 10 to 15 minutes."

This common complaint results from the admission of moisture-laden air into the air-conditioning system through careless installation or servicing procedures. It may also be due to an improper pump down for moisture removal before charging. To correct the condition, a new receiver-drier or accumulator-drier must be installed and the system must be evacuated as long as possible to remove excess moisture before recharging.

If the screens and/or strainers in the receiver-drier become clogged, refrigerant flow will be restricted. Under normal operating conditions, there is no noticeable pressure drop through the receiver-drier. A restriction, however, can result in a considerable pressure drop between the inlet and outlet of the receiver-drier. This condition can easily be detected by feeling both lines. If the outlet line is cooler than the inlet line, then the receiver-drier is clogged and must be replaced. Both inlet and outlet lines should be the same temperature (warm) in a properly operating system.

Installation and Service

Receiver-driers, such as those shown in Figure 8–18, are usually located under the hood of the car. The drier should be placed where it can be kept as cool as possible. At least one independent manufacturer of aftermarket systems mounts the drier inside the evaporator case where it is always surrounded by the cool air of the evaporator.

The inlet and outlet fittings of receiver-driers may be 3/8-inch SAE flare, 5/16-inch barb, O-ring, or block type. Also, receiver-driers may have specially formed inlet and outlet tubing before the fitting provisions. As a rule of thumb, it is recommended to replace a receiver-drier with an exact duplicate. This is particularly important to ensure refrigerant capability with the desiccant.

Manufacturers' recommendations should be followed when mounting a drier. For proper operation, the vertical-type drier must be mounted so that it does not incline more than 15 degrees. The inlet of the drier must be connected to the condenser outlet and the outlet of the drier to the thermostatic expansion valve (TXV).

The word *IN* is generally stamped on the inlet side of the drier. If the inlet is not stamped, an arrow indicating the direction of refrigerant flow should be visible. The service technician must remember that the refriger-

FIGURE 8–18 Typical receiver–driers: (A) O–ring type; (B) Flare type; (C) Barb type. *(Courtesy of BET, Inc.)*

ant flows from the condenser bottom toward the metering device inlet. If the drier is connected in reverse, insufficient cooling will result.

As a result of improper handling or shipping, the internal parts of the drier may become dislodged and cause a partial restriction within the drier. This condition is indicated by a marked temperature change between the inlet and the outlet of the tank. If the restriction is great enough, frosting occurs at the drier outlet.

If the pickup tube is broken because of rough handling, abnormal flashing of the gas occurs in the liquid line. This is the same indication that is evident due to a low charge of refrigerant. In either case, a new drier must be installed.

Screens are located in the inlet of the metering device, the inlet of the compressor, and in the receiver-drier. The screen in the thermostatic expansion valve and compressor can be cleaned or replaced. If the screen in the fixed orifice tube or accumulator-drier becomes clogged, the entire component must be replaced.

ACCUMULATOR

Some air-conditioning installations contain a device known as an accumulator-drier, or more simply, an *accumulator*. The accumulator, Figure 8–19, somewhat resembles a receiver-drier. The accumulator is provided to prevent liquid refrigerant from entering the compressor. It also serves as a tank to store excess liquid refrigerant and it contains a desiccant for drying the refrigerant.

Another name for the accumulator is *suction accumulator* because it is located in the suction line of the air-conditioning system. This device is used in systems that, under certain conditions, may have a flooded evaporator.

FIGURE 8–19 The accumulator is located at the outlet of the evaporator. *(Courtesy of General Motors Corporation Service Operations)*

The accumulator separates the liquid refrigerant from the vapor. In other words, it *accumulates* the liquid.

Refrigerant enters the top of the accumulator and liquid refrigerant falls to the bottom of the tank. Gaseous refrigerant remains at the top of the tank and is moved to the compressor through the U-shaped pickup tube. At the bottom of the tank, the pickup tube contains a small hole or orifice. This orifice allows a very small amount of trapped lubricant and/or liquid refrigerant to return to the compressor.

Recall that a compressor can be damaged by an excess of liquid because it is a *positive displacement pump* and is not designed to compress liquids. The compressor is not damaged because only a controlled amount of liquid is allowed to return to the compressor through the pickup tube orifice.

The characteristics and composition of the desiccant in the accumulator are the same as those for the receiver-drier. The accumulator cannot be serviced. If this device is found to be defective or "wet," the entire unit must be replaced. Air-conditioning systems equipped with an accumulator have an expansion tube that serves as a metering device to the evaporator. If the expansion tube is clogged, it is again necessary to replace the accumulator. The expansion tube is covered in detail in Chapter 11.

When installing or servicing an air-conditioning system, the receiver-drier or accumulator-drier should be the last part connected to the system. Be careful not to let moisture and moisture-laden air enter the system.

WARNING

DO NOT UNCAP THE RECEIVER-DRIER OR ACCUMULATOR-DRIER UNTIL JUST BEFORE THE UNIT IS TO BE INSTALLED. REMEMBER: THE DESICCANT WILL ATTRACT MOISTURE FROM THE SURROUNDING AIR.

SUMMARY

❏   To remove moisture, it is essential to perform a complete evacuation of the system with an efficient vacuum pump.

❏   Whenever a refrigeration system is opened for service, foreign matter can enter the system. Dirt and moisture or other noncondensable materials cause the quality of the refrigerant to deteriorate.

❏   The corrosion of all metal parts due to hydrochloric (HCl) acid (formed from the reaction of moisture and refrigerant) causes small metal particles to slough off the affected components. These particles can stop the flow of refrigerant in the system by clogging the screens that are placed in the system to remove such impurities.

REVIEW

Select the correct answer from the choices given.

1. The moisture content of new refrigerant should not exceed:
    a. 10 ppm.
    b. 20 ppm.
    c. 100 ppm.
    d. 200 ppm.

2. Refrigerant reacts chemically with moisture to:
    a. form harmful deposits on the valve plates.
    b. cause sludge in the compressor crankcase.
    c. block the flow of refrigerant in the evaporator.
    d. form hydrochloric acid in the air-conditioning system.

3. Technician A says that moisture must be removed from new refrigerant and lubricant. Technician B says that air entering an air-conditioning system contains moisture. Who is right?
    a. A only
    b. B only
    c. Both A and B
    d. Neither A nor B

4. Technician A says that moisture may be removed from an air-conditioning system with the use of a vacuum pump. Technician B says that air may be removed from an air-conditioning system with the use of a vacuum pump. Who is right?
    a. A only
    b. B only
    c. Both A and B
    d. Neither A nor B

5. Technician A says that a drying agent is considered an absorbent. Technician B says that a drying agent is known as a desiccant. Who is right?
    a. A only
    b. B only
    c. Both A and B
    d. Neither A nor B

6. The boiling temperature of water (H_2O) at sea level atmospheric pressure is:
    a. 100°F.
    b. 212°C.
    c. both A and B.
    d. neither A nor B.

7. Any moisture content greater than ____ ppm in a motor vehicle air-conditioning (MVAC) system may cause serious damage to the system.
    a. 10
    b. 20
    c. 100
    d. 200

8. Normal sea level atmospheric pressure is about:
    a. 15 psia.
    b. 101 kPa absolute.
    c. both A and B.
    d. neither A nor B.

9. Technician A says that the minimum recommended pump down time is 60 minutes. Technician B says the longer the pump down, the better the moisture removal. Who is right?
    a. A only
    b. B only
    c. Both A and B
    d. Neither A nor B

10. The efficiency of a vacuum pump at higher altitudes:
    a. is reduced.
    b. is increased.
    c. remains the same.
    d. depends on time.

For questions 11 through 14 consider that the air-conditioning system is being properly pumped down and the vacuum pump was run for at least 30 minutes before being turned off.

11. The high-side gauge indicates 0 psig. The low-side gauge indicates 5 psig. The following may be a problem, EXCEPT:
    a. a leak in the air-conditioning system.
    b. a defective low-side gauge.
    c. a defective vacuum pump.
    d. air in the air-conditioning system.

12. The high-side gauge is below 0 psig. The low-side gauge indicates 29 in. Hg and remains steady after 5 minutes. The problem may be:
    a. air in the air-conditioning system.
    b. moisture in the air-conditioning system.
    c. both A and B.
    d. neither A nor B.

13. The high-side gauge is below 0 psig. The low-side gauge indicates 29 in. Hg (plus) but rises to 27 in. Hg after 5 minutes. The problem may be:
    a. residual refrigerant in the air-conditioning system.
    b. a slight leak in the air-conditioning system.
    c. both A and B.
    d. neither A nor B.

14. The high-side gauge indicates 0 psig. The low-side gauge indicates 29 in. Hg (plus). Both gauges remain the same after 5 minutes. The most likely problem is:
    a. a restriction in the air-conditioning system.
    b. a defective high-side gauge.
    c. a defective low-side gauge.
    d. excessive moisture in the air-conditioning system.

15. Which method of moisture and air removal is no longer permitted due to EPA regulations?

    a. Sweep method
    b. Purge method
    c. Both A and B
    d. Neither A nor B

16. What is used to "break" the vacuum in the triple evacuation method?
    a. Dry air
    b. Nitrogen
    c. Both A and B
    d. Neither A nor B

17. The receiver-drier is used in motor vehicle air-conditioning (MVAC) systems having:
    a. a thermostatic expansion valve (TXV).
    b. a fixed orifice tube (FOT).
    c. both A and B.
    d. neither A nor B.

18. Technician A says that the purpose of a pickup tube in a receiver-drier is to ensure that no refrigerant vapor is available at the metering device. Technician B says that the purpose of a pickup tube in an accumulator-drier is to ensure that no refrigerant liquid is available at the compressor. Who is right?
    a. A only
    b. B only
    c. Both A and B
    d. Neither A nor B

19. A desiccant absorbs and holds about _____ drops of moisture at 150°F (65.6°C) per cubic inch (in.^{3}).
    a. 10
    b. 20
    c. 50
    d. 100

20. Technician A says that the drier should be replaced any time there are major repairs made on an air-conditioning system. Technician B says that the desiccant must be compatible with the refrigerant used in the air-conditioning system. Who is right?
    a. A only
    b. B only
    c. Both A and B
    d. Neither A nor B

TERMS

Write a brief description for the following terms:

1. absorb
2. accumulator-drier
3. additives
4. compound gauge
5. debris
6. desiccant
7. efficiency
8. evacuate
9. filter
10. Hg

11. hydrochloric acid
12. leak
13. moisture
14. ppm
15. pump down
16. RH
17. rpm
18. sight glass
19. silica gel
20. strainers

CHAPTER 9

THE REFRIGERATION SYSTEM

OBJECTIVES

On completion and review of this chapter, you should be able to:

- ❏ Identify and discuss the term *air conditioning.*
- ❏ Identify the low and high side of a motor vehicle air-conditioning (MVAC) system.
- ❏ Understand the three basic laws of natural and mechanical refrigeration systems.
- ❏ Define and discuss the term *ton* of refrigeration.
- ❏ Identify the differences between a thermostatic expansion valve (TXV) and a fixed orifice tube (FOT) air-conditioning system.
- ❏ Identify the type of refrigerant used in a motor vehicle air-conditioning (MVAC) system.

INTRODUCTION

Usually what the term *air conditioning* first brings to mind is **cold** fresh air. Actually, a true air-conditioning system automatically controls the temperature, humidity, purity, and circulation of the air. In mobile vehicle air-conditioning (MVAC) applications, air conditioning is a system that cools, dehumidifies, and conditions the air inside the driver and passenger compartments of a vehicle.

THE MECHANICAL REFRIGERATION SYSTEM

The mechanical refrigeration system, depicted in Figure 9–1, installed in a modern vehicle uses a special fluid known as the *refrigerant* to absorb heat inside the vehicle in a component known as an *evaporator.* To do this, the refrigerant changes from a liquid to a vapor as it picks up heat. Because the evaporator is located inside the passenger compartment, air blown over the fins of the evaporator is directed to the driver and passengers for their comfort.

It is then necessary to remove the heat that was absorbed by the refrigerant from the inside of the evapora-

tor. One method of removing this heat is to expel heat-laden refrigerant vapor to the outside **ambient air.** This is not only very expensive, but it is now **illegal.** Current **federal regulations prohibit intentional** venting of any refrigerant into the **atmosphere.** The preferred and accepted method is to recondition the refrigerant for reuse in the system. For this method, the heat alone is removed from the refrigerant and is expelled to the outside ambient air.

The process of reconditioning the heat-laden refrigerant actually begins at the compressor. The function of the compressor is to pressurize the heat-laden vapor until its pressure and temperature are much greater than those of the outside air. The compressor also pumps the heat-laden vapor into the condenser. In the condenser, the vapor gives up its heat and in doing so changes back to a liquid. The condenser is a component of the air-conditioning system that is located outside the passenger compartment, generally in front of the radiator. The ambient air passing over the tubes and fins of the condenser is much cooler than the refrigerant vapor inside the condenser. The refrigerant vapor gives up much of its heat and in doing so changes

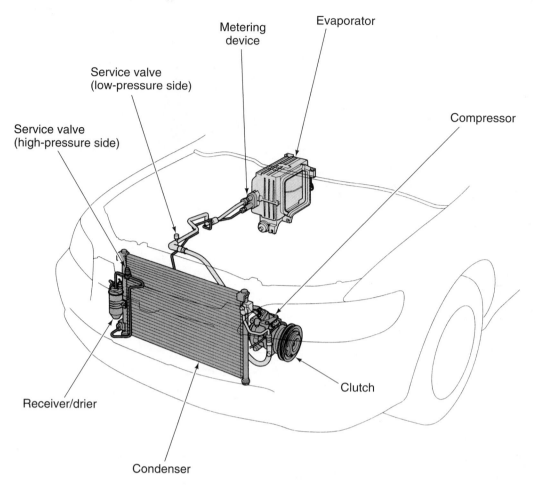

FIGURE 9–1 A late–model air-conditioning system. *(Courtesy of American Honda Motor Co., Inc.)*

back to a liquid. In one type of air-conditioning system, the liquid refrigerant passes from the condenser to the receiver-drier where it is stored until it is needed again by the evaporator. In another type of system, the liquid refrigerant is passed from the condenser directly to the metering device at the inlet of the evaporator.

With either type of air-conditioning system, liquid refrigerant is metered into the evaporator where heat-laden air passing through the evaporator causes the refrigerant to vaporize to repeat the process.

This discussion of a mechanical refrigeration system demonstrates three basic laws of refrigeration that are the bases of all natural and mechanical refrigeration systems.

Law I

To refrigerate is to remove heat. The absence of heat is cold. Heat is ever present.

Law I is illustrated by the refrigeration system of a vehicle. Heat is removed from the air in the passenger compartment of the vehicle, and thus the temperature of the air is lowered.

Law II

Heat is ready to flow or pass to anything that has less heat, Figure 9–2. Nothing can stop the flow of heat; it can only be slowed down. Heat cannot be contained no matter how much insulation is used.

Law II is demonstrated by the special refrigerant used in the air-conditioning system. In this example, heat is ready to flow to anything that contains less heat—to

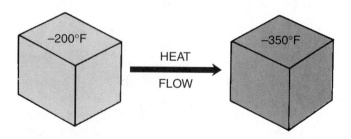

FIGURE 9–2 Heat energy is still available at these low temperatures and will still transfer from the warmer to the colder substance.

the refrigerant in the evaporator and then to the outside (ambient) air through the tubes and fins of the condenser.

Law III

If a change of state is to take place there must be a transfer of heat. If a liquid is to change to a gas, it must take on heat. The heat is carried off in a vapor. If a vapor is to change into a liquid, then it must give up heat. The heat is given up to a less hot surface or medium.

Law III is shown by the liquid refrigerant in the evaporator. That is, as the refrigerant takes on heat, it changes to a vapor. That heat is carried off to be expelled outside the vehicle while changing to a liquid in the condenser.

REFRIGERATION CAPACITY

For many years, the cooling capacity of refrigeration units was rated in **horsepower** (HP). The horsepower is actually a theoretical unit of energy. One horsepower is the amount of energy required to raise 33,000 pounds (1,497 kg) 1 foot (305 mm) in 1 minute, Figure 9–3.

Early refrigeration systems had ratings of 1/4 HP, 1/2 HP, 3/4 HP, 1 HP, and so on. Such a rating, however, was a very inaccurate method of describing the output of an air-conditioning system because the horsepower value generally referred only to the compressor size.

Another term often used to describe the capacity of an air-conditioning system is the *ton*. A ton of refrigeration was generally considered to be equivalent to 1 horsepower. An air-conditioning unit with a rating of 1/2 HP is also said to have the cooling capacity of half a ton of refrigeration.

The value of a ton of refrigeration in Btu/h can be determined if the latent heat of fusion for water (H_2O) is known. The amount of heat required to cause a change in state of 1 pound of ice at 32°F to 1 pound of water (H_2O) at 32°F is 144 Btu.

In applying this value, it must be remembered that a ton of matter, such as water (H_2O), contains 2,000 pounds, Figure 9–4. Because 144 Btu are required to change 1 pound of solid water (H_2O) or ice to a liquid, the equivalent value for 1 ton can be found by multiplying the amount of energy required to change 1 pound by 2,000 pounds.

$$144 \text{ Btu} \times 2,000 \text{ lb.} = 288,000 \text{ Btu}$$

This value is the amount of heat energy (in Btu) required to change the state of 1 ton of ice to 1 ton of liquid in 24 hours. To determine the Btu/h rating for a ton of refrigeration, divide 288,000 by 24, the number of hours in a day.

$$288\,000 \text{ Btu} \div 24 \text{ h} = 12,000 \text{ Btu/h}$$

One ton of refrigeration is thus equivalent to 12,000 Btu/h.

All air-conditioning systems now are rated by Btu. For example, a 3/4-ton air-conditioning system should have a rating of 9,000 Btu; each quarter ton of refrigeration is equivalent to 3,000 Btu.

$$288,000 \text{ Btu} \times 0.25 \text{ ton} = 72,000$$
$$72,000 \div 24 \text{ h} = 3,000 \text{ Btu}$$

The actual Btu rating is, by far, a more accurate method of determining the size of an air-conditioning system. However, because vehicle air-conditioning systems are designed and manufactured for particular

50'

660 LB

FIGURE 9–3 When a horse can lift 660 lb. a height of 50 ft. in 1 min., it has done the equivalent of 33,000 ft.-lb. of work in 1 min., or 1 HP.

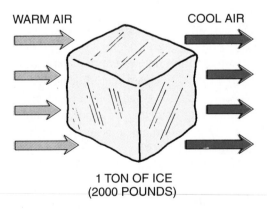

WARM AIR   COOL AIR

1 TON OF ICE
(2000 POUNDS)

FIGURE 9–4 2000 lb of ice requires 144 Btu/lb. to melt. 2,000 lb. × 144 Btu/lb. = 288,000 Btu. When this is accomplished in 24 hours, it is known as a work rate of 1 ton of refrigeration. This is the same as 12,000 Btu/h or 200 Btu/min.

makes and models of automobiles, the Btu ratings are not as important to the technician and are not a part of the decision making for purchase.

Most motor vehicle air-conditioning (MVAC) systems, Figure 9–5, are rated at well over a ton (12,000 Btu) of refrigeration. Because of the tremendous heat load in the vehicle, a unit rated at 6,000 to 8,000 Btu will do a very poor job of keeping the average modern vehicle cool.

For example, the factory-installed air conditioners on many full-size vehicles are rated at about 1.75 tons (21,000 Btu/h). This is equal to about the same amount of cooling capacity that may be required to cool an average two-bedroom house. Of course, a house is better insulated and does not have as great a problem of heat loss by radiation as does a vehicle.

BASIC REFRIGERATION CIRCUIT

The following description of the refrigeration circuit part of the air-conditioning system is given to familiarize the service technician with the general arrangement and function of the components in the system. A good and complete understanding of the overall operation of the air-conditioning system is essential when troubleshooting and working on an air-conditioning system.

> CAUTION
> EYE PROTECTION IS RECOMMENDED WHEN SERVICING AN AIR-CONDITIONING SYSTEM.

Types

There are two basic types of air-conditioning systems found in today's modern vehicle: the thermostatic expansion valve (TXV) type and the fixed orifice tube type **(FOT).**

Study the diagrams of the components of the two types of air-conditioning systems. Figure 9–6 is the diagram of an air-conditioning system that has a thermostatic expansion valve (TXV) as a metering device. Another popular air-conditioning system has a fixed orifice tube (FOT) as a metering device and is shown in Figure 9–7. Note that the TXV system has a receiver-drier but has no accumulator-drier. The FOT system, on the other hand, has an accumulator-drier but has no receiver-drier.

The TXV System

In a TXV system, the compressor pumps heat-laden refrigerant vapor from the evaporator. The refrigerant is compressed and then is sent, under high pressure, to the

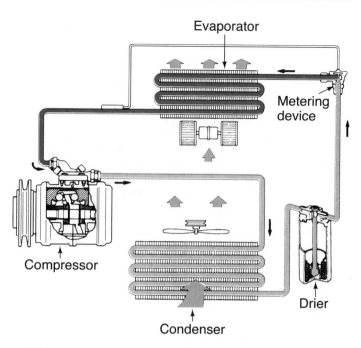

FIGURE 9–5 Typical motor vehicle air-conditioning (MVAC) system.

condenser as a superheated vapor. Because this vapor is much hotter than the surrounding air, it gives up its heat to the outside air flowing through the condenser fins. As the refrigerant vapor gives up its heat, it changes to a liquid. The condensed liquid refrigerant is filtered, dried, and temporarily stored, under pressure, in the receiver-drier until it is needed by the evaporator.

Liquid refrigerant is metered from the receiver-drier into the evaporator by the thermostatic expansion valve. This valve controls the flow of refrigerant into the evaporator. The pressure of the refrigerant is lowered as it flows through the expansion valve. As a result, it begins to boil and changes to a vapor. During this process, the refrigerant picks up heat from the warm air passing through the fins of the evaporator. Thus, the process repeats as this heat is transmitted first to the compressor and then to the condenser for **dissipation.**

The TXV

The thermostatic expansion valve (TXV), illustrated in Figure 9–8, is located at the inlet side of the evaporator. This valve is the controlling device for the system and separates the high side from the low side of the system. A small restriction, or variable orifice, in the valve allows only a small amount of liquid refrigerant to pass through the valve into the evaporator from the drier. The amount of refrigerant passing through the valve depends on the evaporator temperature. The orifice is only about 0.008

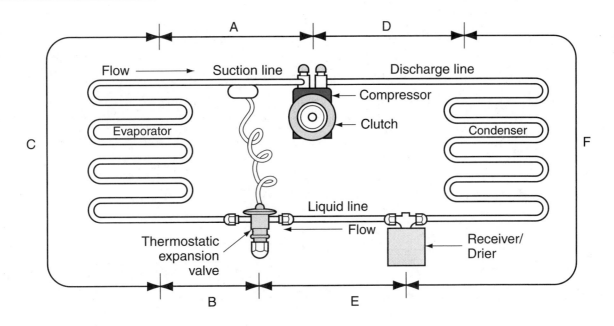

Legend

A Low-pressure vapor
B Low-pressure liquid
C Low-pressure liquid/vapor
D High-pressure vapor
E High-pressure liquid
F High-pressure liquid/vapor

FIGURE 9–6 Thermostatic expansion valve system: (A) low-pressure liquid vapor; (B) low-pressure liquid; (C) low-pressure liquid/vapor; (D) high-pressure vapor; (E) high-pressure liquid; (F) high-pressure liquid/vapor.

inch (0.2 mm) in diameter when wide open. A pin can be raised and lowered in the orifice to change the size of the opening (up to the 0.008-inch diameter). It is evident that only a small amount of refrigerant can enter even when the valve is wide open.

The refrigerant inside the thermostatic expansion valve, and immediately after it, is 100 percent liquid. A small amount of liquid refrigerant, known as **flash gas,** vaporizes immediately after passing through the valve due to the severe pressure drop. All of the liquid refrigerant soon changes state, however. As soon as the pressure drops, the liquid refrigerant begins to boil. All liquid refrig-

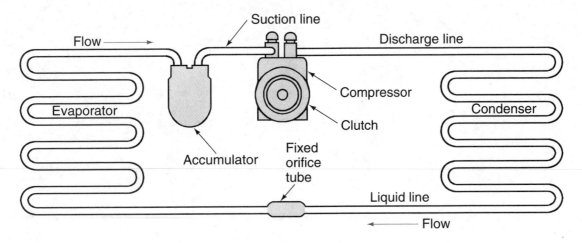

FIGURE 9–7 Orifice tube system.

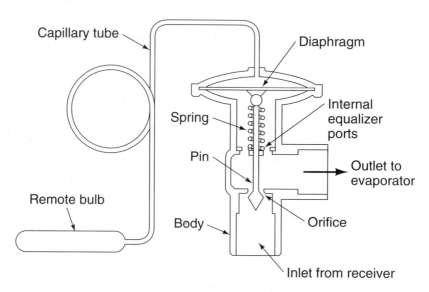

FIGURE 9–8 Typical thermostatic expansion valve (TXV) details.

erant should boil off before reaching the outlet of the evaporator. As it boils, it must absorb heat. This heat is removed from the air passing over the coils and fins of the evaporator. As a result, the air feels cool. Remember, the heat is being removed from the air; cold air is not being created.

The purpose of the thermostatic expansion valve (TXV) is to meter the proper amount of refrigerant into the evaporator for any given condition. Refrigerant that is properly metered into the evaporator is 100 percent liquid just after the thermostatic expansion valve, except for a small amount of flash gas. It should be 100 percent vapor

(gas) before reaching the outlet, or tailpipe, of the evaporator. At the point of total evaporation, the refrigerant is said to be *saturated*. The saturated vapor continues to pick up heat in the evaporator and in the suction line until it reaches the compressor. This refrigerant is said to be *superheated*.

The expansion valve, depicted in Figure 9–9, has a sensing element called a *remote bulb*. This bulb is attached to the evaporator tailpipe to sense outlet temperatures. In this manner, the expansion valve can regulate itself. The thermostatic expansion valve is covered in greater detail in Chapter 12.

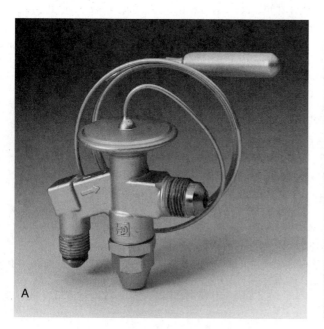
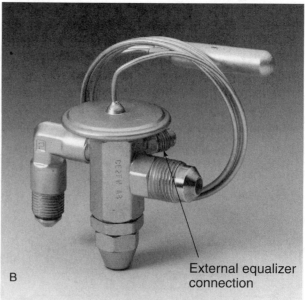

FIGURE 9–9 Typical thermostatic expansion valves (A) internally equalized; (B) externally equalized. *(Courtesy of Parker Hannifin Corporation)*

Receiver-Drier

The receiver-drier, shown in Figure 9–10, is used in systems that have a thermostatic expansion valve as a metering device (see Figure 9–6). This device stores excess refrigerant until it is needed. The receiver-drier (or drier) is a cylindrical metal can with two **fittings** and, in most cases, a sight glass. The drier is located in the high-pressure side of the air-conditioning system between the condenser outlet and the evaporator inlet. In general, the construction of the receiver-drier is such that refrigerant vapor and liquid are separated to ensure that 100 percent liquid is fed to the thermostatic expansion valve. The assembly can be divided into two parts: the receiver and the drier.

The receiver section of the tank is a storage compartment. It holds the proper amount of extra refrigerant required by the system to ensure proper operation. The receiver ensures that a steady flow of liquid refrigerant can be supplied to the thermostatic expansion valve.

The drier section of the tank is simply a bag of desiccant, which is a **chemical** drying agent, that can absorb and hold a small quantity of moisture. The desiccant used in CFC-12 systems may not be compatible with that for HFC-134a systems. Be sure all components of the system are compatible with the refrigerant and refrigeration oil. Use only replacement parts designated for a particular application. Also note that most vendors will not honor warranty claims on new or rebuilt compressors unless the receiver-drier is replaced at the time of service.

A **screen** is placed in the receiver-drier to catch and prevent the circulation of any debris that may be in the system. Although this screen cannot be serviced, many systems have two other filtering screens that can be cleaned or replaced if necessary. These screens are usually located in the metering device (thermostatic expansion valve or fixed orifice tube) inlet and the compressor inlet.

FIGURE 9–10 A typical receiver-drier.

The refrigerant then moves through a rubber hose called the **liquid line** to the evaporator metering device. The liquid line is usually 5/16-inch (7.9 mm) inside diameter (ID), although some may be 1/4-inch (6.3 mm) ID. In some aftermarket installations, the liquid line may be made of copper (Cu), steel, or aluminum (Al). The liquid line can also be a combination of rubber or nylon and copper (Cu), steel, or aluminum (Al). The refrigerant in the liquid line is high-pressure liquid. The receiver-drier is covered in greater detail in Chapter 8.

The FOT System

In an FOT system (see Figure 9–7), the compressor pumps heat-laden refrigerant vapor from the evaporator by way of the accumulator. The refrigerant is compressed and is sent, under high pressure, to the condenser as a superheated vapor. The refrigerant, giving up its heat, changes to a liquid. Liquid refrigerant, under high pressure, is then metered into the evaporator by the FOT. Refrigerant pressure, lowered in the evaporator by the FOT, allows it to pick up heat as it boils off to a vapor. From the evaporator, the refrigerant, which is now mostly vapor, flows to the accumulator.

In the following discussion, the sizes of the system hoses are given, as is the state of the refrigerant in the hoses and in the components. Although the hose sizes may vary in different systems, the state of the refrigerant at various points in all systems is basically the same.

THE FOT

The fixed orifice tube (FOT) is a calibrated restrictor originally used in 1976 in General Motors car lines, followed in 1980 by Ford and, in later years, by Fiat, Jaguar, Mercedes-Benz, and Volvo with GM air-conditioning systems. The FOT is also found in some Audi, Hyundai, and Volkswagen car lines. The FOT replaces the TXV as a means to meter liquid refrigerant into the evaporator. The orifice tube, shown in Figure 9–11, is located in the liquid line between the condenser outlet and evaporator inlet. It meters high-pressure liquid refrigerant into the evaporator as a low-pressure liquid.

The amount of refrigerant entering the evaporator with an orifice tube system depends on the size of the orifice, subcooling of the refrigerant, and the pressure difference (DP) between the inlet and outlet of the orifice device. It is commonly called a fixed orifice tube (FOT) due to its fixed orifice and its tubular shape. The orifice size is from 0.047 inch (1.19 mm) to 0.072 inch (1.83 mm).

Fine-mesh filter screens protect both the inlet and outlet of the orifice tube. If any foreign matter were to

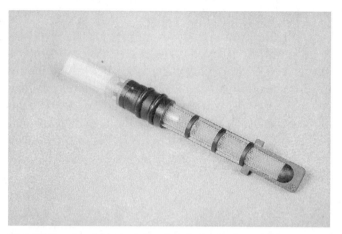

FIGURE 9–11 A typical fixed orifice tube (FOT).

block or even partially block the calibrated orifice, the air-conditioning system either would not function to full efficiency or not function at all.

General Motors orifice tube systems currently use two methods of temperature control. One method, the Cycling clutch orifice tube (CCOT) system, uses a fixed displacement compressor and a pressure- or temperature-actuated cycling switch to turn the compressor's electromagnet clutch off and on. This, in turn, starts and stops the compressor to maintain the desired in-car temperature. The other orifice tube system, the variable displacement orifice tube (VDOT) system, has a variable displacement (VD) compressor. It varies the amount of refrigerant flowing through the system, maintaining the selected in-car temperature without having to cycle the clutch off and on.

Ford Motor Company car lines use the cycling clutch method of temperature control on their orifice tube systems. Either a temperature- or pressure-actuated control may be used to cycle the clutch off and on in order to maintain the desired in-car temperature. Their system is called a fixed orifice tube/cycling clutch (FOTCC) system.

All orifice tube systems have an accumulator located at the evaporator outlet, Figure 9–12. The accumulator prevents unwanted quantities of liquid refrigerant and/or oil from returning to the compressor at any one time. The orifice tube may be found in both CFC-12 and HFC-134a systems. To ensure compatibility, always follow the manufacturer's specifications when replacing this or any other component. Procedures for testing and replacing the orifice tube are found in Chapter 18.

EVAPORATOR

The evaporator is the part of the air-conditioning system where the refrigerant vaporizes as it picks up heat. Heat-laden air is forced through and past the fins and tubes of the evaporator. Heat from the air is picked up by the boiling refrigerant and is carried in the system to the condenser.

Factors that are important in the design of an evaporator include the size and length of the tubing, the number and size of the fins, the number of return bends, and the amount of air passing through and past the fins. The heat load is also an important consideration. *Heat load* refers to the amount of heat, in Btu, to be removed from the vehicle.

The evaporator may have two, three, or more rows of tubing as required to fit inside the evaporator housing

FIGURE 9–12 An accumulator, a pressure vessel designed to withstand normal pressures of an air-conditioning system, is located at the outlet of the evaporator.

and to ensure the required rated capacity, in Btu, of the system. It is essential to ensure that the refrigerant, as it leaves the evaporator on its way to the compressor, is a low-pressure, slightly superheated vapor.

If too much refrigerant is metered into the evaporator it is said to be **flooded.** As a result, the unit will not cool because the pressure of the refrigerant in the evaporator is higher and it does not boil away as quickly. In addition, when the evaporator is filled with liquid refrigerant, the refrigerant cannot vaporize properly. This step is necessary if the refrigerant is to take on heat. A flooded evaporator allows an excess of liquid refrigerant to leave the evaporator, which may result in serious damage to the compressor. Under this condition there is little or no superheat.

If too little refrigerant is metered into the evaporator, the system is said to be **starved.** Again, the unit does not cool because the refrigerant vaporizes, or boils off, too rapidly, long before it passes through the evaporator. Under this condition, the superheat is high to very high.

Accumulator

The accumulator, a tanklike vessel located at the outlet of the evaporator, is an essential part of an orifice tube air-conditioning system. The orifice tube, under certain conditions, will meter more liquid refrigerant into the evaporator that can be evaporated. Excess liquid refrigerant leaving the evaporator would then enter the compressor, causing damage.

To prevent this problem, all liquid and vapor refrigerant and lubricant leaving the evaporator enter the accumulator. The accumulator allows the refrigerant vapor to pass onto the compressor and traps the liquid refrigerant and lubricant. A calibrated orifice, not to be confused with an orifice tube, is included in the U-tube outlet provisions of the accumulator. Its purpose is to allow only a small amount of liquid refrigerant and/or lubricant to return to the compressor with the vapor at any given time. This orifice is generally referred to as an oil *bleed hole,* shown in Figure 9–13, although it also allows small quantities of liquid refrigerant to enter.

In addition, the accumulator contains the desiccant, a chemical drying agent. The desiccant attracts, absorbs, and holds moisture that may have entered the system due to improper or inadequate service procedures. The desiccant is not serviced as a component of the accumulator. If the need for desiccant replacement is indicated, then the accumulator must be replaced as an assembly. The desiccant used in CFC-12 air-conditioning system accumulators may not be compatible with HFC-134a. To be sure of system compatibility, use only replacement parts specifically designated for a particular application.

A fine-mesh screen is placed in the accumulator to catch and prevent the circulation of any debris that may be in the system. Although this screen cannot be serviced, many systems have two others in the system that may be cleaned or replaced. They are located at the compressor inlet and as a part of the orifice tube.

Most vendors will not honor warranty on new or rebuilt compressors unless the accumulator has been replaced at the time of service.

COMPRESSOR

The refrigerant compressor, shown in Figure 9–14, is a pump especially designed to raise the pressure of the refrigerant. According to the laws of physics, when a gas or vapor is compressed (pressure increased), its temperature is also increased proportionately. When its pressure and temperature are increased, the refrigerant condenses more rapidly in the next component, the condenser.

Many different models of current production air-conditioning compressors are found on today's vehicles. Many other models, now discontinued, are still in use. Air-conditioning compressors, depending on design, have one, two, four, five, six, or ten cylinders (pistons).

Several different compressors may be found on some vehicle lines. Different models of a particular compressor may be found on some vehicle lines. For example, a positive- or variable-displacement compressor may be found on the same model year vehicle lines. Other vehicle lines may use several different compressors for

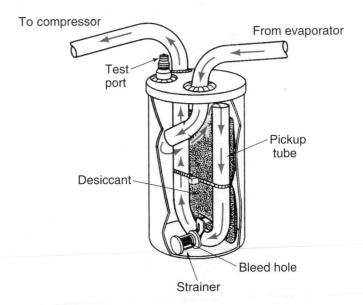

FIGURE 9–13 Cutaway of the accumulator to show the bleed hole to prevent the entrapment of oil.

FIGURE 9–14 Typical compressors: (A) FS–10; (B) HR–6; (C) TR70 Honda; (D) FS–6.

the same model year. Mazda, for example, uses a Diesel Kiki with a nine-groove clutch or a Nippondenso with a single-groove clutch, depending on application.

Regardless of manufacturer and model, most motor vehicle air-conditioning system compressors are of the same design, that is, reciprocating piston. This means that the piston(s) move(s) in a linear motion, back and forth or up and down. The exceptions are rotary and scroll designs. York's rotary vane compressor was introduced, and discontinued, in the early 1980s. Only about fifty thousand were produced, with no appreciable application in the marketplace. Panasonic's rotary vane compressor was introduced in some Ford applications in 1992. The latest design, introduced in 1993, is the scroll compressor.

The following is a brief description of compressor operation. More detailed information is given in Chapter 10, Compressors and Clutches. Service and repair procedures for the most popular compressors are also given in that chapter.

Operation

Each piston of the reciprocating compressor is equipped with a set of suction and discharge valves and valve plates. While one piston is on the intake stroke, the other is on the compression stroke, as illustrated in Figure 9–15. The piston draws in refrigerant through the suction valve and forces it out through the discharge valve. When the piston is on the downstroke, or intake stroke, the discharge valve is held closed by the action of the piston and the higher pressure above it. At the same time, the suction reed valve is opened to allow low-pressure gas to enter. When the piston is on the upstroke, or compression stroke, refrigerant is forced through the discharge valve and the suction valve is held closed by the same pressure.

The compressor separates the low side of the system from the high side. The refrigerant entering the compressor is a low-pressure slightly superheated vapor. When the refrigerant leaves the compressor, it is a high-pressure highly superheated vapor.

Many compressors are equipped with service valves that are used to service the air-conditioning system. The manifold and gauge set is connected into the system at the service valve ports, Figure 9–16. All procedures, such as evacuating and charging the system, are carried out through the manifold and gauge set.

Refrigeration oil is stored in the *sump* of the compressor to keep the crankshaft, connecting rods, and other internal parts lubricated. A small amount of this oil circulates throughout the system with the refrigerant. Internal provisions allow this oil to return to the compressor.

The hose leaving the compressor contains high-pressure refrigerant vapor. It is made of synthetic rubber and is usually 13/32-inch (10.3 mm) inside diameter. This rubber hose often has extended preformed metal (steel or aluminum) ends with fittings. Known as the *hot gas dis-*

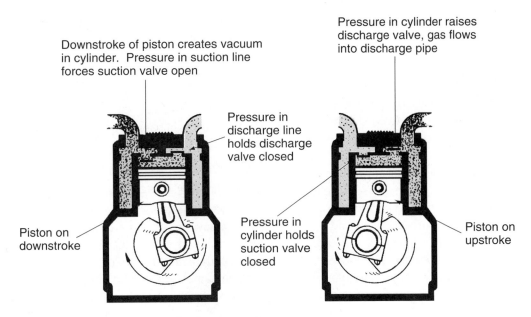

FIGURE 9–15 Operating cycle of reciprocating compressor.

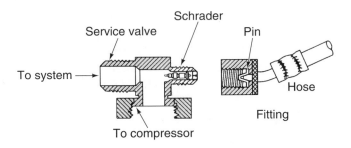

FIGURE 9–16 A typical CFC–12 Schrader-type service valve and service hose connection.

charge line, it connects to the inlet of the condenser. It should be noted that the inlet of the condenser is always at the top—never at the bottom.

HOSES

Refrigerant fluid and vapor hoses may be made of copper (Cu), steel, or aluminum (Al). They are usually made of a synthetic rubber covered with a nylon braid for strength. The inner lining of early systems was typically Buna "N," a synthetic rubber. Buna "N" was not affected by CFC-12, but it is not acceptable for HFC-134a service. Due to efforts to prevent leaks, the lining is now made of nylon, which is compatible with both CFC-12 and HFC-134a, Figure 9–17.

Special consideration must be given for hoses as well as other components used in an HFC-134a air-conditioning system. Many materials that were compatible for a CFC-12 system, such as nitrile or epichlorohydrin, cannot be used for HFC-134a service. For example, O-rings used with fittings in a CFC-12 system, such as nitrile, and those used in an HFC-134a system, such as neoprene, are not interchangeable.

Also, some metals such as copper (Cu) may not be used due to chemical reaction with HFC-134a known as "copper (Cu) plating." It is not known at this time, however, if this will be a problem with system performance.

Standard hose sizes are given a number designation, such as #6, #8, #10, and #12. Size #6 is usually used for the

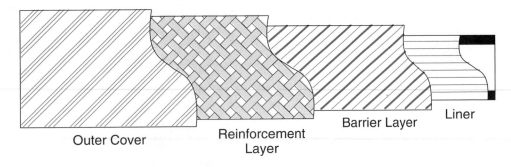

FIGURE 9–17 Construction details of a barrier hose. *(Courtesy of BET, Inc.)*

liquid line, #8 or #10 as the hot gas discharge line, and #10 or #12 as the suction line.

CONDENSER

The purpose of the condenser, depicted in Figure 9–18, is the opposite of that of the evaporator. Heat-laden refrigerant, in the gaseous state, liquefies or condenses in the condenser. To do so, the refrigerant must give up its heat. *Ram air,* or the air passing over the condenser, carries the heat away from the condenser, and the gas condenses. The heat removed from the refrigerant (so that it can change to a liquid), is the same heat that was absorbed in the evaporator to change the refrigerant from a liquid to a gas.

The refrigerant is almost 100 percent vapor when it enters the condenser. A very small amount of vapor may liquefy in the hot-gas discharge line, but the amount is so small that it does not affect the operation of the system.

The refrigerant is not always 100 percent liquid when it leaves the condenser, however. Because only a certain amount of heat can be handled by the condenser at a given time, a small percentage of the refrigerant may leave the condenser in a gaseous state. Again, this condition does not affect the system operation because the next component is the receiver-drier.

As indicated previously, the inlet of the condenser must be at the top. With the inlet at the top, the condensing refrigerant can flow to the bottom of the condenser where it is forced, under pressure, to the receiver-drier through the liquid line.

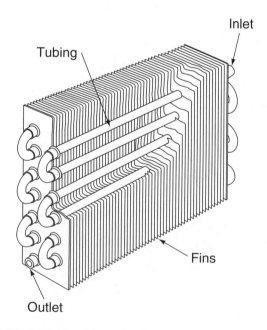

FIGURE 9–18 Condenser details.

The refrigerant in the condenser is a combination of liquid and gas under high pressure. Extreme care must be exercised when servicing this component.

From the condenser, the refrigerant continues to the receiver-drier through the liquid line. At this point, the cycle starts over again. The liquid line from the condenser can be either a rubber or a metal line in a variety of sizes.

THE ROTARY VANE AIR CYCLE (ROVAC) SYSTEM

Based on much the same principle that is used to provide cooling for the passenger compartment of commercial aircraft, several attempts have been made to develop a rotary vane air cycle (ROVAC) air-conditioning system that is suitable for motor vehicle use. The ROVAC system (a trade name), or a similar system, may someday replace the vapor cycle system now in use. It should be noted that an air cycle system is an open system, as opposed to a vapor cycle system, which is a closed system.

As an incentive for developers to find an alternate method of comfort cooling to eliminate refrigerants, U.S. government agencies (such as NASA and the military) have supported the ROVAC efforts. Although in development for many years, a practical system from the standpoint of cost has not yet been produced for use in vehicles.

Operation

The compressor of the ROVAC system is called a *circulator.* The condenser is called a *primary heat exchanger,* and the evaporator is called the *secondary heat exchanger.* A collector in the system serves in a similar manner as an accumulator in a conventional (vapor cycle) system; it separates liquid (hydrocarbon) from the vapor (air). Unlike an accumulator, however, the liquid is retained in the collector and is not metered back into the system by the circulator.

A small amount of liquid oil circulates in the system at all times to provide lubrication for the circulator. The other liquid, comprised of alcohol and hydrocarbons, is vaporized in the secondary heat exchanger as it picks up heat. Conversely, this vapor is changed back to a liquid in the primary heat exchanger as its heat is given up to the outside air.

DIAGNOSIS AND SERVICE

The service technician must enter the air-conditioning system to perform most diagnostic, testing, and service procedures. This is necessary to record the pressures within the system as an aid in determining if there is a problem. This unit deals with the *service valve,* a device that allows the service technician to enter the refrigeration system by mechanical means. The service valve, then, provides a means to connect the manifold and gauge set into the system. Chapter 7 covers the manifold and gauges that are connected to the service valve and used in actual diagnostic and service procedures.

Service Valves

Most air-conditioning systems have two service valves: one on the low side of the system and one on the high side of the system. Some early General Motors, Chrysler, and Ford systems have three valves. The third valve, when found as shown in Figure 9–19, is on the low side of the system.

There are two types of service valves used in the CFC-12 refrigerant system: the Schrader (automatic) valve, depicted in Figure 9–20, and the hand shutoff (manual) valve, shown in Figure 9–21. The service valve, illustrated in Figure 9–22, is used in an R-134a refrigerant system. Although different in appearance and operation, they all serve the same purpose.

Schrader Valve For R-12

The Schrader-type valve is the most popular service valve

for a CFC-12 system. It is very similar to the tire valve in appearance and operation. The Schrader valve, illustrated in Figure 9–23, has only two positions: *cracked* (open) and *back seated* (closed). The normal operating position of this valve is back seated. This valve is cracked or opened by a pin in the end of the manifold hose, as in Figure 9–24, or in a special hose adapter, as in Figure 9–25. Whenever the hose or adapter is screwed onto the Schrader valve, system pressures are impressed on corresponding gauges.

It is important to note that hoses and/or adapters should not be connected to a Schrader access fitting unless the fitting is first connected to the gauge manifold. Conversely, hoses should not be disconnected from the gauge manifold while they are connected to a Schrader service valve. Doing so would result in a loss of refrigerant and could cause personal injury.

The standard low-side Schrader valve fitting is 7/16-20, commonly referred to as 1/4-inch SAE. High-side valve fittings may be the same size as low-side fittings, although many may be 3/8-24. The smaller fitting requires a reducing adapter to connect the high-side hose into the system. Another fitting often found in the high side of the system is the quick connect/disconnect type, depicted in Figure 9–26. This fitting also requires a special adapter to connect the high-side hose into the system.

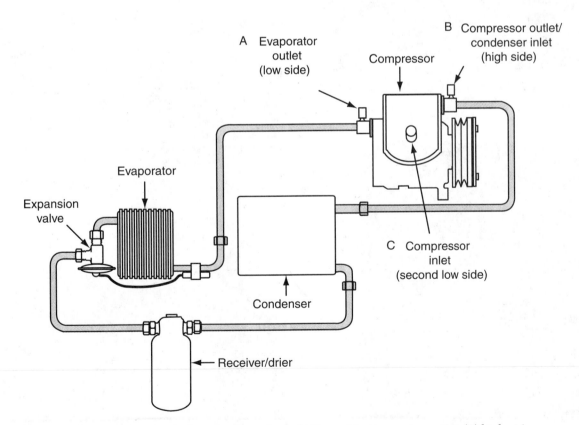

FIGURE 9–19 Early system with three service valves: (A) low-side evaporator outlet, (B) high-side compressor outlet/condenser inlet, and (C) a second low-side compressor inlet.

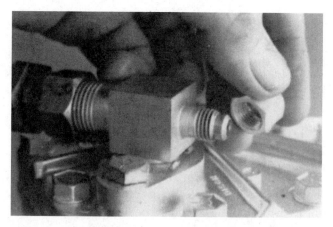

FIGURE 9–20 An "Acorn" protective cap keeps dirt and moisture out of a Schrader–type service valve and, at the same time, helps to prevent leaks. *(Courtesy of BET, Inc.)*

FIGURE 9–21 Remove the protective caps from the service valves.

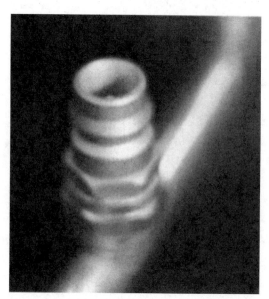

FIGURE 9–22 Typical HFC–134a service fitting. *(Courtesy of BET, Inc.)*

The different size or type of fitting on the high-pressure side of the system helps to prevent reversing the hoses of the manifold and gauge set. Reversing the hoses would impress high-side system pressure on the low-side gauge and could affect gauge accuracy, as well as cause other problems, which are covered later.

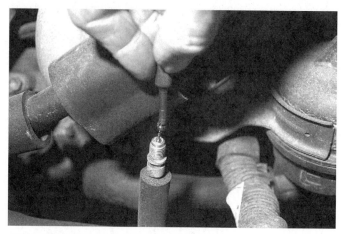

FIGURE 9–23 A typical HFC–134a Schrader valve (core shown being replaced).

Pin

FIGURE 9–24 CFC–12 service hoses equipped with Schrader valve depressor pin.

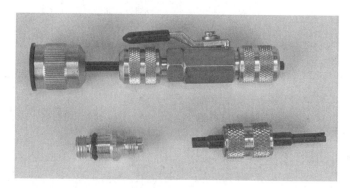

FIGURE 9–25 CFC–12 adapters for Schrader access valves.

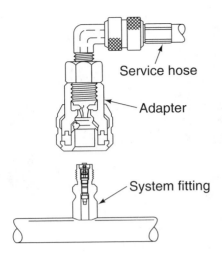

FIGURE 9–26 Quick connect/disconnect fitting and adapter. *(Courtesy of Ford Motor Company)*

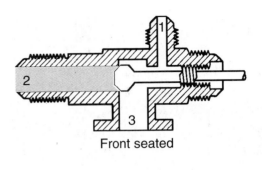

Front seated

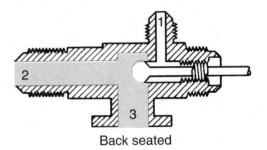

Back seated

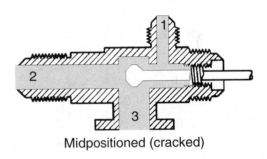

Midpositioned (cracked)

FIGURE 9–27 Service valves in the front–seated(A); back–seated (B); and mid (cracked) position (C).

Hand Shutoff Valve For CFC-12

The hand shutoff service valve is often referred to as a manual valve. Although it is not as common as in past years, it is still found on some systems. This type of service valve is found primarily on York compressors that are used in heavy-duty truck air-conditioning system applications, such as Autocar, Freightliner, Kenworth, and Mack, to name a few. The hand shutoff valve has a 1/4-inch (square end) stem that is used for opening and closing. For this operation, some technicians use pliers or vise grips, but it is recommended that a service valve wrench be used to position this type of valve.

The service valve is back seated when the stem is turned fully counterclockwise (ccw). It is cracked when the stem is turned one or two turns clockwise (cw) off the back-seated position. The valve is front seated when the stem is turned fully clockwise (cw).

The hand shutoff valve is, in effect, a three-position device that can be used for the three functions shown in Figure 9–27.

Front Seated (A). To shut off refrigerant flow: the gauge port (1) is not part of the system (2) but will read compressor (3) pressure.

Back Seated (B). Normal refrigerant operation: the gauge port (1) is not part of the system (2) or compressor (3). Compressor (3) and system (2) are in the circuit.

Mid-position (C). Normal refrigerant operation: the gauge port (1) is part of the system (2) and compressor (3) circuit.

The technician must always back seat the valve, Figure 9–27, before attempting to remove the gauge hose

from the service valves. Failure to back seat the valve will result in a loss of refrigerant. Refrigerant is present at all outlets when the service valve is in the cracked position.

Service Valves For R-134a

Service valve fittings found in HFC-134a systems, shown in Figure 9–28, are much larger and of a different configuration than the service valves found in a CFC-12 system. The HFC-134a system uses quick disconnect Schrader-type fittings that resemble the male end of an air hose coupling. They are not, however, interchangeable. A comparison of fittings for CFC-12 and HFC-134a service is given in Figure 9–29. The different sizes and types of fittings are meant to ensure that the two refrigerants are not accidently mixed.

Observe this word of caution: A fitting found on the fuel rail (line) of some vehicles equipped with fuel injection is the same size as one of the fittings found on an HFC-134a

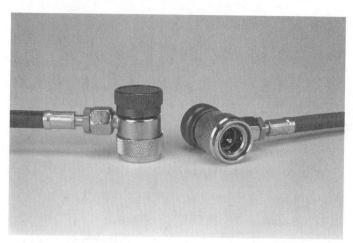

FIGURE 9–28 Typical R–134a service port adapters.

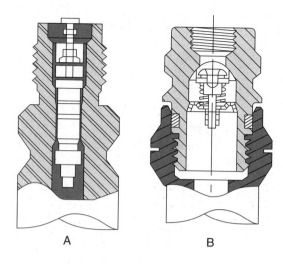

FIGURE 9–29 Comparison of CFC–12 (A) and HFC–134a (B) service valve fittings. *(Courtesy of BET, Inc.)*

air-conditioning system. Although the Schrader-type valve in the fuel system fitting has a greater recess than that found in the air-conditioning system, accidents do happen. Take extreme care not to attach the air-conditioning service hose to the automotive fuel system. Remember, the fuel system contains highly combustible fluids under high pressure. Improper connection, under certain conditions, may create an extreme hazard.

Because of the different type of fitting requirements, the manifold and gauge set for HFC-134a service, covered in Chapter 7, will only connect to the air conditioning system for which it is intended. It will not, for example, match the access fittings used for a CFC-12 air-conditioning system.

SUMMARY

The entire refrigeration cycle exhibits several processes as the refrigerant changes state in various sections of the system. When the pressure of the refrigerant drops in the evaporator, the refrigerant boils. While boiling, the refrigerant picks up heat. The compressor raises the temperature and pressure of the refrigerant so that it condenses in the condenser. At this point, the refrigerant gives up the same heat (in Btu) that it picked up in the evaporator.

The thermostatic expansion valve (TXV) or fixed orifice tube (FOT) controls the flow of refrigerant into the evaporator and thereby separates the high side from the low side of the system. The compressor increases gas pressure and separates the low side from the high side of the system. This is the basic air-conditioning circuit on which all of the other automotive refrigeration circuits are patterned. A good understanding of this simple circuit makes understanding of other circuits much easier.

Service valves, regardless of type or style, require little repair or maintenance. On occasion, however, a service valve may leak. If the leak is through the gauge port

opening (either Schrader or manual type), then a cap with a rubber insert, Figure 9–30, may be used. Permanent repairs should be made the first time the air-conditioning system is opened for service. If the leak is severe or is around the service stem (manual), then it is recommended that the entire service valve be replaced. Whatever the remedy, ensure that the leak has been corrected before returning the vehicle to service.

There is no rule of thumb as to where a service valve will be found in the system. The low-side service valve may be found anywhere from the evaporator outlet to the compressor inlet. On earlier three-valve systems, a service valve is found before (upstream of) the suction pressure regulator device and another low-side service valve is found after (downstream of) the suction pressure regulator device. On either two-valve or three-valve air-conditioning systems, the high-side service valve is usually found anywhere from the compressor outlet to the condenser inlet. On rare occasions it may be found between the condenser outlet and the inlet to the metering device.

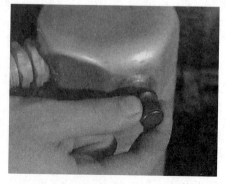

FIGURE 9–30 Removing the protective cap from the accumulator–mounted low–side service valve.

REVIEW

Select the correct answer from the choices given.

1. Technician A says a "ton" of refrigeration is equal to 288,000 Btu/h. Technician B says a "horsepower" of refrigeration is equal to 12,000 Btu/h. Who is right?
   a. A only
   b. B only
   c. Both A and B
   d. Neither A nor B

2. What is the capacity of the average full-size vehicle air-conditioning system?
   a. 18,000 Btu/h
   b. 21,000 Btu/h
   c. 24,000 Btu/h
   d. 12,000 Btu/h

3. Technician A says the absence of "heat" is "cold." Technician B says that "heat" is ever present. Who is right?
   a. A only
   b. B only
   c. Both A and B
   d. Neither A nor B

4. Refrigerant metered into the evaporator:
   a. changes to a vapor.
   b. becomes somewhat superheated.
   c. picks up heat.
   d. all of the above.

5. Technician A says heat is dissipated into the outside ambient air by the condenser. Technician B says heat is dissipated into the outside ambient air by the radiator. Who is right?
   a. A only
   b. B only
   c. Both A and B
   d. Neither A nor B

6. To prevent ice formation on the fins and tubes of the evaporator the temperature should never be allowed to go below:
   a. 30°F (–1.1°C).
   b. 32°F (0°C).
   c. 34°F (1.1°C).
   d. 36°F (2.2°C).

Refer to Figure 9–31 for questions 7 through 12.

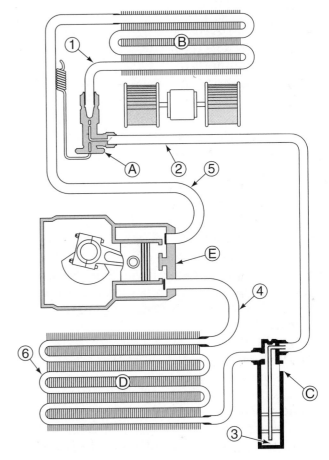

FIGURE 9–31 The refrigeration system.

7. What component is shown as D?
   a. Evaporator
   b. Receiver-drier
   c. Metering device
   d. Condenser

8. What component is shown as C?
   a. Receiver-drier
   b. Accumulator-drier
   c. Compressor
   d. Metering device

9. What component is shown as A?
   a. Fixed orifice tube (FOT)
   b. Thermostatic expansion valve (TXV)
   c. Cycling clutch orifice tube (CCOT)
   d. Automatic expansion valve (AEV)

10. What is the state of the refrigerant at point 4?
    a. Low-pressure vapor
    b. Low-pressure liquid
    c. High-pressure vapor
    d. High-pressure liquid

11. What is the state of the refrigerant at point 2?
    a. Low-pressure vapor
    b. Low-pressure liquid
    c. High-pressure vapor
    d. High-pressure liquid

12. What is the state of the refrigerant at point 3?
    a. Low-pressure vapor
    b. Low-pressure liquid
    c. High-pressure vapor
    d. High-pressure liquid

13. The purpose of the compressor is being discussed. Technician A says it causes the low-pressure vapor to change to a high-pressure liquid. Technician B says it moves refrigerant through the tubes and components of the air-conditioning system. Who is right?
    a. A only
    b. B only
    c. Both A and B
    d. Neither A nor B

14. The state of the refrigerant as it enters the evaporator is being discussed. Technician A says it is all liquid with just a trace of flash gas. Technician B says flash gas is caused by a rapid decrease in pressure. Who is right?
    a. A only
    b. B only
    c. Both A and B
    d. Neither A nor B

15. Which is considered more serious in a TXV system, a flooded evaporator or a starved evaporator? Why?
    a. A flooded evaporator because the refrigerant will not evaporate due to lack of space for expansion.
    b. A flooded evaporator because liquid refrigerant may be returned to the compressor, causing damage.
    c. A starved evaporator because the cooling effect of the evaporator will be drastically reduced.
    d. A starved evaporator because the refrigerant will become superheated and damage the compressor.

16. How many positions does a Schrader-type service valve have?
    a. One
    b. Two
    c. Three
    d. Four

17. The low-side service valve may be found anywhere, EXCEPT:
    a. at the evaporator outlet.
    b. at the compressor inlet.
    c. on the accumulator-drier.
    d. on the receiver-drier.

18. All of the following staterments are true, *except:*
    a. the receiver-drier is located in the liquid line.
    b. the accumulator-drier is located in the suction line.
    c. the evaporator is in the low side of the system.
    d. fixed orifice tubes (FOT) are interchangeable.

19. For a change of state there must be a transfer of heat. Technician A says heat is removed from a substance to cause it to change from a liquid to a vapor. Technician B says adding heat to a substance causes it to change from a solid to a liquid. Who is right?
    a. A only
    b. B only
    c. Both A and B
    d. Neither A nor B

20. A motor vehicle air-conditioning (MVAC) system will automatically control all of the following, EXCEPT:
    a. temperature and humidity of the air.
    b. purity of the air.
    c. engine coolant temperature.
    d. circulation of the air.

TERMS

Write a brief description of the following terms:

1. ambient air
2. atmosphere
3. CCOT
4. chemical
5. cold
6. dissipation
7. federal regulations
8. fittings
9. flash gas
10. flooded
11. FOT
12. horsepower
13. illegal
14. intentional
15. liquid line
16. prohibit
17. screen
18. starved
19. variable displacement
20. VDOT

CHAPTER 10

COMPRESSORS AND CLUTCHES

OBJECTIVES

On completion and review of this chapter, you should be able to:

❏ Discuss and explain the operating principles of a reciprocating compressor.
❏ Discuss and explain the operating principles of a scroll compressor.
❏ Discuss and explain the operating principles of a rotary compressor.
❏ State the purpose and describe the operation of an electromagnetic clutch.
❏ Compare fixed- and variable-displacement compressors.

INTRODUCTION

There are many available makes and models of compressors for motor vehicle air-conditioning (MVAC) **application,** Figure 10–1. Others are on the drawing boards of compressor manufacturers, some of which will be developed and available for use in the near future. The prime consideration for a new compressor design is to reduce weight. Compressors, as well as other components, must be designed to be efficient and durable. Overall vehicle weight is reduced by trimming the weight of individual components. A reduction in overall (gross) vehicle weight provides greater fuel economy; that is, more miles per gallon (kilometers per liter).

FUNCTION

The compressor in the automotive air-conditioning system serves two important functions at the same time. First, it creates a low-pressure condition at the compressor inlet provisions, Figure 10–2, to remove heat-laden refrigerant vapor from the evaporator. This low-pressure condition is essential to allow the refrigerant metering device (thermostatic expansion valve or fixed orifice tube) to admit the proper amount of liquid refrigerant into the evaporator. Second, the compressor compresses the low-pressure refrigerant vapor into a high-pressure vapor, Figure 10–3. This increased pressure raises the heat content of the refrigerant (see Chapter 5). A high pressure with high-heat content is essential if the refrigerant is to give up its heat in the condenser.

Failure of either function will result in a loss or reduction of circulation of the refrigerant within a system. Without proper refrigerant circulation in the system, the air conditioner will either not function properly or not function at all.

DESIGN

Many types of compressors are used in automotive air-conditioning systems. Regardless of type, with a few exceptions, compressors are basically of the same design: **reciprocating** piston. Reciprocating means that the piston moves up and down, to and fro, or back and forth. Two basic methods of driving the piston of a reciprocating compressor are by **crankshaft** or **axial plate.** The axial plate is often called a **swash plate** or **wobble plate.** The exceptions, rotary and **scroll** compressors, which are found on a limited number of vehicle lines, are discussed later in this chapter.

Crankshaft

Driving the piston by crankshaft is an operation very similar to that in an automobile engine. The main difference is that in a compressor the crankshaft drives the piston, whereas in an engine the piston drives the crankshaft, as illustrated in Figure 10–4. The compressor crankshaft is driven, directly or indirectly, off the engine crankshaft by means of pulleys and belts.

Axial Plate

The other method of driving the piston is by axial plate, shown in Figure 10–5. When pressed on the main shaft the plate provides a reciprocation of the piston motion, as illustrated in Figure 10–6. The axial plate is driven by the main shaft off the engine crankshaft, directly or indirectly, by means of pulleys and belts.

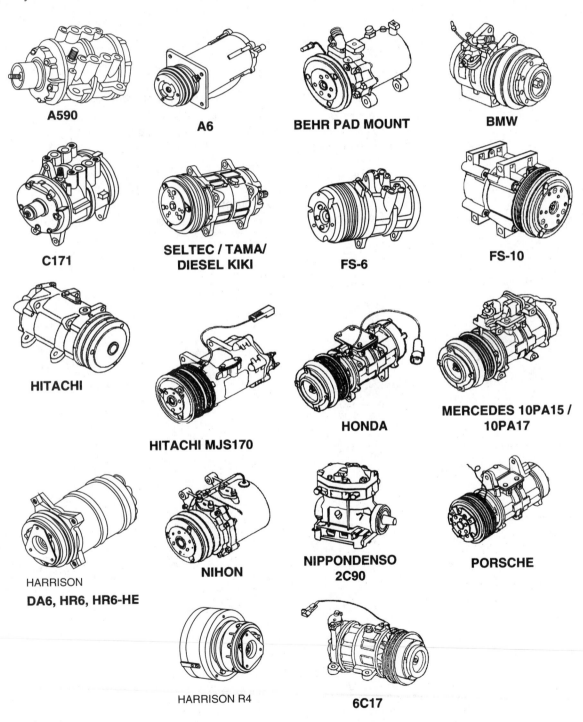

FIGURE 10–1 Some of the many compressors available for vehicle service.

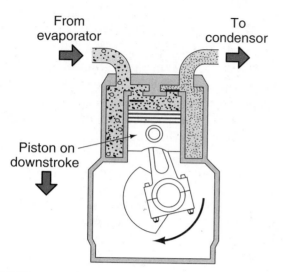

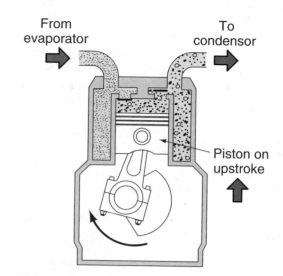

FIGURE 10–2 A low-pressure condition is created on the intake stroke.

FIGURE 10–3 Low-pressure refrigerant vapor is compressed into a high-pressure vapor.

CLUTCH

All automotive air-conditioning compressors have an **electromagnetic** clutch attached to the compressor crank-

shaft or main shaft, as shown in Figure 10–7. This **clutch** provides a means of starting and stopping the compressor. Some compressors are driven by one or two belts off the engine crankshaft and have an idler pulley, Figure

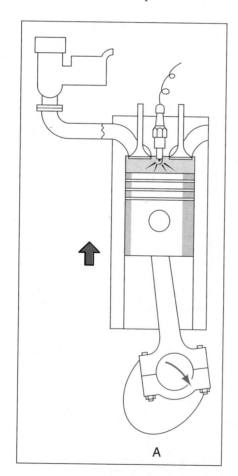

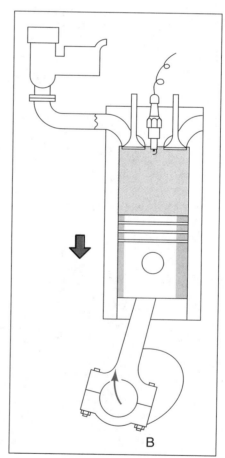

FIGURE 10–4 The burning of air-fuel mixture (A) drives the piston, which, in turn, rotates the crankshaft (B) in an engine.

FIGURE 10–5 Details of compressor axial plate, also referred to as swash plate. *(Courtesy of BET, Inc.)*

10–8, which is used to adjust belt tension. Some use an accessory device, such as an alternator or power-steering pump pulley, to provide a means of adjusting the belt, as shown in Figure 10–9. In some early air-conditioning systems, compressors are driven off the water-pump pulley, which, in turn, is driven by the engine crankshaft pulley, as shown in Figure 10–10. In this application, the belt is tensioned by adjusting the compressor.

Compressors are frequently driven off the crankshaft by a single belt, along with other accessories such as the power steering pump, air pump, alternator, and water pump. This system is known as a *serpentine drive,* and is shown in Figure 10–11. This belt, called V-rib or serpentine, is tensioned by a spring-loaded idler pulley that generally rides on the back (flat) side of the belt.

COMPRESSOR TYPES

There are over 160 currently available makes and models of compressors for motor vehicle air-conditioning systems. The various types include reciprocating piston, scroll, and **rotary vane.**

Reciprocating Piston Compressor

Actually there are three types of reciprocating piston compressors depending on how the pistons are driven: crankshaft, wobble plate, and **scotch yoke.**

The reciprocating piston type of mobile air-conditioning system compressors, depending on design, have one, two, four, five, six, or ten pistons (cylinders).

Single Cylinder. Two **domestic** manufacturers, Tecumseh and York, provided a single-cylinder compressor, which was used on some imports through the mid-1980s.

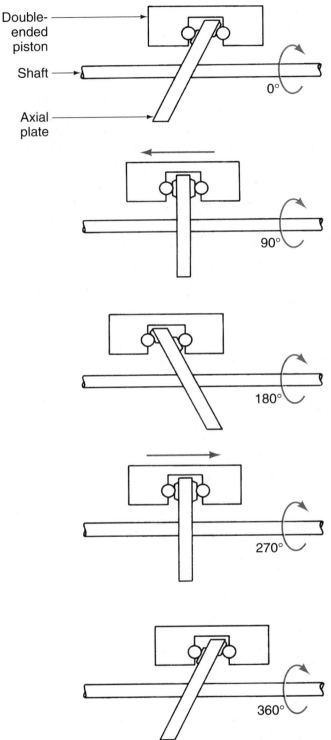

FIGURE 10–6 The piston(s) is (are) moved back and forth or to and fro by an axial plate.

Two Cylinders. A two-cylinder V-type compressor, manufactured by Chrysler Air-Temp, was discontinued in the late 1970s because of its weight.

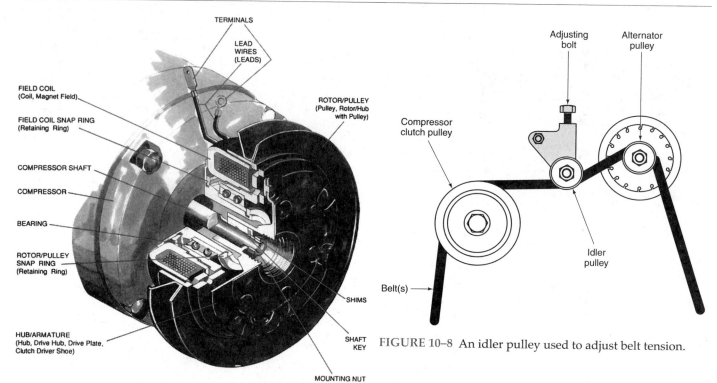

FIGURE 10–7 An electromagnetic clutch provides a means of turning the compressor on and off. (*Courtesy of Warner Electronic*)

FIGURE 10–8 An idler pulley used to adjust belt tension.

Two-cylinder in-line compressors, one type manufactured by both Nippondenso and Tecumseh, and another manufactured by York, were phased out of passenger car service in the early 1980s. These compressors have been replaced with new designs that are lighter in weight and more efficient. The York two-cylinder compressor, Figure 10–12, however, is making a comeback in over-the-road truck air-conditioning systems.

Four Cylinders. A four-cylinder, radial-design compressor is available. The R-4, shown in Figure 10–13, is manufactured by Harrison (Frigidaire) and is found in vehicle lines through the early 1990s. This compressor is available in either standard or light weight versions. A similar compressor, the Tecumseh HR-980, was produced through the late 1980s.

A rotary compressor manufactured by Tecumseh is similar in appearance. Model HR-980 was discontinued during the 1989 model year.

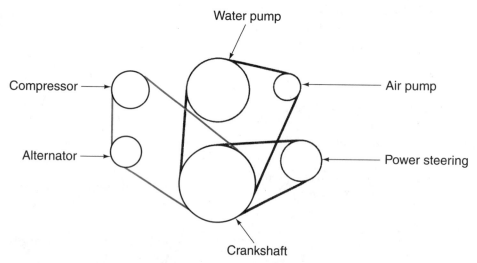

FIGURE 10–9 Compressor driven off the crankshaft pulley via the alternator. The alternator adjustment provides belt tensioning.

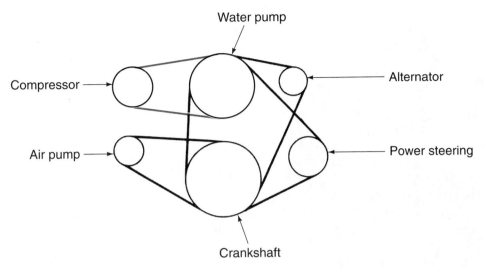

FIGURE 10–10 The compressor is driven off the water–pump pulley, which is driven off the crankshaft pulley. Compressor belt tensioning is accomplished by positioning the compressor. Alternator adjustment provides water–pump belt tensioning.

Five Cylinders. Sanden (Sankyo), Harrison, and Calsonic manufacture a five-cylinder compressor. The Sanden compressor, Figure 10–14 is a positive displacement compressor; the Harrison V-5 compressor in Figure 10–15 is of variable displacement design.

Six Cylinders. At least six six-cylinder compressors are currently available: the Harrison A-6, first put into service in 1962, was replaced with a lighter version, the DA-6, Figure 10–16, in 1982. Two more design changes followed, the "Harrison Redesigned" HR-6, and the "High Efficiency" model, HR6HE. Ford and Chrysler have versions of the six-cylinder compressor originally developed by Nippondenso, Figure 10–17. These compressors are of the axial design.

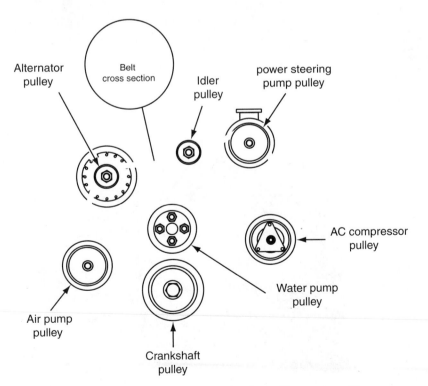

FIGURE 10–11 Single–belt drive system. The belt is tensioned by a spring–loaded idler pulley.

FIGURE 10–12 Two–cylinder York compressor. *(Courtesy of BET, Inc.)*

FIGURE 10–13 Harrison's R4 compressor with V–belt clutch assembly. *(Courtesy of BET, Inc.)*

Seven Cylinders. Harrison manufactures a seven-cylinder variable displacement compressor, Figure 10–18. A seven-cylinder compressor, Figure 10–19, is also manufactured by Honda Air Device Systems (HADS) for use with HFC-134a refrigerant.

Ten Cylinders. A ten-cylinder compressor was introduced by Nippondenso in 1986. This compressor has the same general appearance as the Nippondenso six-cylinder compressor illustrated in Figure 10–17.

Applications

Tecumseh's single-cylinder compressor may be found on some early aftermarket compact car applications, both domestic and **imported**. Nippondenso's two-cylinder compressor is used primarily on some Japanese import

FIGURE 10–14 Sankyo five–cylinder compressor with clutch assembly. *(Courtesy of BET, Inc.)*

FIGURE 10–16 Harrison DA–6 compressor. *(Courtesy of BET, Inc.)*

FIGURE 10–15 Harrison V–5 compressor. *(Courtesy of BET, Inc.)*

FIGURE 10–17 Nippondenso six–cylinder compressor. *(Courtesy of BET, Inc.)*

FIGURE 10–18 Harrison's seven–cylinder variable displacement compressor.

vehicle lines. The Tecumseh and York two-cylinder compressors were often found on intermediate and full-size aftermarket applications, as well as on some early Audi, Ford, Nissan, Porsche, Subaru, and Volkswagen factory-installed systems.

The York two-cylinder compressor is currently used on some heavy-duty trucks as well as on off-the-road equipment. Keihin compressors may be found on some Honda car lines and Hitachi compressors may be found on some Nissan vehicle lines.

The Air-Temp two-cylinder, V-type compressor,

found on some early Chrysler vehicle lines, was discontinued in 1981. Although it is very efficient and durable, it is also heavy. Overall weight is an important factor to be considered in automotive fuel economy. It is still available for use in vehicle restoration where original equipment components are required.

The four-cylinder compressor by Harrison is standard equipment on some General Motors vehicle lines, and may also be found on some Peugeot, Volvo, and Mercedes-Benz vehicles. The Sankyo compressor may be found on Dodge, Plymouth, BMW, Nissan/Datsun, Fiat, Honda, Jeep, Mazda, Porsche, Subaru, Toyota, and Volkswagen vehicles.

Harrison six-cylinder compressors may be found on General Motors, Checker, Ford, Lincoln, Mercury, Audi, Avanti, Jaguar, Mercedes-Benz, Peugeot, Rolls-Royce, and Volvo vehicles.

Sankyo (Sanden) compressors are found on some Dodge, Mazda, Peugeot, Chrysler, Jeep, Renault, Honda, Subaru, and Volkswagen vehicles.

The Nippondenso six-cylinder compressor is found on some Toyota, Ford, Acura, Chevrolet, Honda, Mitsubishi, Mazda, and Mercury vehicles. The Nippondenso ten-cylinder compressor is found on some Acura, Chevrolet, Ford, Honda, Mercury, Mitsubishi, and Toyota vehicles.

A Chrysler-built compressor, Acustar, is found on some Chrysler vehicle lines. This fixed displacement compressor is based on a Nippondenso design.

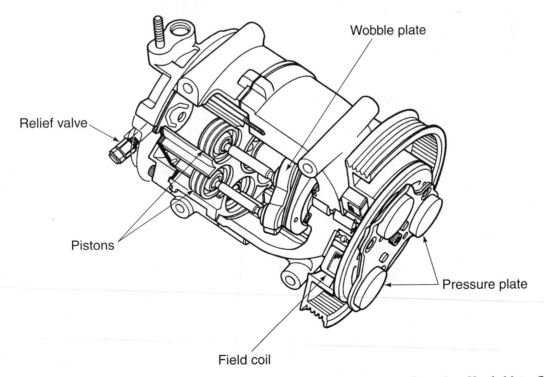

FIGURE 10–19 Honda Air Device Systems (HADS) compressor details. (*Courtesy of American Honda Motor Co. Inc.*)

Some compressors are less commonly used. At the present time, the Harrison variable displacement compressor, known as the V-5, is found only on some General Motors vehicle lines. A Nippondenso-built variable displacement compressor is found on some Chrysler vehicle lines. The Zexel rotary compressor may be found on some Nissan, Toyota, and General Motors vehicle lines. Panasonic's rotary compressor is found on some Ford vehicle lines and a scroll compressor, manufactured by Sanden, is found on some Honda vehicle lines.

COMPRESSOR ACTION

Low-pressure refrigerant vapor is compressed to high-pressure refrigerant vapor by action of the pistons and **valve plates.** For each piston, there is one intake (suction) valve and one outlet (discharge) valve mounted on a valve plate. For simplicity of understanding, a single-cylinder (piston) compressor is discussed here.

By action of the crankshaft, illustrated in Figure 10–20, the piston travels from the top of its stroke to the bottom of its stroke during the first half-revolution of the crankshaft. On the second half-revolution, the piston travels from the bottom of its stroke to the top of its stroke, as shown in Figure 10–21. The first action, top to bottom, is called the *intake* or **suction stroke;** the second action, bottom to top, is called the *compression* or **discharge stroke.**

The piston is fitted with a piston ring, like the one shown in Figure 10–22, to provide a seal between the piston and the cylinder wall. This seal helps to provide a negative (low) pressure on the down or **intake stroke,** and a positive (high) pressure on the up or **exhaust stroke.**

The Intake Stroke. During the intake stroke, a low-pressure area is created atop the piston and below the intake and exhaust valves. The higher pressure atop the intake valve, from the evaporator, allows the intake valve to open to admit heat-laden refrigerant vapor into the cylinder chamber. The discharge valve is held closed during this time period. The much higher pressure atop this valve as opposed to the low pressure below it prevents it from opening during the intake stroke.

The Exhaust Stroke. The exhaust stroke may also be described as the compression stroke. A high-pressure area is created atop the piston and below the intake and exhaust valves. This pressure becomes much greater than that above the intake valve and closes that valve. At the same time, the pressure is somewhat greater than that above the exhaust valve. The pressure difference is great enough to cause the exhaust valve to open. This allows the compressed refrigerant vapor to be discharged from the compressor.

Continuous Action. Piston action is repeated rapidly—once for each revolution of the compressor crankshaft. At road speed, this action may be repeated fifteen hundred or more times each minute for each cylinder of the compressor.

ROTARY VANE COMPRESSOR

The rotary vane compressor in Figure 10–23 provides the greatest cooling capacity per pound of compressor

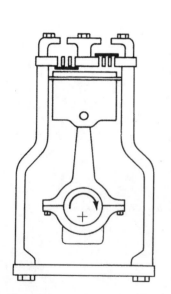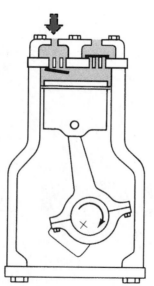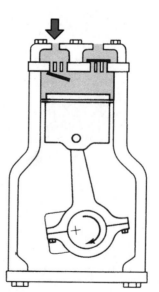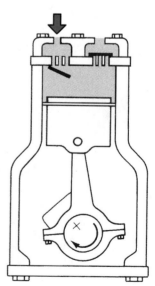

FIGURE 10–20 The piston(s) move(s) down (top to bottom) during the suction (intake) stroke.

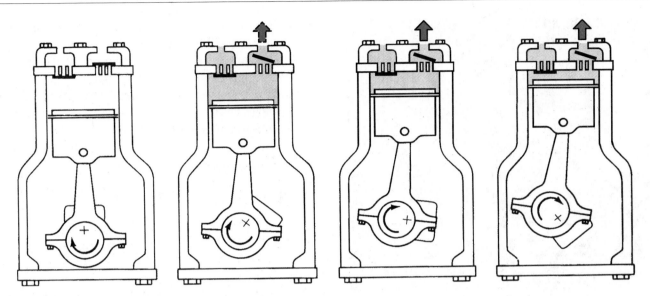

FIGURE 10–21 The piston(s) move(s) up (bottom to top) during the compression (discharge) stroke.

weight. It has no pistons and only one valve called a discharge valve. The discharge valve actually serves as a check valve to prevent high-pressure refrigerant vapor from entering the compressor through the discharge provisions during the off cycle, or when the compressor is not operating. The function of the rotary vane compressor is the same as that of the piston- or reciprocating-type compressor. Its operation, however, is entirely different.

The concept of rotary-type compressors for refrigeration service is not new. Two basic types of rotary vane compressors have been available for nonautomotive refrigeration use for many years: rotating vane and stationary vane. York was first to introduce this compressor to the automotive marketplace, Figure 10–24. Only about fifty thousand York rotary compressors were manufactured in the early 1980s before being discontinued. A rotary vane compressor is found on some Geo Prism, Toyota Corolla, and Toyota Tercel car lines as early as 1989. In 1993, Panasonic introduced a rotary compressor used on some Ford car lines. The Zexel rotary compressor is also used on Nissan's Altima.

FIGURE 10–22 Details of a piston driven by action of the crankshaft in a compressor.

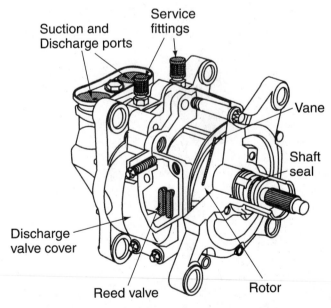

FIGURE 10–23 Cutaway showing the details of a rotary vane compressor. *(Courtesy of Green and Dwiggins, Australian Automotive Air Conditioning, © Nelson, a division of Thomson Learning)*

FIGURE 10–24 York rotary vane compressor. *(Courtesy of BET, Inc.)*

Rotary Compressor Operation

The rotary compressor's shaft turns a rotor assembly with vanes that extend to the wall of the cylinder block, Figure 10–25. A brief explanation of the operation of a typical rotary vane compressor follows.

1. This forms a compression chamber, or several chambers, if there is more than one vane.
2. The rotating vanes then draw in refrigerant vapor through the suction ports.
3. Compression of the refrigerant starts after the vanes have crossed the suction ports, increasing the refrigerant pressure and temperature.
4. The hot vapor is then forced out through the discharge valves to the condenser.

SCROLL COMPRESSORS

Although the scroll compressor, Figure 10–26, was first patented in 1909, it did not meet practical application until introduced by Copeland Corporation in 1988 for use in home air conditioners and heat pumps. Sanden introduced the scroll compressor to the automotive marketplace in 1993. The scroll compressor has but one moving part: the scroll. Many consider its unique design to be a major technological breakthrough in compressor design.

A brief explanation of the operation of a scroll compressor, Figure 10–27, follows.

1. Compression in the scroll compressor is achieved by the interaction of a rotating scroll and a stationary scroll.

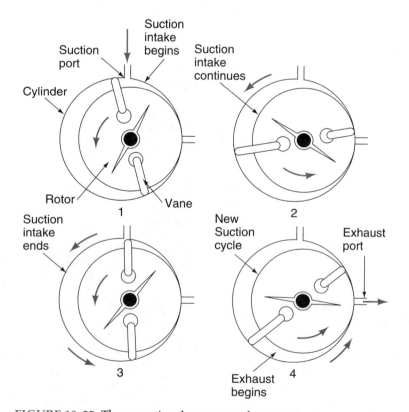

FIGURE 10–25 The operational sequence of a rotary vane compressor.

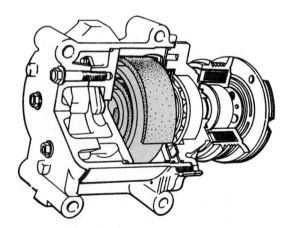

FIGURE 10–26 Scroll compressor details.

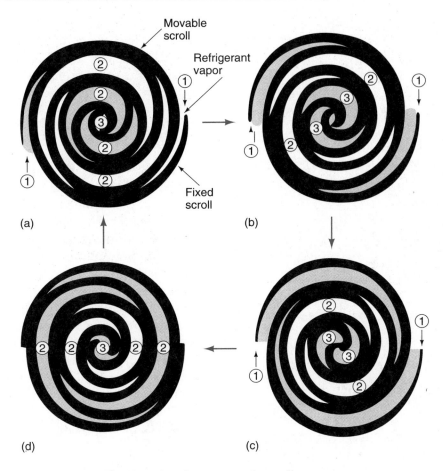

Movable scroll

Refrigerant vapor

Fixed scroll

(a)

(b)

(d)

(c)

FIGURE 10–27 The operational sequence of a scroll compressor.

2. Refrigerant vapor enters the compressor suction port and an outer opening of one of the rotating scrolls.
3. This open passage allows refrigerant vapor to be drawn into the passage of the scroll, which is then sealed off.
4. As the scroll continues to rotate, the passage becomes smaller and the refrigerant vapor is compressed.
5. As the refrigerant vapor is discharged from the compressor discharge port, its temperature and pressure have been increased.

This was a brief explanation of just one vapor passage. During operation, all vapor passages are in various stages of compression at the same time. This provides a nearly continuous suction and discharge pressure at all times.

VARIABLE DISPLACEMENT COMPRESSORS

In 1985, Harrison introduced a variable displacement compressor, designated as model V-5, that is used on some models of General Motors vehicle lines. This compressor, and a later model V-7, can match most any automotive air-conditioning load demand under all conditions. This is accomplished by varying the displacement of the compressor by changing the stroke (displacement) of the pistons.

The five, or seven, axially oriented pistons are driven by a variable-angle wobble plate. The angle of the wobble plate is changed by a bellows-activated control valve located in the rear head of the compressor. This control valve, Figure 10–28 senses suction pressure and controls the wobble plate angle based on crankcase-suction pressure differential. Operation of the control valve is dependent on a difference in pressure, known as delta p (Δ_P).

Whenever the air-conditioning demand is high, suction pressure will be above the control point, and the control valve will maintain a bleed from the compressor crankcase to the suction side. In this case, there is no crankcase-suction pressure differential, and the compressor will have maximum displacement. The wobble plate is at maximum angle, providing greatest stroke (piston travel), Figure 10–29.

Conversely, when the air-conditioning demand is low and the suction pressure reaches the control point,

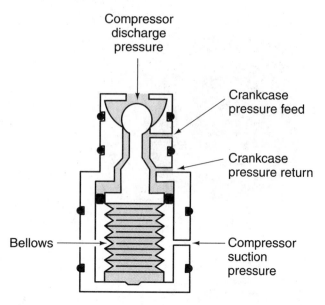

FIGURE 10–28 Variable displacement compressor control valve assembly.

the control valve will bleed discharge gas into the crankcase to the suction plenum, Figure 10–30.

The angle of the wobble plate is actually controlled by a force balance on the five, or seven, pistons. Only a slight increase of the crankcase-suction pressure differential is required to create a force on the pistons sufficient to result in movement of the wobble plate.

Temperature, then, is maintained by varying the

capacity of the compressor, not by cycling the clutch on and off. This action provides a more uniform method of temperature control and, at the same time, eliminates some of the noise problems associated with a cycling clutch system.

SCOTCH YOKE COMPRESSORS

The Scotch yoke compressor design has been used for many years in domestic and commercial refrigeration applications. Opposed pistons are pressed onto opposite ends of a yoke riding on a slider block located on the shaft eccentric of a Scotch yoke compressor. Rotation of the shaft also rotates the yoke, with attached pistons, in a reciprocating motion. The rotating assembly, Figure 10–31, is counterbalanced to ensure smooth operation. A suction **reed valve** is located atop each piston and a reed valve plate is located atop each cylinder. The action of the compressor is like any other piston compressor; low-pressure refrigerant is drawn into the cylinder through the suction valve during the intake stroke. This vapor is compressed as it is forced out, under high pressure, through the discharge valve plate on the exhaust stroke.

DIAGNOSING PROBLEMS AND MAKING REPAIRS

Broken discharge valves in compressors are not an uncommon failure. Broken suction valves and piston rings are not

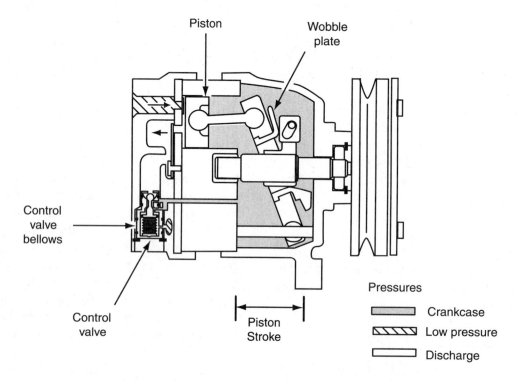

FIGURE 10–29 Variable displacement compressor at maximum displacement.

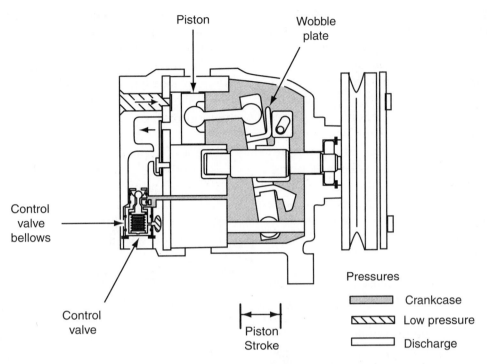

FIGURE 10–30 Variable displacement compressor at minimum displacement.

as common but lead to the same diagnosis. Broken valves and/or rings are easily diagnosed in one- and two-cylinder reciprocating compressors with the use of a manifold and gauge set. The first indication of failure is a higher-than-normal low-side (suction) pressure accompanied by a lower-than-normal high-side (head) pressure.

Valve and ring failures, however, are not as easily diagnosed in four-, five-, and six-cylinder compressors. The first indication of valve or ring failure in these compressors is that the belt(s) will not remain tightened. One

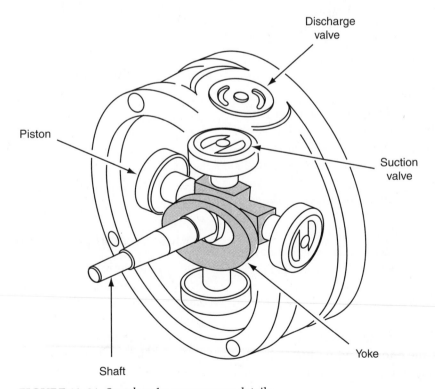

FIGURE 10–31 Scotch yoke compressor details.

defective discharge valve plate in a six-cylinder compressor sets up a vibration that, when not otherwise detected, literally shakes the belt(s) loose. This is true regardless of how well the adjustment provisions are tightened.

Many simple compressor repairs are usually a routine service provided by the automotive air-conditioning technician. These repairs include checking and adding oil, replacing the crankshaft seal, and, in some units, replacing the control valve assembly. More complex repairs are often "shopped out" to a specialty shop with the facilities for semi-mass-rebuilding procedures. Because of the general high cost of labor, one-on-one compressor rebuilding is not economically feasible.

COMPRESSOR FAILURE

Compressor failure, representing almost 30 percent of all vehicle air-conditioning system repairs, is the leading cause of system failure according to a survey conducted in the late 1990s by the Mobile Air Conditioning Society Worldwide (MACSW). The principal cause of compressor failure was found to be leaks, followed by internal mechanical problems. Clutch problems were the least common reasons for compressor failure.

Over half of the compressor failures in this survey were in air-conditioning systems in vehicles that had been retrofitted to HFC-134a from CFC-12 refrigerant. This is most likely because many older compressors designed for CFC-12 refrigerant will not withstand the higher operating pressures associated with HFC-134a refrigerant. The lack of proper lubrication is also noted as a problem with vehicle air-conditioning system compressors, and is generally due to not properly changing the lubricant during retrofit procedures or not checking the lubricant during repair procedures.

Most compressors are designed to function with a **compression ratio** between 5:1 and 7:1. A quick check of a compressor's operating compression ratio can be done by dividing its high-side absolute pressure (psia) by its low-side absolute pressure (psia). Note that one must add 15 to both low-side and high-side gauge readings to obtain the absolute pressure.

For example, assume that the low-side pressure is 30 psig and the high-side pressure is 220 psig. When 15 is added to these values, they become 45 psia and 235 psia.

$$235 \div 45 = 5.2 \text{ or } 5.2{:}1$$

Pressure ratios above 7.5:1 can cause early compressor failure because of the added load on bearings, pistons, and seals. Also, higher operating temperatures generated by the higher operating pressures can cause lubrication breakdown, which results in harmful deposits on the compressor internal assembly.

A study by ACDelco and Delphi Harrison made public by the Mobile Air Conditioning Society (MACS) in late 2000 reveals five major reasons that new replacement compressors fail:

1. Use of refrigerant blends
2. Debris in the air-conditioning system
3. Improper flush agent or flushing procedure
4. Lack of lubrication or improper lubricant
5. Low refrigerant charge
6. Improper trace dye

Use of Refrigerant Blends

Many refrigerant blends contain chemicals that are simply not compatible with some of the compressor components, such as O-rings and seals. To avoid this problem, use only a refrigerant that is designated for a particular compressor application, CFC-12 or HFC-134a.

Debris in the System

Severe compressor failure generally results in debris in the air-conditioning system. Most, but not all, of the debris is removed by properly flushing the system following procedures specified by the compressor manufacturer. It is also important to use an approved flushing agent. The installation of in-line filters following flushing is recommended by General Motors and Ford.

Improper Flush Agent

Use the flush agent recommended by the compressor manufacturer. General Motors recommends that refrigerant be used, claiming that any other cleaning agent breaks down the lubricants and leaves degrading residue, which affects overall system performance. All agree, however, that all flush agent must be removed from the air-conditioning system.

Lack of or Improper Lubricant

The lack of lubricant in a compressor results in early failure due to excessive wear. Because the lubricant circulates in the system with the refrigerant, an ample amount is essential to ensure compressor lubrication at all times. An improper lubricant in an air-conditioning system does not "mix" and circulate with the refrigerant and the results are the same as a lack of lubricant, Figure 10–32.

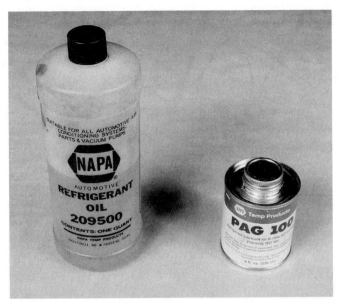

FIGURE 10–32 Be sure that the proper lubricant is added to the air-conditioning system.

Low Refrigerant

Because the lubricant mixes and flows through the air-conditioning system with the refrigerant, a lack of refrigerant will result in a lack of lubricant. Most of the lubricant is often found in the bottom of the receiver-drier or accumulator-drier in systems that are low on refrigerant.

Improper Trace Dye

Lubrication problems are sometimes caused by the use of the wrong trace dye. Note that trace dyes are designated to be compatible with certain types of lubricant. The improper dye may also attack O-rings and compressor shaft seals as well as dilute the lubricant viscosity.

COMPRESSOR REPLACEMENT

A leading compressor rebuilder advises that most rebuilt or remanufactured compressors do not contain lubricant. The proper lubricant must be added per manufacturer's specifications for type and quantity. The warranty may be void unless the following 18-step procedure is followed during replacement compressor installation.

Replacement Procedure

Review all of the information provided with the replacement compressor and follow any specific guidelines given. Use only the proper lubricant—mineral oil for a CFC-12 system and poly alkylene glycol (PAG) or polyol ester (POE) for an HFC-134a air-conditioning system.

Lubricate all O-rings and seals with mineral oil or O-ring lubricant. Torque all fasteners to specifications; do not use an impact wrench. Ensure that the vacuum pump is capable of 29.5 in. Hg at minimum.

1. Recover refrigerant, if any.
2. Remove the old compressor.
3. Transfer any switches or mounting hardware to the replacement compressor.
4. Clean the air-conditioning system with an approved flush solution or install an in-line auxiliary filter.

   NOTE: Do not flush the compressor, metering device, receiver or accumulator, or hoses containing an orifice tube or muffler. Follow the instructions provided with the flush solution or the filter.

5. Replace the receiver-drier or accumulator-drier.
6. Replace the orifice tube or clean the screen in the thermostatic expansion valve (TXV).

   NOTE: If necessary replace the liquid line or TXV, as applicable.

7. Check for proper airflow through the condenser and radiator.
8. Check for proper fan clutch or electric fan operation.
9. Check the replacement compressor clutch air gap.

   NOTE: Refer to manufacturer's specifications.

10. Add the correct type and amount of lubricant per the replacement compressor manufacturer's specifications.

    NOTE: If the replacement compressor already contains lubricant, drain and change it per step 10. Add half the lubricant to the compressor and half to the receiver or accumulator.

11. After installation, turn the compressor shaft (not clutch rotor) 10–12 turns to ensure that lubricant has not collected in the valve plate area atop the pistons.
12. Check the compressor clutch electrical circuit to ensure the proper voltage.

    NOTE: Refer to the manufacturer's specifications.

13. Leak check the system and inspect for irregularities.
14. Use HFC-134a or CFC-12 refrigerant, as applicable, only.

   NOTE: Use of any other type of refrigerant will void the warranty.

15. Evacuate the air-conditioning system for a minimum of:
   a. 45 minutes if the ambient temperature is above 80°F (26.7°C).
   b. One hour if a dual-air-conditioning system or if ambient temperature is below 80°F (26.7°C).
16. Charge the air-conditioning system with the proper refrigerant, Figure 10–33.
17. Hold a performance test and/or return the air-conditioning system to service.

COMPRESSOR SERVICE

The information for typically servicing some of the popular compressors is given in Chapter 11. Service for other types of compressor is similar. It is generally recommended that the manufacturer's service manuals be followed for specific repairs. Procedures are given in Chapter 11 for servicing Harrison's six-cylinder R4 and V-5; Nippondenso; Panasonic; Sanden; Tecumseh's HR-980; and York's Vane Rotary shaft oil seal, replacing lubricant and rebuilding or replacing the clutch assembly.

SUMMARY

❏ Reciprocating compressors have a piston or pistons that draw low-pressure heated refrigerant vapor into a chamber, increase its heat content and pressure, and "pump" it out as a high-pressure, high-temperature vapor.

FIGURE 10–33 Be sure to use the proper refrigerant.

❏ A scroll compressor draws low-pressure heated refrigerant vapor through its suction port into a continuously rotating scroll where its pressure and temperature are increased. It is then forced out through its discharge port as a high-pressure, high-temperature vapor.
❏ In a rotary compressor, a rotating vane draws in low-pressure heated refrigerant vapor through the suction port and increases its temperature and pressure before forcing it out through the discharge port. It is discharged as a high-temperature, high-pressure vapor.
❏ An electromagnetic clutch is used to engage and disengage (turn on and off) a compressor, as desired, in present applications of automotive air-conditioning systems.

REVIEW

Select the correct answer from the choices given.

1. Technician A says that a reciprocating compressor pumps refrigerant as a vapor. Technician B says that a scroll compressor pumps refrigerant as a liquid. Who is right?
   a. A only
   b. B only
   c. Both A and B
   d. Neither A nor B

2. The low-side gauge of an R-12 system indicates 55 psig (379 kPa) and the high-side gauge indicates 130 psig (896 kPa). Technician A says the problem may be an indication of a low charge of refrigerant. Technician B says the problem may be a blown head gasket or broken discharge valve. Who is right?
   a. A only
   b. B only
   c. Both A and B
   d. Neither A nor B

3. Compressor function is being discussed. Technician A says a compressor creates a low-pressure condition at the inlet. Technician B says a compressor compresses a low-pressure vapor into a high-pressure vapor. Who is right?
   a. A only
   b. B only
   c. Both A and B
   d. Neither A nor B

4. If a refrigerant is to give up its heat, there must be:
   a. high pressure.
   b. high heat.
   c. both A and B.
   d. neither A nor B.

5. All of the following are current compressor designs, EXCEPT:
   a. wobble plate.
   b. stationary plate.
   c. reciprocating.
   d. rotary vane.

6. A reciprocating-type compressor's pistons are driven off :
   a. a crankshaft.
   b. a wobble plate.
   c. both A and B.
   d. neither A nor B.

7. All automotive compressors are driven directly or indirectly off the:
   a. water pump pulley.
   b. accessory pulley.
   c. alternator pulley.
   d. crankshaft pulley.

8. Reciprocating piston-type compressors may have _____ cylinders.
   a. one, two, four, five, six, seven, or ten
   b. two, four, six, eight, ten, or twelve
   c. one, three, five, six, seven, or ten
   d. one, two, four, six, eight, or ten

9. Refrigerant _____ is drawn into the compressor on its _____ stroke.
   a. vapor/exhaust
   b. vapor/intake
   c. liquid/exhaust
   d. liquid/intake

10. What prevents high-pressure refrigerant from returning to the low side of the system through a reciprocating compressor?
    a. Check valve
    b. Piston action
    c. Discharge valve
    d. Suction valve

11. Technician A says the exhaust stroke of a compressor is also referred to as the discharge stroke. Technician B says the discharge valve is closed during the exhaust stroke of the compressor. Who is right?
    a. A only
    b. B only
    c. Both A and B
    d. Neither A nor B

12. Who was first to introduce a rotary vane compressor for motor vehicle air-conditioning (MVAC) system service?
    a. Tecumseh
    b. Zexel
    c. Panasonic
    d. York

13. Which compressor design has the fewest moving parts?
    a. Reciprocating
    b. Scroll
    c. Rotary
    d. Scotch yoke

14. What is variable in a variable displacement compressor?
    a. Piston bore
    b. Piston stroke
    c. Suction pressure
    d. Control valve

15. The principal cause of compressor failure is due to:
    a. leaks.
    b. a defective clutch.
    c. broken reed valve(s).
    d. defective bearing(s).

16. What is the compression ratio of a compressor that pumps a low-side pressure of 28 psig and a high-side pressure of 213 psig?
    a. 7.6:1
    b. 6.7:1
    c. 5.3:1
    d. 3.5:1

17. Technician A says that a lack of refrigerant will result in a lack of lubricant in an air-conditioning system. Technician B says that most of the lubricant is found in the bottom of the evaporator in a system low on refrigerant. Who is right?
    a. A only
    b. B only
    c. Both A and B
    d. Neither A nor B

18. All of the following are refrigeration lubricants, EXCEPT:
    a. PAG.
    b. mineral oil.
    c. POE.
    d. paraffin oil.

19. What may be used to lubricate O-rings?
    a. Special O-ring lube
    b. Mineral oil
    c. Both A and B
    d. Neither A nor B

20. All of the following are true statements, *except:*
    a. an electromagnetic clutch is used on all compressors as a means of temperature control.
    b. a pressure control may be used to cycle a compressor clutch on and off for temperature control.
    c. a control thermostat may be used to cycle a compressor clutch on and off for temperature control.
    d. a control valve may be used in a variable displacement compressor to provide a means of temperature control.

TERMS

Write a brief description of the following terms:

1. application
2. axial plate
3. clutch
4. compression ratio
5. crankshaft
6. discharge stroke
7. domestic
8. electromagnetic
9. exhaust stroke
10. imported

11. intake stroke
12. reed valve
13. reciprocating
14. rotary vane
15. Scotch yoke
16. scroll
17. suction stroke
18. swash plate
19. valve plates
20. wobble plate

CHAPTER 11

COMPRESSOR SERVICE

OBJECTIVES

On completion and review of this chapter, you should be able to:

- ❏ Service the shaft oil seal and clutch assembly of the Harrison four-, five-, six-, and seven-cylinder compressors.
- ❏ Service the shaft oil seal and clutch assembly of the Nippondenso compressor.
- ❏ Service the shaft oil seal and clutch assembly of the Panasonic compressor.
- ❏ Service the shaft oil seal and clutch assembly of the Sankyo compressor.
- ❏ Service the shaft oil seal and clutch assembly of the Tecumseh HR-980 type of compressor.
- ❏ Check and correct the lubricant level of the Harrison, Nippondenso, Sankyo, Panasonic, and Tecumseh types of compressors.

INTRODUCTION

Special tools are required for most service procedures covered in this chapter. The tools and materials listed are adequate for all service procedures given in this chapter. All of the tools and materials listed, however, are not required for all of the procedures. Depending on the vehicle application, the tools required for compressor service may be English, metric or a combination of both.

The procedures in this chapter assume that the compressor has been removed from the vehicle unless otherwise stated. If the compressor to be repaired is in the vehicle, it should be removed for ease of service. Before removing the compressor, ensure that the air-conditioning system is free of refrigerant. If there is refrigerant in the air-conditioning system then it should be recovered using the proper equipment.

WARNING
IT IS A VIOLATION OF FEDERAL LAW TO INTENTIONALLY VENT ANY TYPE OF REFRIGERANT TO THE AMBIENT ATMOSPHERE.

Tools

• Arbor press • Basic hand tool set • Compressor: Clutch pulley support, Clutch tool set, Service tool set, Shaft seal service set, Holding fixture, Pressure test fitting, O-ring remover/installer, Shaft seal and shaft seal seat remover/installer, and Gasket scraper • Detector, Leak • Dipstick, Oil • Drill rod, 5/16" • Gauge: Air gap, Angle, and Nonmagnetic feeler • Graduated measure • Hammer: Ball peen, Plastic, and Soft face • Indicator, Dial with magnetic base • Manifold and gauge set with service hoses • Meter:

Digital Volt-ohmmeter (DVOM), Ammeter, and Thermometer • Pliers: Slip joint and internal and external snap ring • Pulley puller, two- and three-jaw • Punch and chisel set • Safety glasses or goggles • Screwdriver set • Wrenches: Adjustable 8" and 10", Allen set, Combination, 1/4" and 3/8" Drive socket set, 9/16" Thinwall socket, Spanner, and Torque in.-lb. and ft.-lb. (N•m)

Materials

• Brass washers: six 10 mm • Clean shop rags • Clutch bearing • Compressor: Clutch bearing, Shaft seal assembly, Control valve, Gaskets and O-rings, Pulley bearings, Retainer rings, and Oil filler plug O-ring. • Lubricant: Mineral oil, PAG or POE lubricant • O-ring lubricant • Mineral spirits • Refrigerant: CFC-12 or HFC-134a • Thread lock

SERVICING THE HARRISON SIX-CYLINDER COMPRESSOR

Service procedure for bench testing and repairing the Harrison six-cylinder compressor follows the typical general procedures required to service all models, such as A-6, DA-6, HR-6, and HD6/HT6. These procedures are given in eight parts, as follows:

PART	PROCEDURE	PAGE
1	Checking and adding lubricant	182
2a	The A-6 clutch	183
2b	The clutch, all except A-6	184
3a	The A-6 shaft seal	188
3b	The shaft seal, all except A-6	191
4	The rear head, all except A-6	191
5	The front head, all except A-6	193
6	The cylinder assembly, except A-6A	193
7	The center seal, except A-6	194
8	Rebuilding the A-6 compressor	194

PART 1
CHECKING AND ADDING LUBRICANT

The design of Harrison's six-cylinder compressors requires a different **lubricant** checking procedure than that used for some other types of compressors. These compressors, new or rebuilt, are generally shipped fully charged with lubricant: model A-6 with 11 ounces (325 mL), and models DA-6, HR-6, and DT6/HD6 with 8 ounces (237 mL). Compressors for CFC-12 refrigerant service are charged with 525 viscosity mineral oil and those for HFC-134a are charged with PAG lubricant.

Procedure

1. Clean the external surfaces of the compressor so that it is free of oil and grease.

2. Remove the drain plug located in the compressor sump, A-6 only.
3. To drain the lubricant:
   a. A-6: Place the compressor in a horizontal position, drain hole facing downward, over a graduated container.
   b. Other models: Place the compressor in a vertical position with the suction/discharge ports facing downward over a graduated container.
4. Allow several minutes to drain the compressor. Measure and note the amount of lubricant removed. Discard the old lubricant.
   a. If the quantity removed is less than 4 ounces (118 mL) and the system shows no signs of a greater loss, add 6 ounces (177 mL) of new lubricant.
   b. If the quantity removed is greater than 4 ounces (118 mL) and the system shows no signs of a greater loss, then add the same amount of new lubricant.
   c. If a major component of the system is also replaced, see the following "Service Notes" for the additional amount of lubricant to be added for the component.
   d. If the compressor is replaced with a rebuilt unit, then replace the lubricant in the amount as indicated in steps 3 or 4 and then add 1 ounces (29.6 mL) more.

Service Notes

If the compressor shows that foreign matter is present or that the lubricant contains chips or metallic particles when it is removed and drained, then it should be flushed. The receiver-drier or accumulator-drier, as applicable, should also be replaced after the air-conditioning system has been flushed. The compressor inlet and thermostatic expansion valve screens should be cleaned as well. If the compressor is used in an orifice tube system, replace the orifice tube.

With the exception of a flushed system, add lubricant as shown to any system that has had a major component replaced.

Component	A-6 System	All other
Accumulator	Replace with same amount of lubricant drained, plus 3 oz (88.7 mL). If no lubricant drained, add 3 oz (88.7 mL) to new accumulator.	Same as A-6
Compressor (flushed)	11 oz (325 mL)	8 oz (237 mL)
Condenser	1 oz (29.6 mL)	Same as A-6
Evaporator	3 oz (88.7 mL)	2 oz (59.1 mL)

Disregard any loss of lubricant due to the changing of a line, hose, or muffler unless the component contained a measurable amount of lubricant. If this is the case, then add the same amount and type of clean refrigeration lubricant as removed from the component. Use only refrigeration grade mineral oil in a CFC-12 refrigerant air conditioning system and only PAG or POE lubricant, as specified, in an HFC-134a refrigerant air conditioning system.

FIGURE 11–1 Lubricant replacement chart for Harrison six–cylinder compressors.

If the system has been flushed, add a full charge of clean refrigeration lubricant to the six-cylinder compressor system, as specified in the chart of Figure 11–1.

PART 2A
SERVICING THE DA-6 COMPRESSOR CLUTCH

This service procedure covers servicing the Harrison A-6 compressor clutch assembly. Procedures for servicing compressor models HR-6 and HD6/HT6 compressor clutch assemblies are covered in Part 2B.

Procedure

1. Mount the compressor in a **holding fixture** and secure the fixture in a vise.
2. Using a drive plate holding tool and a 9/16-inch thin wall socket, remove the locknut from the shaft.
3. Using snap ring pliers, remove the clutch hub retaining ring. Remove the spacer under the ring.
4. Remove the hub and drive plate with the hub and drive plate remover tool, as shown in Figure 11–2.
5. Using the snap ring pliers, remove the pulley and bearing snap ring retainer, as illustrated in Figure 11–3.

6. Place a puller pilot over the crankshaft. Remove the pulley using a pulley puller, as shown in Figure 11–4.

    NOTE: The puller pilot must be in place. Placing the puller against the crankshaft will damage the internal assembly.

7. If the pulley bearing is to be replaced, use a sharp tool, such as a small screwdriver, to remove the wire retaining ring.
8. From the rear of the pulley, press or drive the bearing out.
9. Scribe the location of the coil **housing** in relation to the compressor body to ensure proper alignment during reassembly.
10. Using snap ring pliers, remove the coil housing retainer ring, as in Figure 11–5.
11. Lift off the coil housing assembly.
12. Check the coil for loose connections or cracked insulation.
13. Briefly, connect the coil to a 12-volt battery with an ammeter in series. If the coil draws more than 3.2 amperes at 12 volts, then it should be replaced.
14. Inspect the hub and drive plate. Look for signs of looseness between the hub and drive plate. See the note following step 15.
15. Check the pulley and bearing assembly.

FIGURE 11–2 Removing the hub and drive plate assembly. *(Courtesy of General Motors Corporation, Service Operations)*

FIGURE 11–3 Removing the pulley retaining ring. *(Courtesy of General Motors Corporation, Service Operations)*

FIGURE 11–4 Removing the pulley. *(Courtesy of General Motors Corporation, Service Operations)*

FIGURE 11–5 Removing the coil housing retainer ring. *(Courtesy of General Motors Corporation, Service Operations)*

NOTE: If the frictional surface of the pulley or drive plate shows signs of warpage due to excessive heat, then that part should be replaced. Slight scoring, as shown in Figure 11–6, is normal. If either assembly is heavily scored, then it should be replaced.

16. Check the pulley bearing for signs of excessive noise, binding, or looseness. Replace the bearing, steps 20 and 21, if necessary.
17. Note the original position of the coil housing by the scribe marks.
18. Slip the coil housing into place.
19. Replace the snap ring to secure the housing.
20. If bearing replacement is necessary, press out the old bearing.
21. Press a new bearing into the pulley, as illustrated in Figure 11–7, and replace the wire retaining ring.
22. Using the proper tool, as shown in Figure 11–8, press or drive the pulley and bearing assembly on the compressor front head.
23. Install the retainer snap ring.

PART 2B
SERVICING THE SIX-CYLINDER COMPRESSOR CLUTCH

The following procedure for servicing the Harrison DA-6 compressor is applicable to later models such as HR-6 and HD6/HT6. There are two clutch plate and hub assemblies,

FIGURE 11–6 Scoring of the drive and driven plates is normal. Do not replace for this condition. *(Courtesy of General Motors Corporation, Service Operations)*

This ridge of tool is up when installing bearing

FIGURE 11–7 Installing the pulley and drive plate bearing. *(Courtesy of General Motors Corporation, Service Operations)*

FIGURE 11–8 Installing the pulley and drive plate on the compressor. *(Courtesy of General Motors Corporation, Service Operations)*

also referred to as a clutch driver, without a torque cushion and with a thin torque cushion. Removal and replacement procedures are the same for both types, although a different clutch holding tool is used for each type.

Procedure

1. Clamp the compressor into an appropriate holding fixture.

2. Hold the clutch hub with the clutch holding tool and remove the shaft nut, as shown in Figure 11–9.
3. Use the clutch plate and hub installer/remover tool to remove the clutch hub, as illustrated in Figure 11–10.
4. Remove the **shaft key** and set aside for reassembly.
5. Use the snap ring pliers to remove the rotor snap ring, as shown in Figure 11–11.

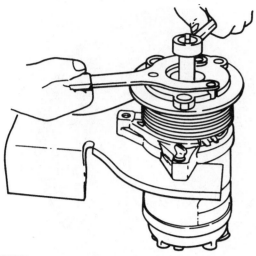

FIGURE 11–9 Remove the shaft nut while holding the clutch hub. (*Courtesy of General Motors Corporation, Service Operations*)

6. Use the pulley rotor and bearing guide over the compressor shaft and insert the puller in the rotor slots to remove the rotor, as shown in Figure 11–12.
7. Mark the clutch coil **terminal** location on the compressor front head for ease in reassembly.
8. Install the puller pilot on the front head.
9. Install the puller and tighten the forcing screw against the pilot to remove the clutch coil, as in Figure 11–13.
10. Attach the rotor and bearing puller tool (less forcing screw) to the rotor. Place the puller atop a solid flat surface, as illustrated in Figure 11–14.
11. With the rotor bearing remover tool and universal handle, drive the bearing out of the rotor hub.
12. Remove the rotor and bearing puller tool from the rotor.

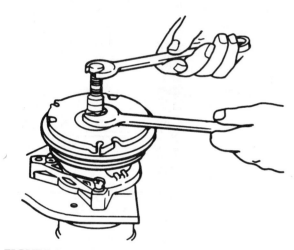

FIGURE 11–10 Remove the clutch hub. (*Courtesy of General Motors Corporation, Service Operations*)

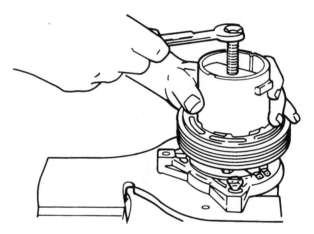

FIGURE 11–12 Remove the rotor using the rotor remover tool. (*Courtesy of General Motors Corporation, Service*

FIGURE 11–11 Use snap ring pliers to remove the snap ring. (*Courtesy of General Motors Corporation, Service Operations*)

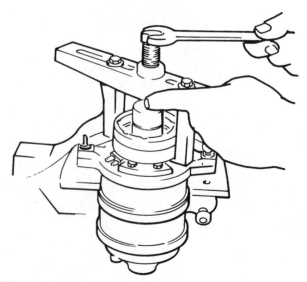

FIGURE 11–13 Remove the clutch coil. (*Courtesy of General Motors Corporation, Service Operations*)

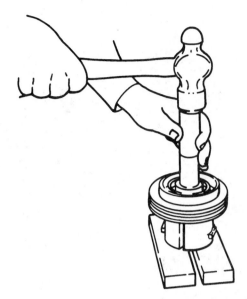

FIGURE 11–14 Place the puller atop a solid flat surface and, using the rotor bearing remover tool, drive the bearing out of the rotor. (*Courtesy of General Motors Corporation, Service Operations*)

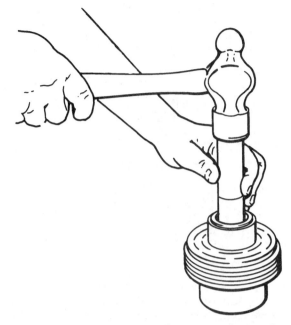

FIGURE 11–15 Place the rotor on the support block and, using the bearing installer tool, drive the bearing fully into the rotor. (*Courtesy of General Motors Corporation, Service Operations*)

13. Place the rotor on the support block, as in Figure 11–15, to support the rotor during bearing installation.
14. Align the bearing with the hub **bore.** Using the puller and bearing installer tool, drive the bearing fully into the hub.
15. Position the bearing staking guide and staking pin tool in the hub bore.
16. Strike the staking pin with a hammer. Form three stakes 120 degrees apart. The staked metal should not touch the outer race of the bearing. *Take care not to damage the bearing* during the staking procedure.
17. Remove the staking tools and support block.
18. Place the clutch coil on the front head of the compressor. Note the location of the electrical terminals as marked during removal of the coil.
19. Assemble the clutch coil installer, puller crossbar, and bolts atop the clutch coil, as shown in Figure 11–16.
20. Turn the forcing screw of the crossbar to force the clutch coil onto the front head of the compressor. Make sure the clutch coil and clutch coil installer tool remain in-line during this procedure.
21. After the clutch coil is fully seated, use a 1/8-inch punch and stake the front head at 120-degree intervals to hold the coil in the proper position.
22. Position the rotor and bearing assembly on the front head.
23. Assemble the puller pilot, crossbar, and bolts atop the rotor and bearing assembly, as shown in Figure 11–17.
24. Tighten the forcing screw to force the rotor and

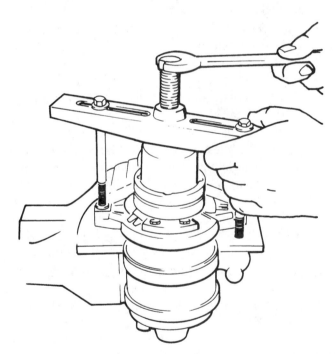

FIGURE 11–16 Assemble the clutch coil installer. (*Courtesy of General Motors Corporation, Service Operations*)

bearing assembly onto the compressor front head. Make sure the assembly stays in-line during this procedure.
25. Use the snap ring pliers to install the rotor and bearing assembly retainer snap ring (see Figure 11–11).
26. Install the shaft key into the hub key groove. The

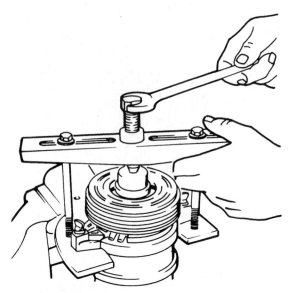

FIGURE 11–17 Assemble the puller pilot, crossbar, and bolts atop the rotor and bearing assembly. (*Courtesy of General Motors Corporation, Service Operations*)

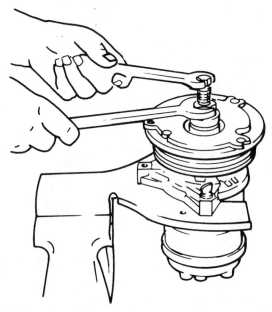

FIGURE 11–18 Install the clutch plate and hub assembly. (*Courtesy of General Motors Corporation, Service Operations*)

key should protrude about 1/8 inch (3.2 mm) out of the keyway.

27. Align the shaft key with the shaft keyway and position the clutch plate and hub assembly onto the compressor shaft.
28. Install the drive plate installer and bearing onto the clutch plate and hub assembly. The forcing tip of the installer must be flat or the end of the shaft/axial plate assembly will be damaged.
29. Using wrenches, as shown in Figure 11–18, force the clutch plate and hub assembly onto the compressor shaft. Remove the tool from time to time to ensure that the key is still in place in the keyway. The key should be even with or slightly above the clutch hub when the hub is fully seated.
30. Using a nonmagnetic feeler gauge, check the air gap. The air gap should be between 0.015 inch (0.38 mm) and 0.025 inch (0.64 mm).
31. Install the shaft nut and torque it to 8–16 ft.-lb. (11–22 N•m).

PART 3A
REPLACING THE A-6 SEAL

Careful handling of the seal parts, as shown in Figure 11–19, is important. Neither the seal face nor the seal seat should be touched with the fingers because of the etching effect of the acid normally found on the fingers.

Procedure

1. Using a 9/16-inch thin wall socket wrench and

clutch hub holding tool, remove the shaft nut; refer to Figure 11–20.
2. Using snap ring pliers, remove the clutch hub **retainer ring.**
3. Remove the spacer under the retainer ring.
4. Remove the plate assembly using a clutch hub and drive plate puller, as shown in Figure 11–21.
5. Remove the **shaft seal** seat retainer ring using internal snap ring pliers, as in Figure 11–22.
6. Remove the seal seat using a shaft seal seat remover, as shown in Figure 11–23. (Use the appropriate tool.)
7. Remove the shaft seal using the shaft seal remover, as illustrated in, Figure 11–24.
8. Remove the shaft seal seat O-ring using an O-ring remover (a wire with a hook on the end), as shown in Figure 11–25.

NOTE: Take care not to scratch the mating surfaces.

9. Ensure that the inner bore of the compressor is free of all foreign matter. Flush the area with clean refrigeration mineral oil.
10. Wet the seal seat O-ring with refrigerant mineral oil or O-ring lubricant and place it on the seal installer tool; slide the O-ring into place, then remove the tool.
11. Coat the shaft seal liberally with refrigeration mineral oil and place it on the shaft seal installer tool, as illustrated in Figure 11–26. Slide the shaft seal into place in the bore. Rotate the seal clockwise

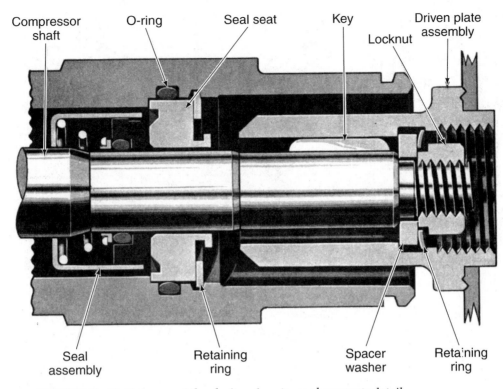

Compressor shaft   O-ring   Seal seat   Key   Locknut   Driven plate assembly

Seal assembly   Retaining ring   Spacer washer   Retaining ring

FIGURE 11–19 Cutaway of the shaft seal cavity to show parts detail.

(cw) until it seats on the flats provided. Rotate the tool counterclockwise (ccw) and remove it.

12. Wet the shaft seal seat with refrigerant mineral oil and place it on the remover/installer tool. Slide the shaft seal seat into position and remove the tool.

13. Install the shaft seal seat snap ring. Note that the beveled edge of the snap ring must face the outside of the compressor.

14. Before replacing the clutch hub and drive plate, the seal should be checked for leaks, steps 15 through 19.

15. Install the test fitting and connect the manifold and gauge set to the test ports.

16. Connect the center service hose to a refrigerant source.

17. Pressurize the compressor to 50 psig (344.8 kPa).

18. With a leak detector, check the shaft seal area for escaping refrigerant.

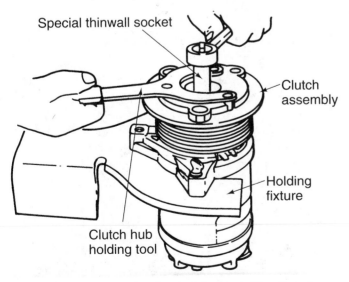

Special thinwall socket

Clutch assembly

Holding fixture

Clutch hub holding tool

FIGURE 11–20 Removing the shaft locknut. *(Courtesy of General Motors Corporation, Service Operations)*

Clutch plate and hub assembly remover

Clutch plate and hub assembly

FIGURE 11–21 Removing the hub and drive plate assembly. *(Courtesy of General Motors Corporation, Service Operations)*

FIGURE 11–22 Removing the shaft seal retainer. (*Courtesy of General Motors Corporation, Service Operations*)

FIGURE 11–23 Removing the seal seat. (*Courtesy of General Motors Corporation, Service Operations*)

FIGURE 11–24 Removing the shaft seal assembly. (*Courtesy of General Motors Corporation, Service Operations*)

19. If only a small leak is detected, rotate the crankshaft a few turns to seat the seal, then recheck the seal area for leaks. If the leak is heavy, or if it persists, the seal must be removed and checked for defects.
20. Place the drive key into the crankshaft keyway, as illustrated in Figure 11–27. Allow about 3/16 inch (4.8 mm) of the key to protrude over the end of the keyway.
21. Align the key with the keyways of the drive plate, then slide the drive plate into position.
22. Using a hub and drive plate installer, press the

plate on the crankshaft, as shown in Figure 11–28.

NOTE: Take care not to force the drive key into the shaft seal. Occasional rotation of the drive plate during assembly ensures that it is seated properly.

23. A clearance of 0.030 inch ± 0.010 inch (0.76 mm ± 0.25 mm) should exist between the drive plate and the rotor, as illustrated in Figure 11–29.
24. Replace the spacer and clutch hub retainer ring. The

FIGURE 11–25 Removing the seal seat O–ring. (*Courtesy of General Motors Corporation, Service Operations*)

FIGURE 11–26 Installing the seal seat O–ring and the shaft seal. (*Courtesy of General Motors Corporation, Service Operations*)

FIGURE 11–27 Drive plate key installed in keyway. *(Courtesy of General Motors Corporation, Service Operations)*

Clearance 0.030" (0.762 mm)

FIGURE 11–29 Checking the air gap. *(Courtesy of General Motors Corporation, Service Operations)*

Special tools

FIGURE 11–28 Installing the driven plate. *(Courtesy of General Motors Corporation, Service Operations)*

FIGURE 11–30 Secure the front head in the holding fixture. *(Courtesy of BET, Inc.)*

retainer ring is installed using the snap ring pliers.
25. Replace the shaft nut.

PART 3B
REPLACE THE SHAFT SEAL, ALL EXCEPT A-6

Procedure

1. Secure the front head in the holding fixture, as shown in Figure 11–30.
2. Carefully insert the cylinder and shaft assembly into the front head, with the shaft end down.
3. Lubricate the center O-ring with clean mineral refrigeration oil or **O-ring** lubricant and position it at the center cylinder O-ring groove.
4. With the discharge crossover O-ring in place (see step 5), insert a 5/16-inch (7.9 mm) drill rod through the discharge crossover passage in both cylinder halves.
5. Using both hands, carefully press the cylinder halves together.

6. Remove the drill rod.
7. Reassemble the rear head, valve plates, **gasket,** and O-ring, following the procedures given in Part 4.
8. Install the clutch coil and the clutch assembly as outlined in Part 2B.

PART 4
REAR HEAD SERVICE, EXCEPT A-6

Procedure

1. Drain the lubricant from the compressor into a graduated container. Note the quantity drained and discard it.
2. Remove the clutch and coil assembly as outlined in Part 2B.
3. Mark the location and note the alignment of the front head, cylinder assembly, and rear head, as shown in Figure 11–31.

Alignment marks

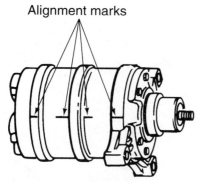

FIGURE 11–31 Mark the location of the front head, cylinder assembly, and rear head. *(Courtesy of General Motors Corporation, Service Operations)*

NOTE: This is important to ensure proper reassembly.

4. Remove the six compressor through-bolts and gaskets. Discard the gaskets.
5. With a plastic hammer and wooden block, tap around the edge of the rear head to disengage it from the cylinder assembly, as in Figure 11–32.
6. Separate the rear head, head gasket, valve plates, and O-ring. Discard the gasket and O-ring.
7. Inspect the head and valve plates. Discard any that are defective.
8. Secure the front head and cylinder assembly in the holding fixture, as illustrated in Figure 11–33.
9. Install two guide pins in the front head and cylinder assembly, as shown in Figure 11–34. Insert the guide pins with the small diameter end *up*, as shown.
10. Liberally lubricate a new O-ring with clean mineral refrigeration oil or O-ring lubricant. Install the O-ring in the rear cylinder O-ring groove.
11. Install the suction reed valve plate over the guide pins.
12. Install the discharge valve plate over the guide pins.

NOTE: Check for proper position of the valve plates.

13. Install the rear head gasket over the guide pins.

NOTE: Be sure it is in the proper position.

14. Carefully install the rear head over the guide pins. The alignment mark on the rear head should match the alignment mark on the cylinder assembly.
15. Using both hands, press down on the rear head to force it over the O-ring. Recheck the alignment marks.
16. Remove the compressor from the holding fixture and place it on the workbench.

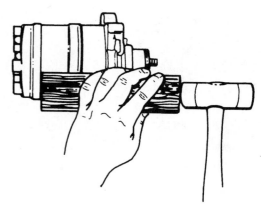

FIGURE 11–32 Disengage the rear head from the cylinder assembly. *(Courtesy of General Motors Corporation, Service Operations)*

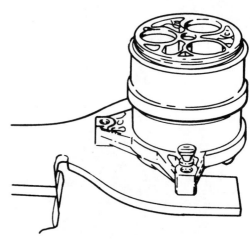

FIGURE 11–33 Secure the front head and cylinder assembly in the holding fixture. *(Courtesy of General Motors Corporation, Service Operations)*

12 o'clock

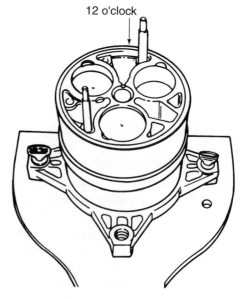

FIGURE 11–34 Install two guide pins in the front head and cylinder assembly. *(Courtesy of General Motors Corporation, Service Operations)*

17. Install new through-bolt gaskets on the six through-bolts.
18. Install four bolts into the compressor. After all four bolts have been threaded into the rear head, remove the two guide pins.
19. Install the other two through-bolts and alternately torque all six to 72–84 in.-lb. (8–10 N•m).
20. Replace the same quantity of lubricant as was removed. Use only 525 viscosity mineral oil for CFC-12 service or PAG lubricant for HFC-134a service.
21. Replace the clutch and coil assembly as outlined in Part 2B.

PART 5
FRONT HEAD SERVICE, EXCEPT A-6

Procedure

1. Drain the compressor lubricant into a **graduated container.** Note the quantity of lubricant removed, then discard it.
2. Remove the clutch and coil assembly as outlined in Part 2B.
3. Remove the compressor shaft seal assembly, as outlined in Part 3B.
4. Mark the location and note the alignment of the front head, cylinder assembly, and rear head (refer to Figure 11–31).

   NOTE: This step is important to ensure proper reassembly.

5. Remove the six through-bolts and discard the brass **washers.**
6. Using a plastic hammer, tap the front head to disengage it from the cylinder assembly.
7. Remove the front head, valve plates, O-ring, and head gasket. Discard the O-ring and head gasket.
8. Inspect the head and valve plates. Replace any that are found to be defective.
9. Rest the rear head and cylinder assembly on the support block, as shown in Figure 11–35.
10. Install two guide pins into the rear head and cylinder assembly. Insert the guide pins with the small diameter down, as shown.
11. Liberally lubricate the new O-ring with clean mineral oil or O-ring lubricant and install the O-ring in the front cylinder O-ring groove.
12. Install the suction reed valve plate over the guide pins.
13. Install the discharge valve plate over the guide pins. Check for proper position of the valve plates before proceeding.

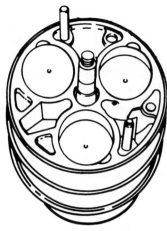

FIGURE 11–35 Rest the rear head and cylinder assembly on the support block. (*Courtesy of General Motors Corporation, Service Operations*)

14. Install the front head gasket over the guide pins.
15. Note the position of the alignment marks and carefully install the front head over the guide pins.
16. Using both hands, press down on the front head to force it over the O-ring. Recheck the alignment marks to ensure proper alignment of the compressor assembly.
17. Place new brass washers on the six through-bolts.
18. Install four of the through-bolts into the compressor.
19. After all four bolts have been threaded into the rear head, remove the two guide pins.
20. Install the other two through-bolts and alternately torque all six bolts to 72–84 in.-lb. (8–10 N•m).
21. Install the new shaft seal as outlined in Part 3.
22. Replace the same quantity of lubricant as was removed. Use mineral oil for CFC-12 service or PAG lubricant for HFC-134a service.
23. Replace the clutch and coil assembly as outlined in Part 1.

PART 6
COMPRESSOR CYLINDER ASSEMBLY, EXCEPT A-6

Procedure

1. Follow steps 1 through 6 of Part 4 to remove the rear head assembly.
2. Follow steps 3, 6, and 7 of Part 5 to remove the front head assembly.

   NOTE: If removal was to replace the assembly center seal, refer to Part 7 for procedure and reassembly. If removal was to replace the cylinder assembly, proceed with step 3.

3. Set the rear head on the support block and insert two guide pins, as shown in Figure 11–36. Install the rear head gasket in the proper position.
4. Install the rear discharge valve plate over the guide pins. Check for proper positioning of the valve plate.
5. Install the rear suction valve plate over the guide pins. Check to ensure the proper positioning of the valve plates.
6. Lubricate the rear head O-ring liberally with clean mineral oil or O-ring lubricant and install the O-ring in the rear cylinder O-ring groove.
7. Carefully lower the cylinder assembly over the guide pins to the rear head.
8. Using both hands, press the cylinder and shaft assembly down into the rear head.
9. Follow steps 11 through 23 of Part 5 to replace the front head assembly.

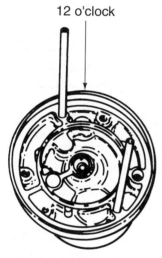

12 o'clock

FIGURE 11–36 Set the rear head on the support block with two guide pins. *(Courtesy of General Motors Corporation, Service Operations)*

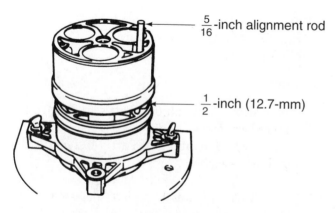

$\frac{5}{16}$-inch alignment rod

$\frac{1}{2}$-inch (12.7-mm)

FIGURE 11–37 Separate the two cylinder halves no more than 1/2 inch (12.7 mm). *(Courtesy of General Motors, Service Operations)*

PART 7
CENTER CYLINDER SEAL, EXCEPT A-6
Procedure

1. Follow steps 1 through 6 of Part 4 to remove the rear head assembly.
2. Follow steps 3, 5, and 6 of Part 5 to remove the front head assembly.
3. Using a wooden block and plastic hammer, tap around the rear cylinder half to separate the two cylinder sections, as shown in Figure 11–37. Do not separate the sections more than 0.5 in. (12.7 mm).
4. Remove and discard the center cylinder assembly O-ring.
5. Check to be sure that the small O-ring between the two cylinder halves is in place. It may stick to the front half or be in the rear half recess. It need not be replaced unless it is missing.

PART 8
REBUILD THE HARRISON A-6 COMPRESSOR

Rebuilding the General Motors Harrison Division A-6 compressor, shown in Figure 11–38, is considered a major service operation. The compressor must be removed from the vehicle and placed on a clean workbench, preferably one covered with a piece of clean white paper.

The following procedures are based on the use of the proper service tools and on the condition that an adequate stock of service parts is on hand.

Procedure

1. Mount the compressor in a holding fixture and secure the fixture in a vise.
2. Using a drive plate holding tool and the 9/16-inch thin wall socket, remove the locknut from the shaft.
3. Using the snap ring pliers, remove the clutch hub retaining ring. Remove the spacer under the ring.
4. Remove the hub and drive plate with the hub and drive plate remover tool, as shown in Figure 11–39.
5. Using the snap ring pliers, remove the pulley and bearing snap ring retainer, as illustrated in Figure 11–40.
6. Place a puller pilot over the crankshaft. Remove the pulley using a pulley puller, as shown in Figure 11–41.

NOTE: The puller pilot must be in place. Placing the puller against the crankshaft will damage the internal assembly.

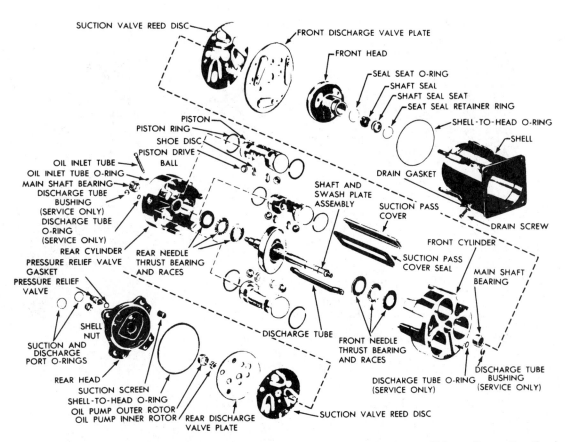

FIGURE 11–38 Exploded view of General Motors six–cylinder compressor. *(Courtesy of General Motors Corporation, Service Operations)*

7. If the pulley bearing is to be replaced, use a sharp tool, such as a small screwdriver, to remove the wire retaining ring.

8. From the rear of the pulley, press or drive the bearing out.

9. Scribe the location of the coil housing in relation to the compressor body to ensure proper alignment during reassembly.

10. Using snap ring pliers, remove the coil housing retainer ring, as shown in Figure 11–42.

FIGURE 11–39 Removing the hub and drive plate assembly. *(Courtesy of General Motors Corporation, Service Operations)*

FIGURE 11–40 Removing the pulley retaining ring. *(Courtesy of General Motors Corporation, Service Operations)*

FIGURE 11–41 Removing the pulley. (*Courtesy of General Motors Corporation, Service Operations*)

11. Lift off the coil housing assembly.
12. Follow the procedure outlined in Part 3A.
13. Remove the oil sump plug. Remove the compressor from the holding fixture and drain the lubricant from the compressor into a graduated container.
14. After noting the quantity of lubricant drained, discard the lubricant.
15. Return the compressor to the holding fixture with the rear head up.
16. Remove the four nuts from the shell studs. Remove the rear head, as shown in Figure 11–43. *Take* care not to scratch or nick the Teflon® surface.
17. Remove the oil pump drive and driven gears. Set the gears aside in their original position to ensure proper assembly.
18. Remove the rear head-to-shell O-ring and discard.
19. Using two screwdrivers, as shown in Figure 11–44, carefully pry up the rear discharge valve plate assembly. Lift out the valve plate assembly.
20. Carefully lift out the suction reed valve.
21. Remove the oil inlet tube and O-ring with the oil inlet tube remover, as in Figure 11–45. *To avoid damage to the tube and shell, do not omit this step.*
22. Carefully remove the compressor assembly from the holding fixture and lay it on its side.
23. Gently tap the front head casting with a soft hammer to slide the internal assembly and head out of the shell.

NOTE: Do not attempt to remove the internal assembly without removing the front head with it to avoid serious damage to the Teflon surface.

24. Place the internal assembly and the front head on the support block, as shown in Figure 11–46.
25. Carefully remove the front head. *Use extreme caution to prevent damage to the Teflon-coated surface of the*

FIGURE 11–42 Removing the coil housing retainer ring. (*Courtesy of General Motors Corporation, Service Operations*)

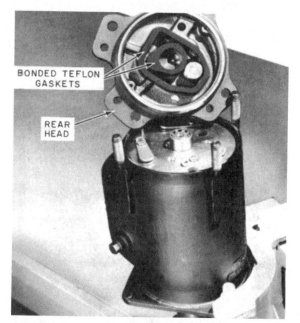

FIGURE 11–43 Rear head removal. (*Courtesy of General Motors Corporation, Service Operations*)

front head.

26. Remove the discharge and suction valve plates from the internal assembly.
27. Remove and discard the front O-ring gasket.
28. Replace the six-cylinder internal assembly.
29. Examine the front and rear discharge and suction valve plates for damaged or broken valves. Replace any damaged valves.

FIGURE 11–44 Removing the rear discharge valve plate. *(Courtesy of General Motors Corporation)*

FIGURE 11–45 Removing the oil inlet tube and O–ring. *(Courtesy of General Motors Corporation, Service Operations)*

FIGURE 11–46 Remove the internal assembly (held on the support block) with the front head. *(Courtesy of General Motors Corporation, Service Operations)*

30. Examine the Teflon surfaces on the front and rear heads. The heads must be replaced if damaged, nicked, or scratched.
31. Examine the suction screen in the rear head. Clean or replace the screen if it is clogged or damaged.
32. Examine the oil pump gears. Replace both gears if either gear shows signs of damage or wear.
33. Inspect the seal and seal seat for damage of any kind.
34. Replace the seal assembly if damage is noted. A new seal is suggested when the compressor is rebuilt.
35. Check the coil for loose connections or cracked insulation. If the coil is checked with an ammeter, the reading should be no more than 3.2 amperes at 12 volts.
36. Check the pulley and bearing assembly.

NOTE: If the frictional surface of the pulley shows signs of warpage due to excessive heat, then the pulley should be replaced. Slight scoring, as shown in Figure 11–47, is normal. The assembly should be replaced if it is heavily scored.

37. Check the pulley bearing for signs of excessive noise or looseness. Replace if necessary.
38. Inspect the hub and drive plate. Refer to the note following step 36.
39. Place the internal mechanism, oil pump end down, in the support block.
40. Place the suction valve plate and then the discharge valve plate in place over the dowel pins.
41. Carefully locate the front head in place over the dowel pins, as illustrated in Figure 11–48. *Take care not to damage the Teflon surface.*
42. Place the front O-ring in position. This O-ring is properly located between the head and discharge valve.
43. Locate the lubricant pickup tube hole with the center of the shell, as shown in Figure 11–49.
44. Slide the shell over the internal assembly.
45. Gently tap the shell in place with a soft hammer. *Do not pinch or distort the O-ring.*
46. Hold the internal mechanism securely in the shell and invert the assembly. Place the assembly with the front end down in the holding fixture. Secure the fixture in a vise.

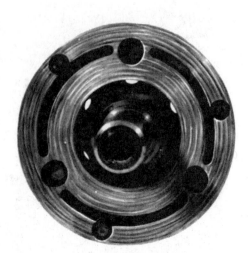

FIGURE 11–47 Scoring of the drive and driven plates is normal. Do not replace for this condition. *(Courtesy of General Motors Corporation, Service Operations)*

47. Ensure that the internal mechanism is in the proper position and drop the lubricant pickup tube in place, as shown in Figure 11–50. Some units are equipped with an O-ring. If so equipped, make sure the O-ring is in place.

48. Install the suction valve plate and the discharge valve plate in their proper positions.

49. Install the oil pump gears. Install the rear O-ring gasket.

50. Position the rear head casting so it is aligned with the dowel pins, as in Figure 11–51. *Note the position in which the outer oil pump gear must be placed to prevent damage to the Teflon surface of the head.*

51. Position the oil pump and slide the rear head into place.

52. Install the four hex nuts and torque them to 19–23 ft.-lb. (25.7–31.1 N•m).

53. Follow the procedures outlined in Part 3A to install the seal assembly.

54. Leak test the compressor.

55. Note the original position of the coil housing by the scribe marks.

56. Slip the coil housing into place.

57. Replace the snap ring to secure the housing.

58. Press the new bearing into the pulley, as in Figure 11–52, and replace the wire retaining ring (if bearing replacement is necessary).

59. Using the proper tool, shown in Figure 11–53, press or drive the pulley and bearing assembly on the compressor front head.

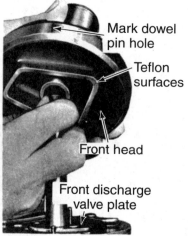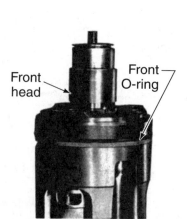

FIGURE 11–48 Installing the front reed valve, head, and O–ring. *(Courtesy of General Motors Corporation, Service Operations)*

60. Install the retainer snap ring.
61. Follow the procedures outlined in Part 2A.
62. Turn the clutch hub by hand to ensure that it (or the internal assembly) is not dragging or binding.

63. Spin the pulley to ensure that it is not dragging or binding. It should turn freely.
64. Refill the compressor with 525 viscosity mineral oil for CFC-12 service or PAG lubricant for HFC-134a service.

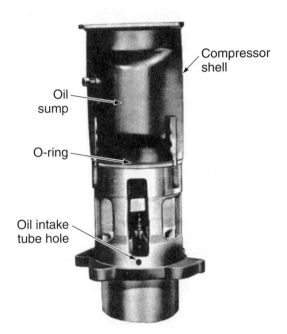

FIGURE 11–49 Replace the shell over the internal assembly. *(Courtesy of General Motors Corporation, Service Operations)*

FIGURE 11–52 Installing the pulley and drive plate bearing. *(Courtesy of General Motors Corporation, Service Operations)*

FIGURE 11–50 Installing the oil intake (pickup) tube. *(Courtesy of General Motors Corporation, Service Operations)*

FIGURE 11–51 Installing the rear head. *(Courtesy of General Motors Corporation, Service Operations)*

FIGURE 11–53 Installing the pulley and drive plate on the compressor. *(Courtesy of General Motors Corporation, Service Operations)*

SERVICING THE HARRISON R4 COMPRESSOR

This service procedure has three parts: Part 1 for servicing the seal assembly, Part 2 for checking and adding lubricant, and Part 3 for servicing the clutch.

PART 1
REPLACING THE SEAL

Procedure

1. Remove the shaft nut using a 9/16-inch thinwall socket wrench and clutch hub holding tool, as shown in Figure 11–54.
2. Remove the plate using a clutch hub and drive plate puller, as illustrated in Figure 11–55.

   NOTE: If the shaft key is not removed in step 2, remove the shaft key.

3. Remove the shaft seal seat retainer ring using the snap ring pliers.
4. Remove the seal seat using a shaft seal seat remover.

FIGURE 11–54 Using a 9/16–inch thinwall socket and hub holding tool, remove the shaft nut.

FIGURE 11–55 Using the clutch hub and drive plate puller, remove the hub and drive plate. If the shaft key did not come out, remove the key.

5. Remove the shaft seal using the shaft seal remover, as shown in Figure 11–56.
6. Remove the shaft seal seat O-ring using an O-ring remover. Take care not to scratch the mating surfaces.
7. Ensure that the inner bore of the compressor is free of all foreign matter. Flush the area with clean mineral oil.
8. Place the seal seat O-ring on the installer tool and slide the O-ring into place, as illustrated in Figure 11–57. Then, remove the tool.
9. Coat the shaft seal liberally with mineral oil and place it on the shaft seal installer tool.
10. Slide the tool into place in the bore. Rotate the seal clockwise (cw) until it seats on the flats provided. Rotate the tool counterclockwise (ccw) and remove it.
11. Place the shaft seal seat on the remover/installer tool, as shown in Figure 11–58.
12. Slide the shaft seal seat into position and remove the tool.
13. Install the shaft seal seat snap ring.

    NOTE: The beveled edge of the snap ring must face the outside of the compressor.

14. Check the seal for leaks before replacing the clutch hub and drive plate.
    a. Install the test fitting on the compressor.
    b. Connect the manifold and gauge set to the test ports.
    c. Connect the center hose to a refrigerant source.
    d. Briefly, open the high- and low-side manifold hand valves to pressurize the compressor.
15. Use a leak detector to check the shaft seal area for escaping refrigerant.
    a. If a small leak is detected, rotate the crankshaft a few turns by hand to seat the seal; then recheck the seal area for leaks.

FIGURE 11–56 Using an O–ring remover, remove the shaft seal seat O-ring. Take care not to scratch the mating surfaces.

FIGURE 11–57 Place the seal seat O–ring on the installer tool and slide the O–ring into place. Remove the tool.

b. If a heavy leak is detected, or if it persists, the seal must be removed and checked for defects.

16. Place the drive key into the clutch plate keyway.

NOTE: About 3/16 inch (4.8 mm) of the key should be allowed to protrude over the end of the keyway (see Figure 11–72).

17. Align the key with the keyways of the drive plate and compressor crankshaft. Then slide the drive plate into position.
18. Using a hub and drive plate installer, press this part on the crankshaft.

NOTE: Take care not to force the drive key into the shaft seal. Occasional rotation of the drive plate during assembly ensures that it is seated properly.

19. Using a nonmagnetic feeler gauge, check the drive plate to rotor clearance.

NOTE: Clearance should be 0.030 inch ±0.010 inch (0.76 mm, ±0.25 mm).

20. Replace the shaft nut.

PART 2
CHECKING AND ADDING LUBRICANT

Harrison R4 compressors are factory charged with 5.5–6.5 fluid ounces (163–192 mL) of lubricant.

If the compressor shows signs that foreign matter is present or that the lubricant contains chips or metallic particles when it is removed and drained, then the system should be flushed. The receiver-drier, desiccant, or accumulator-drier (as applicable) should be replaced after the system is flushed. The compressor inlet and/or TXV inlet screens should be cleaned as well.

FIGURE 11–58 Liberally coat the shaft seal with refrigeration oil and place it on the shaft seal installer tool. Slide the shaft into place in the bore. Rotate the seal clockwise until it seats on the flats provided. Rotate the tool counterclockwise and remove it.

If the system is flushed, add a full 6 ounces (177 mL) of clean refrigeration lubricant to the compressor. With the exception of a flushed system, add lubricant as shown in Figure 11–59 to any system that has had the major component replaced.

Procedure

1. Clean the external surface of the compressor so that it is free of oil and grease.
2. Position the compressor with the shaft end up over a graduated container.
3. Drain the compressor. Allow it to drain for at least 10 minutes. Measure and note the amount of lubricant removed, and then discard the old lubricant.

Component	R-4 Compressor
Accumulator-drier	Replace with same amount as drained, plus 3 oz. (88.7 mL). If no luricant was drained, add 2 oz. (59.1 mL) to the new accumulator.
Condenser	1 oz. (29.6 mL)
Evaporator	2 oz. (59.1 mL)
Receiver-drier	1 oz. (29.6 mL)

Disregard any loss of lubricant due to the changing of a line, hose, or muffler unless the component contains a measurable amount of lubricant. If this is the case, then add the same amount of new clean lubricant as was removed from the component.

FIGURE 11–59 Add lubricant for component replacement.

4. Add new lubricant in the same amount as the lubricant drained.

   NOTE: If the replacement compressor is new, drain it as outlined in steps 2 and 3, then add new lubricant in the amount drained from the old compressor.

5. If a major component is also replaced, see the introduction of this procedure for the addition of lubricant for the component.

6. If the loss of refrigerant occurs over an extended period of time, then add 3 fluid ounces (88.71 mL) of new lubricant. Do not exceed a total of 6.5 ounces (192 mL) of lubricant.

PART 3
SERVICING THE COMPRESSOR CLUTCH

Procedure

1. Using the clutch hub holding tool and a 9/16-inch thinwall socket, remove the retaining nut from the compressor shaft.
2. Remove the clutch plate and hub assembly using the clutch plate and hub assembly remover tool.
3. If the shaft key was not removed in step 2, remove the shaft key.
4. Mark the location of the clutch coil terminals to ensure proper reassembly.
5. Remove the rotor and bearing assembly retaining ring using the snap ring pliers, as shown in Figure 11–60.
6. Install the rotor bearing and puller guide over the end of the compressor shaft, as illustrated in Figure

11–61. The guide should seat on the front head of the compressor.
7. Remove the clutch rotor and assembly parts using a puller, as in Figure 11–62
8. Using a cold chisel and hammer, bend the tabs of the six pulley rim mounting screw lockwashers flat; refer to Figure 11–63.
9. Loosen and remove all six screws using a 7/16-inch, 6-point box wrench.

FIGURE 11–61 Positioning the rotor and bearing puller guide. *(Courtesy of General Motors Corporation, Service Operations)*

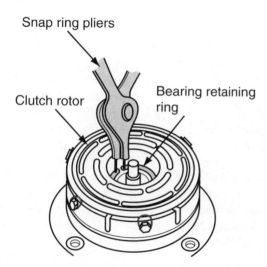

FIGURE 11–60 Remove the bearing and rotor retaining ring.

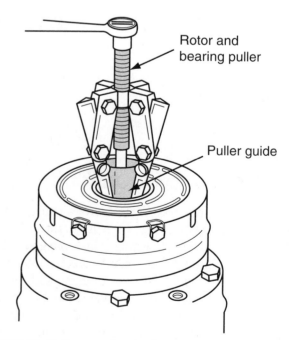

FIGURE 11–62 Remove the clutch rotor.

FIGURE 11–63 Bending the locking tabs to allow removal of the capscrews. (*Courtesy of BET, Inc.*)

FIGURE 11–64 Separating the pulley rim from the rotor. (*Courtesy of BET, Inc.*)

10. Separate the pulley rim from the rotor, as illustrated in Figure 11-64.
11. Visually check the coil for loose connections or cracked insulation.
12. Briefly, connect the coil to a 12-volt battery with an ammeter connected in series. If the coil draws more than 3.2 amperes at 12 volts, then it should be replaced.
13. Inspect the clutch plate and hub assembly. Check for signs of looseness between the plate and hub. See the note following step 14.
14. Check the rotor and bearing assembly.

NOTE: If the frictional surface of the clutch plate or rotor shows signs of warpage due to excessive heat, then that part should be replaced. Slight scoring, as shown in Figure 11–65, is normal; if either assembly is heavily scored, however, it should be replaced.

15. Check the bearing for signs of excessive noise, binding, or looseness. Replace the bearing if necessary.
16. Place the rotor and bearing assembly, split side down, atop two soft wood blocks, as shown in Figure 11–66.

FIGURE 11–65 Scoring of the drive and driven plates is normal. Do not replace for this condition. (*Courtesy of General Motors Corporation, Service Operations*)

17. Using the bearing remover/installer, with a hammer, drive the bearing from the rotor; refer to Figure 11–67. The bearing may also be removed with the arbor press.

18. Turn the rotor over with the frictional surface resting on a block of soft wood.

19. Using the bearing remover/installer, with a hammer, drive the bearing into the rotor. To ensure the alignment of the bearing outer surface into the rotor inner surface, the use of an arbor press, shown in Figure 11–68, is recommended.

NOTE: Ensure that pressure is exerted on the outer bearing race during insertion. If pressure is exerted on the inner bearing race (by either method of insertion), premature failure of the bearing will result.

20. Use a prick punch to stake the bearing into the rotor, as illustrated in Figure 11–69.
21. With the coil in place, join the pulley rim to the rotor.
22. Replace and/or tighten the six retaining screws using a 7/16-inch, 6-point box wrench.
23. Using a cold chisel and hammer, bend the tabs of the six mounting screw lockwashers up against a flat of each of the screws (one tab for each screw), as shown in Figure 11–70.
24. Position the assembly on the front head of the compressor.
25. Using the rotor assembly installer with a universal handle, shown in Figure 11–71, drive the assembly into place. Before the assembly is fully seated, ensure that the coil terminals are in the proper location and the three protrusions on the rear of the coil housing align with the locator holes in the front head.
26. Use snap ring pliers to install the retainer ring.
27. Clean the frictional surfaces of the clutch plate and rotor, if necessary.
28. Insert the key into the slot (keyway) of the hub. Do not insert the key into the compressor crankshaft slot (keyway).

FIGURE 11–66 Place the rotor atop two soft wood blocks. *(Courtesy of BET, Inc.)*

FIGURE 11–67 Driving the old bearing from the rotor. *(Courtesy of BET, Inc)*

FIGURE 11–68 Pressing the new bearing into the rotor. *(Courtesy of BET, Inc.)*

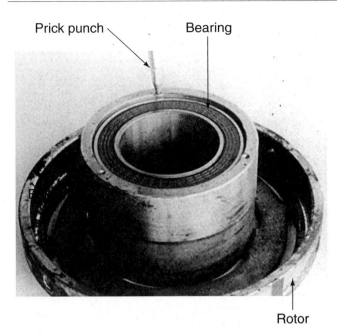

Prick punch   Bearing

Rotor

FIGURE 11–69 Using a prick punch to stake the bearing into the rotor. *(Courtesy of BET, Inc.)*

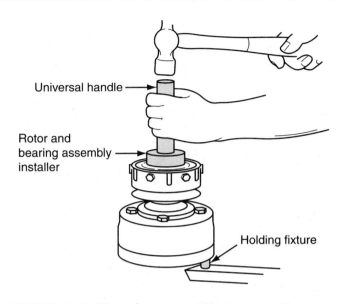

Universal handle

Rotor and bearing assembly installer

Holding fixture

FIGURE 11–71 Using the rotor and bearing assembly installer, drive the clutch rotor assembly into place on the front head of the compressor.

NOTE: The key should protrude about 3/16 inch (4.7 mm) below the hub, as in Figure 11–72.

29. Place the clutch plate and hub assembly onto the compressor shaft by matching the key of the hub to the keyway of the shaft.

30. Using a clutch plate and hub installer, press this part on the crankshaft. Do not hammer it into position.

31. Use a nonmagnetic feeler gauge to ensure an air gap of 0.020–0.040 inch (0.508–1.016 mm) between the frictional surfaces.

32. Replace the shaft nut and torque to 8–12 ft.-lb. (10.8–16.3 N•m).

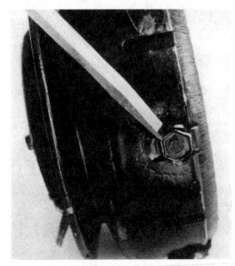

FIGURE 11–70 Bend the tabs against the capscrews to prevent loosening during use. *(Courtesy of BET, Inc.)*

$\frac{3}{16}$ in (4.7mm)

FIGURE 11–72 The key, inserted into the keyway of the hub, should protrude 3/16 inch (4.7 mm). *(Courtesy of BET, Inc.)*

SERVICING THE HARRISON V-5 AND V-7 COMPRESSORS

The Harrison V-5 and V-7 compressors, like many other models, require special service tools. This service procedure, given in five parts, depicts the V-5 compressor. Procedures for the V-7 compressor are similar. Part 1 is for clutch service, Part 2 is for compressor shaft seal service, Part 3 is for rear head service, Part 4 is for front head service, and Part 5 is for **control valve** service.

PART 1
CLUTCH SERVICE

Procedure

1. Clamp the compressor into an appropriate holding fixture.
2. Hold the clutch hub with the clutch holding tool and remove the shaft nut, as illustrated in Figure 11–73.
3. Use the clutch plate and hub installer/remover tool to remove the clutch hub, as shown in Figure 11–74.
4. Remove the shaft key and set it aside for reassembly.
5. Use the snap ring pliers to remove the rotor snap ring, as in Figure 11–75.
6. Use the pulley rotor and bearing guide over the compressor shaft and insert the puller in the rotor slots to remove the rotor, as shown in Figure 11–76.
7. Mark the clutch coil terminal location on the compressor front head for ease in reassembly.
8. Install the puller pilot on the front head.
9. Install the puller and tighten the forcing screw against the pilot to remove the clutch coil, as illustrated in Figure 11–77.

10. Attach the rotor and bearing puller tool (less forcing screw) to the rotor. Place the puller atop a solid flat surface, as illustrated in Figure 11–78.

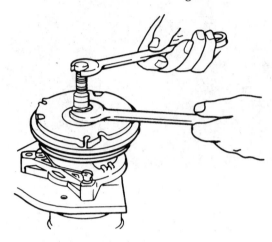

FIGURE 11–74 Remove the clutch hub. (*Courtesy of General Motors Corporation, Service Operations*)

FIGURE 11–75 Use snap ring pliers to remove the snap ring.

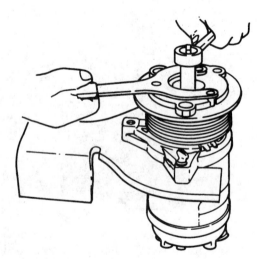

FIGURE 11–73 Remove the shaft nut while holding the clutch hub. (*Courtesy of General Motors Corporation, Service Operations*)

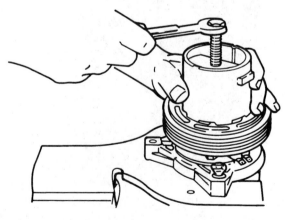

FIGURE 11–76 Remove the rotor using the rotor remover tool. (*Courtesy of General Motors Corporation, Service Operations*)

11. Drive the bearing out of the rotor hub with the rotor bearing remover tool and universal handle.

12. Remove the rotor and bearing puller tool from the rotor.

13. Place the rotor on the support block, as in Figure 11–79, to support the rotor during bearing installation.

14. Align the bearing with the hub bore. Using the puller and bearing installer tool, drive the bearing fully into the hub.

15. Position the bearing staking guide and staking pin tool in the hub bore.

16. Strike the staking pin with a hammer. Form three stakes 120 degrees (120°) apart. The staked metal should not touch the outer race of the bearing. Take care not to damage the bearing during the staking procedure.

17. Remove the staking tools and support block.

18. Place the clutch coil on the front head of the compressor. Note the location of the electrical terminals as marked during removal of the coil.

19. Assemble the clutch coil installer, puller crossbar, and bolts atop the clutch coil, as shown in Figure 11–80.

20. Turn the forcing screw of the crossbar to force the clutch coil onto the front head of the compressor. Make sure the clutch coil and clutch coil installer tool remain in-line during this procedure.

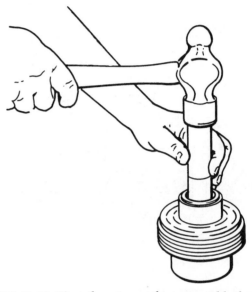

FIGURE 11–79 Place the rotor on the support block and, using the bearing installer tool, drive the bearing fully into the rotor. (*Courtesy of General Motors Corporation, Service Operations*)

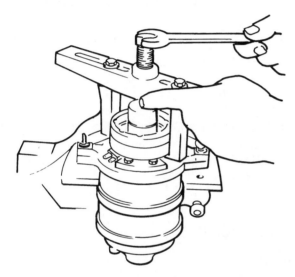

FIGURE 11–77 Remove the clutch coil. (*Courtesy of General Motors Corporation, Service Operations*)

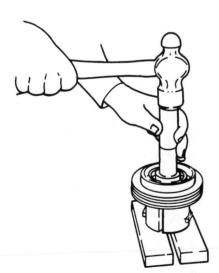

FIGURE 11–78 Place the puller atop a solid flat surface and, using the rotor bearing remover tool, drive the bearing out of the rotor. (*Courtesy of General Motors Corporation, Service Operations*)

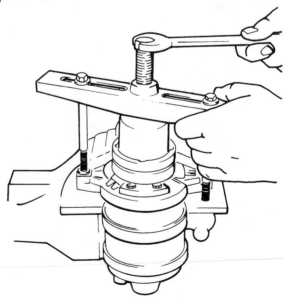

FIGURE 11–80 Assemble the clutch coil installer and puller crossbar atop the clutch coil. (*Courtesy of General Motors Corporation, Service Operations*)

21. After the clutch coil is fully seated, use a 1/8-inch punch and stake the front head at 120-degree (120°) intervals to hold the coil in the proper position.
22. Position the rotor and bearing assembly on the front head.
23. Assemble the puller pilot, crossbar, and bolts atop the rotor and bearing assembly, as shown in Figure 11–81.
24. Tighten the forcing screw to force the rotor and bearing assembly onto the compressor front head.

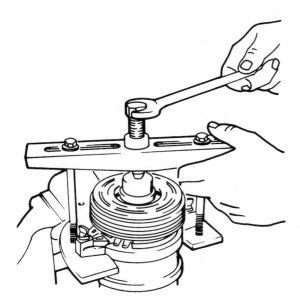

FIGURE 11–81 Assemble the puller pilot, crossbar, and bolts atop the rotor and bearing assembly. *(Courtesy of General Motors Corporation, Service Operations)*

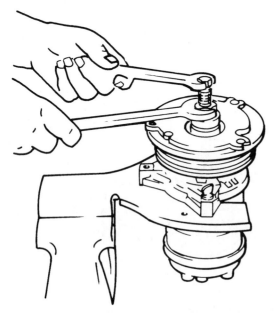

FIGURE 11–82 Install the clutch plate and hub assembly. *(Courtesy of General Motors Corporation, Service Operations)*

Make sure the assembly stays in-line during this procedure.
25. Use the snap ring pliers to install the rotor and bearing assembly retainer snap ring (see Figure 11–75).
26. Install the shaft key into the hub key groove. The key should protrude about 1/8 inch (3.2 mm) out of the keyway.
27. Align the shaft key with the shaft keyway and position the clutch plate and hub assembly onto the compressor shaft.
28. Install the drive plate installer and bearing onto the clutch plate and hub assembly, as shown in Figure 11–82. The forcing tip of the installer must be flat or the end of the shaft/axial plate assembly will be damaged.
29. Using wrenches, as shown, force the clutch plate and hub assembly onto the compressor shaft. Remove the tool from time to time to ensure that the key is still in place in the keyway. The key should be even with or slightly above the clutch hub when the hub is fully seated.
30. Using a nonmagnetic feeler gauge, check the air gap. The air gap should be between 0.015 inch (0.38 mm) and 0.025 inch (0.64 mm).
31. Install and torque the shaft nut to 8–16 ft.-lb. (11–22 N•m).

PART 2
SHAFT SEAL SERVICE

Procedure

1. Remove the clutch hub as outlined in Part 1, steps 1 through 3.

   NOTE: It is not necessary to remove the clutch rotor for seal service.

2. Remove the shaft key and set aside for reassembly.
3. Clean the inside of the seal cavity to prevent any dirt or foreign matter from entering the compressor when the seal is removed.
4. Engage the knurled tangs of the seal seat remover and installer tool into the seal by turning the tool clockwise (cw).
5. With a rotary motion, as illustrated in Figure 11–83, remove the seal from the cavity.
6. Remove the O-ring from the seal cavity using the O-ring remover tool.
7. Clean and check the seal cavity for nicks and/or burrs.
8. Lubricate the new O-ring with clean mineral oil or

FIGURE 11–83 With a rotary motion, remove the seal.

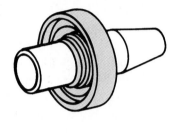

FIGURE 11–84 Install the seal protector in the seal. *(Courtesy of General Motors Corporation, Service Operations)*

O-ring lubricant and attach it to the O-ring installer.

9. Install the O-ring into the seal cavity. The lower recess is for the O-ring seal.

10. Rotate the tool to seat the O-ring into its recess and remove the installer tool.

11. Lubricate the new seal with clean mineral oil and attach it to the seal remover/installer tool.

12. Install the seal protector, as shown in Figure 11–84, in the seal and place over the compressor shaft.

13. Push the seal into place with a rotary motion.

14. Using snap ring pliers, install the new snap ring to retain the seal. Use the sleeve from the remover/installer tool to press on the snap ring until it snaps into its groove.

15. With a clean, lint-free cloth, remove the excess lubricant from the seal cavity.

16. Replace the clutch hub by following the procedures outlined in Part 1, steps 26 through 31.

PART 3
REAR HEAD SERVICE

Procedure

1. Drain the lubricant from the compressor into a graduated container. Note the quantity drained and discard the lubricant.

2. Remove the clutch and coil assembly, as outlined in Part 1.

3. Mark the location, as in Figure 11–85, and note the alignment of the front head, cylinder assembly, and rear head. This is important to ensure proper reassembly.

4. Remove the six compressor through-bolts and gaskets. Discard the gaskets.

5. With a plastic hammer and wooden block, tap around the edge of the rear head to disengage it from the cylinder assembly, as shown in Figure 11–86.

6. Separate the rear head, head gasket, valve plates, and O-ring. Discard the gasket and O-ring.

7. Inspect the head and valve plates. Discard any that are defective.

8. Place the rear head on a clean, flat surface with the control valve facing the 6 o'clock position, as illustrated in Figure 11–87.

9. Insert the guide pins in the mounting holes at the 5 and 11 o'clock positions, small end facing down.

10. Install the discharge valve plate over the guide pins.

NOTE: The elongated hole should be at the upper left (11 o'clock) guide pin, as in Figure 11–88.

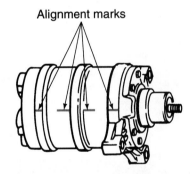

FIGURE 11–85 Mark the location of the front head, cylinder assembly, and rear head to ensure proper reassembly. *(Courtesy of General Motors Corporation, Service Operations)*

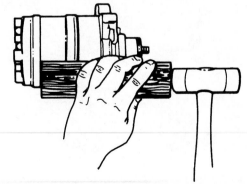

FIGURE 11–86 Disengage the rear head from the cylinder assembly. *(Courtesy of General Motors Corporation, Service Operations)*

FIGURE 11–87 Place the rear head on a flat surface with the control valve facing 6 o'clock. *(Courtesy of BET, Inc.)*

FIGURE 11–88 Insert the assembly guide pin at the 11 o'clock position. *(Courtesy of BET, Inc.)*

11. Install the suction valve plate over the guide pins (see note for step 10).

12. Remove the 5 o'clock guide pin before proceeding.

13. Lubricate the new O-ring with clean mineral oil or O-ring lubricant and install it in the cylinder O-ring groove.

14. Carefully install the front head and cylinder assembly over the guide pin.

15. Using both hands, press the front head and cylinder assembly down and into the rear head.

16. Add the new gaskets to the six through-bolts. Install five bolts into the assembly.

17. Ensure that three or four through-bolts are securely screwed into the rear head. Then, remove the 11 o'clock guide pin.

18. Insert the other through-bolt and torque all bolts to 72–84 in.-lb. (8–10 N•m).

19. Leak check the assembly. Attach the pressure test fitting and pressurize the compressor to 50 psig (344.8 kPa) with refrigerant. Check the front and rear O-rings, shaft seal, and through-bolt gaskets.

20. Depressurize the compressor and remove the testing connector.

21. Install the clutch coil following the procedures outlined in Part 1, steps 18 through 21.

22. Replace the lubricant with the same quantity as was removed.

PART 4
FRONT HEAD SERVICE

Procedure

1. Drain the compressor lubricant into a graduated container. Note the quantity and discard the lubricant.

2. Remove the clutch and coil assembly as outlined in Part 1.

3. Remove the compressor shaft seal assembly as outlined in Part 2.

4. Mark the location and note the alignment of the front head, cylinder assembly, and rear head. (See Figure 11–85.) This step is important to ensure proper reassembly.

5. Remove the six through-bolts and discard the washers.

6. Using a plastic hammer, tap the front head to disengage it from the cylinder assembly.

7. Remove the front head, valve plates, O-ring, and head gasket. Discard the O-ring and head gasket.

8. Inspect the head and valve plates. Discard any that are found to be defective.

9. Rest the rear head and cylinder assembly on the support block.
10. Install two guide pins into the rear head and cylinder assembly. Insert the guide pins with the small diameter down.
11. Liberally lubricate the new O-ring with clean mineral oil or O-ring lubricant and install the O-ring in the front cylinder O-ring groove.
12. Install the suction reed valve plate over the guide pins.
13. Install the discharge valve plate over the guide pins. Check for proper position of valve plates before proceeding.
14. Install the front head gasket over the guide pins.
15. Note the position of the alignment marks and carefully install the front head over the guide pins.
16. Using both hands, press down on the front head to force it over the O-ring. Recheck the alignment marks to ensure proper alignment of the compressor assembly.
17. Install new gaskets on all six through-bolts.
18. Install four through-bolts into the compressor. After all four bolts have been threaded into the rear head, remove the two guide pins.
19. Install the other two through-bolts and alternately torque all six bolts to 72–84 in.-lb. (8–10 N•m).
20. Install the new shaft seal as outlined in Part 2.
21. Replace the same quantity and type of lubricant as was removed.
22. Replace the clutch and coil assembly as outlined in Part 1.

PART 5
CONTROL VALVE SERVICE
Procedure

1. Use snap ring pliers to remove the control valve retaining ring.
2. Remove the control valve assembly.
3. Lubricate the O-ring(s) with clean mineral oil or O-ring lubricant.
4. Use thumb pressure to push the control valve into the compressor.
5. Use the snap ring pliers to install the snap ring. Be sure the snap ring is properly seated in the ring groove.

SERVICING THE NIPPONDENSO SIX-CYLINDER COMPRESSOR

Servicing the Nippondenso six-cylinder compressor, shown in Figure 11–89, is limited to the following: shaft seal, covered in Part 1; checking and adding lubricant, covered in Part 2; servicing the clutch, covered in Part 3; and rebuilding the compressor, covered in Part 4.

PART 1
REPLACING THE SHAFT SEAL

Seal replacement for the Nippondenso compressor is somewhat different from most other compressors in that the front head assembly must first be removed.

Procedure

1. Remove the clutch and coil assemblies as outlined in Part 3.
2. Remove the shaft key, Figure 11–90, using the shaft key remover tool.
3. Remove the felt lubricant absorber and retainer from the front head cavity.
4. Clean the outside of the compressor with pure mineral spirits and air dry.

   NOTE: Do not submerge the compressor into mineral spirits.

5. Drain the compressor lubricant into a graduated measure as outlined in Part 2.
6. Remove the six through-bolts from the front head. Use the proper tool; some require a 10 mm socket and others require a 6 mm Allen wrench.
7. Discard the six brass washers, if equipped, and retain the six bolts.
8. Gently tap the front head with a plastic hammer to free it from the compressor housing.
9. Remove and discard the head-to-housing O-ring and the head-to-valve plate gasket.
10. Place the front head on a piece of soft material, such as cardboard, with the cavity side up.
11. Use the shaft seal seat remover to remove the seal seat, as illustrated in Figure 11–91.

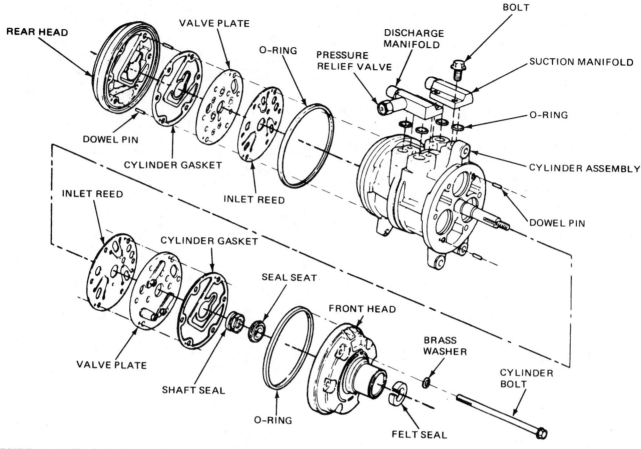

FIGURE 11–89 Exploded view of Nippondenso compressor. (*Courtesy of Ford Motor Company*)

12. Remove the shaft seal cartridge, using both hands, as shown in Figure 11–92.
13. Liberally coat all seal parts, compressor shaft, head cavity, and gaskets with clean refrigeration lubricant.
14. Carefully install the shaft seal cartridge, as shown in Figure 11–93, making sure to index the shaft seal on the crankshaft slots.
15. Install the seal seat into the front head using the

seal seat installer, as shown in Figure 11–94.
16. Install the head-to-valve plate gasket over the **alignment pins** in the compressor housing.
17. Install the head-to-housing O-ring.
18. Carefully slide the head onto the compressor housing, ensuring that the alignment pins engage in the head.

FIGURE 11–90 After removing the clutch and coil assemblies, use the shaft key remover (not shown) to remove the shaft key.

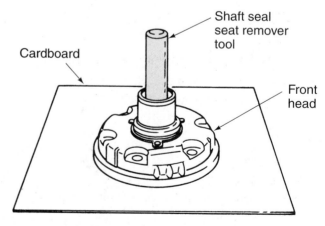

FIGURE 11–91 Using the shaft seal seat remover tool to remove the seal seat from the front head. (*Courtesy of Ford Motor Company*)

19. Using six new brass washers, if required, install the six compressor through-bolts.
20. Using a 10 mm socket or 6 mm Allen wrench, as required; tighten the bolts to a 260 in.-lb. (29.4 N•m) torque.

FIGURE 11–92 Remove the shaft seal.

FIGURE 11–93 Liberally coat all seal parts, compressor shaft, head cavity, and gaskets with clean refrigeration oil. Carefully install the shaft seal cartridge, making sure to index the shaft seal on the crankshaft slots.

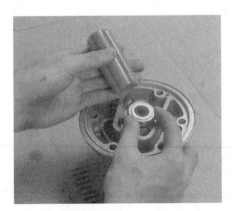

FIGURE 11–94 Install the seal seat into the front head using the seal seat installer.

NOTE: Use an alternate pattern, as illustrated in Figure 11–95, when torquing the bolts.

21. Replace the lubricant with clean mineral oil for a CFC-12 system or PAG lubricant for an HFC-134a system, as outlined in Part 2.
22. Install the crankshaft key using a drift.
23. Align the ends of the felt and its retainer, as in Figure 11–96, and install them into the head cavity. Be sure the felt and retainer are fully seated against the seal plate.
24. Replace the clutch and coil assemblies, as outlined in Part 3.

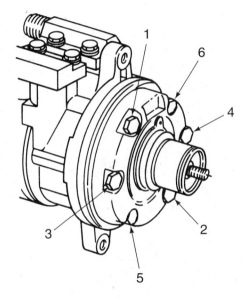

FIGURE 11–95 Use an alternate pattern when torquing bolts. Torque sequence for tightening bolts is shown. *(Courtesy of Ford Motor Company)*

FIGURE 11–96 Align openings in the felt and retainer. *(Courtesy of Ford Motor Company)*

PART 2
CHECKING AND/OR ADDING LUBRICANT

The Nippondenso compressor is factory charged with 13 ounces (384 mL) of 500 Saybolt universal viscosity (SUV) refrigeration lubricant. It is not recommended that the lubricant level be routinely checked unless there is evidence of a severe loss.

The following procedure assumes that the suction and discharge service valves have been removed from the compressor.

Procedure

1. Drain the compressor lubricant through the suction and discharge service ports into a graduated container.
2. Rotate the crankshaft one revolution to ensure that all lubricant has been drained.
3. Note quality (inspect drained lubricant for brass or other metallic particles, which indicate a compressor failure) and quantity. Record the amount of lubricant that was removed, in ounces (oz.) or milliliters (mL).
4. Discard the old lubricant as required by local regulations.
5. Add lubricant as follows:
   a. If the amount of lubricant drained was 3 ounces (89 mL) or more, then add an equal amount of clean lubricant of the proper type.
   b. If the amount of lubricant drained was less than 3 ounces (89 mL), then add 5–6 ounces (148 to 177 mL) of clean lubricant of the proper type.
6. If the compressor is to be replaced, drain all of the lubricant from the new or rebuilt compressor and replace the lubricant as outlined in step 5A or 5B, as applicable.

NOTE: Lubricant is added into the suction and/or discharge port(s). Rotate the compressor crankshaft at least five revolutions by hand after adding lubricant.

PART 3
SERVICING THE CLUTCH

The Nippondenso compressor may be equipped with either a Nippondenso or Warner clutch assembly. Although these two clutches are similar in appearance, their parts are not interchangeable. Complete clutch assemblies are, however, interchangeable on this compressor.

The apparent difference in the two clutches is that the Nippondenso pulley, depicted in Figure 11–97, has two narrow single-row bearings that are held in place with a wire snap ring. The Warner clutch, shown in Figure 11–98, has a single wide double-row bearing that is staked or crimped in place.

The service procedures of the two clutches are similar. The same special tools are used for serving either clutch with one exception: the clutch, pulley support, used for removing the bearing(s), differs between the Nippondenso and Warner clutches.

Procedure

1. Remove the hub nut.
2. Use the hub remover and remove the clutch hub, as shown in Figure 11–99.

NOTE: The shaft/hub key need not be removed. Take care not to lose the **shim** washer(s).

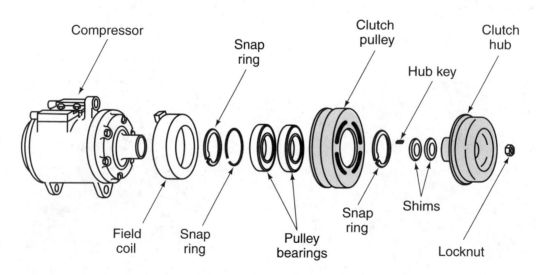

FIGURE 11–97 Exploded view of a Nippondenso compressor with a Nippondenso clutch.

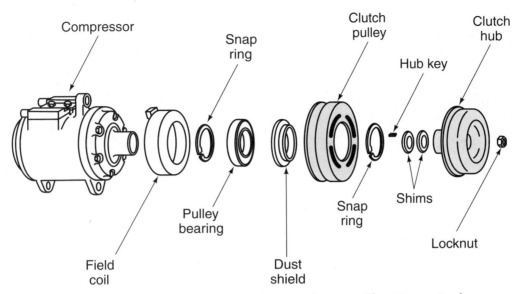

FIGURE 11–98 Exploded view of a Nippondenso compressor with a Warner clutch.

3. Use the snap ring pliers to remove the pulley retainer snap ring.
4. With the shaft protector in place, as shown in Figure 11–100, remove the pulley and bearing assembly with the three-jaw puller.

   NOTE: Make certain that the puller jaws are firmly and securely located behind the pulley to avoid damage.

5. Use the snap ring pliers to remove the **field coil** retaining snap ring, as illustrated in Figure 11–101.
6. Note the location of the coil electrical connector and lift the field coil from the compressor.

   NOTE: In the case of a Nippondenso clutch, use a small screwdriver and remove the bearing retaining snap ring before proceeding.

7. Support the pulley with the proper clutch pulley support, as in Figure 11–102 (see the introductory statement to this Part 3 procedure).
8. Drive out the bearing(s) using a hammer and bearing remover.
9. Lift out the dust shield and retainer or leave them in place. Make sure the dust shield is in place *before* installing the bearing.
10. Install new bearing(s) using the bearing installer and the hammer, as shown in Figure 11–103. Bearing(s) must be fully seated in the rotor.
11. Replace the wire snap ring if servicing a Nippondenso compressor . If servicing a Warner compressor, stake the bearing in place using the prick punch and the hammer.

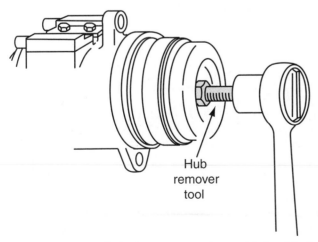

FIGURE 11–99 Use the hub remover to remove the clutch hub.

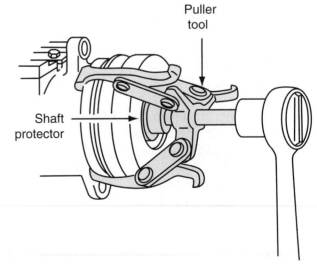

FIGURE 11–100 With the shaft protector in place, use a three–jaw puller to remove the pulley and bearing assembly.

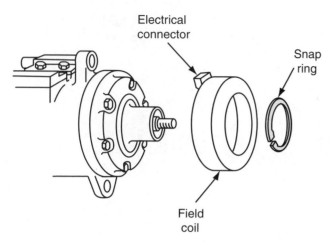

FIGURE 11–101 The field coil is held in place with the snap ring. Note the location of the electrical connector and use snap ring pliers to remove the snap ring, then lift off the field coil.

12. Hook up a 12-volt battery to the clutch coil with an ammeter in series and a voltmeter in parallel.
13. Record the voltage and amperage.

NOTE: The clutch coil should draw approximately 60 volt-amperes (5 amperes at 12 volts).

14. Replace the coil if:
   a. the draw is in excess of 5 amperes, indicating turn-to-turn shorting.
   b. there is no draw (0 ampere), indicating an open coil. First ensure that electrical connections are secure.

NOTE: Before reassembly, clean all parts, including the pulley bearing surface and the compressor front head, with pure mineral spirits.

15. Install the field coil. Be sure the locator pin on the compressor engages with the hole in the clutch coil.
16. Install the snap ring. Be sure the beveled edge of the snap ring faces out.
17. Slip the rotor and bearing assembly squarely on the head. Using the bearing remover/pulley installer tool, shown in Figure 11–104, gently tap the pulley on the head.
18. Install the rotor and bearing snap ring. The beveled edge of the snap ring must face out.
19. Install shim washers and/or be sure they are in place. Check the shaft and hub key to ensure proper seating.
20. Align the hub keyway with the key in the shaft. Press the hub onto the compressor shaft using the

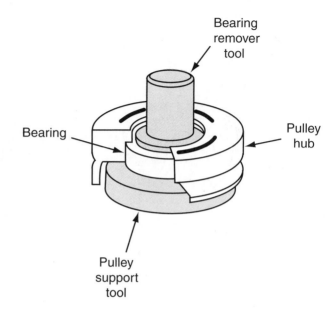

FIGURE 11–102 Use a hammer (not shown) and bearing remover tool to drive out the bearing after placing the hub on a pulley support.

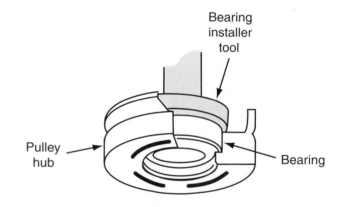

FIGURE 11–103 Use a bearing installer and hammer to drive the new bearing(s) into the pulley hub.

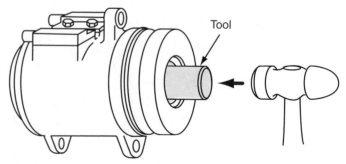

FIGURE 11-104 Gently tap the pulley assembly onto the compressor head.

hub replacer tool, shown in Figure 11–105. Do not drive (hammer) the hub on; to do so will damage the compressor.

21. Using a nonmagnetic feeler gauge, check the air gap between the hub and rotor, shown in Figure 11–106.

NOTE: The air gap should be 0.021–0.036 inch (0.53–0.91 mm).

22. Turn the shaft (hub) one-half turn and recheck the air gap. Change the shim(s) as necessary to correct the air gap.

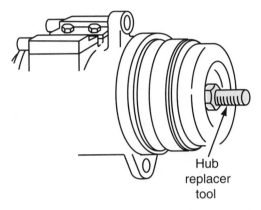

FIGURE 11-105 Press the hub onto the shaft using the hub replacer tool.

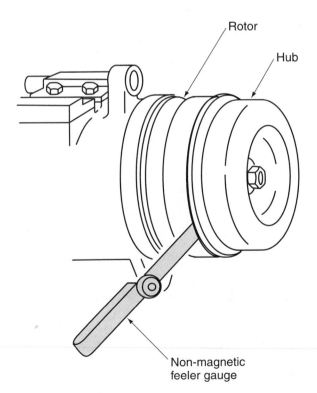

FIGURE 11-106 Check the air gap between the hub and rotor using a nonmagnetic feeler gauge.

23. Install the locknut and tighten to 10–14 ft.-lb. (13.6–19.0 N•m).
24. Recheck the air gap. See steps 21 and 22.

PART 4
REBUILDING THE NIPPONDENSO COMPRESSOR

Rebuilding the Nippondenso compressor is limited to replacing the valve plates, O-rings, and gaskets.

Procedure

1. Remove the clutch and clutch coil assemblies, as outlined in Part 3.
2. Using the shaft key remover, remove the shaft key, as shown in Part 1, Figure 11–90.
3. Remove the felt oil absorber and retainer from the front head cavity.
4. Clean the outside of the compressor with pure mineral spirits and air dry. Do not submerge the compressor in the mineral spirits.
5. Drain the compressor lubricant into a graduated measure. Refer to Part 2 for this procedure.
6. Remove the six through-bolts from the front head of the compressor. Use the proper tool; some require a 10 mm hex socket and others require a 6 mm Allen wrench.
7. Discard the six brass washers, if equipped. Retain the six through-bolts for reassembly.
8. Gently tap the front head with a plastic hammer to free it from the compressor housing.
9. Remove and discard the head-to-housing O-ring and the head-to-valve plate gasket.

NOTE: If replacing the seal assembly, proceed with step 10. If not, proceed with step 12.

10. Place the front head on a piece of soft material, such as cardboard or a cup.
11. Use the shaft seal seat remover to remove the seal seat.
12. Using both hands, as in Figure 11–107, remove the shaft seal cartridge.
13. Tap the rear head with a plastic hammer and remove the head.
14. Discard the head-to-housing O-ring.
15. Remove the valve plate from the head using the valve plate remover, as illustrated in Figure 11–108.
16. Discard the head-to-valve plate gasket.
17. Tap on the compressor body lugs, shown in Figure 11–109, to separate the front and rear housings.

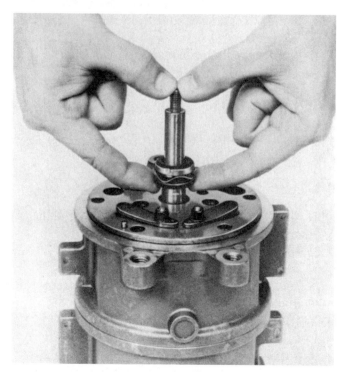

FIGURE 11-107 Removing the shaft seal. *(Courtesy of Ford Motor Company)*

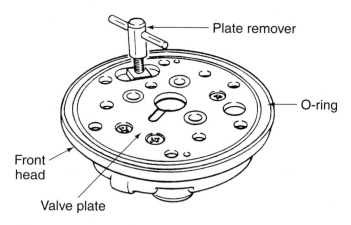

FIGURE 11-108 Use the special tool to remove the valve plate from the head. *(Courtesy of Ford Motor Company)*

NOTE: Separate the housings no more than 1 inch (25.4 mm), as illustrated in Figure (11–110.

18. Inspect both suction valve plates for damage.
19. Inspect both discharge valve plates for damage.

NOTE: Replace any valve plate found to be damaged.

20. Inspect all mating surfaces for nicks and/or burrs.
21. Inspect for brass or metallic material in the compressor piston bores or body.

NOTE: Replace the compressor if brass or metallic material is found.

22. Liberally coat all O-rings and gaskets with clean mineral oil or O-ring lubricant.
23. Position the front-to-rear housing O-ring and slide the two housings together.
24. Install the rear head to the valve plate gasket. Install the discharge valve plate and the suction valve plate.

NOTE: Make sure the gaskets and valve plates are aligned with the alignment pins in the rear head.

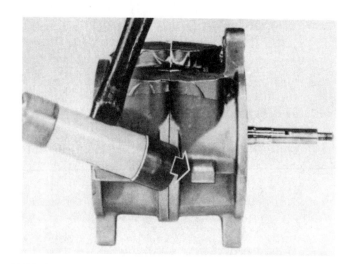

FIGURE 11-109 Separating the front and rear housings. *(Courtesy of Ford Motor Company)*

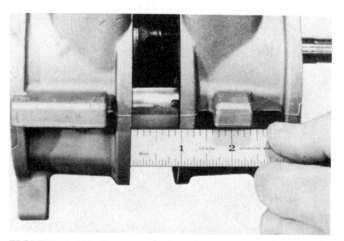

FIGURE 11-110 Separate the housings no more than 1 inch (25.4 mm). *(Courtesy of Ford Motor Company)*

25. Install the rear head O-ring and mount the rear head and valve plate assembly to the compressor housing. The rear head alignment pins must engage in corresponding holes in the compressor housing.
26. Position the compressor on the rear head and install the suction valve plate, the discharge valve plate, and the head-to-valve plate gasket.

   NOTE: Make sure the valve plates and gasket are aligned with the alignment pins in the compressor housing.

27. Install the compressor shaft seal cartridge, as in Figure 11–111, making sure to index the shaft seal on the crankshaft slots.

   NOTE: If the seal seat was removed, proceed with step 7. If not, proceed with step 8.

28. Install the seal seat into the front head using the seal seat installer. Refer to Figure 11–94.
29. Install the front head-to-housing O-ring and carefully slide the head onto the compressor body.
30. Using six new brass washers, if required, install the six compressor through-bolts.

31. Using a 10 mm hex socket or 9 mm Allen wrench, as required, tighten the bolts to 260 in.-lb. (29 N•m). Use an alternate pattern, shown in Figure 11–112, when tightening bolts.
32. Replace the lubricant as outlined in Part 2.
33. Install the crankshaft key using a drift.
34. Align the ends of the felt and its retainer and install into the head cavity. Be sure the felt and retainer are fully seated against the seal plate.
35. Replace the clutch and coil assemblies as outlined in Part 3.

   NOTE: Before installing the compressor on the car, check for sticking and/or binding conditions, as follows.

36. Insert the socket wrench with torque handle into the shaft locknut and turn clockwise (cw). The suction and discharge service valves should be removed for this test.
37. Note the torque required to turn the compressor crankshaft. The maximum torque required should be 7 ft.-lb. (9.5 N•m) or less.

   NOTE: If a greater torque is required, determine the cause and correct it.

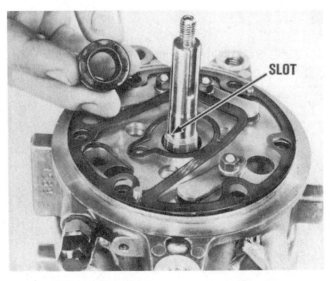

FIGURE 11-111 Index flat on the shaft seal with the slot on the crankshaft. (*Courtesy of Ford Motor Company*)

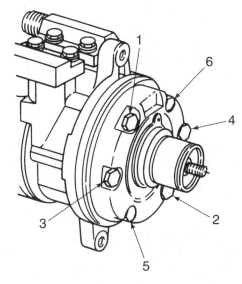

FIGURE 11-112 Use an alternate pattern when torquing bolts. The torque sequence for tightening bolts is shown. (*Courtesy of Ford Motor Company*)

SERVICING THE NIPPONDENSO TEN-CYLINDER COMPRESSOR

The Nippondenso six- and ten-cylinder compressors are similar in appearance. Unlike the Nippondenso six-cylinder compressor, the ten-cylinder compressor shaft seal may be serviced without removing the front head.

Service to the ten-cylinder compressor, however, is limited to the clutch and shaft seal. It is not advisable to attempt internal repairs.

These procedures are given in two parts: Part 1 for clutch service and Part 2 for seal service.

PART 1
CLUTCH SERVICE

Procedure

1. While holding the clutch plate with a strap wrench or clutch plate holding tool, remove the shaft nut with a 12 mm socket.
2. Use the clutch plate remover, shown in Figure 11–113, to remove the clutch plate.
3. Remove the clutch plate shim(s) and set them aside for later reassembly.
4. Use the snap ring pliers, as in Figure 11–114, to remove the pulley retaining snap ring.
5. Use the plastic hammer, depicted in Figure 11–115, to tap the pulley off the compressor. Take care not to damage the pulley.
6. Remove the clutch coil ground wire from the compressor.
7. Use the snap ring pliers to remove the clutch coil retaining snap ring.

8. Lift the coil from the compressor.
9. Assemble the pulley with the pulley support and bearing remover on a hydraulic press, as shown in Figure 11–116. Make sure the tools and the pulley are in alignment.
10. Press the bearing from the pulley rotor. Discard the bearing.
11. Assemble the pulley with the pulley support, the bearing replacer tool, and the new bearing, as shown in Figure 11–117. Make certain that the tools, the pulley, and the bearing are in alignment.
12. Carefully press the bearing into the pulley.

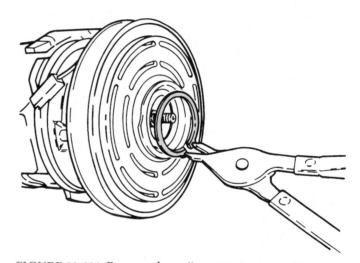

FIGURE 11-114 Remove the pulley retaining snap ring.

FIGURE 11-113 Use clutch plate remover to remove the clutch plate.

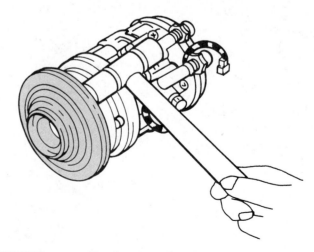

FIGURE 11-115 Use the plastic hammer to tap the pulley off the rotor.

13. Place the clutch coil into position and secure it with a snap ring. Be sure the snap ring is seated.
14. Replace the clutch coil ground wire.
15. Use the plastic hammer to gently tap the pulley onto the compressor. Take care not to damage the pulley.
16. Replace the pulley snap ring. Make sure it is fully seated.
17. Replace the clutch plate shim(s).
18. Replace the clutch plate.

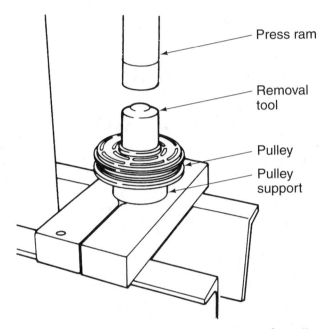

FIGURE 11-116 Use a hydraulic press to remove the pulley bearing.

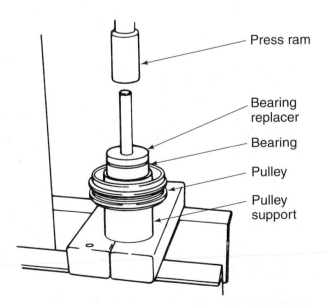

FIGURE 11-117 Use a hydraulic press to replace the pulley bearing.

19. Use a nonmetallic feeler gauge and check the clutch plate clearance. Clearance should be 0.016–0.028 inch (4–7 mm). If necessary, add or subtract shim(s) to obtain the proper clearance.
20. Replace the compressor shaft nut and torque to 10–14 ft.-lb. (13.6–19.0 N•m).
21. Recheck the air gap. Adjust it if necessary. See step 19.

PART 2
SHAFT SEAL SERVICE

Procedure

1. Remove the clutch and coil assembly, as outlined in Part 1.
2. Remove the felt seal from the seal cavity.
3. Remove the thrust plate snap ring using the O-ring pliers.
4. Use the shaft key remover, shown in Figure 11–118, to remove the shaft key. Set the key aside for reuse during the reassembly.
5. Insert the shaft thrust plate remover/installer into the seal cavity to engage the thrust plate.
6. Hold down on the holder ring and pull out on the T-handle to remove the thrust plate. Discard the thrust plate.
7. Insert the shaft thrust seal remover/installer into the seal cavity and engage the shaft thrust seal. Press against the seal while turning the tool clockwise (cw).
8. Pull the tool out of the seal cavity to remove the seal. Discard the shaft thrust seal.
9. Install the new seal on the thrust seal remover/installer. Engage the seal to the tool by turning the tool or seal clockwise (cw) while applying pressure.
10. Coat the seal and seal cavity liberally with clean refrigeration lubricant.

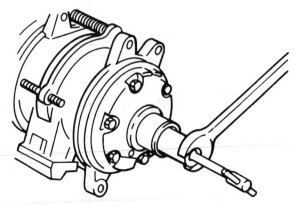

FIGURE 11-118 Use the shaft key remover to remove the shaft key.

11. Insert the seal into the seal cavity until it is fully seated.
12. Rotate the tool counterclockwise (ccw) to release the tool from the seal. Remove the tool from the seal cavity.
13. Use the thrust plate remover/installer to install the new thrust plate.

14. Replace the thrust plate snap ring. Be sure the snap ring is fully seated.
15. Replace the felt seal.
16. Replace the shaft key.
17. Replace the clutch and coil assembly, as outlined in Part 1.

SERVICING THE PANASONIC VANE-TYPE COMPRESSOR

The main components of the Panasonic vane-type compressor are the rotor with three vanes, a sludge control valve, discharge valve, and a thermal protector. The following services may be performed: checking and adjusting the TH lubricant level described in Part 2; servicing the clutch, as outlined in Part 3; servicing the thermal protector, sludge control, and discharge valve, as described in Part 4; and servicing the compressor, as outlined in Part 5.

Part 1A covers typical procedures for removing and Part 1B for replacing a Panasonic vane-type compressor.

CAUTION
WEAR SAFETY GLASSES OR GOGGLES WHEN HANDLING REFRIGERANT OR SERVICING A BATTERY.

PART 1A
REMOVE THE COMPRESSOR
Procedure

1. Ensure that all refrigerant has been properly recovered from the air-conditioning system.

   NOTE: It is unlawful to vent refrigerant to the atmosphere.

2. Carefully disconnect the battery ground cable.
3. Loosen the belt tensioner and remove the drive belt.
4. Raise the front of the vehicle to gain under-engine clearance if necessary.

   CAUTION
   SUPPORT THE VEHICLE WITH APPROPRIATE STANDS.

5. Disconnect the clutch cycling switch wire as well as the clutch coil wire.

6. Remove the low- and high-side service valve fitting bolts and discard the O-rings or gaskets.
7. Remove the compressor mounting bolts.
8. Remove the compressor from the engine.

PART 1B
REPLACE THE COMPRESSOR
Procedure

1. With the compressor in position, install the mounting bolts.
2. Position the low- and high-side service fittings with new O-rings or gaskets in place.

   NOTE: Lubricate the O-rings or gaskets with clean mineral oil or O-ring lubricant.

3. Be certain that the fittings are in proper position and install the bolts.
4. Reconnect the cycling clutch switch wire and the clutch coil wire.
5. Remove the supports and lower the vehicle.
6. Replace the compressor drive belt.
7. Reconnect the battery ground cable.

   CAUTION
   WEAR SUITABLE EYE PROTECTION.

8. Leak test the system as outlined in Chapter 7.
9. Evacuate the system, as outlined in Chapter 8.
10. Charge the system, as outlined in Chapter 19.

PART 2
CHECKING AND ADJUSTING
THE LUBRICANT LEVEL

A new Panasonic rotary vane compressor contains 6.78 ounces (200 mL) of a special paraffin-based refrigeration

FIGURE 11-119 Remove the Allen holding bolt (A). Remove the clutch armature (B). (*Courtesy of BET, Inc.*)

lubricant, designated as YN-9. Too much lubricant will reduce the system capacity and too little lubricant will result in insufficient lubrication. It is therefore necessary to adjust the lubricant when replacing the compressor, as follows:

Procedure

1. Remove the defective compressor from the vehicle, as outlined in Part 1A.
2. Drain the lubricant from the defective compressor into a calibrated container and note the amount removed. Allow the compressor to drain thoroughly.
3. Drain the lubricant from the replacement compressor into a second calibrated container. Allow the compressor to drain thoroughly.
4. Add the same amount of clean refrigeration lubricant to the replacement compressor as was removed.
5. Add an additional 0.68 ounce (20 mL) of lubricant.
6. Install the compressor on the vehicle, as outlined in Part 1B.

PART 3
SERVICING THE CLUTCH ASSEMBLY

Procedure

1. Use an Allen wrench to remove the clutch armature Allen bolt.
2. Remove the clutch armature, as illustrated in Figure 11–119A and 11–119B.
3. Remove the shim(s) and set aside.

FIGURE 11-120 Remove/replace the rotor snap ring. (*Courtesy of BET, Inc.*)

4. Remove the clutch rotor and pulley snap ring using internal snap ring pliers, as in Figure 11–120.
5. Remove the clutch rotor and pulley.
6. Remove the three clutch field coil screws using a screwdriver, as illustrated in Figure 11–121.
7. Remove the clutch field coil.
8. Replace the clutch field coil and secure it with three screws.
9. Replace the clutch rotor and pulley and secure them with an internal snap ring.
10. Replace the shim(s) removed in step 3.
11. Replace the clutch armature and secure it with an Allen bolt.

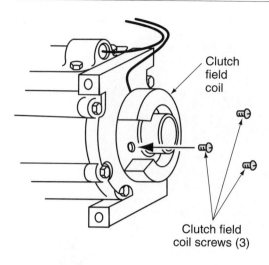

FIGURE 11-121 Using a screwdriver, remove the three clutch field coil screws.

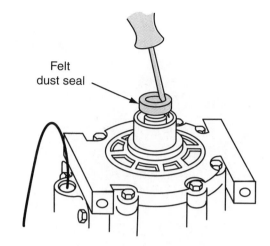

FIGURE 11-122 Remove the felt dust seal.

PART 4
SERVICING THE SHAFT SEAL

Procedure

1. Remove the clutch, as outlined in Part 3.

   NOTE: It is not necessary to remove the clutch coil for seal service.

2. Remove the felt dust seal from the seal cavity, as illustrated in Figure 11–122.
3. Remove the shaft seal snap ring using internal snap ring pliers, as in Figure 11–123.
4. Remove the seal seat using the seal remover, as shown in Figure 11-124.
5. Remove the shaft seal using the seal remover/installer, as in Figure 11-125.
6. Coat all seal parts with clean refrigeration lubricant.
7. Install the shaft seal using the remover/installer tool.
8. Install the shaft seal using the seal remover tool.
9. Replace the shaft seal snap ring.
10. Replace the felt dust seal.
11. Replace the clutch assembly, as outlined in Part 3.

PART 5
SERVICING THE COMPRESSOR

The thermal protector, sludge control, and discharge valve are the only components that may be serviced in the Panasonic rotary compressor. It is not necessary to remove the clutch assembly or shaft seal assembly for this service.

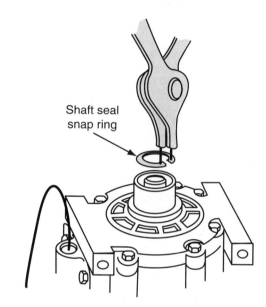

FIGURE 11-123 Remove (replace) the shaft seal snap ring.

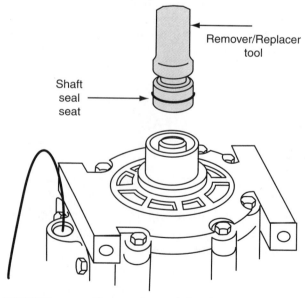

FIGURE 11-124 Remove (install) the shaft seal seat.

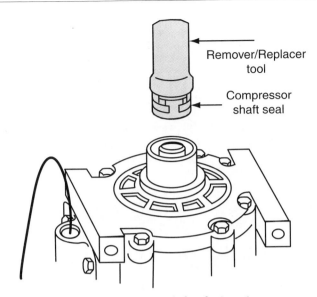

FIGURE 11-125 Remove (install) the shaft seal.

FIGURE 11-126 Remove the six hex nuts. (*Courtesy of BET, Inc.*)

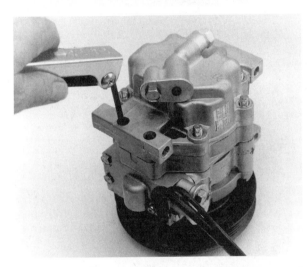

FIGURE 11-127 Remove the two Allen bolts. (*Courtesy of BET, Inc.*)

Procedure

1. Drain the lubricant from the compressor, as outlined in Part 2.
2. Remove the six housing cover hex nuts at the rear of the compressor, as in Figure 11–126.
3. Remove two Allen bolts from the rear of the compressor, as shown in Figure 11–127.
4. Lift off the housing cover.
5. Remove three lubricant control valve bolts, as shown in Figure 11–128.
6. Remove the lubricant control valve.
7. Remove and discard the lubricant control valve gasket, as shown in Figure 11–129.
8. Remove the two compression springs and spring stoppers, shown in Figure 11–130.
9. Using a screwdriver, as illustrated in Figure 11–131, remove the thermal protector holddown bracket screw.
10. Using a socket wrench, remove the four thermal protector housing bolts.
11. Remove the thermal protector housing.
12. Remove and discard the thermal protector housing gasket.
13. Use internal snap ring pliers to remove the thermal protector snap ring retainer, as shown in Figure 11–132.
14. Push the thermal protector out of the housing, as shown in Figure 11–133.
15. Use a socket wrench to remove the two discharge valve stopper bolts, as in Figure 11–134.
16. Remove the discharge valve stopper and the discharge valve, as shown in Figure 11–135.
17. Replace the discharge valve and stopper.

18. Secure them with the two bolts removed in step 15.
19. Install the thermal protector and secure it with the snap ring.
20. Replace the thermal protector housing with a new gasket and secure it with four capscrews, as in Figure 11–136.
21. Secure the thermal protector holddown bracket with the screw removed in step 9.
22. Replace the two compression springs and spring stoppers.
23. With a new gasket in place, install the lubricant control valve with three bolts (see Figure 11–128).
24. Install the housing cover, as illustrated in Figure 11–137.
25. Secure the housing cover with two Allen bolts and six hex nuts.
26. Replace the lubricant following the procedures as outlined in Part 2.

FIGURE 11-128 Remove/replace the three oil control valve bolts. *(Courtesy of BET, Inc.)*

FIGURE 11-130 Remove/replace the two compression springs and spring stoppers. *(Courtesy of BET, Inc.)*

FIGURE 11-129 Remove the gasket. *(Courtesy of BET, Inc.)*

FIGURE 11-131 Remove the thermal protector bracket hold-down screw. *(Courtesy of BET, Inc.)*

CAUTION
ALWAYS WEAR SAFETY GLASSES AND OTHER APPROPRIATE SAFETY EQUIPMENT WHEN WORKING ON MOTOR VEHICLE AIR-CONDITIONING (MVAC) SYSTEMS.

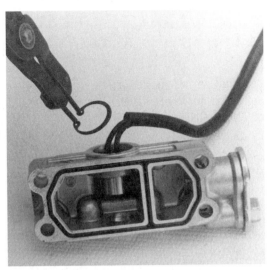

FIGURE 11-132 Remove/replace the thermal protector snap ring. *(Courtesy of BET, Inc.)*

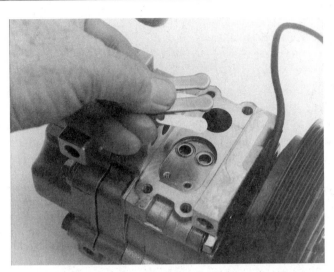

FIGURE 11-135 Remove/replace the discharge valve stopper and valve. *(Courtesy of BET, Inc.)*

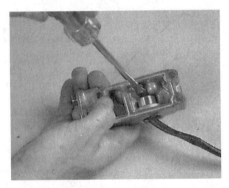

FIGURE 11-133 Push the thermal protector out of the housing.

FIGURE 11-136 Replace the thermal protector housing.

FIGURE 11-134 Remove the two discharge valve stopper bolts. *(Courtesy of BET, Inc.)*

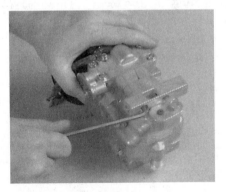

FIGURE 11-137 Install the housing cover.

SERVICING THE SANKYO COMPRESSOR

Rebuilding the Sankyo compressor, Figure 11–138, is limited to replacing the shaft seal (Part 1); servicing the clutch (Part 3); and repairing the rear head and/or valve plate (Part 4). The proper lubricant level of the Sankyo compressor is checked following the procedures outlined in Part 2.

PART 1
REPLACING THE COMPRESSOR SHAFT SEAL

In general, when replacing the compressor shaft **oil seal** it is advisable to remove the compressor from the vehicle. However, if there is sufficient clearance in front of the compressor, the shaft seal may be serviced without removal from the vehicle. It must first be isolated from the system and all refrigerant must be recovered.

Procedure

1. Using a 3/4-inch hex socket and **spanner** wrench, shown in Figure 11–139, remove the crankshaft hex nut.
2. Remove the clutch front plate, as in Figure 11–140, using the clutch front plate puller.
3. Remove the shaft key and spacer shims and set aside.
4. Using the snap ring pliers, as shown in Figure 11–141, remove the seal retaining snap ring.
5. Remove the seal seat, as shown in Figure 11–142, using the seal seat remover and installer.
6. Remove the seal, as shown in Figure 11–143, using the seal remover tool.
7. Remove the shaft seal seat O-ring, as shown in Figure 11–144, using the O-ring remover.

8. Discard all parts removed in steps 5, 6, and 7.
9. Clean the inner bore of the seal cavity by flushing it with clean mineral oil.
10. Coat the new seal parts with clean mineral oil. Do not touch the carbon ring face with the fingers. Normal body acids will etch the seal and cause early failure.
11. Install the new shaft seal seat O-ring. Make sure it is properly seated in the internal groove. Use the remover tool to position the O-ring properly.
12. Install the seal protector on the compressor crankshaft. Lubricate the part liberally with clean mineral oil.

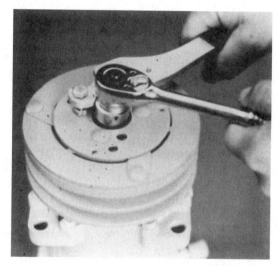

FIGURE 11-139 Removing the crankshaft hex nut. *(Courtesy of Sankyo)*

FIGURE 11-138 Sankyo five-cylinder compressor with clutch assembly. *(Courtesy of BET, Inc.)*

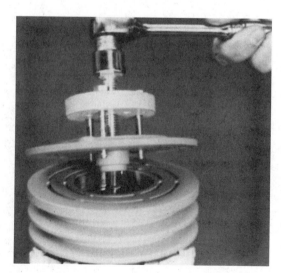

FIGURE 11-140 Removing the front clutch plate. *(Courtesy of Sankyo)*

FIGURE 11-141 Removing the seal seat retainer ring. *(Courtesy of Sankyo)*

FIGURE 11-143 Removing the seal. *(Courtesy of Sankyo)*

FIGURE 11-142 Removing the seal seat. *(Courtesy of Sankyo)*

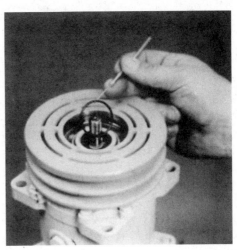

FIGURE 11-144 Removing the O-ring. *(Courtesy of Sankyo)*

13. Place the new shaft seal in the seal installer tool and carefully slide the shaft seal into place in the inner bore. Rotate the shaft seal clockwise (cw) until it seats on the compressor shaft flats.
14. Rotate the tool counterclockwise (ccw) to remove the seal installer tool.
15. Remove the shaft seal protector.
16. Place the shaft seal seat on the remover/installer tool and carefully reinstall it in the compressor seal cavity.
17. Replace the seal seat retainer.
18. Reinstall the spacer shims and shaft key.
19. Position the clutch front plate on the compressor crank shaft.
20. Using the clutch front plate installer tool, a small hammer, and an air gap gauge, reinstall the front plate.
21. Draw down the front plate with the shaft nut. Use the air gap gauge for *go* at 0.016 inch (0.4 mm) and

no-go at 0.031 inch (0.79 mm); refer to Figure 11–145.
22. Using the torque wrench, tighten the shaft nut to a torque of 25–30 ft.-lb. (33.0–40.7 N•m).

PART 2
CHECKING THE LUBRICANT LEVEL

The compressor lubricant level should be checked at the time of installation and after repairs are made when it is evident that there has been a loss of lubricant. The Sankyo compressor is factory charged with 7 fluid ounces (207 mL) of Suniso 5GS lubricant. Only this lubricant, or an equivalent, should be added to the system.

The system must be purged of refrigerant before the lubricant level is checked. A special angle gauge and dipstick are used to check the lubricant level.

FIGURE 11-145 Reinstalling the front clutch plate and checking the air gap. (*Courtesy of Sankyo*)

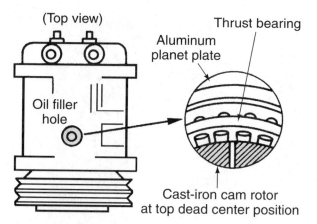

FIGURE 11-146 Position the rotor to top dead center (TDC). (*Courtesy of Sankyo*)

Procedure

1. Position the angle gauge tool across the top flat surfaces of the two mounting ears.
2. Center the bubble and read the inclination angle.
3. Remove the oil filler plug. Rotate the clutch front plate to position the rotor at the top dead center (TDC), as shown in Figure 11–146.
4. Face the front of the compressor. If the compressor angle is to the right, rotate the clutch front plate counterclockwise (ccw) by 120 degrees (120°). If the compressor angle is to the left, rotate the plate clockwise (cw) by 120 degrees (120°), as in Figure 11–147.

   NOTE: The dipstick tool for this procedure is marked in eight increments. Each increment represents 1 ounce (29.6 mL) of lubricant.

5. Insert the dipstick until it reaches the STOP position marked on the dipstick.
6. Remove the dipstick and count the number of increments of lubricant.
7. Compare the compressor angle and the number of increments with the table in Figure 11–148.
8. If necessary, add lubricant to bring it up to the proper level.

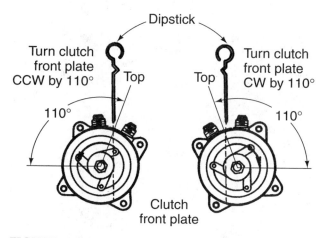

FIGURE 11-147 Rotate the clutch front plate. (*Courtesy of Sankyo*)

Inclination Angle In Degrees	Acceptable Oil Level In Increments
0	6-10
10	7-11
20	8-12
30	9-13
40	10-14
50	11-16
60	12-17

FIGURE 11-148 Dipstick reading versus inclination angle. (*Courtesy of Sankyo*)

NOTE: Do not overfill. Use only clean refrigeration lubricant of the proper type and grade.

9. Check that the rubber O-ring is in place. Reinstall the oil filler plug. Tighten the plug to a torque of 8–9 ft.-lb. (10.8–12.2 N•m).

NOTE: If the oil filler plug leaks when leak testing the system, do not overtighten the plug. Remove the plug and replace the O-ring.

PART 3
SERVICING THE CLUTCH ROTOR AND COIL

This procedure presumes that the compressor has not been removed from the vehicle, and that ample clearance is provided in front of the compressor for clutch service.

Procedure

1. Loosen the compressor and/or idler pulley and remove the belt(s).
2. Use a 3/4-inch hex socket and spanner wrench to remove the crankshaft hex nut.
3. Remove the clutch front plate, using the clutch front plate puller.
4. Using the snap ring pliers, remove the internal and external snap rings, as illustrated in Figures 11–149A and 11–149B.
5. Using the pulley puller, shown in Figure 11–150, remove the rotor assembly.
6. If the clutch coil is to be replaced, remove the three retaining screws and the clutch field coil. Omit this step if the coil is not to be replaced.
7. Using the snap ring pliers, remove the bearing retainer snap ring.
8. From the back (compressor) side of the rotor, knock out the bearing using the bearing remover tool and a soft hammer.
9. From the front (clutch face) side of the rotor, install the new bearing using the bearing installer tool and a soft hammer. Take care not to damage the bearing with hard blows of the hammer.
10. Reinstall the bearing retainer snap ring.
11. Reinstall the field coil (or install a new field coil, if necessary) using the three retaining screws.
12. Align the rotor assembly squarely with the front compressor housing.
13. Using the rotor two-piece installer tools and a soft hammer, carefully drive the rotor into position until it seats on the bottom of the housing, as in Figure 11–151.

A

B

FIGURE 11-149 Removing the internal snap ring (A). Removing the external snap ring (B). (Courtesy of Sankyo)

FIGURE 11-150 Removing the rotor assembly. (Courtesy of Sankyo)

14. Reinstall the internal and external snap rings using the snap ring pliers.
15. Align the slot in the hub of the front plate squarely with the shaft key.
16. Drive the front plate on the shaft using the installer tool and a soft hammer. Do not use unnecessary hard blows.

FIGURE 11-151 Drive the front plate onto the shaft. (Courtesy of *Sankyo*)

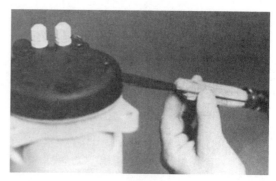

FIGURE 11-152 Removing the rear head. (*Courtesy of BET, Inc.*)

17. Check the air gap with *go* and *no-go* gauges.
18. Replace the shaft nut and tighten it to a torque of 25–30 ft.-lb. (33.9–40.7 N•m) using the torque wrench.
19. Replace the belt(s) and tighten to 90–110 ft.-lb. (122–149.1 N•m) tension.

PART 4
REBUILDING THE SANKYO COMPRESSOR

Rebuilding the Sankyo compressor is limited to those procedures previously covered, replacing the shaft seal, servicing the clutch, and adjusting the proper lubricant level. This procedure, then, may be used if it is determined that the valve plate, rear head, or valve plate gaskets are defective.

Procedure

1. Remove the five screws from the cylinder head using a 13 mm hex socket wrench.
2. Remove the head and valve plate assembly from the cylinder block by tapping lightly with a soft hammer on the gasket scraper, which is placed between the valve plate and the cylinder head; refer to Figure 11–152.
3. To remove the valve plate, as in Figure 11–153, insert the gasket scraper between the valve plate and the cylinder block. Do not damage the mating surfaces.
4. Carefully remove all gasket material, as in Figure 11–154, from the mating surfaces. Do not nick or

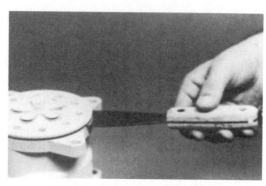

FIGURE 11-153 Removing the valve plate. (*Courtesy of BET, Inc.*)

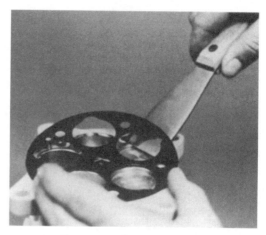

FIGURE 11-154 Removing the gasket material. (*Courtesy of BET, Inc.*)

scratch the surfaces.
5. Apply a thin coat of clean mineral oil or gasket lubricant to all gaskets and mating surfaces.
6. Install the valve plate gasket on the cylinder block. The alignment pin ensures that the gasket is installed properly.
7. Place the valve plate into position. The alignment pin must pass through the hole provided in the valve plate.

8. Install the head gasket on the valve plate. Check for the proper alignment of the gasket.

9. Reinstall the cylinder head and check for the proper alignment.

10. Install the five head screws and tighten them to a snug fit.

11. Tighten the five screws to a torque of 22–25 ft.-lb. (29.8–33.8 N•m). Tighten the screws in the sequence shown in Figure 11–155. Do not under- or overtighten these screws.

12. Add 1 or 2 ounces (29.6 or 59.2 mL) of lubricant to the compressor to compensate for any loss that occurs as a result of this repair.

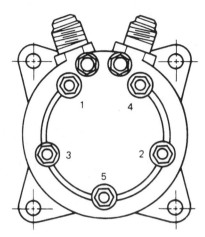

FIGURE 11-155 Rear head torque sequence. *(Courtesy of Sankyo)*

SERVICING THE TECUMSEH HR-980 COMPRESSOR

The Tecumseh HR-980 compressor, used on some Ford vehicle lines, requires special tools for servicing. Because the internal assembly is not accessible, service is limited to shaft seal and clutch repairs.

This compressor is factory charged with 8 ounces (236.6 mL) of 500 viscosity refrigeration lubricant. In a balanced system, approximately 3–4 ounces (88.7–118.3 mL) of lubricant will be found in the compressor, 2–3 ounces (59.2–88.7 mL) in the evaporator, 1 ounce (29.6 mL) in the condenser, and 1 ounce (29.6 mL) in the accumulator.

This procedure is given in three parts: Part 1 for checking and adding lubricant, Part 2 for servicing the clutch, and Part 3 for replacing the shaft seal.

PART 1
CHECKING AND ADDING LUBRICANT

Procedure

1. If replacing the compressor: Drain the new or rebuilt compressor and replace the old lubricant with 4 ounces (118.3 mL) of clean refrigeration lubricant.

2. If replacing the evaporator: Add 3 ounces (88.7 mL) of clean refrigeration lubricant to the new evaporator before installing.

3. If replacing the condenser: Add 1 ounce (29.6 mL) of clean refrigeration lubricant to the new condenser before installing it.

4. If replacing the accumulator: Drain the lubricant from the old accumulator through the pressure switch fitting into a graduated container. Discard the old lubricant.

5. Add the same amount of clean refrigeration lubricant to the new accumulator, plus 1 ounce (29.6 mL). If no lubricant was drained from the old accumulator, add 1 ounce (29.6 mL) of lubricant to the new accumulator.

PART 2
CLUTCH SERVICE (FIGURE 11–156)

Procedure

1. Remove the retaining nut.

2. Use the hub remover tool and remove the clutch hub from the compressor shaft, as shown in Figure 11–157. Remove and retain the shim(s).

3. Remove the clutch pulley retaining nut using a spanner wrench.

4. Remove the pulley and bearing assembly from the compressor by hand. If the assembly cannot be removed by hand, use the shaft protector and pulley remover, as illustrated in Figure 11–158.

5. Remove the field coil from the compressor.

6. Clean the front of the compressor to remove any dirt and/or corrosion.

7. Place the pulley on the clutch pulley support, as shown in Figure 11–159.

8. Use the pulley replacer tool to drive out the bearing.

9. Turn the pulley over, with the flat side atop a clean board.

10. Position the new bearing in the bearing bore of the pulley, as in Figure 11–160, and use the pulley bearing replacer to seat the bearing.

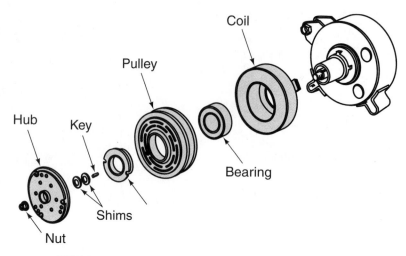

FIGURE 11-156 An exploded view of the clutch assembly.

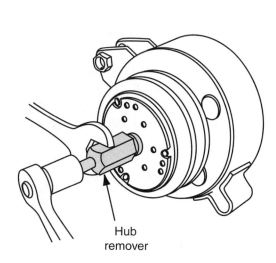

FIGURE 11-157 Remove the clutch hub.

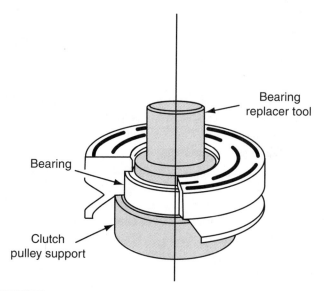

FIGURE 11-159 Pulley placed atop the clutch pulley support for service.

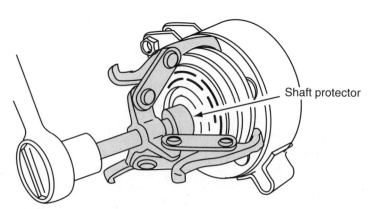

FIGURE 11-158 Use the shaft protector and three-jaw puller to remove the pulley assembly if it cannot be removed by hand.

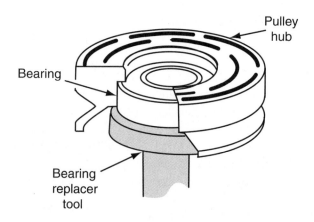

FIGURE 11-160 Use a bearing replacer to seat the new bearing.

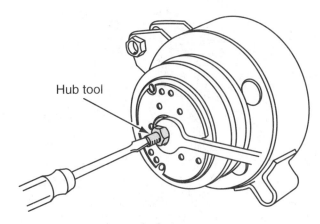

FIGURE 11-161 Replace the hub.

11. Stake the new bearing. Use a blunt drift or punch at three equally spaced places inside the bore. Do not use the same places that were used to retain the old bearing.
12. Install the field coil. The slots of the coil should fit over the housing lugs. The electrical connector should be toward the top of the compressor.
13. Install the pulley and bearing assembly on the front of the compressor. If properly aligned, the assembly should slide on.

    NOTE: If difficult, use the pulley replacer and tap lightly with a plastic hammer. Do not use unnecessary force.

14. Apply a drop of thread lock to the threads of the pulley retainer nut.

15. Install the pulley retainer nut and tighten to 65–70 ft.-lb. (88–94 N•m) using the spanner and torque wrench.
16. Make certain that the key is aligned with the key-way of the clutch hub; install the hub and shim(s) onto the compressor shaft. Use the hub replacer, as shown in Figure 11–161.

    NOTE: Do not drive the hub onto the shaft, as compressor damage will result.

17. Install the nut and tighten to 10–14 ft.-lb. (14–18 N•m).
18. Check the air gap at three equally spaced intervals (120 degrees) around the pulley. Record the measurements.
19. Rotate the compressor pulley one-half turn (180 degrees) and repeat step 17. The air gap should be 0.021–0.036 inch (0.53–0.91 mm). If the air gap is greater or less than these specifications, add or remove shims (step 16) to bring the air gap into specifications.

PART 3
REPLACE THE SHAFT SEAL (FIGURE 11–162)

Procedure

1. Remove the clutch and coil, as outlined in Part 2.
2. Remove the key from the compressor shaft.
3. Carefully pry the dust shield from the compressor using a small screwdriver, as shown in Figure 11–163.

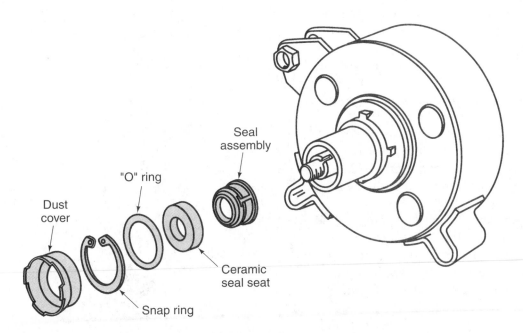

FIGURE 11-162 Exploded view of the shaft seal assembly.

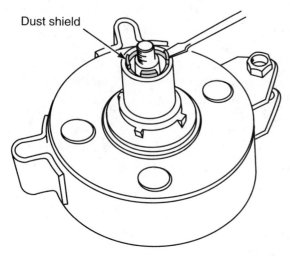

Dust shield

FIGURE 11-163 Pry the dust shield from the seal cavity.

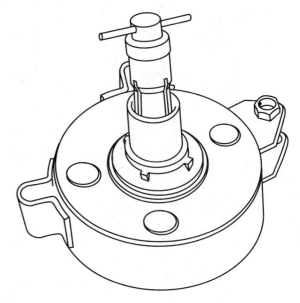

FIGURE 11-164 Insert the seal seat tool into the seal cavity.

NOTE: Take care not to damage the end of the compressor housing.

4. Remove the seal snap ring retainer using the internal snap ring pliers.
5. Clean the inside of the seal cavity to prevent entry of foreign material when the seal is removed.
6. Insert the shaft seal seat tool, as in Figure 11–164, and engage the seal. Tighten the outer sleeve to expand the tool in the seal seat.
7. Pull on the tool, while rotating it clockwise (cw), to remove the seal seat.
8. Use the O-ring remover and remove the O-ring, as shown in Figure 11–165.
9. Insert the seal assembly tool into the compressor, as shown in Figure 11–166. While forcing the tool downward, rotate it counterclockwise (ccw) to engage the tangs of the seal.
10. Pull the seal from the compressor and remove the seal from the tool.
11. Check the inside of the compressor to ensure that all surfaces are free of nicks and burrs.
12. Coat the O-ring liberally with clean mineral oil or O-ring lubricant and insert it into the cavity using the O-ring installer, O-ring sleeve, and O-ring guide.
13. With the O-ring in place, remove the tools from the cavity.
14. Coat the shaft seal liberally with clean mineral oil and carefully engage the seal with the seal remover/replacer tool, as illustrated in Figure 11–167.
15. Carefully place the seal over the shaft and, while rotating it, slide the seal down the shaft until the assembly engages the flats and is in place.
16. Rotate the tool to disengage it from the seal. Remove the tool from the cavity.

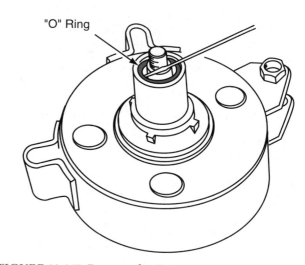

"O" Ring

FIGURE 11-165 Remove the O-ring.

FIGURE 11-166 Insert the seal on the remover/replacer tool.

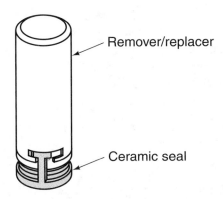

Remover/replacer

Ceramic seal

FIGURE 11-167 Engage the seal in the remover/replacer tool.

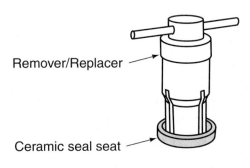

Remover/Replacer

Ceramic seal seat

FIGURE 11-168 Engage the seal seat with the remover/replacer tool.

17. Coat the seal seat liberally with clean mineral oil and engage the seal seat with the remover/replacer tool, as in Figure 11–168.
18. Carefully insert the seal seat onto the compressor shaft with a clockwise (cw) rotation. Take care not to disturb the O-ring installed in step 12.
19. Disengage the tool from the seal seat and remove the tool.
20. Install the snap ring using snap ring pliers. The flat side of the snap ring must be against the seal seat. Do not bump or tap the snap ring into place; to do so may damage the seal seat, which is made of a ceramic material.
21. Place the compressor in a horizontal position and install the pressure test fitting and manifold gauge set.
22. Pressurize the compressor to 50 psig (344.75 kPa).

NOTE: The Environmental Protection Agency (EPA) allows Refrigerant-22 to be used for leak testing. HCFC-22 leaks are easily detected using either a halogen or halide leak detector.

23. Leak test the compressor. If the shaft seal leaks, then temporarily install the shaft nut and rotate the compressor shaft several turns by hand.

NOTE: If the leak persists, remove the seal and inspect it to determine the cause of a leak.

24. Remove the nut. Recover the test refrigerant and remove the test fitting.
25. Install the dust shield.
26. Replace the clutch and coil, as outlined in Part 2.

REVIEW

Select the correct answer from the choices given.

1. All of the following are lubricants approved for motor vehicle air-conditioning (MVAC) system use, *except:*
   a. Mineral oil.
   b. POG.
   c. PAG.
   d. POE.

2. Technician A says that the lubricant does not have to be checked in a rebuilt compressor. Technician B says that the lubricant does not have to be checked in a new compressor. Who is right?

   a. A only
   b. B only
   c. Both A and B
   d. Neither A nor B

3. Technician A says that mineral oil may be used to lubricate an O-ring before installation. Technician B says a special lubricant is available that may be used to lubricate an O-ring before installation. Who is right?
   a. A only
   b. B only
   c. Both A and B
   d. Neither A nor B

4. Technician A says that metallic particles in the lubricant may be expected whenever a compressor fails. Technician B says the lubricant in a failed compressor will have a pungent odor. Who is right?
   a. A only
   b. B only
   c. Both A and B
   d. Neither A nor B

5. When flushing an air-conditioning system, all of the following components should be flushed EXCEPT:
   a. the accumulator-drier.
   b. the condenser.
   c. the evaporator.
   d. the hoses and fittings.

6. The air gap of the clutch is checked with:
   a. a brass feeler gauge.
   b. an aluminum feeler gauge.
   c. both A or B.
   d. neither A nor B.

7. A clutch current draw of 3 amperes at 12 volts is:
   a. excessive.
   b. expected.
   c. minimum.
   d. not sufficient.

8. Technician A says one must take care not to install a shaft seal backward. Technician B says one must take care not to install a snap ring backward. Who is right?
   a. A only
   b. B only
   c. Both A and B
   d. Neither A nor B

9. A torque wrench may be scaled in:
   a. in.-lb. (N•m).
   b. ft.-lb. (N•m).
   c. both A and B.
   d. neither A nor B.

10. Technician A says that rotating a compressor several times will often stop a new shaft seal leak. Technician B says a leaking new shaft seal will stop leaking within 5 minutes of compressor operation. Who is right?
    a. A only
    b. B only
    c. Both A and B
    d. Neither A nor B

11. Technician A says the lubricant in all compressors may be drained through the low- and high-side service ports. Technician B says most compressors have an oil plug that may be removed to drain the lubricant. Who is right?
    a. A only
    b. B only
    c. Both A and B
    d. Neither A nor B

12. If alignment positions are at 1 o'clock and 4 o'clock, they are ___ degrees apart.
    a. 120
    b. 90
    c. 60
    d. 30

13. Which of the following manufacture(s) a ten-cylinder compressor?
    a. Harrison
    b. Nippondenso
    c. Tecumseh
    d. Panasonic

14. Technician A says a bent valve plate may be straightened using an arbor press. Technician B says an arbor press may be used to remove or replace clutch bearings. Who is right?
    a. A only
    b. B only
    c. Both A and B
    d. Neither A nor B

15. In what order should the bolts, shown in Figure 11–169, be tightened?
    a. 1-4-2-5-3-6
    b. 1-3-5-2-4-6
    c. 1-2-3-4-5-6
    d. 1-6-2-5-3-4

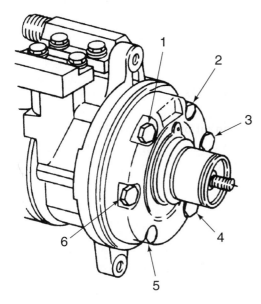

FIGURE 11-169 Figure for question 15.

16. A bearing should be staked in a DA-6 clutch hub at:
    a. 90-degree intervals.
    b. 120-degree intervals.
    c. 180-degree intervals.
    d. none of the above.

17. Technician A says that mineral oil is generally used in a compressor designated for CFC-12 service. Technician B says that PAG lubricant is generally used in a compressor designated for HFC-134a service. Who is right?
    a. A only
    b. B only
    c. Both A and B
    d. Neither A nor B

18. Technician A says that CFC-12 may be used to pressure test a compressor. Technician B says that HFC-134a may be used to pressure test a compressor. Who is right?
    a. A only
    b. B only
    c. Both A and B
    d. Neither A nor B

19. The clutch air gap may be adjusted:
    a. by removing or adding shims.
    b. by torquing the compressor shaft nut.
    c. by repositioning the retainer snap ring.
    d. The air gap cannot be adjusted.

20. Technician A says the felt found in the front head of some compressors is there to absorb oil. Technician B says the felt is placed in the front head of some compressors to absorb compressor noise. Who is right?
    a. A only
    b. B only
    c. Both A and B
    d. Neither A nor B

TERMS

Write a brief description of the following terms:

1. alignment pins
2. bore
3. control valve
4. digital
5. DVOM
6. field coil
7. gasket
8. graduated container
9. holding fixture
10. housing
11. lubricant
12. oil seal
13. O-ring
14. retainer ring
15. shaft key
16. shaft seal
17. shim
18. spanner
19. terminal
20. washers

CHAPTER 12

METERING DEVICES

OBJECTIVES

On completion and review of this chapter, you should be able to:

❑ Understand and discuss the functions of the thermostatic expansion valve (TXV).
❑ Understand and discuss the functions of the fixed orifice tube (FOT).
❑ Compare the functions and performances of the TXV and FOT.
❑ Identify the internally and externally equalized TXV.
❑ Determine the differences between FOT and TXV air-conditioning systems by sight.

INTRODUCTION

The control of the amount of refrigerant that is allowed to enter the evaporator, under varying heat-load conditions, is the job of the metering device. Basically, two types of metering devices are used in automotive air-conditioning applications: the thermostatic expansion valve **(TXV)** and the fixed **orifice** tube **(FOT).**

The thermostatic expansion valve (TXV or TEV) is also called the **H-valve** or block valve. The TXV or H-valve, diagrammed in Figure 12–1, is usually found outside the evaporator case at the inlet provisions of the evaporator core. The fixed orifice tube (FOT), shown in Figure 12–2, may be an internal part of the inlet provisions of the evaporator or may be found inside the liquid line anywhere between the condenser outlet and the evaporator inlet. The TXV on some systems is found inside the valves-in-receiver (VIR). On other systems, it is part of the combination or "combo" valve shown in Figure 12–3. These metering devices are discussed in this chapter.

THE THERMOSTATIC EXPANSION VALVE

Two types of thermostatic expansion valves are commonly used: the internally equalized valve and the externally equalized valve, Figure 12–4. Many factory-installed air conditioners use externally equalized valves; aftermarket manufacturers commonly use internally equalized valves. Pressure drop across the evaporator dictates which type of valve to use. Figure 12–5 illustrates the typical location of the thermostatic expansion valve in an air-conditioning system.

OPERATION OF THE THERMOSTATIC EXPANSION VALVE

Figure 12–6 illustrates the construction of a thermostatic expansion valve (TXV). The TXV has an orifice with a needle-type valve and **seat** to provide variable metering. The needle is actuated by a **diaphragm** that is controlled by three factors:

❑ Evaporator pressure exerted on the bottom of the diaphragm that tends to close the valve
❑ **Superheat** spring pressure against the bottom of the **needle valve** that tends to close the valve
❑ Pressure of the **inert** liquid/vapor in the **remote bulb** via the **capillary tube** against the top of the diaphragm that tends to open the valve

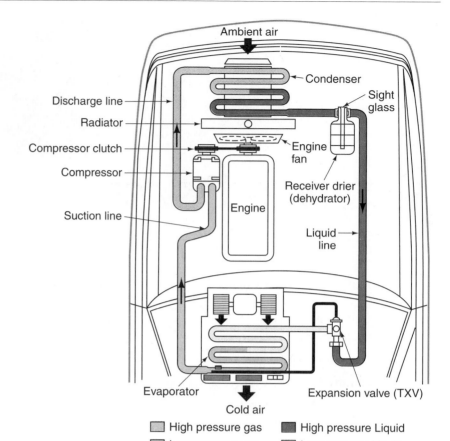

FIGURE 12–1 A typical thermostatic expansion valve (TXV) system. *(From Green/Green and Dwiggins, Australian Automotive Air Conditioning, © Nelson, a division of Thomson Learning)*

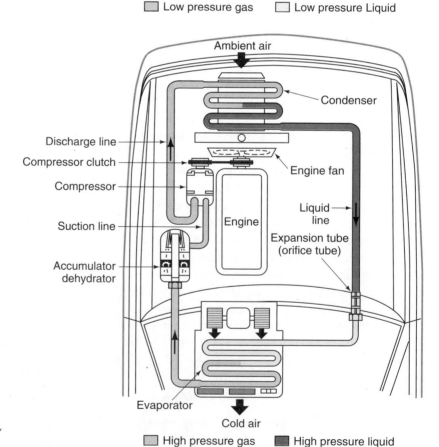

FIGURE 12–2 A typical fixed orifice tube (FOT) system. *(From Green/Green and Dwiggins, Australian Automotive Air Conditioning, © Nelson, a division of Thomson Learning)*

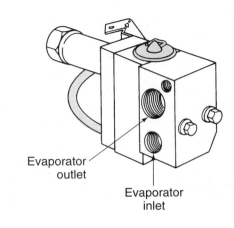

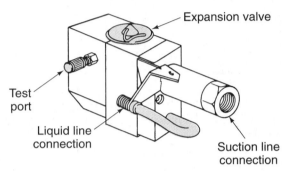

FIGURE 12–3 Ford suction throttling valve and expansion valve assembly (combination valve).

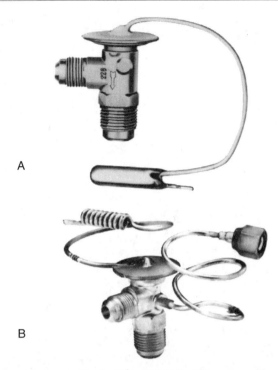

FIGURE 12–4 Typical thermostatic expansion valves: (A) internally equalized; (B) externally equalized. *(Courtesy of Singer Controls Division, Schiller Park, IL)*

Remote Bulb

Several types of inert liquids are used in the remote bulb, Figure 12–7. For example, assume that the fluid in the bulb is the same as that used in the system. Because the same fluid or refrigerant is used, each exerts the same pressure, assuming that the temperature of each fluid is the same.

Under normal design considerations, the liquid refrigerant entering the evaporator **boils** off by picking up heat and is in vapor (gas) form by the time it exits the evaporator coil. In fact, the refrigerant should be all vapor before reaching the end of the evaporator coil, and the vapor should become slightly superheated. Although the superheated vapor is somewhat warmer than the temperature at which evaporation took place, its pressure has not changed. The remote bulb of the expansion valve is clamped onto the suction line, sometimes referred to as the "tailpipe." In this location, the bulb senses the warmer temperature at the evaporator outlet. The temperature of the inert fluid within the remote bulb increases, and its corresponding pressure is exerted on top of the diaphragm.

The increased pressure of the inert fluid exerted on top of the diaphragm is greater than the combination of the evaporator pressure and the superheat spring pressure. As a result, the needle moves away from the seat in

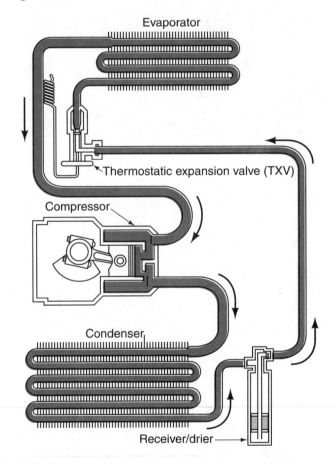

FIGURE 12–5 Relation of the thermostatic expansion valve to the air-conditioning system.

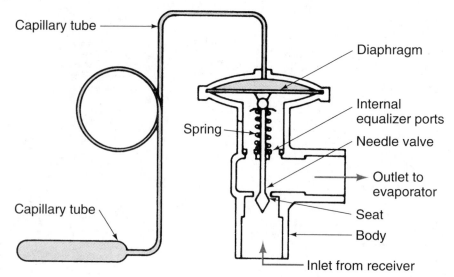

FIGURE 12–6 Thermostatic expansion valve details.

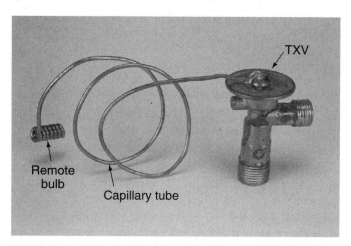

FIGURE 12–7 Typical O-ring type of TXV with remote bulb.

the orifice. The needle valve opens until the superheat spring pressure and the evaporator pressure are great enough to balance the remote bulb pressure.

For example, when the needle valve is closed, it does not allow enough refrigerant to enter the evaporator. Thus, the evaporator pressure remains low and the suction vapor is warm. Accordingly, the suction vapor has high superheat. This condition causes a positive pressure on top of the diaphragm, which opens the needle valve.

When the needle valve is open, too much refrigerant is allowed to enter the evaporator. As a result, the evaporator pressure rises and the suction vapor is cool; that is, it has too little (or no) superheat. This condition reduces pressure on top of the diaphragm and creates a positive pressure under it, which closes the needle valve.

This condition continues until the three pressures of the thermostatic expansion valve are in balance.

Under stable conditions, the evaporator remains fully operational under all load conditions in the manner just described.

The TXV, then, has three main functions:

- ❑   It **throttles.**
- ❑   It **modulates.**
- ❑   It **controls.**

Throttling Action

The expansion valve separates the high side from the low side of the air-conditioning system. Because there is a pressure drop across the valve, the flow of refrigerant is **restricted,** or throttled. The state of the refrigerant entering the valve is a high-pressure liquid. The refrigerant leaving the valve is a low-pressure liquid. This drop in refrigerant pressure results in little change in the state of the refrigerant. A small percentage of refrigerant vaporizes due to the severe drop in pressure. This vaporization is referred to as "flash gas."

Modulating Action

The TXV is designed to meter the proper amount of liquid refrigerant into the evaporator, as required under varying **heat-load** conditions, to maintain the proper cooling action. The amount of refrigerant required at any given time varies with different heat loads. The TXV modulates from the wide-open position, as shown in Figure 12–8A, to the closed position, shown in Figure 12–8B. The valve constantly seeks a balance between these two positions to ensure the proper metering of refrigerant under all load conditions.

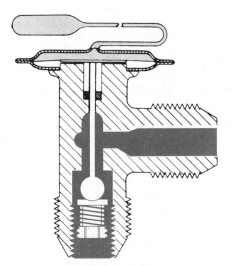 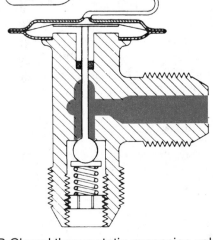

A Open thermostatic expansion valve   B Closed thermostatic expansion valve

FIGURE 12–8 Open thermostatic expansion valve (A) and closed thermostatic expansion valve (B).

Controlling Action

The expansion valve is designed to change the amount of liquid refrigerant metered into the evaporator in response to load or heat changes. More refrigerant is required by the evaporator as the load increases. As the load is decreased, however, the valve closes and less refrigerant is metered into the evaporator core. This controlling action of the thermostatic expansion valve maintains proper refrigerant metering into the evaporator under varying heat-load conditions.

Superheat

The liquid refrigerant delivered to the evaporator coil should vaporize, or evaporate, completely before it reaches the coil outlet. The point at which it is completely vaporized is known as its *saturation point.*

Liquid refrigerant boils (vaporizes) at a low temperature: approximately –21.6°F (–29.8°C) for CFC-12 and –15.7°F (–26.5°C) for HFC-134a at sea-level atmospheric pressure (14.697 psia). The actual boiling point of the refrigerant in the evaporator is determined by its pressure. Refer to the temperature-pressure charts in Chapter 5 for the boiling point of refrigerant based on its pressure. The temperature-pressure charts illustrate that **saturated** vapor is still cold and capable of absorbing more heat. The cold vapor flowing through the remainder of the coil continues to absorb heat and becomes superheated. In other words, the temperature of the refrigerant increases above the point at which it evaporates or vaporizes.

For example, an evaporator operating at a suction pressure of 28.5 psig (196.5 kPa) has a saturated liquid temperature of 30°F (–1.1°C), according to the temperature-pressure chart in Figure 12–9. As the refrigerant vaporizes (due to the absorption of heat from the evaporator), the temperature of the vapor rises until the temperature at the coil outlet, or **tailpipe,** reaches 35°F (1.67°C). Thus, the difference between the inlet and the outlet temperatures is 5°F (2.7°C).

The difference in temperature, 5°F (2.7°C) in this illustration, is known as superheat. All expansion valves

TEMPERATURE-PRESSURE CHART
(Evaporator temperature range)

TEMPERATURE		PRESSURE	
°F	°C	psig	kPa
20	–6.6	21	144.7
22	–5.5	22.4	154.4
24	–4.4	23.8	164.1
26	–3.3	25.3	174.4
28	–2.2	26.8	184.7
30	–1.1	28.5	196.5
32	0	30	206.8
34	+1.1	31.7	218.5
36	+2.2	33.4	230.2
38	+3.3	35.1	242
40	+4.4	36.9	254.4

FIGURE 12–9 Temperature-pressure chart, TXV range.

are adjusted at the factory to operate under the superheat conditions present in the particular type of air-conditioning system for which they are designed. When an expansion valve is being replaced, it is important to use a valve of the proper size and superheat range. Although many thermostatic expansion valves look the same, they differ greatly in their applications.

The TXV as a Control Device

The TXV consists of seven major parts, as shown in Figure 12–10:

- ❏ Valve body
- ❏ Valve seat
- ❏ Valve diaphragm
- ❏ Pushrod(s)
- ❏ Valve stem and needle
- ❏ Superheat spring with **adjuster**
- ❏ Capillary tube with remote bulb

As indicated previously, the remote bulb is fastened to the outlet, or tailpipe, of the evaporator. The bulb senses the tailpipe's temperature and activates the diaphragm in the valve through the capillary tube. In this manner, the proper amount of refrigerant is metered into the evaporator core.

For example, a high evaporator tailpipe temperature means that the evaporator is *starved* for refrigerant. This condition is indicated by an increase in the superheated vapor leaving the evaporator. As a result, the low-side pressure gauge indicates lower-than-normal readings.

The increased heat at the tailpipe causes an increase in the pressure exerted on the diaphragm by the expanding gases in the remote bulb through the capillary tube. The diaphragm, in turn, forces the pushrods down against the valve stem and the needle valve, which causes the needle valve to be pushed off its seat. In this way, more refrigerant is metered into the evaporator.

When the tailpipe temperature is low, there is less pressure exerted on the remote bulb, capillary tube, and diaphragm, resulting in the needle valve being seated. In this case, the flow of refrigerant into the evaporator is restricted.

Equalizer. As stated previously, thermostatic expansion valves are either internally or externally equalized. The term **equalized** refers to provisions made for exerting evaporator pressure to the bottom side of the diaphragm. In an internally equalized valve there is a drilled passage

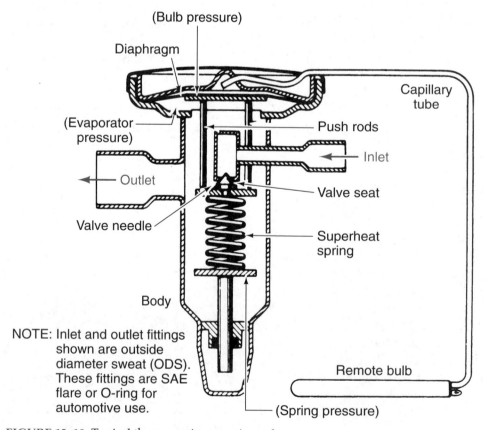

FIGURE 12–10 Typical thermostatic expansion valve.

from the evaporator side of the needle valve to the underside of the diaphragm. An externally equalized valve functions in the same manner, but it can pick up evaporator pressure at the outlet of the evaporator.

To overcome the effect of a pressure drop in larger evaporators, the externally equalized TXV is used. The external equalizer tube is connected to the tailpipe of the evaporator and runs to the underside of the diaphragm in the expansion valve. This arrangement impresses the pressure of the tailpipe, through the equalizer tube, to the underside of the diaphragm. The use of an external equalizer eliminates the effect of the pressure drop across the evaporator coil. The superheat setting, therefore, depends only on the adjustment of the superheat spring tension.

Screen

The thermostatic expansion valve (TXV) is equipped with a screen, depicted in Figure 12–11, in the inlet side of the valve. This screen can be cleaned if it becomes clogged, but it should be replaced if it becomes too clogged for cleaning. The receiver/drier should also be replaced if the screen requires cleaning or replacement. The screen must not be omitted from the system.

PERFORMANCE TESTING THE TXV

It must be noted that it is not always possible to do a **performance** test of the thermostatic expansion valve (TXV or TEV) in the vehicle. If this is the case, the TXV must be removed from the vehicle for bench testing, as outlined in Part 2 of this procedure. The procedure in Part 1 is given for those TXVs that can be tested in the vehicle. If it is necessary to remove the TXV, follow the procedures given in Part 3.

Tools

• Manifold and gauge set • Service hose set • Set of hand tools • Coupler, 1/4″ FF • Tee, 1/4″ MF • Safety glasses/goggles • Test cap, 1/4″ FF (drilled to 0.026″) • Adapters: TXV inlet × 1/4″ MF, TXV outlet × 1/4″ MF, TXV equalizer × 1/4″ MF

Materials

• Container of ice and water • Salt • Container of 125°F (52°C) water • Nitrogen

NOTE: A constant pressure source of 70 psig (483 kPa) is required to perform the bench test. Other suitable pressure sources, aside from nitrogen, include dry air or carbon diox-

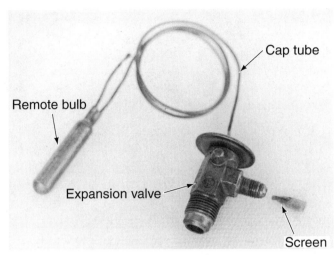

FIGURE 12–11 Typical thermostatic expansion valve (TXV) screen.

ide. These are dangerously high-pressure materials. They must have suitable pressure-regulating valves to reduce the pressure to a safe level.

Part 1 Procedure

This procedure is given for testing the thermostatic expansion valve (TXV or TEV) in the vehicle.

1. Attach the manifold and gauge set.
2. Start the engine and adjust the engine speed to 1,000-1,200 rpm.
3. Adjust all of the air-conditioning controls to maximum (MAX) cooling.
4. Operate the system for 10-15 minutes.

   NOTE: Follow steps 5 through 9 for abnormally low-side gauge readings or steps 10 through 12 for abnormally high low-side gauge readings.

5. Observe the low-side gauge reading. If it is abnormally low, place a warm (125°F or 52°C) rag around the TXV body, Figure 12–12.
6. Observe the low-side gauge. If the pressure rises to normal, or near normal, then moisture in the system is indicated (see the following note). If the pressure does not rise, proceed with Step 7.

   NOTE: To correct moisture in the system, replace the desiccant (receiver-drier or accumulator-drier). Evacuate, charge, and retest the system.

FIGURE 12–12 Place a warm rag around the TXV body.

FIGURE 12–13 Place the remote bulb in a warm rag.

7. Remove the TXV remote bulb from the evaporator outlet and warm it in the hand or in a warm (125°F or 52°C) rag, Figure 12–13.

8. Observe the low-side gauge pressure. If the pressure rises, then the remote bulb was probably improperly placed (see the following note). If the pressure does not rise, proceed with step 9.

NOTE: To correct the rise in pressure, reposition the remote bulb, insulate, and retest the system.

9. If the low-side gauge reading is abnormally low and steps 6 through 9 do not correct the problem, remove the TXV from the system for bench testing (see Part 2).

NOTE: Observe the inlet screen when removing the TXV. If it is clogged, the receiver-drier or desiccant must be replaced after servicing or cleaning the TXV.

10. If the low-side gauge reading is abnormally high, remove the remote bulb from the evaporator outlet and place it in an ice-water bath.

NOTE: Rock salt added to the ice water will lower its temperature to 32°F (0°C).

11. If the pressure falls to normal or near normal, the problem may be:
    a. A lack of insulation at the remote bulb. Reinsulate the area and retest the system.
    b. An improperly placed remote bulb. Reposition the remote bulb and retest the system.

12. If the pressure does not fall to normal or near normal, remove the TXV and bench test as outlined in Part 2.

13. Turn off all of the air-conditioning controls.

14. Reduce the engine speed to idle and stop the engine.

15. Remove the manifold and gauge set.

Part 2 Bench Test the TXV

This procedure is given to bench test the thermostatic expansion valve that is removed from the system; see the procedure in Part 3 to remove and replace the TXV.

1. Close the high- and low-side manifold hand valves.

2. Remove the low-side service hose at the manifold.

3. Install a 1/4-inch female flare coupler to the low-side manifold.

4. Install a 1/4-inch male flare tee to the flare coupler at the low side manifold.

5. Reinstall the low-side manifold hose to the 1/4-inch flare tee.

6. Install a 1/4-inch test cap, drilled to 0.026 inch (0.66 mm), to the 1/4-inch tee.

7. Install the TXV inlet × 1/4-inch MF adapter to the inlet of the TXV.

8. Install the TXV outlet × 1/4-inch MF adapter to the outlet of the TXV.

9. Connect the low-side manifold hose to the expansion valve outlet.

10. Connect the high-side manifold hose to the expansion valve inlet.

11. Prepare the nitrogen cylinder and regulator for service.

12. Adjust the nitrogen regulator to exactly 70 psig (483 kPa).

13. Fill an insulated container with cracked ice and add water. Use a thermometer to determine when the temperature is 32°F (0°C). If necessary, add salt and stir the mixture.

14. Heat water in a second container until it reaches 125°F (52°C).

NOTE: Figure 12–14 diagrams the thermostatic expansion valve test connections for an internally equalized thermostatic expansion valve (TXV) only. If an externally equalized expansion valve is to be tested, another fitting must be added before the test cap, as shown in Figure 12–15. The external equalizer is connected to this fitting. If an externally equalized expansion valve is to be tested, the following additional tools are required: one 1/4-inch female flare coupler and one 1/4-inch male flare tee.

15. Place the thermostatic expansion valve remote bulb into the container of water that was heated to 125°F (52°C).
16. Open the high-side gauge manifold hand valve and adjust it to exactly 70 psig (483 kPa).
17. Read the low-side gauge.

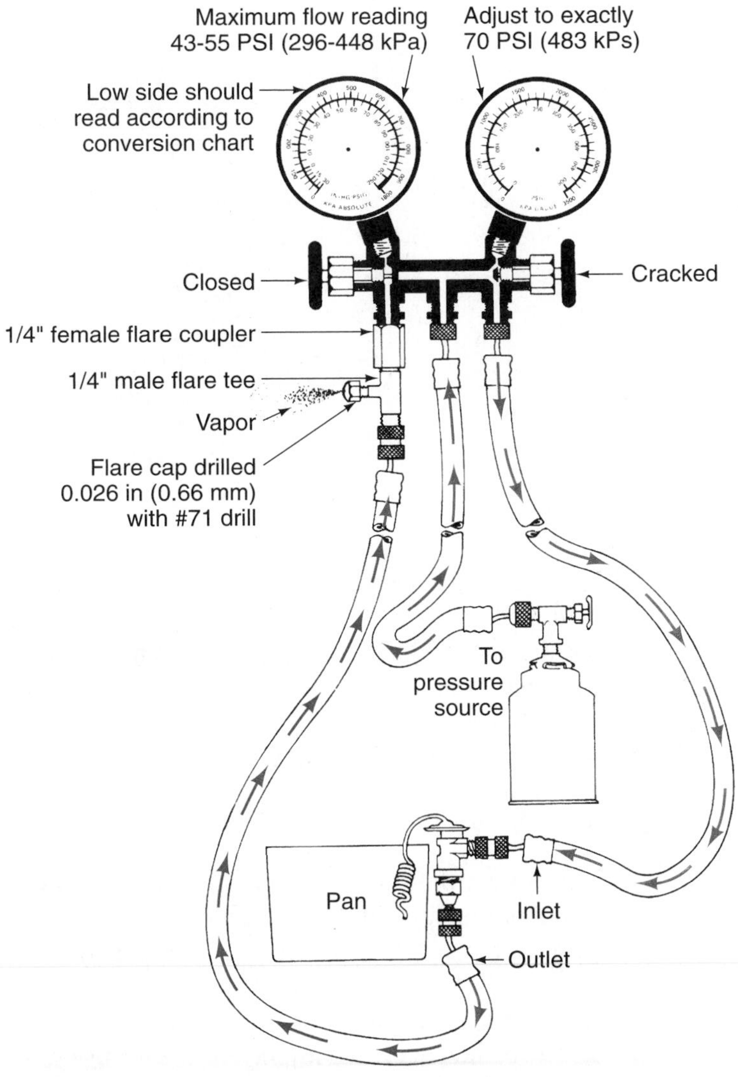

NOTE: The maximum flow test should be 43-55 psig (296-379 kPa). Readings over 55 psig (379 kPa) indicate a flooding valve. A reading under 43 psig (296 kPa) indicates a starving valve.

18. Place the thermal bulb into the container of water at 32°F (0°C).
19. Open the high-side gauge manifold hand valve and adjust it to exactly 70 psig (483 kPa).
20. Read the low-side gauge.

NOTE: Refer to the conversion charts in Figure 12–16 or 12–17, as applicable, for the proper low-side reading. The low-side gauge must be within the limits specified in the conversion chart if the valve is to pass the minimum flow test.

FIGURE 12–14 Test connections for the internally equalized valve.

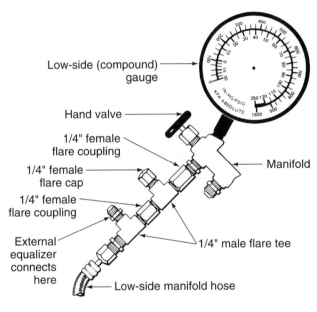

FIGURE 12–15 Modified version of Figure 12–14 for use with externally equalized valves.

CAUTION

ALWAYS WEAR APPROPRIATE SAFETY GLASSES OR GOGGLES WHEN WORKING WITH HIGH PRESSURE GASES SUCH AS NITROGEN.

Conversion Chart	
Superheat Setting ° F	Pounds Per Square Inch Gauge
5	23 lb to 26 lb
6	22 1/4 lb to 25 1/4 lb
7	21 1/2 lb to 24 1/2 lb
8	21 lb to 24 lb
9	20 1/4 lb to 23 1/4 lb
10	19 1/2 lb to 22 1/2 lb
11	19 lb to 22 lb
12	18 lb to 21 lb
13	17 1/2 lb to 20 1/2 lb
14	17 lb to 20 lb
15	15 1/2 lb to 18 1/2 lb

FIGURE 12–16 Conversion chart for TXV testing (English). For absolute (psia) ass 14.7 to the pressure indicated.

Superheat Setting ° C	Kilopascals (Gauge)
2.8	158.6-179.3
3.3	153.4-174.1
3.9	148.2-168.9
4.4	144.8-165.5
5.0	139.6-160.3
5.6	134.5-155.1
6.1	131.0-151.7
6.7	124.1-144.8
7.2	120.7-141.3
7.8	117.2-137.9
8.3	106.9-127.6

FIGURE 12–17 Conversion chart for TXV testing (metric). For absolute kPa, add 101.4 to the pressure indicated.

Removing and Replacing the TXV

This procedure is given in two parts: Part 3A for servicing the standard TXV and Part 3B for servicing the H-valve TXV. This procedure assumes that there is no refrigerant remaining in the air-conditioning system.

Part 3A Servicing the Standard TXV

1. Remove the insulation tape from the remote bulb.
2. Loosen the clamp to free the remote bulb.
3. Disconnect the external equalizer, if so equipped.
4. Remove the liquid line from the inlet of the TXV.
5. Remove and discard the O-ring, if so equipped.
6. Inspect the inlet screen (Figure 12–18).
   a. If clogged, clean or replace the screen. Skip to step 13.
   b. If not clogged, proceed with step 7.
7. Remove the evaporator inlet fitting from the outlet of the TXV.
8. Remove and discard the O-ring, if so equipped.
9. Remove the holding clamp (if provided on the TXV) and carefully remove the TXV from the evaporator.
10. Carefully position the new TXV on the evaporator.
11. Insert new O-ring(s) on the evaporator inlet, if so equipped.
12. Attach the evaporator inlet to the TXV outlet. Tighten to the proper torque (see the manufacturer's specifications).
13. Install new O-rings on the liquid line fitting, if so equipped.
14. Attach the liquid line to the TXV inlet. Tighten to the proper torque (see the manufacturer's specifications).

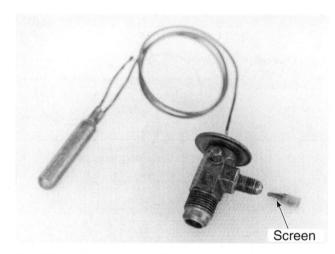

FIGURE 12–18 Inspect the TXV inlet screen.

15. Reconnect the external equalizer tube, if so equipped.
16. Position the remote bulb and secure it with a clamp.
17. Tape the remote bulb to prevent it from sensing ambient air.
18. Proceed with step 13 of Part 3B.

Part 3B H-Valve TXV

1. Disconnect the wire connected to the pressure cutout or pressure differential switch, as applicable.
2. Remove the bolt from the line sealing plate found between the suction and liquid lines.
3. Carefully pull the plate from the H-valve.
4. Cover the line openings to prevent the intrusion of foreign matter.
5. Remove and discard the plate to the H-valve gasket.

6. Remove the two screws from the H-valve.
7. Remove the H-valve from the evaporator plate.
8. Remove and discard the H-valve to the evaporator plate gasket.
9. Install a new H-valve with gasket, Figure 12–19.
10. Replace the two screws. Torque to 170–230 in.-lb. (20–26 N•m).
11. Replace the line plate to the H-valve gasket.
12. Hold the line assembly in place and install the bolt. Torque to 170–230 in.-lb. (20–26 N•m).
13. Replace the access panels and any hardware that was previously removed.
14. Reconnect the battery, following the instructions given in the manufacturer's shop manual.
15. Leak test, and evacuate and charge the system with refrigerant.
16. Hold a performance test, if required.

Part 4 Cleaning the TXV

NOTE: If the TXV fails to pass either one or both of the tests given in Part 1 or Part 2, cleaning may be attempted on some valves. Otherwise, a new valve must be used. Although each valve is different in structure, the following steps may be used as a cleaning guide.

1. Remove the diaphragm, capillary, and remote bulb assembly.
2. Remove the superheat adjusting screw. Carefully count the number of turns required to remove the screw. Knowing the number of turns aids in locating the proper position when reassembling the valve.
3. Remove the superheat spring and the valve seat. Remove the valve and pushrod(s).
4. Clean the valve and all of the parts in *clean* mineral spirits. Let the parts drain and then blow-dry them.
5. Reverse steps 1 through 4 and reassemble the valve.
6. Check the expansion valve for maximum and minimum flow, as outlined in this procedure.
7. If the valve fails to pass the maximum and minimum flow tests, attempt to adjust the superheat spring setting.
8. If the valve fails the test repeatedly, a new valve, like the one shown in Figure 12–20, must be installed. No further repair is possible.

OTHER VALVE TYPES

Basically, the internally and externally equalized thermostatic expansion valves (TXVs) and the fixed orifice tube (FOT) have prevailed as metering devices in motor vehicle air-conditioning (MVAC) system applications.

FIGURE 12–19 A typical H-valve.

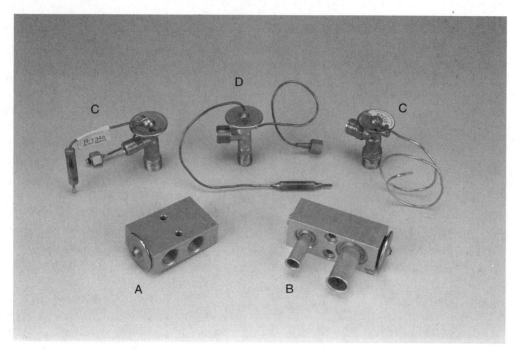

FIGURE 12–20 Thermostatic expansion valves: (A) H-valve; (B) block type; (C) internally equalized; (D) externally equalized.

Throughout the years there have been many devices used in MVAC systems to regulate temperature by regulating the evaporator pressure. Many of these devices, such as the evaporator pressure regulator (EPR) and suction throttling valve (STV), are seldom encountered by today's automotive technician. The latest of these devices was called a valves-in-receiver and is briefly discussed next.

Valves-in-Receiver (VIR)

On some early vehicles a capsulized thermostatic expansion valve (TXV) is found in the *valves-in-receiver* (VIR) or in the later version *evaporator equalizer valves-in-receiver* (EEVIR), as illustrated in Figure 12–21. The VIR or EEVIR also contains the receiver with desiccant and a capsulized suction pressure regulator known as a *positive-operated suction throttling valve* (POASTV) or, more simply, a POA valve. The VIR was last used in 1983 on an Audi 5000.

The TXV is capsulized and placed inside the VIR assembly with the POA valve. Its function and purpose are similar to that of the standard TXV: to meter the proper amount of refrigerant into the evaporator under varying heat-load conditions.

Combination Valve

The *combination valve*, also called a *combo valve*, is found on some early Ford vehicles. The combo valve is similar in operation to the VIR in that it contains a thermostatic

FIGURE 12–21 Typical evaporator equalizer valves-in-receiver (EEVIR or VIR). *(Courtesy of BET, Inc.)*

expansion valve (TXV) and a suction pressure regulator known as a *suction throttling valve* (STV). However, the combo valve, as shown in Figure 12–22, does not include a receiver or desiccant as does the VIR.

Other Valves

In certain Chrysler vehicles the thermostatic expansion valve, as shown in Figure 12–23 is called the H-valve. Some

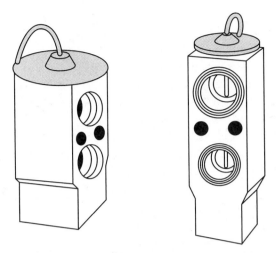

FIGURE 12–22 Typical HFC-134a combination valve.

Ford vehicles use a similar valve, called a *block valve*. Like the combination valve, the H-valve and the block valve do not have capillary tubes. Unlike the combination valve, however, the H-valve and the block valve do not have suction pressure regulators as part of their assembly.

FIXED ORIFICE TUBE

The orifice tube is a calibrated restrictor located in the liquid line that is used as a means of metering liquid refrigerant into the evaporator. Its purpose is to meter

high-pressure liquid refrigerant into the evaporator as a low-pressure liquid.

The amount of refrigerant that enters the evaporator in an orifice tube air-conditioning system depends on three factors: the size of the orifice, subcooling of the refrigerant, and the pressure differential (Δp) between the inlet and outlet of the orifice device. The orifice tube is frequently referred to as a fixed orifice tube (FOT) because of its tubular shape and fixed orifice. It is available in sizes ranging from 0.047 inch (1.19 mm) to 0.072 inch (1.83 mm), depending on application. The color of the orifice tube generally designates its size as well as its application. Although similar in appearance, FOTs are not generally interchangeable. General Motors' color codes may be bronze, white, or white/red. The color codes given in Chart 1 are for Ford's FOTs.

COLOR	ORIFICE SIZE (diameter)
Blue or Black	0.067 inch (1.70 mm)
Red	0.062 inch (1.57 mm)
Orange	0.057 inch (1.45 mm)
Brown	0.053 inch (1.35 mm)
Green	0.047 inch (1.19 mm)

CHART 1: Colors identify orifice size for Ford FOT

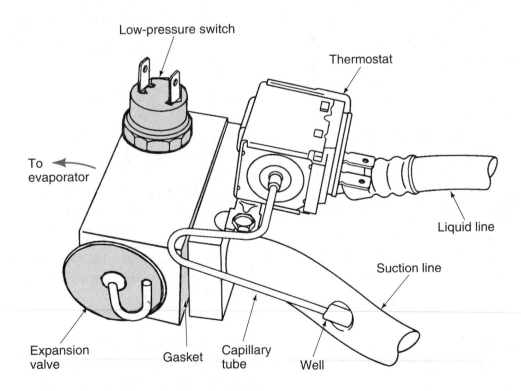

FIGURE 12–23 Chrysler's H-valve assembly.

A fine-mesh filter screen, Figure 12–24, protects the inlet and outlet of the orifice tube. If foreign matter blocks, or partially blocks, the orifice, the air-conditioning system will not function to full efficiency. If the blockage is severe enough, the system may not function at all. The inlet of the early-series fixed orifice tube used by General Motors does not contain a screen. However, beginning in 1976, a screen has been included in the later-series fixed orifice tube. In either case, replacement of the FOT is suggested if the screen or tube becomes clogged. The accumulator should also be replaced to prevent a recurrence of the clogged FOT. Although this screen may often be cleaned, it is generally recommended that it be replaced if found to be clogged.

O-rings are used on the FOT body, Figure 12–25, to provide a seal to prevent refrigerant from bypassing the orifice. The FOT should be replaced if any O-rings are found to be defective.

The fixed orifice tube, with few exceptions, is located in a cavity in the liquid line or at the inlet connection provisions of the evaporator and is, in general, easily accessible. However, some vehicles have an inaccessible FOT located inside the liquid line. If found to be defective, the liquid line must be replaced or a repair kit must be used to replace a section of the liquid line containing the FOT. The procedures for replacing the FOT are included in this chapter.

Some "aftermarket engineers" replace the standard original equipment manufacturer's (OEM) fixed orifice tube with an aftermarket "Smart Variable Orifice Valve" (VOV). The Smart VOV utilizes system pressure to move a metering piston relative to a fixed opening in the sleeve. This is claimed to compensate for reduced compressor output at idle speeds, thereby increasing the air-conditioning system's cooling performance. The Smart VOV manufacturer claims that it is a "drop-in" replacement for ineffective OEM orifice tubes and that it can offer a "dramatic improvement on factory R-134a systems." Before making any changes to an automotive air-conditioning system, however, it is strongly suggested that the manufacturer's recommendations be followed.

Beginning in 1999, the factory-installed air-conditioning system in some Jeep models has been equipped with a variable orifice tube. The Jeep design, however, is slightly different from the aftermarket VOVs. Since then other vehicle manufacturers have followed Jeep's lead.

There are two methods of temperature control used on General Motors' orifice tube systems. One method, called a cycling clutch orifice tube (CCOT) system, uses a fixed displacement compressor. A pressure- or temperature-actuated cycling switch is used to turn the compressor's electromagnetic clutch on and off. This action starts and stops the compressor to maintain the desired in-vehicle temperature.

The other orifice tube system used by General Motors has a variable displacement (VD) compressor, which regulates the quantity of refrigerant that flows through the system to maintain the selected in-vehicle temperature. This system is called the variable displacement orifice tube (VDOT) system and eliminates the requirement to cycle a clutch on and off to maintain the desired in-vehicle temperature conditions.

Many Ford vehicles use the cycling clutch method of temperature control on their orifice tube systems. Either a temperature- or pressure-actuated control may be used to cycle the clutch on and off to maintain the selected in-vehicle temperature conditions. Ford's system is called a fixed orifice tube/cycling clutch (FOTCC) system.

When the engine is stopped and/or the air-conditioning system is shut down, the pressure of the refrigerant within the system will equalize. When this occurs, the flow of refrigerant may be detected as a faint hissing sound for 30–60 seconds.

All orifice tube air conditioning systems have an accumulator located at the evaporator outlet, Figure 12–26. The accumulator prevents unwanted quantities of

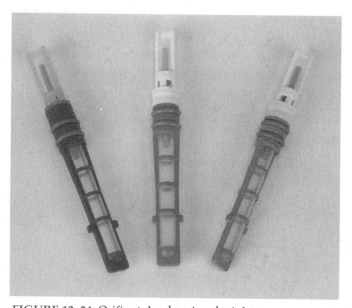

FIGURE 12–24 Orifice tube showing the inlet screen.

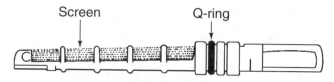

FIGURE 12–25 A typical fixed orifice tube.

FIGURE 12–26 Accumulator located at the evaporator outlet.

liquid refrigerant and/or lubricant from returning to the compressor at any time.

TESTING AND/OR REPLACING THE FOT

Remember that orifice tubes are not interchangeable although the same service tool may be used to remove and replace them. When replacing the orifice tube, it is most important that the correct replacement be used.

Some vehicles have a nonaccessible orifice tube in the liquid line. Its exact location, which could be anywhere between the condenser outlet and evaporator inlet, is determined by a circular depression or three indented notches in the metal portion of the liquid line. An orifice tube replacement kit is used to replace this type of orifice tube and 2.5 inches (63.5 mm) of the metal liquid line.

Testing the fixed orifice tube is reasonably simple. There are only two basic problems that may be found: moisture in the system freezing at the orifice tube or a clogged tube.

This service procedure is given in three parts: Part 1 covers testing the FOT, Part 2 covers replacing the accessible FOT, and Part 3 covers replacing the nonaccessible FOT.

Tools

• Manifold and gauge set • FOT remover/installer • Service hose set • Service wrench set • Tubing cutter

Materials

• Orifice tube • Refrigerant • Accumulator • Refrigeration oil • O-rings

Part I Testing the FOT

1. Attach the manifold and gauge set.

2. Start the engine and adjust the speed to 1,000–1,200 rpm.
3. Adjust all of the air-conditioning controls to maximum (MAX) cooling.
4. Operate the system for 10–15 minutes.
5. Observe the low-side gauge. An abnormally low-side gauge reading indicates that the orifice tube is not metering a sufficient amount of refrigerant into the evaporator.

NOTE: It may be necessary to bypass the low-pressure switch to prevent clutch cycling on a CCOT system. Bypass the switch only long enough to perform the test. Use steps 6 through 8 to determine if the problem is caused by moisture or by a restriction in the air-conditioning system.

6. Place a warm rag (125°F) (52°C) around the fixed orifice tube.
7. Observe the low-side gauge. If the pressure rises to normal or near normal, moisture in the system is indicated.
8. If moisture is found in the system, the accumulator must be replaced. If moisture is not indicated, the orifice tube is probably clogged and must be replaced. Follow the procedure outlined in Part 2 of this procedure.

Part 2 Removing and Replacing the FOT

As in Part 1, it must be noted that orifice tubes are not interchangeable although the same service tool may be used to remove and replace them. When replacing the orifice tube, it is most important that the correct replacement part be used.

Some vehicles have a nonaccessible orifice tube in the liquid line. Its exact location, which could be anywhere between the condenser outlet and the evaporator inlet, is determined by a circular depression or three indented notches in the metal portion of the liquid line. An orifice tube replacement kit is used to replace this type of orifice tube and 2.5 inches (63.5 mm) of the metal liquid line.

This service procedure assumes that all the refrigerant has been removed from the air-conditioning system. It is given in two parts: Part 3A for servicing the accessible FOT and Part 3B for replacing the nonaccessible FOT.

Part 3A Servicing the Accessible FOT

The procedure is given in two parts: removing the FOT,

steps 1 through 9; and replacing the FOT, steps 10 through 14.

NOTE: It may not be necessary to disconnect the battery for this procedure.

1. Using the proper open-end or flare-nut wrenches, remove the liquid line connection at the inlet of the evaporator to expose the FOT, Figure 12–27.
2. Remove and discard the O-ring(s) from the liquid line fitting, if so equipped.
3. Pour a very small quantity of clean refrigeration oil into the FOT well to lubricate the seals, Figure 12–28.

WARNING

USE A LUBRICANT THAT IS COMPATIBLE WITH THE SYSTEM REFRIGERANT, THAT IS, MINERAL OIL FOR CFC-12 AND PAG FOR HFC-134A.

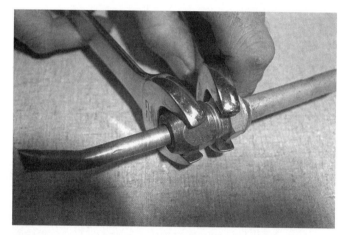

FIGURE 12–27 Using the proper open-end or flare-nut wrenches, remove the liquid line from the evaporator inlet fitting.

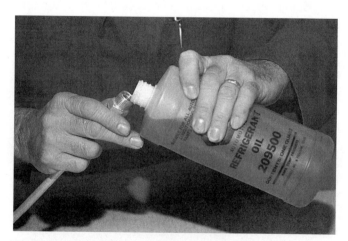

FIGURE 12–28 Pour a small quantity of refrigeration lubricant into the orifice tube well to lubricate the O-rings.

4. Insert the FOT removal tool onto the FOT, Figure 12–29.
5. Turn the T-handle of the tool slightly clockwise (cw), only enough to engage the tool onto the tabs of the FOT.
6. Hold the T-handle and turn the outer sleeve or spool of the tool clockwise to remove the FOT. Do not turn the T-handle, Figure 12–30.

NOTE: If the FOT breaks during removal, proceed with step 7. If it does not break, proceed with step 9.

7. Insert the extractor into the well and turn the T-handle clockwise until the threaded portion of the tool is securely inserted into the brass portion of the broken FOT, Figure 12–31.
8. Pull the tool. The broken FOT should slide out, Figure 12–32.

NOTE: The brass tube may pull out of the plastic body. If this happens, remove the brass tube from the puller and reinsert the puller into the plastic body. Repeat steps 7 and 8.

9. Liberally coat the new FOT with clean refrigeration oil, Figure 12–33. (See the note following step 3.)
10. Place the FOT into the evaporator cavity and push it in until it stops against the evaporator tube inlet dimples, Figure 12–34.
11. Install a new O-ring, if so equipped.
12. Replace the liquid line and tighten to the recommended torque, Figure 12–35. (Refer to the manufacturer's specifications.)
13. Replace the accumulator.

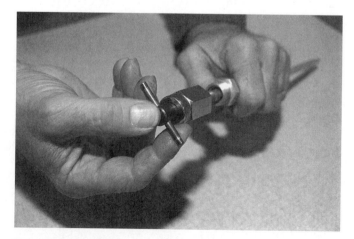

FIGURE 12–29 Insert an orifice tube removal tool and turn the T-handle slightly clockwise to engage the orifice tube.

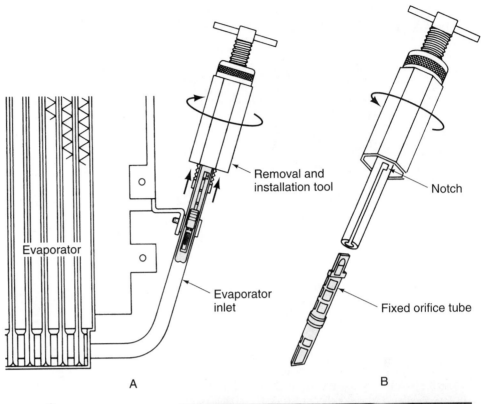

FIGURE 12–30 Remove the fixed orifice tube by turning the outer sleeve only. Do not turn the handle. The handle is turned only enough to engage the notch of the tool onto the fixed orifice tube, detail B. *(Courtesy of General Motors Corporation, Service Operations)*

A

B

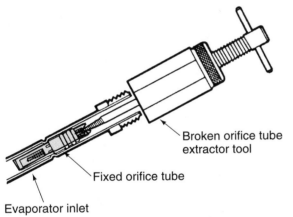

FIGURE 12–31 Using an extractor tool to remove a broken fixed orifice tube.

FIGURE 12–33 Coat the new orifice tube liberally with clean refrigeration lubricant.

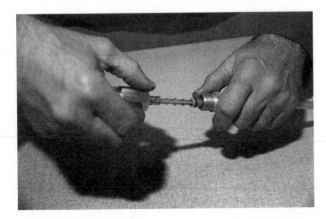

FIGURE 12–32 Pull the tool. The orifice tube should slide out.

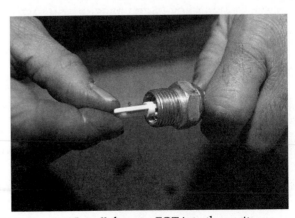

FIGURE 12–34 Install the new FOT into the cavity.

FIGURE 12–35 Connect the liquid line to the evaporator and tighten the nut with the proper wrenches.

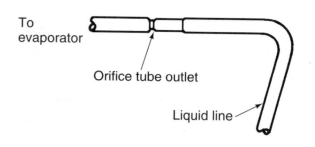

To evaporator

Orifice tube outlet

Liquid line

From condenser

FIGURE 12–36 Locating the orifice tube.

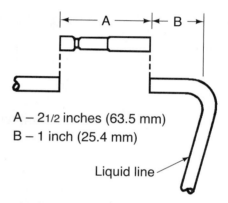

A – 2 1/2 inches (63.5 mm)
B – 1 inch (25.4 mm)

Liquid line

FIGURE 12–37 Cut out the old orifice tube.

Part 3B Servicing the Nonaccessible FOT

1. Disconnect the battery, following the manufacturer's instructions.
2. Remove the liquid line from the evaporator inlet. Remove and discard the O-rings, if so equipped.
3. Remove the liquid line from the condenser outlet. Remove and discard the O-rings, if so equipped.

   NOTE: Note how the liquid line was routed, steps 2 and 3, so it can be replaced in the same manner.

4. Locate the orifice tube. The outlet side of the orifice tube can be identified by a circular depression or three notches, Figure 12–36.
5. Use a tube cutter to remove a 2.5-inch (63.5 mm) section of the liquid line, Figure 12–37.

   NOTE: Allow at least 1 inch (25.4 mm) of exposed tube at either side of any bend.

   CAUTION
   DO NOT USE EXCESSIVE PRESSURE ON THE FEED SCREW OF THE TUBE CUTTER TO AVOID DISTORTING THE LIQUID LINE TUBING. IF A HACKSAW MUST BE USED, FLUSH BOTH ENDS OF THE CUT LIQUID LINE WITH CLEAN REFRIGERATION OIL TO REMOVE CONTAMINANTS, SUCH AS METAL CHIPS.

6. Slide a compression nut onto each section of the liquid line.
7. Slide a compression ring onto each section of the liquid line with the taper portion toward the compression nut.

8. Lubricate the two O-rings with clean refrigeration oil and slide one onto each section of the liquid line.

   WARNING
   USE A LUBRICANT THAT IS COMPATIBLE WITH THE SYSTEM REFRIGERANT, THAT IS, MINERAL OIL FOR CFC-12 AND PAG FOR HFC-134A.

9. Attach the orifice tube housing, with the orifice tube inside, to the two sections of the liquid line, Figure 12–38.
10. Hand tighten both compression nuts. Be sure to note the flow direction indicated by the arrows. The flow should be toward the evaporator or away from the condenser.
11. Hold the orifice tube housing in a vise or other suitable fixture to tighten the compression nuts.

    NOTE: Be sure the hose bends are in the same position as when they were removed from the air-conditioning system for ease in replacing the liquid line.

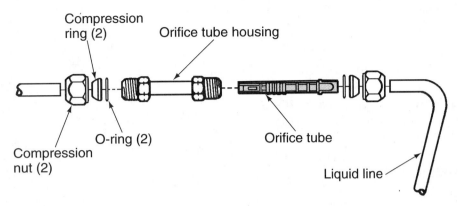

FIGURE 12–38 Exploded view of the new orifice tube assembly.

12. Tighten each compression nut to a torque of 65-70 ft.-lb. (87–94 N•m).
13. Insert new O-ings on both ends of the liquid line, if so equipped.
14. Install the liquid line:
    a. Attach the condenser end of the liquid line to the condenser and tighten to the proper torque.
    b. Repeat step 14a for the evaporator end of the liquid line.
15. Leak test, and evacuate and charge the system.

SUMMARY

If the expansion valve is removed from the air-conditioning system for cleaning or other service, it may be bench checked for proper operation before it is reinstalled. This chapter covers a method for bench checking the TXV for efficiency. This procedure saves time as well as refrigerant, which is otherwise lost through a defective valve.

Thermostatic expansion valves are provided with flange, flare, or O-ring fittings on each side. The compari-son among the fittings is shown in Figure 12–39. Although the valves shown may have the same ratings, they cannot be interchanged because the types of fittings do not mate with each other.

The previous descriptions of the thermostatic expansion valve and the expansion tube make it clear that these devices are more sensitive to foreign materials than are any other parts of the air-conditioning system. This fact makes it essential to keep the system as free as possible of contaminants during service procedures.

To prevent the vital parts of the expansion valve from sticking or becoming corroded, the air conditioner should be operated for short periods during the months that normal operation is not practical. In this manner, the internal parts of the TXV, as well as the compressor, are lubricated and are kept operating freely. It must be noted, however, that some air-conditioning systems are equipped with a low ambient temperature switch and cannot be operated during cold weather. If this is the case, periodic operation of the system during cold weath-er is impossible.

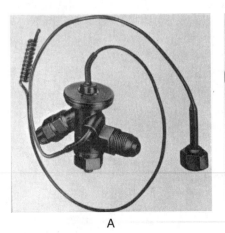
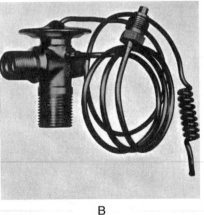

A   B   C

FIGURE 12–39 Typical expansion valve types: (A) flare; (B) O-ring; (C) flange.

REVIEW

Select the correct answer from the choices given.

1. The state of the refrigerant immediately after the metering device is being discussed. Technician A says that the refrigerant is a vapor with traces of liquid. Technician B says that the refrigerant is a liquid with traces of vapor. Who is right?
   a. A only
   b. B only
   c. Both A and B
   d. Neither A nor B

2. A TXV-equipped air-conditioning system is being discussed. Technician A says a starved evaporator may be serious because it can result in high superheat. Technician B says a flooded evaporator may be serious because it can result in liquid slugging of the compressor. Who is right?
   a. A only
   b. B only
   c. Both A and B
   d. Neither A nor B

3. All of the following are metering devices, EXCEPT:
   a. a TXV.
   b. a VOV.
   c. a FOT.
   d. a PMV.

4. An FOT-equipped air-conditioning system is being discussed. Technician A says a starved evaporator is serious because it results in poor cooling and high superheat. Technician B says a flooded evaporator is serious because it results in poor cooling and liquid slugging of the compressor. Who is right?
   a. A only
   b. B only
   c. Both A and B
   d. Neither A nor B

5. Technician A says the ideal superheat for a motor vehicle air-conditioning (MVAC) system is 10°F–20°F (5.6°C–11.1°C). Technician B says that flash gas is a contributing factor for superheat. Who is right?
   a. A only
   b. B only
   c. Both A and B
   d. Neither A nor B

6. Technician A says the temperature difference between the evaporator inlet and outlet is known as subcooling. Technician B says the temperature difference between the evaporator outlet and compressor inlet is known as superheat. Who is right?
   a. A only
   b. B only
   c. Both A and B
   d. Neither A nor B

7. Technician A says that a TXV-equipped air-conditioning system has a desiccant. Technician B says that an FOT-equipped air-conditioning system has a desiccant. Who is right?
   a. A only
   b. B only
   c. Both A and B
   d. Neither A nor B

8. All of the following are functions of the TXV, EXCEPT:
   a. it meters the flow of refrigerant into the evaporator.
   b. it controls the refrigerant pressure in the evaporator.
   c. it provides a dividing line between high and low side.
   d. it prevents freeze-up of the evaporator coil.

9. A clogged screen in a metering device will cause:
   a. low suction pressure.
   b. high suction pressure.
   c. high head pressure.
   d. engine overheating.

10. The remote bulb of a TXV is fastened to the:
    a. evaporator outlet.
    b. liquid line.
    c. TXV outlet.
    d. evaporator inlet.

11. Most FOTs may be serviced by:
    a. cleaning the screen.
    b. cleaning the tube.
    c. either A or B
    d. neither A nor B

12. All of the following may be found in an FOT-equipped air-conditioning system, EXCEPT:
    a. an auxiliary drier.
    b. a discharge line muffler.
    c. an accumulatot-drier.
    d. an external equalizer.

13. An FOT may be identified by:
    a. its size.
    b. its color.
    c. its shape.
    d. all of the above.

14. The VOV was first used by:
    a. Ford.
    b. Jeep.
    c. General Motors.
    d. Chrysle.r

15. Technician A says that the fixed orifice tube meters a fixed amount of refrigerant into the evaporator. Technician B says that the thermostatic expansion valve meters a fixed amount of refrigerant into the evaporator. Who is right?
    a. A only
    b. B only
    c. Both A and B
    d. Neither A nor B

16. The superheat may be adjusted in:
    a. an FOT.
    b. a TXV.
    c. both A and B.
    d. neither A nor B.

17. All of the following may be found in a TXV-equipped air-conditioning system, EXCEPT:
    a. an auxiliary drier.
    b. an internal equalizer.
    c. an accumulator-drier.
    d. an external equalizer.

18. All of the following designate a type of FOT, EXCEPT:
    a. CCOT.
    b. FOTCC.
    c. CCFOT.
    d. VOV.

19. The evaporator pressure is equalized in the TXV by provisions of:
    a. an internal equalizer.
    b. an external equalizer.
    c. Both A and B.
    d. Neither A nor B.

20. Technician A says that if the inlet screen of the TXV is clogged, it may be cleaned. Technician B says that if the inlet screen of the FOT is clogged, it may be replaced. Who is right?
    a. A only
    b. B only
    c. Both A and B
    d. Neither A nor B

TERMS

Write a brief description of the following terms:

1. adjuster
2. boils
3. capillary tube
4. diaphragm
5. equalized
6. FOT
7. heat load
8. H-valve
9. inert
10. modulates

11. needle valve
12. orifice
13. performance
14. remote bulb
15. restricted
16. saturated
17. seat
18. superheat
19. tailpipe
20. TXV

CHAPTER 13

BASIC ELECTRICITY REVIEW

OBJECTIVES

On completion and review of this chapter, you should be able to:

- ❏ Understand and use the laws of electricity, such as Ohm's law.
- ❏ Discuss the differences among insulators, conductors, and semiconductors.
- ❏ Define voltage, current, and resistance.
- ❏ Explain the basic concepts of capacitance and inductance.
- ❏ Discuss the differences between AC and DC currents.
- ❏ Illustrate and explain series, parallel, and series-parallel circuits.
- ❏ Discuss the basic theory of semiconductors.
- ❏ Explain the theory of electromagnetism.
- ❏ Describe the electrical components and protection devices used in motor vehicle air-conditioning (MVAC) systems.
- ❏ Explain the operation of electrical and electronic components used in mobile air-conditioning systems.
- ❏ Understand circuit defects such as opens, shorts, and grounds.
- ❏ Understand, explain, and use wiring diagrams, schematics, and symbols.

INTRODUCTION

The electrical and electronic systems of today's modern vehicles are not as complicated or difficult to understand as they may first appear. A good understanding of the principles and laws that govern electrical circuits is a great asset in diagnosing electrical and electronic system problems. The purpose of this chapter is to discuss the laws that dictate electrical behavior, how circuits operate, the differences among the types of circuits, and how to apply Ohm's law.

This chapter also provides a review of electrical theory and components that are associated with vehicle air-conditioning systems. The technician must understand the operation of the various electrical components as well as the way they affect electrical system operation to be able to accurately and quickly diagnose electrical system failures.

Also, it is important that the technician understand, and be able to use, the various test equipment designed for electrical system diagnosis. This chapter provides an introduction to the various types of test equipment used for troubleshooting and diagnosing electrical systems. The various types of defects that cause system failure are discussed in this chapter as well.

ELECTRICITY

A leading electronic dictionary defines *electricity* as "the phenomena associated with dynamic and static electric charges, such as electrons." Because electricity is a form of energy that cannot be seen, it is important for an automotive technician to understand electrical behavior. Electrical behavior is confined to definite laws that produce predictable results and effects. A review of the laws of electricity begins with a short study of atoms and the atomic structure.

ATOMIC STRUCTURE

As discussed in Chapter 5, an atom is made up of electrons in tightly bound fixed **orbits** around a nucleus, Figure 13–1. The electrons, which are free to move within their orbits around the nucleus, are negatively charged. As such, they repel each other when they get close and therefore stay as far away as possible from each other without leaving their orbits.

Atoms like to have the same number of electrons and protons to be in balance. In order to remain in balance, an atom will give up an electron to another atom or it may pull an electron in from another atom. A specific number of electrons are in each of the electron orbit paths around the nucleus of the atom. Regardless of the matter, the orbit nearest the nucleus has but two electrons. The second orbit has as many, but no more than, eight electrons; the third orbit has as many, but no more than, eighteen electrons; and the fourth holds up to, but no more

than, thirty-two electrons. The number of electrons in an atom determines the number of orbits.

Conductors and Insulators

It is important to know about electrons and their orbits to understand the atomic structure of matter. For example, examine the atomic structure of copper (Cu), a metal most commonly used as a conductor of electricity. A copper (Cu) atom contains twenty-nine electrons; two in the first orbit, eight in the second, and eighteen in the third orbit. The remaining electron is found in the outer orbit, Figure 13–2. The outer orbit, sometimes called the shell, is referred to as the **valence ring.**

Conductors.   Because there is only one electron in the valence ring of a copper (Cu) atom, copper wire is a good **conductor** of electricity. Other good conductors, such as silver (Ag) and gold (Au), have only one or two electrons in their valence rings. The valence rings of these atoms can be made to give up their electrons with little effort.

Electricity is the movement of electrons from one atom to another, Figure 13–3. Atoms that have but one electron in their valence ring allow it to easily move to the valence ring of another atom nearby. A wire made up of millions of copper atoms therefore becomes a good

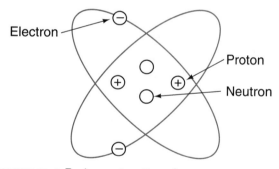

FIGURE 13–1 Basic construction of an atom.

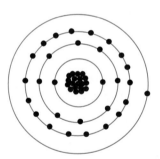

FIGURE 13–2 Basic construction of a copper atom.

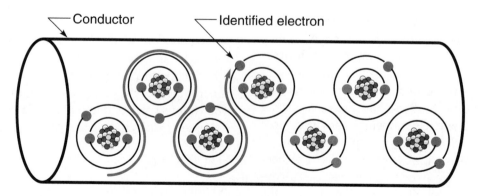

FIGURE 13–3 Electrons moving from atom to atom.

conductor of electricity. To have a current flow of electricity in a wire, one electron is added to one of the copper atoms. That atom will then shed its original electron to another atom, which will shed its original valence electron to another, and so on. As the electrons move from atom to atom, a force is released, which may be used to light lamps, run motors, energize relays, and so on. Therefore, there is electricity as long as the electrons are moving in the conductor.

Insulators. **Insulators** are materials in which electrons do not flow easily. A good insulator atom has seven or eight electrons in its valence ring. They are tightly held around the nucleus and cannot be easily moved. An insulator prevents electron flow and covers the outside of most conductors to keep the moving electrons within the conductor.

To summarize, the number of electrons in the valence ring determines whether an atom is a good conductor or a good insulator. Some atoms are neither good insulators nor good conductors. These are called **semiconductors.**

If the outer orbit contains one to three electrons, the atom is a conductor; if it contains five to eight electrons, the atom is an insulator. If it contains four electrons, the atom is a semiconductor. Keep in mind, however, that everything conducts electricity.

The electrons in the atoms of a conductor can be freed from their outer orbits by several forces such as by chemical reaction, friction, heat, light, magnetism, and pressure. When electrons are moved out of their orbit, they form an electrical current under proper conditions.

The human body is a conductor of electricity. When performing service on an electrical system, be aware that an electrical shock may be possible. Although the shock itself may be harmless, the consequences of the reaction to the shock may cause injury. For example, when working under the hood of a vehicle one might accidentally come into contact with a spark plug. The resulting shock, although not pleasant, would probably not cause permanent damage. The shock, however, could cause one to involuntarily jump back, causing the head to strike the open hood. This **involuntary** reaction, could result in a serious injury.

Electricity Defined

A better definition of electricity is the movement of electrons through a conductor, Figure 13–4. Electrons are attracted to protons. Because there is an excess of electrons on one end of the conductor and an excess of protons on the other end, there are many electrons attracted to the protons. This attraction of the electrons toward the protons is called electrical pressure. The amount of this electrical pressure is determined by the number of electrons that are attracted to the protons. A better term for this electrical pressure is *electromotive force* (EMF). It is the EMF that attempts to push an electron out of its orbit and toward the excess protons. If an electron is freed from its orbit, the atom acquires a positive charge because it now has one more proton than it has electrons. The unbalanced atom or ion attempts to return to its balanced state so it will attract electrons from the orbit of other balanced atoms. This starts a chain reaction as one atom captures an electron and another releases an electron. As this action continues to occur, electrons will flow through the conductor, approaching the speed of light (186,000 miles per second). A stream of free electrons forms and an electrical current is started. This does not mean that a single electron travels the full length of the conductor. It does mean, however, that the overall effect is that of electrons moving in one direction. The strength of the flow of electrons depends on the potential difference (voltage) and an opposition to flow known as **resistance** .

The three elements of electricity are voltage, current, and resistance. How these three elements interrelate governs the behavior of electricity. When the technician understands the laws that govern electricity, then understanding the function and operation of automotive electrical systems is much easier.

Voltage. *Voltage* can be defined as an electrical pressure. It is this pressure, called electromotive force (EMF), that causes the movement of electrons in a conductor. Voltage is the force of attraction between the positive and negative charges. An electrical pressure difference is created when

Conductor

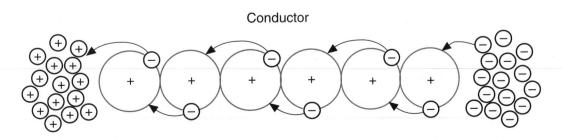

FIGURE 13–4 As electrons flow in one direction from one atom to another, an electrical current is developed.

there is a mass of electrons at one point in the circuit and a lack of electrons at another point in the circuit. The battery or alternator in a vehicle is used to apply EMF.

The amount of pressure applied to a circuit is measured in volts, Figure 13–5. If a voltmeter were connected across the two terminals of a battery it should read 12.6 volts, indicating that there is a potential difference of 12.6 volts between the two battery terminals.

In a circuit that has current flowing, voltage will exist between any two points in that circuit where current is flowing through resistance, Figure 13–6. The only time voltage does not exist is when the resistance is zero. In Figure 13–6, the voltage potential between points A and C and B and C is 12.6 volts. However, the pressure difference is zero between points A and B, so the voltmeter will indicate 0 volt.

Current. **Current** is defined as the rate of electron flow, Figure 13–7, and is measured in amperes. Current is a measurement of the number of electrons passing through any given point of the circuit in one second.

The current in a circuit will increase as the voltage is increased, provided the resistance of the circuit remains constant.

Current is the movement of electrons through a conductor. Negatively charged electrons move toward something that is positively charged because there is a potential difference. Electron theory states that because electrons are negatively charged, current flows from the most negative (–) point to the most positive (+) point within an electrical circuit. In other words current flow is from negative (–) to positive (+), a theory that is widely accepted in the electronic industry.

The second current flow theory, called the conventional theory, states that current flows from positive (+) to negative (–). The basic reason for this theory is that although electrons move toward the protons, the energy or force that is released as the electrons move begins at the point where the first electron moved to the most positive charge. As electrons continue to move in one direction, the released energy moves in the opposite direction. This is the oldest theory and serves as the basis for most electrical diagrams in the automotive industry.

To somewhat simplify the matter, a third theory has been developed to explain the mystery of current flow. This theory, referred to as the hole-flow theory, is actually based on both the electron theory and the conventional theory.

The automotive technician may find references to either one of these theories. Whichever it is does not matter as long as there is a good understanding of what cur-

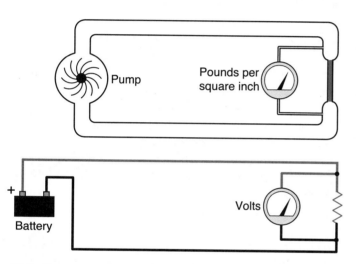

FIGURE 13–5 Voltage in an electrical circuit can be compared to pressure in a water system.

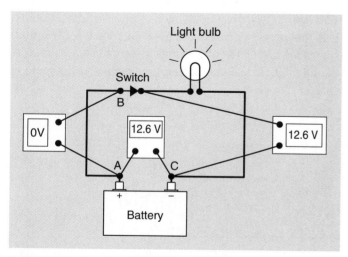

FIGURE 13–6 A simplified light circuit illustrating voltage potential.

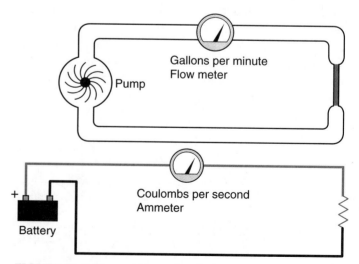

FIGURE 13–7 Current in an electrical circuit can be compared to the flow rate in a water system.

rent flow is and what affects it. From this understanding, basic knowledge is developed relative to how circuits function and how to troubleshoot and repair them. This text presents current flow as moving from positive to negative and electron flow as moving from negative to positive, Figure 13–8. Regardless of the theory, understand that current flow is the result of the movement of electrons.

Resistance. The third component in electricity is resistance—the opposition of current to flow. Resistance is measured in ohms, often designated by the Greek letter omega (Ω). The type, size, length, and temperature of the material used as a conductor determine its resistance. An electrical component that uses electricity to operate, such as a blower motor, has a greater amount of resistance than the conductor.

A complete electrical circuit consists of a power source, a load or resistance unit, and conductors.

A load (resistance) is required to change electrical energy into another form of energy such as light, heat, or mechanical movement. There is resistance in any load

device of a circuit, such as a clutch coil, blower motor, or relay (Figure 13–9). The basic characteristics that determine the amount of resistance in any part of a circuit are:

- ❏ The atomic structure of the material. The fewer the electrons in the outer valence ring, the greater the resistance of the conductor.
- ❏ The length of the conductor. The longer the conductor, the higher its resistance.
- ❏ The diameter of the conductor. The smaller the cross-sectional area of the conductor, the higher its resistance.
- ❏ Temperature. A change in the temperature of the conductor causes a change in its resistance.
- ❏ Physical condition of the conductor. If the conductor is damaged by nicks or cuts, its resistance is increased because its diameter is decreased.

There may be unwanted resistance in a circuit such as that caused by a corroded connection or a broken conductor. This additional resistance can cause the load com-

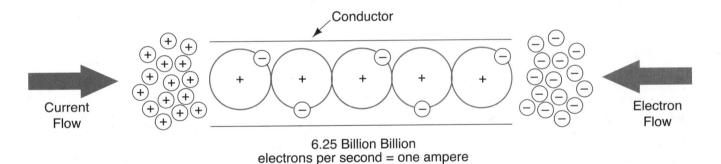

FIGURE 13–8 The rate of electron flow is called current and is measured in amperes.

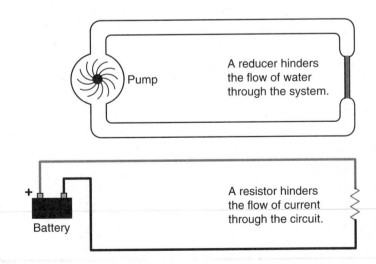

FIGURE 13–9 A resistor in an electrical circuit can be compared to a reducer in a water system.

ponent, such as a relay, to operate at reduced efficiency or not to operate at all. The following principles predict the impact of excessive resistance in a circuit:

- ❏ Voltage always drops as current flows through a resistance.
- ❏ An increase in resistance causes a decrease in current.
- ❏ Resistance changes electrical energy into heat energy.

OHM'S LAW

One must understand **Ohm's law** to understand how electrical circuits work. Ohm's law states that it takes one volt (V) of electrical pressure to push one ampere (A) of electrical current through one ohm (Ω) of electrical resistance. This law is expressed mathematically as: $1V = 1A \times 1\,\Omega$

An easy way to remember the formula of Ohm's law is to draw a circle and divide it into three parts, as shown in Figure 13–10. Cover the value to be calculated and the needed formula is all that shows.

This formula is most often expressed as: $E = I \times R$, Figure 13–11. Whereas E represents EMF or voltage, I represents the intensity of the electron flow current, and R represents resistance. This formula may be used to find one electrical value when the other two are known. For example, if there were 3A of current and 4Ω of resistance in a circuit, there would be 12V of electrical pressure.

$$E = 3A \times 4\Omega \ \ or \ \ E = 3 \times 4 \ \ or \ \ E = 12V$$

If the voltage and resistance are known, the current can be quickly calculated by using Ohm's law. Because $E = I \times R$, the formula becomes $I = E/R$. Consider a 12V circuit with 6Ω of resistance. Apply the following formula:

$$I = E/R \ \ or \ \ I = 12/6 \ \ or \ \ I = 2\,A$$

The same logic can be used to calculate resistance

when the voltage and current are known, $R = E/I$ (Figure 13–12).

To determine the current in a circuit, cover the I in the circle to reveal the formula, $I = E/R$, as shown in Figure 13–13. To apply the formula:

$$I = E/R \ \ or \ \ I = 12/3 \ \ or \ \ I = 4A$$

Based on Ohm's law current will increase if there is a decrease in resistance and the voltage is not changed. Likewise, if there is an increase in resistance, current will decrease. Ohm's law explains what happens when a circuit

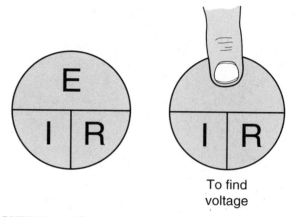

FIGURE 13–11 The formula $E = 1 \times R$.

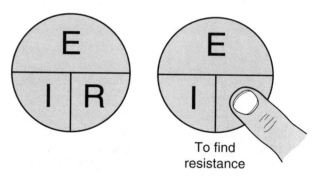

FIGURE 13–12 The formula $R = E/I$.

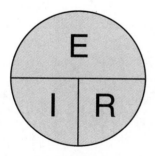

FIGURE 13–10 Chart for finding values of voltage, current, and resistance.

FIGURE 13–13 The formula $I = E/R$.

is changed or when something goes wrong in a circuit. For example, refer to Figure 13–14A; a 12V circuit with a 3Ω lamp. This circuit will have 4A of current flowing through it. If a 1Ω resistor is added, as shown in Figure 13–14B, there will be a total resistance of 4Ω. Because of the increased resistance, current will decrease to 3A. The lamp will receive less current and will not be as bright as it was before adding the additional resistance.

Another factor to consider is voltage drop. Before adding the 1Ω resistor, the 12V source voltage was dropped by the lamp. With the additional resistance, the voltage drop of the lamp was reduced to 9V. The remaining 3V were dropped through the 1Ω resistor. By using Ohm's law when the circuit current was 4A, the lamp had 3Ω of resistance. To find the voltage drop, multiply the current by the resistance.

$$E = I \times R \quad \text{or} \quad E = 4 \times 3 \quad \text{or} \quad E = 12V$$

When adding the resistor to the circuit, the lamp still had 3Ω of resistance; however, the current in the circuit decreased to 3A. Again, determine the voltage drop by multiplying the current by the resistance.

$$E = I \times R \quad \text{or} \quad E = 3 \times 3 \quad \text{or} \quad E = 9V$$

The voltage drop of the additional resistor is calculated in the same way: $E = I \times R$ or $E = 3V$. The total voltage drop of the circuit is the same for both circuits; however, the voltage drop at the lamp changed. This will also cause the lamp to be dimmer.

Electrical Power

Electrical power is expressed in watts. A **watt** (W) is equal to 1 volt (V) multiplied by 1 ampere (A). There is another mathematical formula that expresses the relationship among voltage, current, and power. It is simply: $P = E \times 1$. Power (P) measurements are measurements of the rate at which electricity is doing work.

A good example of power is the household light bulb, which is rated by wattage. A watt (W) is a unit of electrical power (P). A 100W bulb, for example, is brighter and uses more electricity than a 60W bulb. Although a technician is seldom concerned with wattage, an understanding of electrical power is helpful in understanding electrical circuits.

The light bulb, or lamp, of Figure 13–14A, had a 12V drop at 4A of current. To calculate the power used by the lamp multiply the voltage by the current.

$$P = E \times I \quad \text{or} \quad P = 12V \times 4A \quad \text{or} \quad P = 48W$$

The bulb produced 48W of power. In the other example in which the resistor was added, the lamp dropped 9V at 3A of current. The power of the bulb is calculated as done previously.

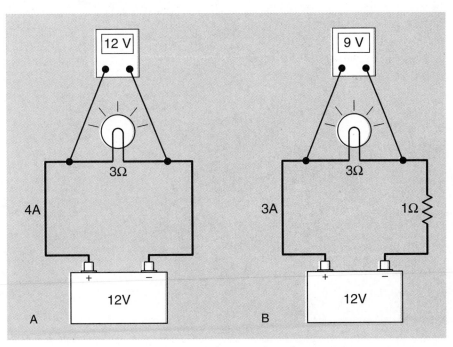

FIGURE 13–14 The light circuit shown with normal circuit values and with added resistance in series.

$$P = E \times I \quad or \quad P = 9V \times 3A \quad or \quad P = 27W$$

This lamp produced 27W of power, which was a little more than half of the original. It would, then, be about half as bright. The key in understanding why is to remember that the lamp did not change; the circuit did.

Capacitance

A **capacitor** is used in an electrical circuit to temporarily store an electrical charge, Figure 13–15. A capacitor is made up of two metal plates that are separated by an insulator or dielectric material. The insulator is generally wax paper, Teflon, or other such material. Capacitors do not consume power. Power stored in a capacitor is returned to the circuit when it is discharged. Because a capacitor stores voltage, it also absorbs voltage changes in the circuit. By providing this temporary storage, damaging high-voltage **spikes** may be controlled.

A capacitor is generally connected in parallel across the circuit, Figure 13–16. Capacitors operate on the principle that opposite charges attract each other, and that there is a potential voltage between any two oppositely charged points. When a switch is closed, the protons at the positive (+) battery terminal will attract some of the electrons on one plate of the capacitor away from the area near the dielectric material. As a result, the atoms of the positive (+) plate become unbalanced because there are more protons than there are electrons in the atom. This plate now has a positive (+) charge because of the shortage of electrons, Figure 13–17. The positive (+) charge of this plate will attract electrons on the other plate. The dielectric material keeps the electrons on the negative (–) plate from crossing over to the positive (+) plate, resulting in a storage of electrons on the negative (–) plate. The movement of electrons to the negative (–) plate and away from the positive (+) plate is an electrical current.

Current flows "through" the capacitor until the

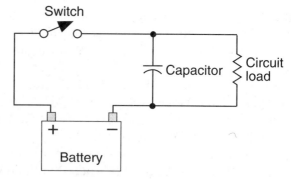

FIGURE 13–16 A capacitor connected to a circuit.

voltages across the capacitor and across the battery are equal. Current flow through a capacitor is only the effect of the electron movement onto the negative (–) plate and away from the positive (+) plate. Electrons do not actually pass through the capacitor from one plate to another. The charges on the plates do not move through the electrostatic field. They are stored on the plates as static electricity. When the charges across the capacitor and battery are equalized, there is no potential difference and no more current will flow through the capacitor, Figure 13–18. As the capacitor charges to the battery voltage, current flow through the resistor increases. When the capacitor and battery voltages are equal, maximum current flows through the load, Figure 13–19.

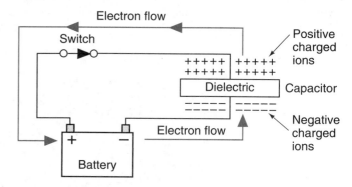

FIGURE 13–17 The positive plate sheds its electrons.

FIGURE 13–15 Capacitors that can be found in automotive electrical circuits.

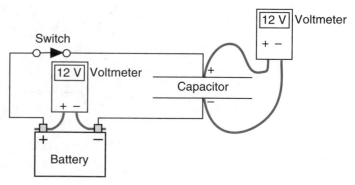

FIGURE 13–18 A capacitor that is fully charged.

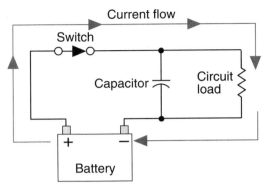

FIGURE 13–19 Current flow with a fully charged capacitor.

When the switch is opened, current flow from the battery through the resistor is stopped. However, the capacitor has a storage of electrons on its negative (–) plate. Because the negative (+) plate of the capacitor is connected to the positive (+) plate through the resistor, the capacitor now acts as a battery. The capacitor will discharge the electrons through the resistor until the atoms of the positive (+) and negative (–) plates return to a balanced state.

If a high-voltage spike occurs in the circuit, the capacitor will absorb the spike before it damages the circuit components. It takes time to change the charge of a capacitor. Electrons must move before the voltage changes. When the spike occurs, the capacitor starts to charge to its spike voltage, but the spike usually dissipates before the voltage across the capacitor charges much. A capacitor can also be used to stop current flow immediately when a circuit is opened, such as when a compressor clutch is de-energized. A capacitor also stores a high-voltage charge to be discharged when a circuit needs the voltage, such as in some air bag systems.

Types of Currents

There are two types of electrical current: direct current (**DC**) and alternating current (AC). The type of current flow is determined by the direction it flows and by the type of voltage that drives it.

Direct Current. Direct current (DC) in a vehicle, which is produced by either a battery or an alternator*, is the same throughout the circuit and flows in the same direction. Voltage and current are constant after the switch is turned on. Most of the electrically controlled units in a vehicle require direct current for operation.

* An alternator produces alternating current (AC), which is then rectified to direct current (DC).

Alternating Current. An alternating current (AC) is produced anytime a conductor is rotated in a magnetic

field. In an AC circuit, voltage and current do not remain constant. Alternating current changes directions from positive to negative and from negative to positive. The voltage in an AC circuit starts at zero and rises to a positive value. It falls back to zero and goes to a negative value, then returns to zero, Figure 13–20.

Electrical Circuits

The electrical term *continuity* refers to a circuit being continuous without interruption. For current to flow, the electrons must have a continuous path from the source voltage to the load component and back through the source. A simple circuit in a vehicle is made up of five parts:

- ❏ Power source (battery)
- ❏ Control device
- ❏ Protective device
- ❏ Conductors (wires and body metal)
- ❏ Load (motors, relays, coils, etc.)

A basic circuit, such as the one shown in Figure 13–21, includes a source (battery), a control device (switch) to turn the circuit on and off, a protection device (fuse), and a load (lamp). In this illustration, current

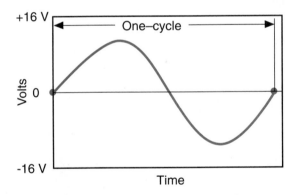

FIGURE 13–20 Alternating current reverses direction.

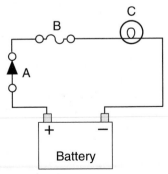

FIGURE 13–21 A basic electrical circuit including (A) a switch, (B) a fuse, and (C) a lamp.

flows from the battery through the lamp and returns to the battery when the switch is closed. The circuit is complete when the switch is closed. Opening and closing the switch to control electrical flow would be the same if the switch were installed in the **ground** side of the lamp.

Types of Circuits

Basically, there are three different types of electrical circuits. They are:

- ❑ **Series circuit**
- ❑ **Parallel circuit**
- ❑ Series-parallel circuit

The following is a brief explanation of each type of electrical circuit.

Series Circuit. A series circuit consists of one or more loads with only one path for current to flow. If any of the components in the circuit fails, the entire circuit will not function. All of the current that comes from the battery must pass through each load, then back to the battery.

The total resistance of a series circuit is calculated by adding all of the resistances together. For example, Figure 13–22 shows a series circuit with three lamps—one having 2Ω resistance, and two having 1Ω resistance each. The total resistance of this circuit, then, is 4Ω.

$$2 + 1 + 1 = 4$$

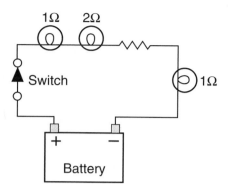

FIGURE 13–22 The total resistance in a series circuit is the sum of all resistances in the circuit.

The typical characteristics of a series circuit are:

- ❑ The total resistance is the sum of all resistances.
- ❑ The current through each resistor is the same.
- ❑ The current is the same throughout the circuit.
- ❑ The voltage drop across each resistor is different if the resistor values are different, Figure 13–23.
- ❑ The sum of the voltage drop of each resistor equals the source voltage.

Parallel Circuit. In a parallel circuit each path of current flow has separate resistances. Current can flow through more than one parallel leg at the same time in a parallel circuit, Figure 13–24. In this type of circuit, the failure of a component in one parallel path generally

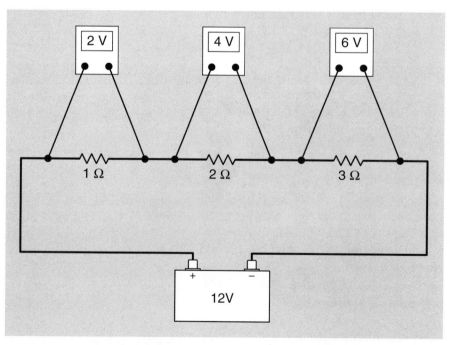

FIGURE 13–23 The voltage drop across each resistor in series is different if their resistance values are different.

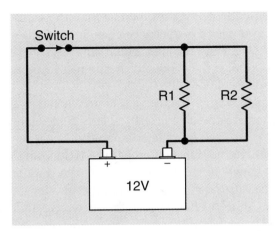

FIGURE 13–24 In a parallel circuit, current can flow through more than one parallel leg at a time.

does not affect the components in other paths of the circuit, except if the component is shorted or otherwise interrupts the circuit protective device.

The total resistance of a parallel circuit with two paths for current flow is calculated by using the following formula:

$$R_T = \frac{R_1 \times R_2}{R_1 + R_2}$$

In Figure 13–24 if R_1 had a value of 3Ω and R_2 had a value of 6Ω, the total resistance can be found by using the following formula:

$$R_T = \frac{R_1 \times R_2}{R_1 + R_2} \text{ or } \frac{3 \times 6}{3 + 6} \text{ or } \frac{18}{9} \text{ or } R_T = 2\Omega$$

Based on this calculation, it can be determined that the total circuit current is 6A (12V divided by 2Ω = 6A). Using Ohm's law and a basic understanding of electricity, other things about this circuit can be quickly determined.

The applied voltage to each path of this circuit is 12V; therefore, there must be a 12V drop in each path. The voltage drop across R_1 is 12V, and the voltage drop across R_2 is also 12V. Using the voltage drops, find the current that flows through each path.

Because load R_1 has 3Ω and drops 12V, the current that flows through it must be 4A (I = E/R).

Load R_2 has 6Ω and drops 12V; therefore, the current flow is 2A (I = E/R). The total current flow through the circuit is 6A (4A + 2A = 6A).

The total resistance can be calculated by using Ohm's law for the whole circuit:

$$R = E/I \text{ or } R_T = 12/6 \text{ or } R_T = 2\Omega$$

The total resistance in a parallel circuit is always less than its lowest individual resistance, because current has more than one path to follow. If more parallel resistors are added, then more circuits are added, causing the total resistance to decrease. If all of the resistances in the parallel circuit are equal, apply the following formula:

$$R_T = \frac{\text{One Resistor Value}}{\text{Number of Resistors}}$$

To determine the current in a parallel circuit, each path is treated as an individual circuit. Applied voltage is the same to all paths. To determine the branch current, simply divide the source voltage by the branch resistance:

$$I = E \div R$$

The total resistance of a circuit with more than two paths can be calculated by using the following formula:

$$R_T = \frac{1}{\dfrac{1}{R_1} + \dfrac{1}{R_2} + \dfrac{1}{R_3} + \dfrac{1}{R_4}}$$

It is often easier to calculate the total resistance of a parallel circuit by using the total current. Begin by finding the current through each path of the parallel circuit; then add them together to find the total current. Use basic Ohm's law to calculate the total resistance as follows:

$$R = E/I \text{ or } R = 12/6.5A \text{ or } R = 1.85\Omega$$

The basic characteristics of a parallel circuit are:

- ❑ The voltage applied to each parallel leg is the same.
- ❑ The voltage dropped across each parallel path is the same; however, if the leg contains more than one resistor, the voltage drop across each of them will depend on the resistance of each resistor in that path.
- ❑ The total resistance of a parallel circuit is always less than the resistance of any of its path.
- ❑ The current flow through the path is different if the resistance is different.
- ❑ The sum of the current in each path equals the total current of the parallel circuit.

Series-Parallel Circuit. A series-parallel circuit has some loads that are in series with each other and some that are in parallel, Figure 13–25. To calculate the total resistance in this type of circuit, calculate the equivalent series loads of the parallel branches first. Next, calculate the series resistance and add it to the equivalent series load. For example, the parallel portion of the circuit has two branches with 4Ω resistance each, and the series portion has a single load of 10Ω.

The current flow through each parallel path is calculated by using the resistance of each leg and voltage drop across that leg. To do this, first find the voltage drops. Because all 12V are dropped by the circuit, some are dropped by the parallel circuit and the rest by the resistor in series. Because the circuit current is 1A, the equivalent resistance value of the parallel circuit is 2Ω, and the resistance of the series resistor is 10Ω. Use Ohm's law to calculate the voltage drop of the parallel circuit:

$$E = I \times R \quad \text{or} \quad E = 1 \times 2 \quad \text{or} \quad E = 2V$$

Two volts are dropped by the parallel circuit. This means 2 volts are dropped by each of the 4Ω resistors. Using the voltage drop, current flow through each parallel leg can be calculated as follows:

$$I = E/R \quad \text{or} \quad I = 2/4 \quad \text{or} \quad I = 0.5\,A$$

Because the resistance on each path is the same, each path has 0.5A current flowing through it. The sum of the amperages is equal to the current of the circuit (0.5 + 0.5 = 1).

It is important to realize that the actual or measured values of current, voltage, and resistance may be somewhat different from the calculated values. The change is caused by the effects of heat on the resistances. As the voltage pushes current through a resistor, the resistor heats up, changing the electrical energy into heat energy. This heat may cause the resistance to increase or decrease, depending on the material it is made of. The best example of a resistance changing electrical energy into heat energy is a light bulb (lamp). A lamp gives off light because the conductor inside heats up and glows when current flows through it.

Although a technician seldom has the need to calculate the values in an electrical circuit, the ability to use Ohm's law will help determine what is wrong when a circuit does not function properly. A technician uses electrical meters to measure current, voltage, and resistance. If a measured value is not within specifications, the technician should understand what is wrong and be prepared to explain why.

Using Ohm's Law

Generally, vehicle electrical systems are connected in parallel. Actually, each electrical system is made up of many series circuits that are wired in parallel. This allows each electrical component to work independently of each other. When one device is turned on or off, the function of other components will not be affected.

For example, refer to Figure 13–26, which shows a 12V circuit with one 3Ω lamp. When the switch that controls the lamp is closed, current flows and the lamp lights up. Four amperes will flow through the circuit and the lamp.

$$I = E/R \quad \text{or} \quad I = 12/3 \quad \text{or} \quad I = 4A$$

Add a 6Ω lamp in parallel to the 3Ω lamp, as shown in Figure 13–27. When the switch for the new lamp is

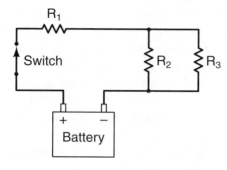

FIGURE 13–25 Series-parallel circuit: R_1 is in series with the parallel R_2 and R_3.

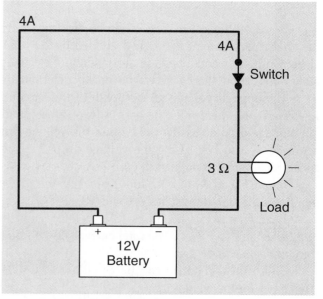

FIGURE 13–26 A simple light circuit.

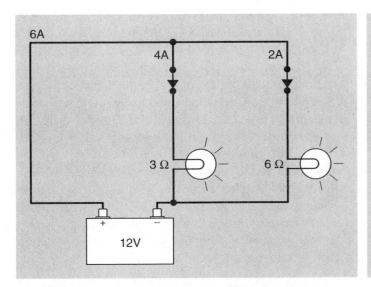

FIGURE 13–27 Two light bulbs connected in parallel.

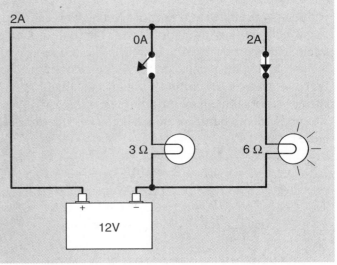

FIGURE 13–28 Two light bulbs connected in parallel; one switched on, the other switched off.

closed, 2A will flow through that lamp. The 3Ω lamp is still receiving 12V and has 4A flowing through it; it will operate in the same way and with the same brightness as it did before adding the 6Ω lamp. The only thing that has changed is circuit current; it is now 6A (4 + 2 = 6).

Leg 1: $I = E/R$ or $I = 12/3$ or $I = 4A$
Leg 2: $I = E/R$ or $I = 12/6$ or $I = 2A$

When the switch to the 3Ω lamp, Figure 13–28, is opened, the 6Ω lamp works with the same brightness as it did before opening the switch. In this case two things have happened: the 3Ω lamp no longer is lit, and the circuit current dropped 2A.

Another lamp is added in parallel, Figure 13–29, except a 1Ω lamp and switch were added in parallel to the circuit. With the switch for the new lamp closed, 12A will flow through that circuit. The other lamps are the same and have the same brightness as before. Again, the only thing that has changed is the total circuit current, which is now 18A.

Leg 1: $I = E/R$ or $I = 12/3$ or $I = 4A$
Leg 2: $I = E/R$ or $I = 12/6$ or $I = 2A$
Leg 3: $I = E/R$ or $I = 12/1$ or $I = 12A$
        Total current = 4A + 2A + 12A or 18A

When the switch for any of these lamps is opened or closed, the only thing that happens is that the lamps are either turned off or on, and the total current through the circuit changes. Notice that the total circuit current goes up as more parallel paths are added.

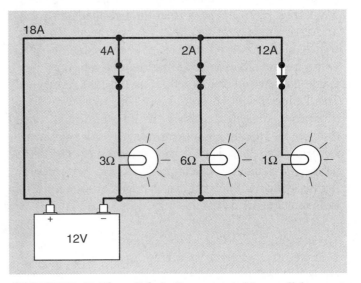

FIGURE 13–29 Three light bulbs connected in parallel.

The oft used statement, "Current always takes the path of least resistance," is somewhat misused. As illustrated in these circuits, current flows to all of the lamps regardless of their resistances. The resistances with lower values will draw higher currents, but all of the resistances will receive the current they allow. A better statement may be, "Higher amounts of current will flow through lower resistances"—an important point to remember when diagnosing electrical problems.

Ohm's Law Is Always True. The previous examples of Ohm's law have illustrated that when resistance decreases, current increases. If a 0.6Ω lamp were used instead of the 3Ω lamp, Figure 13–30, the other lamps will work with the same brilliance as they did before; however, 20A of current will flow through the 0.6Ω lamp, raising the total circuit current to 34A. Lowering the resistance on the one path of the parallel circuit does only one thing: it greatly increases the current through the circuit. High current such as this may burn the wires that carry the 20A or 34A lamp but it would not affect the wires to the other lamps.

Leg 1: $I = E/R$ or $I = 12/0.6$ or $I = 20A$
Leg 2: $I = E/R$ or $I = 12/6$ or $I = 2A$
Leg 3: $I = E/R$ or $I = 12/1$ or $I = 12A$
Total current $= 20A + 2A + 12A$ or $34A$

An increase in resistance to one of the parallel paths should cause a decrease in current. In Figure 13–31, a 1Ω resistor was added after the 1Ω lamp. This resistor is in series with the lamp, and the total resistance of that path is now 2Ω (1 + 1 = 2). The current through that path is 6A. The other lamps were not affected by the change. The only change to the whole circuit was in the total circuit current, which is now 12A. The added resistance lowered the total circuit current and changed the way the 1Ω lamp works. This lamp will now drop only 6V. The other 6V will be dropped by the added resistor. The 1Ω lamp will be much dimmer than before; its power rating has been reduced from 144W to 36W. The additional resistance caused the lamp to be dimmer. The lamp itself was not changed, only the resistance of that path has changed. The dimness is caused by the circuit, not the bulb.

Leg 1: $I = E/R$ or $I = 12/3$ or $I = 4A$
Leg 2: $I = E/R$ or $I = 12/6$ or $I = 2A$

Leg 3: $I = E/R$ or $I = 12/1+1$ or $I = 12/2$ or $I = 6A$
Total current $= 4A + 2A + 6A$ or $12A$

If a resistance of 0.333Ω is added at a point that is common to all of the parallel paths, Figure 13–32, to the negative connection at the battery, the circuit current will decrease, causing a change in the operation of the lamps in the circuit. The total resistance of the lamps, in parallel, is 0.667Ω.

The total resistance of the circuit is 1Ω (0.667 + 0.333), which means the circuit current is now 12A. Because there will be a voltage drop across the 0.333Ω resistor, each of the parallel legs will have less than the source voltage. To find the amount of voltage dropped by the parallel circuit, multiply the amperage by the resistance:

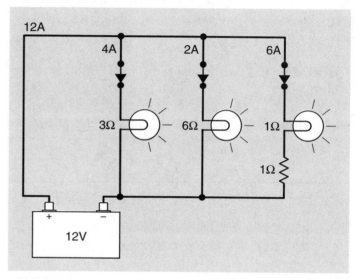

FIGURE 13–31 A series circuit contained in a leg of a parallel circuit.

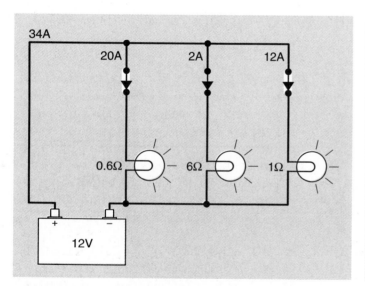

FIGURE 13–30 Three light bulbs connected in parallel.

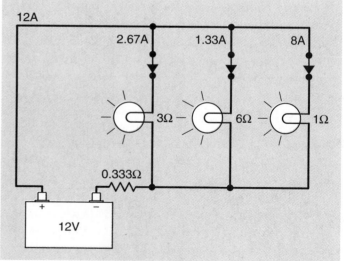

FIGURE 13–32 A resistor in series with a parallel circuit.

$12A \times 0.667\Omega = 7.992$ or 8V.

That means that 8V will be dropped by the parallel circuit. The remaining 4V will be dropped by the 0.333Ω resistor ($12 \times 0.333 = 3.996$). The amount of current through each path can be calculated by taking the voltage drop and dividing it by the resistance of the leg.

Leg 1: $I = E/R$ or $I = 8/3$ or $I = 2.667A$
Leg 2: $I = E/R$ or $I = 8/6$ or $I = 1.333A$
Leg 3: $I = E/R$ or $I = 8/1$ or $I = 8A$
Total current = 2.667 + 1.333 + 8 or 12A

The added resistance affected the operation of all of the lamps because it was added at a point that was common to all of the lamps. All of the lamps would be dimmer, and circuit current would be lower.

Semiconductors

All electrical materials are classified as conductors, insulators, or semiconductors. Semiconductors include diodes, transistors, and silicon-controlled rectifiers. Semiconductors are often called solid-state devices because they are constructed of a solid material. The most common materials used in the construction of semiconductors are silicon or germanium. Both of these materials, classified as crystals, have four electrons in their outer orbits, Figure 13–33. Because of their crystal-type structure, each atom shares an electron with four other atoms. As a result each atom will have eight electrons in its outer orbit. All the orbits are filled and there are no free electrons, thus the material falls somewhere between conductor and insulator; hence the term *semiconductor*.

Magnetism Principles

Magnetism is a force used to produce most of the electrical power in the world. It is also the force used to create the electricity to recharge a vehicle's battery, make a starter work, and produce signals for various operating systems. A magnet is a material that attracts ferrous materials, such as iron (Fe) and steel.

There are two types of magnets used in modern vehicles: permanent magnets and electromagnets. Permanent magnets are magnets that do not require an external force or power to maintain a magnetic field. Electromagnets, on the other hand, depend on electrical current flow to produce and, in most cases, retain their magnetic field.

Permanent Magnets. All magnets have a north (N) polarity and a south (S) polarity. A magnet, if reasonably balanced and allowed to hang freely, will align itself with Earth's north and south poles. The end that faces Earth's north is called the north-seeking pole; the other end is called the south-seeking pole. Like poles of two magnets repel each other, and unlike poles of two magnets attract each other. This natural phenomenon is shown in Figure 13–34.

A strong magnet produces many more lines of force than does a weak magnet. The invisible lines of force emerge in the magnet at the north and south poles. Inside the magnet, the lines of force extend from pole to pole, Figure 13–35.

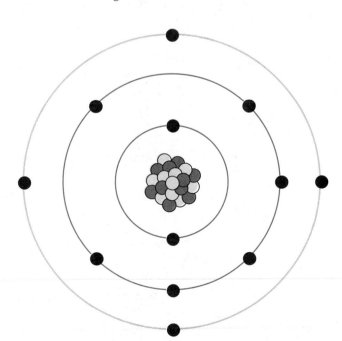

FIGURE 13–33 Semiconductors contain four valence electrons.

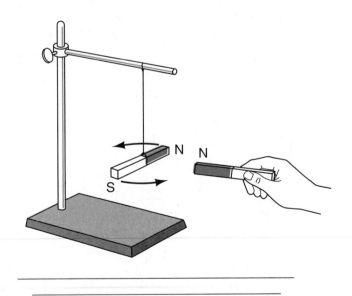

FIGURE 13–34 Unlike poles attract; like poles repel.

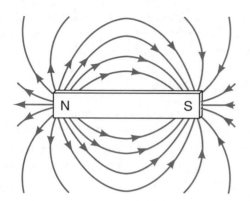

FIGURE 13–35 Lines of force extend from pole to pole.

The magnetic field of force is all of the space, outside the magnet, that contains the invisible lines of magnetic force. There is no known insulator for magnetic lines of force; they penetrate all substances. The lines of force, however, may be deflected by other magnetic materials or by another magnetic field, Figure 13–36.

Electromagnets. André-Marie Ampère observed that current flowing in the same direction through two nearby wires will cause the wires to be attracted to one another, Figure 13–37. He also noted that if current flow in one of the wires is reversed, the wires will repel each other, Figure 13–38. In addition, he noted that if current is passed through a coiled wire, the magnetic field combines to form a large magnetic field that has true north and south poles. Looping the wire doubles the magnetic field flux density where the wire runs parallel to itself. The illustration in Figure 13–39 shows how these lines of force join and add to each other.

As loops are added, the magnetic fields from each loop join and increase the flux density. To make the magnetic field even stronger, a soft iron core can be placed in the center of the coil, Figure 13–40. Soft iron is a material

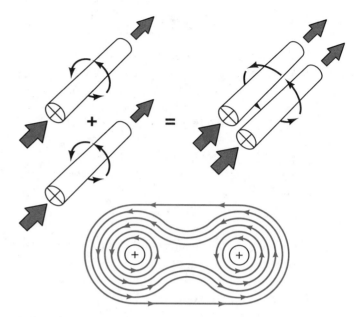

FIGURE 13–37 Lines of force combine if current flow in two conductors is in the same direction.

that has high permeability and provides an excellent conductor for the magnetic field that travels through the center of the wire coil.

The strength of an electromagnetic coil is affected by the following factors:

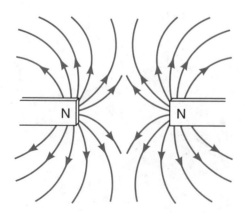

FIGURE 13–36 A magnetic field may be deflected by another magnetic field.

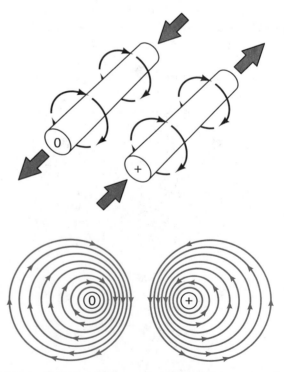

FIGURE 13–38 Lines of force repel if current flow is in the opposite direction in one of the two conductors.

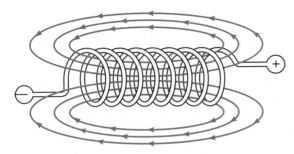

FIGURE 13–39 Lines of force join in a coil.

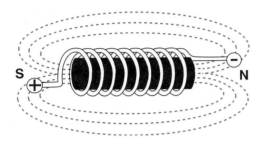

FIGURE 13–40 The addition of an iron core concentrates the flux density.

❑ The amount of current flowing through the wire
❑ The number of windings or turns
❑ The size, length, and type of core material
❑ The direction and angle at which the lines of force are cut

The strength of the magnetic field is measured in ampere-turns:

$$\text{ampere-turns} = \text{amperes} \times \text{number of turns}$$

The magnetic field strength is measured by multiplying the current flow in amperes through a coil by the number of complete turns of wire in the coil, Figure 13–41. For example, a 1,000-turn coil with 1 ampere of current would have a field strength of 1,000 ampere-turns. This coil would have the same field strength as a coil with 100 turns and 10 amperes of current.

Theory of Induction

Electricity can be produced by magnetic induction, which occurs when a conductor is moved through a magnetic field, Figure 13–42, or when a magnetic field is moved across a conductor. A difference of potential is set up between the ends of the conductor and a voltage is induced. This voltage exists only when the magnetic field or the conductor is in motion.

The induced voltage can be increased by either

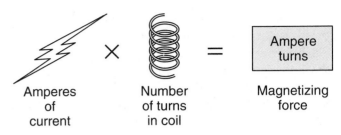

FIGURE 13–41 Formula for determining magnetic force.

increasing the speed in which the magnetic lines of force cut the conductor, or by increasing the number of conductors that are cut. This is the principle of the operation of a charging system.

A common induction device in a vehicle is the ignition coil. The coil reaches a point of saturation as current is increased. The magnetic lines of force, representing stored energy, collapse when the applied voltage is removed. This causes the lines of force to collapse, and the magnetic energy is returned as electrical energy.

Voltage induced in the wires of a coil when current is first connected or disconnected is called self-induction. The resulting current flows in the opposite direction of the applied current and tends to reduce the magnetic force. Self-induction, governed by Lenz's law, states, "An induced current flows in a direction to oppose the magnetic field that produced it."

Self-induction voltage spikes, with the exception of the ignition coil, are generally not wanted in modern vehicles that have delicate voltage-sensitive, computer-controlled circuits. When a switch is opened, for example, self-induction tends to continue to supply current in the same direction as the original current. As the magnetic field collapses, it induces voltage in the wire. Self-induction, then, can cause an electrical arc to occur across an opened switch. The arcing may momentarily bypass the switch and allow the circuit that was turned off to operate for a short period of time. This arcing will also burn the contacts of the switch.

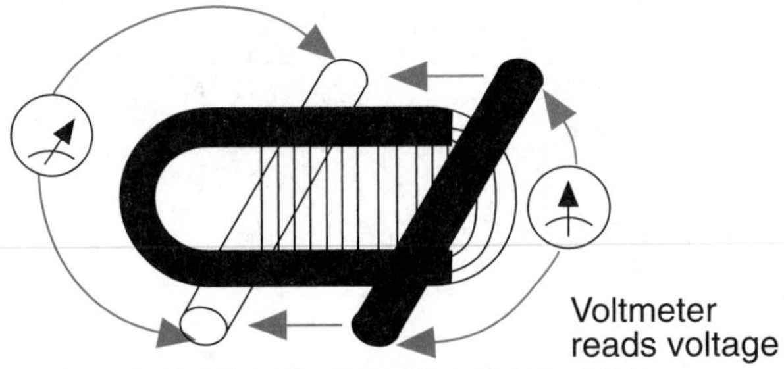

Conductor movement

Voltmeter reads voltage

FIGURE 13–42 Moving a conductor through a magnetic field induces an electrical potential difference.

Self-induction is commonly found in electrical components that contain a coil, such as a compressor clutch coil. In order to reduce the arcing across contacts, a capacitor or **clamping diode** may be connected to the circuit. The capacitor absorbs the high voltage and prevents arcing across the contacts. Diodes are semiconductors that allow current to flow in only one direction. A clamping diode, Figure 13–43, that is connected in parallel to the coil will prevent current flow from the self-induction coil to the switch.

Magnetic induction is also the basis for a generator and many of the sensors used on today's vehicles. In a generator, often referred to as an alternator, a magnetic field rotates inside a set of conductors. As the magnetic field crosses the conductors, a voltage is induced that is proportional to the speed of the rotating field, the strength of the field, and the number of conductors the field cuts through.

Magnetic sensors are commonly used to measure engine, vehicle, or shaft speeds. These sensors typically use a permanent magnet, and rotational speed is determined by the passing of blades or teeth in and out of the magnetic field. As a tooth moves in and out of the magnetic field, the strength of the magnetic field changes and a voltage signal is induced. This signal is sent to a control device, where it is interpreted.

Induced Voltage Spikes

As the number of electronic components and systems in vehicles increased, so did the problems associated with induced voltage spikes, or electromagnetic interference (EMI). The low-power delicate integrated circuits (ICs) used on modern vehicles are sensitive to the signals produced as a result of EMI. Voltage spikes are produced whenever current in a conductor is turned on and off. The static electricity created by friction, such as from the tires in contact with the road or from fan belts in contact with the pulleys, also causes unwanted EMI.

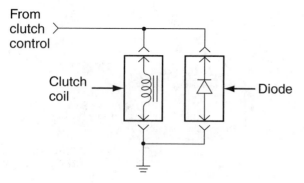

FIGURE 13–43 A diode in the clutch circuit prevents electrical spikes.

From clutch control

Clutch coil

Diode

The computer requires messages to be sent over circuits in order to communicate with other computers, sensors, control devices, and actuators. If any of these signals is disrupted, the engine and/or accessories may turn off. EMI can disrupt the vehicle's computer systems by inducing false messages to the computer.

EMI can be suppressed by any one of the following methods:

- Adding a resistance to the conductors such as to the high-voltage system secondary circuit
- Connecting a capacitor in parallel and a choke coil in series with the circuit
- Shielding the conductor or load components with a metal or metal-impregnated plastic
- Increasing the number of paths to ground by using designated ground circuits providing a very low resistance path to ground
- Adding a clamping diode in parallel or an isolation diode in series to the component

CIRCUIT PROTECTION DEVICES

Vehicle electrical circuits are protected against high current flow that exceeds the circuit's capacity. Excessive current generally is the result of a decrease in the circuit resistance, such as when a component or wire becomes shorted. A short is an unintentional low resistance path in which excessive current will flow. When the current of a circuit reaches a predetermined maximum, a circuit protection device will open and interrupt current flow, preventing damage to the circuit's wires and/or components.

Some late-model vehicles may use a thermistor as a protection device in some of their circuits. A thermistor changes its resistance in proportion to changes in heat; its resistance increases with an increase in temperature. As current flow is increased, the resistance in the circuit is increased. The increased resistance, in turn, lowers the current flow, thereby protecting the circuit.

Fuses

The most common protection device is the fuse, Figure 13–44. It is a simple device that contains a calibrated metal strip that melts if the current flowing through it exceeds its rating.

If the metal strip melts, a "blown fuse" results. The cause of the excessive current must be found and corrected before a new fuse, of the same rating, is installed.

Fuses are generally located in a central fuse block under the dash or a in power distribution box under the

FIGURE 13-44 Fuses and circuit breakers.

hood. They may also be found in relay boxes and electrical junction boxes. The fuse box may also be found behind a kick panel, in the glove box, or in a variety of other places. Fuse ratings and information relative to the circuits can be found in the vehicle owner's manual as well as in the manufacturer's service manual.

Fuses are rated by amperage and voltage. Do not install a fuse with a higher rating into a circuit than that specified by the equipment manufacturer. To do so may damage or destroy the circuit.

There are three types of fuses: glass or ceramic, blade type, and bullet or cartridge type. Glass and ceramic fuses are generally found in older vehicles. They are also sometimes found in an in-line fuse holder generally connected in series with an accessory component.

Glass Fuse. A glass fuse consists of a small glass cylinder with metal end caps. The visible metal fuse element, inside the glass tube, connects the two caps. The rating of the fuse, in voltage and amperes, is usually marked on one of the caps.

Ceramic Fuse. The ceramic-type fuse is of similar construction, except that the fuse element is encased inside the ceramic tube and is not visible. Also known as a cartridge-type fuse, it is found in many European vehicles. These fuses may be made of plastic as well as of a ceramic material. They generally have pointed ends and the metal strip rounds from end to end.

Blade-type Fuse. A blade-type fuse is found in a flat plastic housing and is available in three different physical sizes: mini, standard, and maxi. The plastic housing is formed around two male blade-type connectors. The fuse element, a metal strip, connects to the connectors inside

the plastic housing. The rating, in amperes, is marked on top of the plastic housing.

Fusible Link A vehicle may have one or more fusible links to provide circuit protection for the main power wires before they are divided into smaller circuits at the fuse box. The fusible links are usually located at a main connection near the battery or starter solenoid, Figure 13-45. The current capacity of a fusible link is determined by its size; it is usually four wire sizes smaller than the circuit it is protecting.

In wire sizing, the smaller the wire, the larger its number. For example, a circuit that uses 14-gauge wire would require an 18-gauge fusible link for protection.

Maxi-Fuse. Instead of a fusible link, some vehicles have a maxi-fuse to divide the electrical system into smaller circuits. If, for any reason, a fusible link fails, many electrical systems may be affected. By using a maxi-fuse to divide the electrical system into smaller circuits, the consequence of a circuit defect will not be as severe. The maxi-fuse also makes the technician's task of troubleshooting and diagnosing a circuit failure easier.

Several maxi-fuses may be used instead of a single fusible link.

Circuit Breaker

A circuit breaker is used in a circuit that is susceptible to routine overloads. There are three types of circuit breakers: manual resetting, semiautomatic resetting, and automatic resetting. The manual resetting circuit breaker is reset by pressing a button after the cause of interruption has been found. The semiautomatic reset type must be disconnected from the power source to be reset.

The automatic or self-resetting circuit breaker resets itself after it has cooled sufficiently. This type of circuit

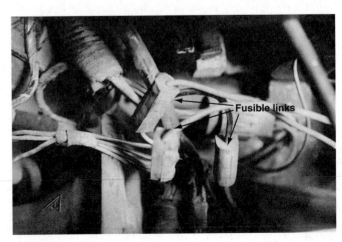

FIGURE 13-45 Fusible links.

breaker, Figure 13–46, uses a bimetallic strip that reacts to excessive current.

When an overload or circuit defect occurs, causing an excessive amount of current draw, the current flowing through the bimetallic strip causes it to heat. As the strip heats, it bends and opens a set of electrical contacts. After the contacts are opened, current no longer flows and the strip cools and closes the electrical contacts. If the cause of the excessive current has not been corrected, however, the contacts will again open. This action will continue as long as the circuit is overloaded.

Some accessories are protected by an electronic circuit breaker (ECB). This type of circuit breaker has a positive temperature coefficient and greatly increases its resistance when excessive current passes through it. The excessive current heats the ECB and its resistance increases, which, in turn, causes a decrease in current. The ECB will reset after a few seconds when current flow is interrupted.

ELECTRICAL COMPONENTS

Electrical circuits require different components, depending on the type of work they do and how they perform it. A blower motor, for example, may be wired directly to the battery, but it will run until the battery drains. A switch will provide control of the blower motor. If more than one speed is desired, however, a speed control device will also be needed.

Several components may be used in an electrical circuit to achieve the desired results. These components include switches, relays, and resistors.

Switches

Mechanical switches are the most common means of providing control of electrical current flow in an air-conditioning system, Figure 13–47. A switch controls the on and off operation of a circuit or directs the flow of current through various circuits. The electrical contacts inside a switch complete a circuit to carry the current when it is closed. When the electrical contacts are open, current will not flow and the circuit is interrupted.

A normally open (NO) switch will not allow current flow in its normal position. Its electrical contacts are open until the switch is acted on by an outside force, which closes them to permit current flow. A normally closed (NC) switch will allow current flow when it is in its normal position. Its electrical contacts are closed until the switch is acted on by an outside force, which opens them to interrupt current flow.

Some switches are designed to be momentary contact switches. A momentary contact switch usually has a spring that holds the contacts open, or closed, until an outside force is applied to close, or open, them. A good example of a NO momentary switch is the horn button on most vehicles.

The simplest switch is the single-pole, single-throw (SPST) switch. Figure 13–48 shows a schematic symbol of this switch controlling an on and off operation of a single circuit.

Some electrical systems may have a single-pole, double-throw (SPDT) switch. This type of switch has one input circuit with two output circuits. The input is directed to either output, depending on the selected switch position, Figure 13–49.

Relay

Circuits may contain an electromagnetic switch, which is more commonly referred to as a relay, Figure 13–50. A relay requires low current to control a high-current circuit. The relay coil has a very high resistance and draws very low current to produce a magnetic field. The magnetic field draws in an armature that has a movable contact, which closes onto a stationary contact. Normally open (NO) relay electrical contacts are closed by the electromagnetic field, whereas normally closed (NC) relay electrical contacts are

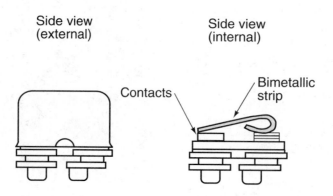

FIGURE 13–46 A self–resetting circuit breaker uses a bimetallic strip that opens if current draw is excessive.

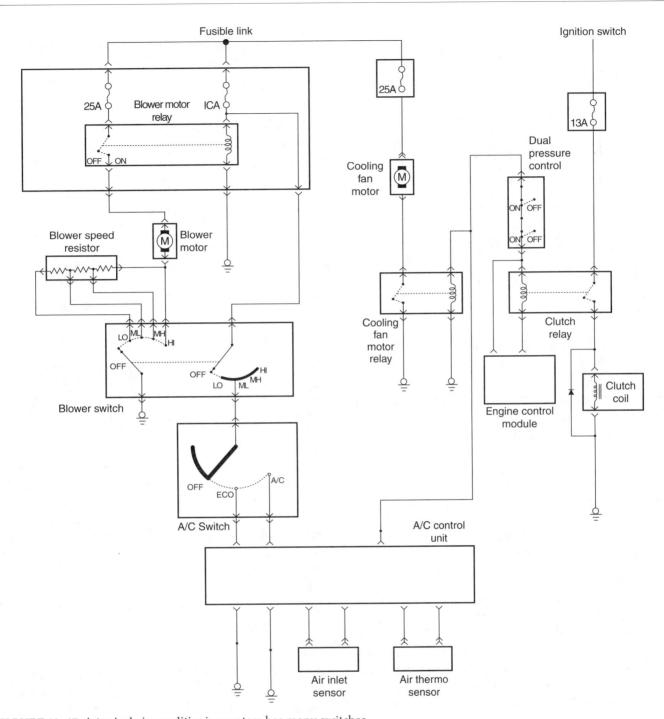

FIGURE 13-47 A typical air-conditioning system has many switches.

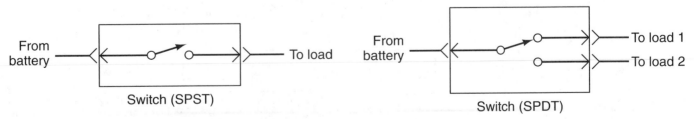

FIGURE 13-48 A typical SPST switch shown in its open position.

FIGURE 13-49 A typical SPDT switch.

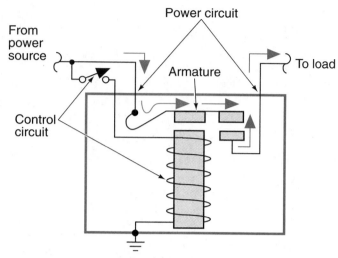

FIGURE 13–50 A relay uses electrical current to create a magnetic field to draw the contact points together to close the circuit.

opened by the electromagnetic field. The contacts are designed to carry the high current required to operate the load component. When current is applied to the coil, the contacts close and heavy battery current flows to the load component that is being controlled.

The illustration in Figure 13–51 shows a relay application in a blower motor circuit. The relay electrical contacts are closed and opened by supplying current to the relay coil. This provides a high-current electrical path from the source to the blower motor. The advantage is that this high current does not pass through the control switch.

Solenoid

A solenoid is an electromagnetic device that operates in much the same manner as a relay. The primary difference is that a solenoid has a movable iron core. Solenoids are used to do mechanical work, such as switch electrical, vacuum, and liquid circuits. The iron core inside the coil of the solenoid is spring loaded. When current flows through the coil, the magnetic field created around the coil attracts the core and moves it into the coil. To do work, the core is attached

to a mechanical linkage, which causes something to move. When current flow through the coil stops, the spring pushes the core back to its original position.

Buzzers

A buzzer is similar in construction to a relay, except for the internal wiring. Current is supplied to the coil through the normally closed contact points. When voltage is applied to the buzzer, current flows through the contact points to the coil. When the coil is energized, the contact arm is attracted to the magnetic field. As soon as the contact arm is pulled down, the current flow to the coil opens, and the magnetic field is dissipated. The contact arm closes again, closing the circuit to the coil. This opening and closing action occurs very rapidly. It is this movement that generates the vibrating signal.

Resistors

A resistor is an electrical device used to provide a fixed or variable resistance in an electrical circuit. Two types of resistors used in vehicle air-conditioning system applications are the stepped resistor and the variable resistor.

Stepped Resistor. A stepped resistor is often used, with a multiposition switch, to control the blower motor speed, Figure 13–52. By changing switch position, the resistance is increased or decreased within the blower motor circuit. If the current flows through a low resistance, higher current flows to the motor, increasing its speed. If the switch is placed in the low-speed position, additional resistance is added to the circuit. Less current flows to the motor, which causes it to operate at a reduced speed.

Variable Resistor. The most common types of variable resistors are the rheostat and potentiometer. A rheostat, Figure 13–53, has one terminal connected to the fixed end of a resistor and a second terminal con-

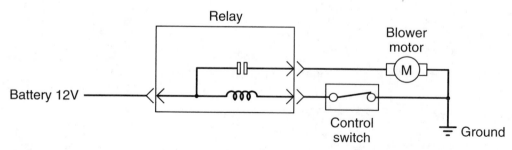

FIGURE 13–51 Relay used in a blower motor circuit.

nected to a movable contact called a wiper. Changing the position of the wiper on the resistor increases or decreases the resistance.

When a potentiometer is installed into a circuit, one wire is connected to a power source and the second wire is connected to the opposite end of the resistor and to ground. A third wire is connected to the wiper contact,

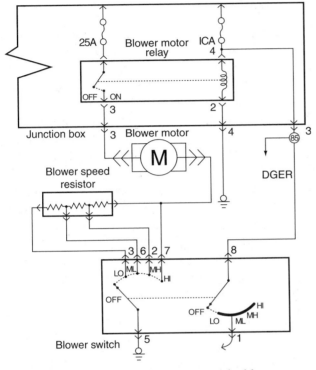

FIGURE 13–52 A stepped resistor is used for blower motor speed control.

FIGURE 13–53 Rheostat, a variable resistor used to provide infinite fan/blower–speed control. (Courtesy of BET, Inc.)

Figure 13–54. The wiper senses a variable voltage drop as it is moved over the resistor. Because the current always flows through the same amount of resistance, the total voltage drop measured by the potentiometer is very stable. The potentiometer, therefore, is a common type of input sensor for the vehicle's onboard computers.

Diode

The diode is the simplest of the semiconductor devices. A diode allows current flow in one direction only. It can function as a switch, acting as an insulator or a conductor, depending on the direction of voltage bias.

The action of a diode depends on which side receives a positive voltage. In DC circuits, positive pressure, or voltage, always comes from the positive side of the battery. Therefore, a diode's action depends on whether the anode or the cathode is connected to the positive side of the battery. When positive voltage is present on the P-side or anode, the diode is forward biased and current will flow through it, Figure 13–55. When positive voltage is present at the cathode or N-side, the

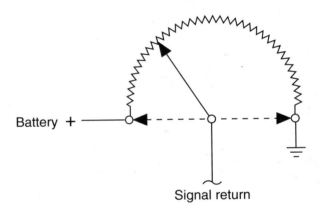

FIGURE 13–54 A potentiometer is used to send a voltage signal from the switch's wiper.

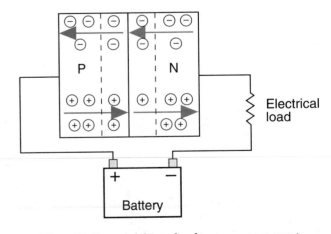

FIGURE 13–55 Forward–biased voltage causes current.

diode is reverse biased and current flow is prevented, Figure 13–56. A diode is a conductor when it is forward biased and an insulator when it is reverse biased. Because a diode is a semiconductor there will always be a voltage drop across it.

Zener Diodes

A diode that is reverse biased will not conduct current. However, if the reverse voltage is increased, a voltage level will be reached at which the diode will conduct in the reverse direction. Reverse current can destroy a simple *People-needed*-type diode.

A zener diode, Figure 13–57, is designed to operate in reverse bias at the breakdown region. A large current flows in reverse bias at the point where the breakdown voltage is reached. This prevents the voltage from climbing any higher, making the zener diode an excellent component for regulating voltage.

The zener diode that is rated at 15 volts will not conduct in reverse bias when the voltage is below 15 volts. At 15 volts it will conduct in reverse bias, but the voltage will not increase over 15 volts as long as a resistor is in series with the diode to limit current.

Light-Emitting Diode

A light-emitting diode (LED) has a small lens built into it so that light can be seen when current flows through it, Figure 13–58. When the LED is forward biased, there is current flow and light radiates from the junction of the diode. Normally, an LED requires 1.5 to 2.2 volts to light.

Clamping Diode

A voltage surge or spike is produced whenever current flow through a coil, such as that used in a compressor clutch, is interrupted. This voltage spike is the result of the magnetic field collapsing around the coil. The movement of the field across the windings induces a very high voltage spike, which can damage delicate electronic components as it flows through the electrical system wiring. In some circuits a capacitor can be used as a "shock absorber" to prevent component damage due to this surge. A clamping diode is often used in vehicle electrical circuits to prevent voltage spikes. Installing a clamping diode in parallel with the coil provides a bypass for the electrons during the time the circuit is open.

An example of the use of a clamping diode on an air-conditioning compressor clutch is shown in Figure 13–59. Because the clutch operates by electromagnetism, opening the clutch coil circuit produces a high voltage spike that, if

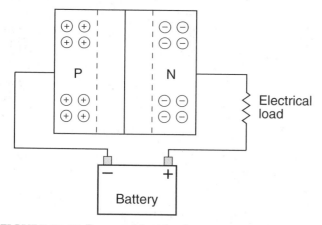

FIGURE 13–56 Reverse–biased voltage prevents current.

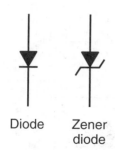

FIGURE 13–57 Diode and zener diode symbols.

FIGURE 13–58 A light–emitting diode uses a lens to emit the generated light; (B) the symbol for an LED.

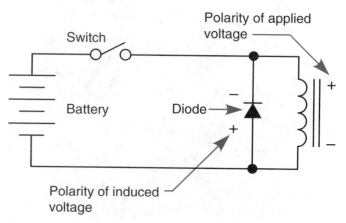

FIGURE 13–59 A diode is used to prevent induced voltage spikes.

left unchecked, could damage the onboard computers. The clamping diode, connected in the circuit is in reverse bias, when the clutch is energized. When the switch is opened the voltage across the clutch coil reverses; the diode conducts and shorts out the voltage, thereby preventing the voltage spike from reaching the computers.

Transistor

A transistor, Figure 13–60, is a solid-state device used to control current flow in the circuit. It can be used to allow a predetermined amount of current to flow or to resist this flow. The two principal uses for transistors are as switches and amplifiers. The average life of a transistor is about 100 years, unless abused.

There are two types of transistors: NPN and PNP. They are not interchangeable. It may be noted that the NPN type is the most commonly used transistor in automotive electronics.

Current can be controlled through a transistor. A transistor, therefore, can be used as a very fast electrical switch. It is also possible to control the amount of current flow through the collector. This is because the output current is proportional to the amount of current that flows through the base leg.

Photo Transistor. A small lens is used to focus incoming light onto the sensitive portion of the photo transistor. When light strikes the transistor, holes and free electrons form. These increase current flow through the transistor according to the amount of light. The stronger the light intensity, the more current that will flow. This type of transistor is often used as a sun sensor in automatic temperature control circuits.

Thyristors

The most common type of thyristor used in automotive applications is the silicon-controlled rectifier (SCR). Like

FIGURE 13–60 Transistors are used in motor vehicle applications.

the transistor, the SCR has three legs. However, it consists of four regions that are arranged PNPN. The three legs of the SCR are called the anode or P-terminal; the cathode or N-terminal; and the gate, which is one of the center regions.

The SCR will also block any reverse current from flowing from the cathode to the anode. Because current can flow only in one direction through the SCR, it can rectify AC current to DC current.

INTEGRATED CIRCUITS

An integrated circuit (IC) is a complex circuit of many resistors, transistors, diodes, capacitors, and other electronic devices that are formed onto a tiny silicon chip, Figure 13–61. As many as 30,000 transistors can be placed on a chip as small as 1/4-inch (6.35 mm) square.

Integrated circuits are constructed by photographically reproducing circuit patterns onto a silicon wafer. The process begins with a large-scale drawing of the circuit, which is reduced until it is the actual size of the circuit.

The small size of the chip has made it possible for vehicle manufacturers to have several computer-controlled systems in a vehicle without requiring a lot of space.

Use caution when working with and around integrated circuits (IC). They are easily damaged by static electricity. Antistatic wrist straps should be worn to reduce the possibility of destroying the integrated circuit. Do not connect or disconnect an IC to the circuit with the power on. The arc produced may damage the chip. Do not use an ohmmeter to test with unless specific instructions are followed.

CIRCUIT DEFECTS

Vehicle electrical problems may be classified as one of three types: an open, a short, or a high resistance. Each of these problems will cause a component to operate incorrectly or not at all. An understanding of what each of these problems will do to an electrical circuit is a key to the proper diagnosis of any electrical problem.

Open

An **open** is an intentional or unintentional break in the electrical circuit, Figure 13–62. An open may be caused by turning a switch to its off position, a broken wire, a burned-out component such as a relay coil, a disconnected wire or connector, or anything else that interrupts electrical flow in a circuit. In an open circuit, current will not flow and its components will not function. Because there is no current flow, there will be no voltage drops in the circuit. The source voltage is available everywhere in

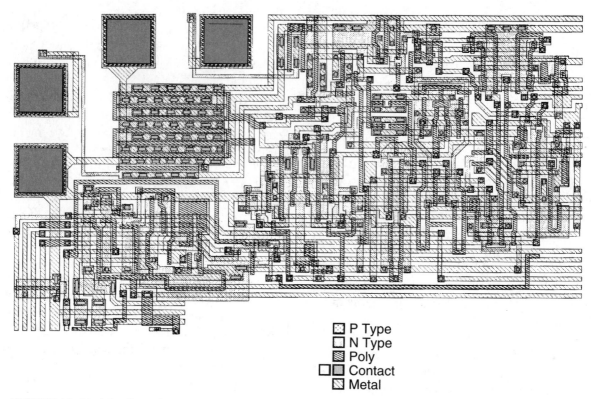

☒ P Type
☐ N Type
▨ Poly
☐▨ Contact
▨ Metal

FIGURE 13–61 An enlarged illustration of an integrated circuit with thousands of transistors, diodes, resistors, and capacitors. The actual size can be less than 1/4 inch square.

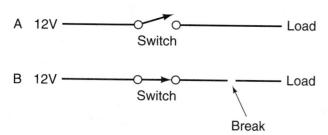

FIGURE 13–62 Two examples of an open circuit; (A) open switch; (B) broken wire.

the circuit up to the point where the open has occurred. Source voltage is even available after a load, if the open has occurred after that point.

Nearly all opens are caused by a break in the continuity of the circuit, which may occur anywhere in the circuit.

An open circuit resulting from a blown fuse or open circuit breaker will cause the circuit to be inoperable. The cause of the problem, however, is the excessive current that blew the fuse. This is most often a short that caused a circuit to open.

Short

A **short** is the result of an unwanted path for current to flow. Shorts cause an increase in current flow that can burn

wires or components. Sometimes two circuits become shorted together, and one circuit will supply unwanted power to another circuit. Improper wiring or damaged wiring insulation are the major causes of short circuits. A short is often an unwanted circuit to ground, Figure 13–63, providing a low-resistance path for current to travel.

High Resistance

High-resistance problems cause the current flow to be lower than normal, and prevent the components in the circuit from operating properly. The components in high-resistant circuits receive less voltage and, therefore, operate at less efficiency or do not operate at all.

The common causes for high resistance are loose connections or corroded connectors. The resistances become additional loads in the circuit that prevent full voltage to the normal loads in the circuit. High resistance can occur before or after a load.

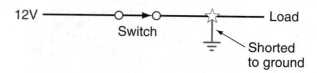

FIGURE 13–63 A short is most often an unwanted circuit to ground.

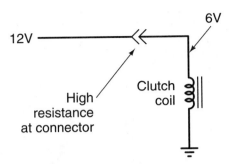

FIGURE 13–64 High resistance in the clutch circuit causes low voltage at the coil.

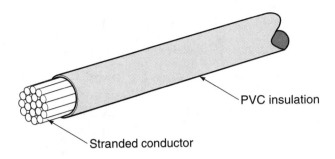

FIGURE 13–65 Stranded primary wiring.

Note the circuit illustrated in Figure 13–64, a simple compressor clutch circuit with unwanted resistances at the connector to the coil. Normally this 4Ω coil would have 3A of current flowing through it and would drop 12V. With the added resistance the clutch coil would only drop 6 or 8 volts and would not develop a magnetic field of sufficient strength to engage the clutch. The clutch may also chatter and/or slip due to the reduced available voltage.

WIRING AND CIRCUIT DIAGRAMS

Vehicles are equipped with a half-mile or more of wires in different gauge sizes and lengths, with color traces for ease of identification. Locating the cause of an electrical problem can be a difficult task if one does not have a good understanding of wiring systems and diagrams.

This chapter shows how to read a wiring diagram, how to interpret the symbols used, and how terminals are used. It is also important to know how to determine the correct type and size of wire to use to carry the anticipated current. It is possible to cause an electrical problem by simply using the wrong gauge size of wire. A technician should have an understanding of the three resistance factors of wire: its length, diameter, and temperature.

Vehicle Wiring

Primary wiring refers to the conductors that carry low voltage, such as the lighting circuits. Secondary wiring refers to the conductors used to carry high voltage, such as spark plug wires. The insulation of primary wires is usually thin when compared with secondary wires, which have extra thick insulation.

Most primary wiring conductors used in vehicles are made of several strands of copper (Cu) twisted together and covered with an insulation such as polyvinyl chloride (PVC), Figure 13–65. Copper (Cu) is most desirable because of its low resistance and because it may be easily connected using crimp or solder connectors. Other types of conductor materials that may be used

in vehicles include silver (Ag), gold (Au), aluminum (Al), and tin-plated brass.

A solid copper (Cu) wire may be used in low-voltage, low-current circuits where flexibility is not required, such as in a relay coil. Solid wire should not be used where high voltage, high current, or flexibility is required, unless used by the vehicle manufacturer.

Stranded wire is preferred because it is more flexible than solid wire. PVC insulation is used because it can withstand the effects of corrosion and extremes of temperature. Also, PVC insulation is not harmed by battery acid, antifreeze, or gasoline.

Wire Sizes

Some consideration for a margin of safety should be taken into account when selecting a wire size. The three major factors that determine the proper size of wire to be used are:

- The diameter of the wire must be large enough, for its length, to carry the required current for the load.
- The wire must be able to withstand the anticipated vibration.
- The wire must be able to withstand the anticipated amount of heat exposure.

Wire size is based on the diameter of the conductor. The larger its diameter, the less its resistance. The two common size standards used to designate wire size is the American Wire Gauge (AWG) and the metric systems.

The AWG assigns a number to the wire based on its diameter, Figure 13–66. The higher the number, the smaller the wire. An 18 gauge wire, for example, is smaller than a 14 gauge wire. Most vehicle electrical systems use 14, 16, and 18 gauge wires. An exception are battery cables, which are generally 2, 4, or 6 gauge.

In the metric system, wire size designation is determined by the cross-sectional area of the wire. Metric wire size is expressed in square millimeters

American Wire Gauge Table		
Gauge	Dia/Mils	Circular/Mils
000 (3/0)	410	167,800
00 (2/0)	365	133,100
0	325	105,500
2	258	66,370
4	204	41,740
6	162	26,250
8	128	16,510
10	102	10,380
12	81	6,530
14	64	4,107
16	51	2,583
18	40.3	1,624
20	32	1,022
22	25.4	642

FIGURE 13–66 American Wire Gauge (AWG) table.

Metric Size (mm^2)	AWG (Gauge) Size	Ampere Capacity
0.5	20	4
0.8	18	6
1.0	16	8
2.0	14	15
3.0	12	20
5.0	10	30
8.0	8	40
13.0	6	50
19.0	4	60

FIGURE 13–67 Approximate AWG to metric equivalents.

(mm^2). In the metric system the smaller the number, the smaller the wire. A comparison of metric to AWG wire sizes is given in Figure 13–67.

Both the wire diameter and length affect its resistance. For example, 10 feet of 16 gauge wire is capable of conducting 20 amperes with minimal voltage drop. However, if the conductor is 15 feet, a larger 14 gauge wire will be required; for 20 feet, 12 gauge wire will be required. The additional wire size is needed to prevent voltage drops in the wire.

An increase in temperature creates a proportional increase in resistance. A wire having a known resistance of 0.03 ohm per 10 feet at 70°F, when exposed at 170°F may increase its resistance to 0.04 ohms per 10 feet. Wires that are to be installed in areas of high temperatures, such as the engine compartment, must be of sufficient size so that the resistance increase will not affect the operation of the load component. It is also important that the insulation be capable of withstanding high temperatures.

Terminals and Connectors

Terminal connections are used to connect the wires from the voltage source to the load. A modern vehicle may have as many as 500 separate electrical circuit connections. The terminals used to make these connections must perform with very low voltage drop. A loose or corroded connection can cause an unwanted voltage drop that will result in poor operation of the load component.

Terminals may be crimped or soldered to the conductor. The terminal makes the electrical connection and must be capable of withstanding the stress of normal vibration. The illustration in Figure 13–68 shows several different types of terminals used in a vehicle electrical system.

To reduce the number of connectors in an electrical system, a common connection is often used, Figure 13–69. If there are several electrical components physically close to each other, a common connection or splice eliminates the use of separate connectors for each component wire.

Printed Circuits

Most instrument panels and control panels use printed circuit boards (pcb) as circuit conductors. A printed circuit is made of a thin phenolic or fiberglass board on which copper (Cu), or another conductive material, has been bonded. Unwanted portions of the conductive metal are then etched away by acid. The remaining strips of conductors provide the circuit path for the panel lights and other circuitry. An edge connector joins the printed circuit board to the vehicle wiring harness.

When it is necessary to perform repairs on or around the printed circuit board, do not touch the surface of the printed circuit with your fingers. The acid present in normal body oils can damage the surface. If the printed circuit board needs to be cleaned, use a commercial cleaning solution designed for electrical use. If this solution is not available, it is possible to clean the board by lightly and gently rubbing the surface with a pencil eraser.

Wiring Harness

Vehicle manufacturers use wiring harnesses to reduce the number of wires hanging loose under the hood or dash. The wiring harness is made up of insulated wires that are bundled into separate assemblies and are joined together by multiple-pin connectors, some which have more than sixty individual wire terminals.

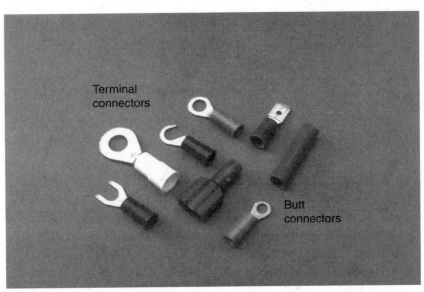

FIGURE 13–68 Different terminals and connectors.

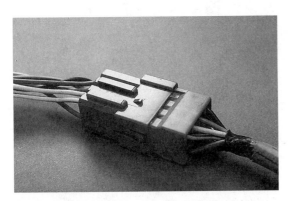

FIGURE 13–69 Multiple–circuit hard shell connector.

There are several individual wiring harnesses in a vehicle. The engine compartment harness and the under-dash harness are examples of a harness, Figure 13–70. Most wiring harnesses are encased in flexible tubing for protection and to provide quick installation. The tubing has a seam that can be opened to accommodate the installation or removal of wires from the harness. The seam closes after the wires are installed, and will close even more when the tubing is bent.

Wiring Protective Devices

Wire protection devices prevent damage to the wiring by maintaining proper wire routing and retention. Special clips, retainers, straps, and supplementary insulators provide additional protection to the conductor over what the insulation itself is capable of providing. Whenever the technician must remove one of these devices to per-

form a repair, it is important that the device be reinstalled to prevent additional electrical problems.

Do not use metal clamps to secure wires to the frame or body of a vehicle. The metal clamp may cut through the insulation and cause a short to ground. Use plastic clips only.

Wiring Diagrams

One of the most important tools for diagnosing and repairing electrical problems is the wiring diagram. The diagram identifies the wires and connectors in each circuit of a vehicle. It also shows where different circuits are interconnected, where they receive their power, where the ground is located, and the colors of the different wires. All of this information is critical to properly troubleshoot and diagnose electrical problems.

A wiring diagram may cover many pages and show the wiring of the entire vehicle, or it may cover an individual circuit on a single page. A single-circuit diagram is also referred to as a block diagram. Both types of diagrams show the wire color codes and connectors.

A wiring diagram generally shows the wires, connections to switches and other components, and the types of connectors used throughout the circuit. Total vehicle wiring diagrams may be spread out over many pages of a service manual or displayed on a single large page that unfolds out of the manual. A system wiring diagram is actually a portion of the total vehicle diagram where a system and all related circuitry are shown on a single page. System diagrams are easier to use than vehicle diagrams because there is less information to sort through.

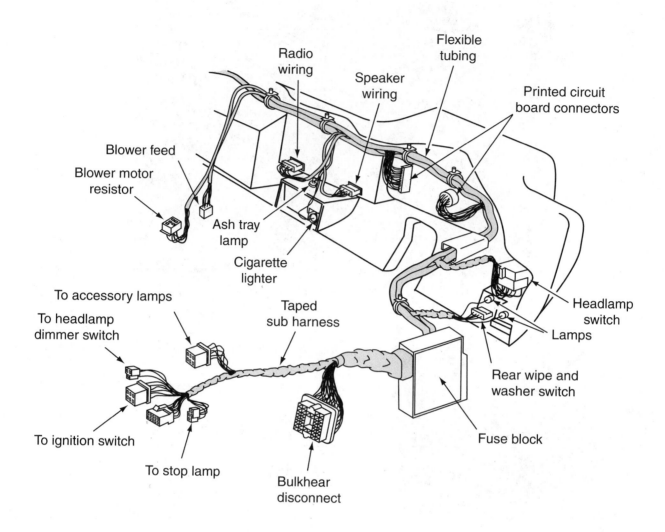

FIGURE 13–70   Complex wiring harness.

ELECTRICAL SYMBOLS

Rather than show an actual drawing of the components, most wiring diagrams use symbols to represent them. Often, a symbol displays the basic operation of the component. Many different symbols have been used in wiring diagrams throughout the years. Figure 13–71 illustrates some of the more popular ones. Although symbols may vary with each manufacturer, they are somewhat similar. Technicians should be familiar with symbols but need not memorize them because most wiring diagrams include a "legend" to help interpret the symbols.

Color Codes and Circuit Numbering

Most of the wires in a vehicle are covered with a colored PVC insulation. Colors are used to identify wires and electrical circuits with the aid of a corresponding color reference on a wiring diagram. Most wiring diagrams also include circuit numbers or letters and numbers to help identify a specific circuit. Both types of coding identifications make tracing circuits easier for troubleshooting and diagnosis of electrical problems. Not all manufacturers, however, use the same method of identifying wires and circuits.

Most wiring diagrams list the appropriate color coding used by the manufacturer. Be sure you understand the

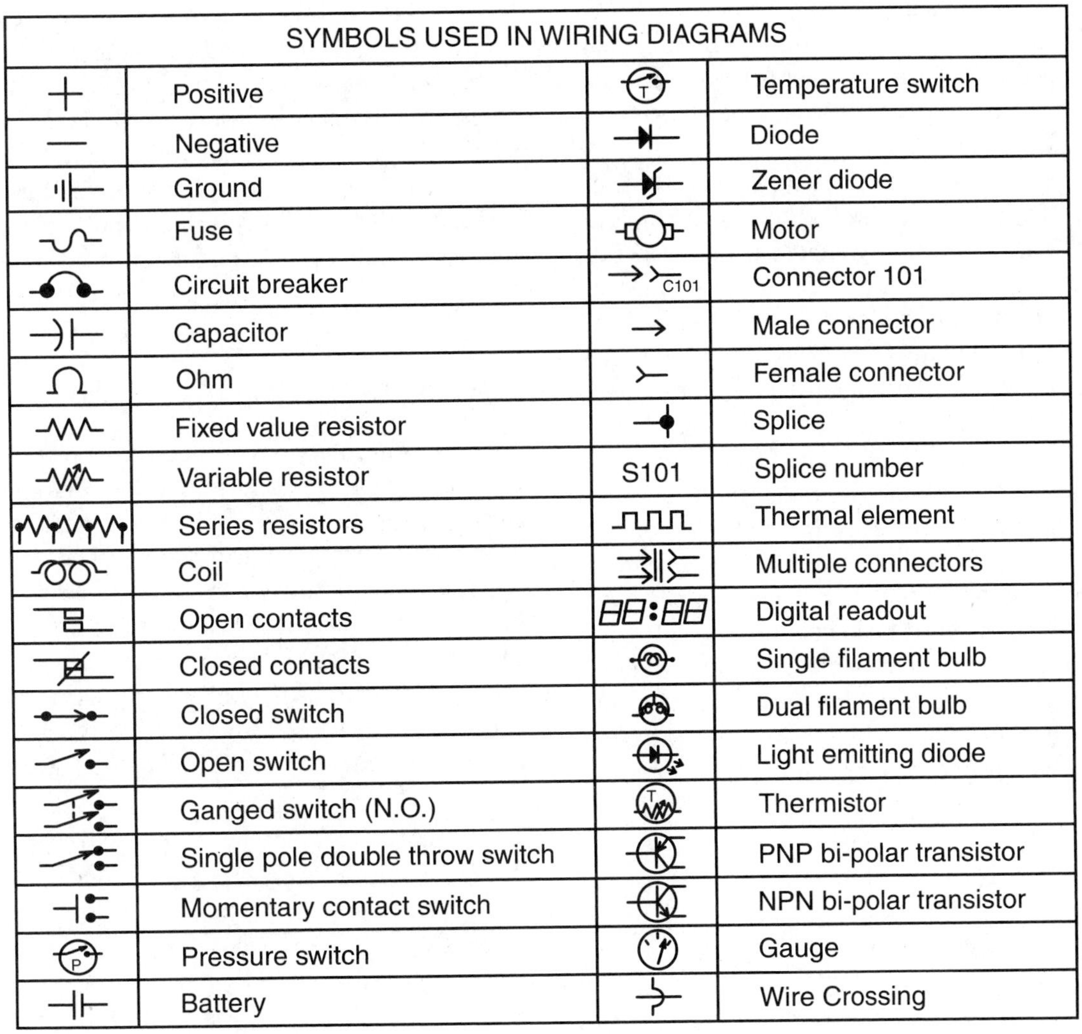

SYMBOLS USED IN WIRING DIAGRAMS			
+	Positive		Temperature switch
—	Negative		Diode
	Ground		Zener diode
	Fuse		Motor
	Circuit breaker		Connector 101
	Capacitor		Male connector
Ω	Ohm		Female connector
	Fixed value resistor		Splice
	Variable resistor	S101	Splice number
	Series resistors		Thermal element
	Coil		Multiple connectors
	Open contacts	88:88	Digital readout
	Closed contacts		Single filament bulb
	Closed switch		Dual filament bulb
	Open switch		Light emitting diode
	Ganged switch (N.O.)		Thermistor
	Single pole double throw switch		PNP bi-polar transistor
	Momentary contact switch		NPN bi-polar transistor
	Pressure switch		Gauge
	Battery		Wire Crossing

FIGURE 13–71 Typical schematic symbols.

color code reference before looking for a wire by color. A wire leading to the same component in a Ford may be green, whereas it may be yellow in a Chevrolet, for example.

In most color codes, the first group of letters designates the base color of the insulation, and a second group of letters, if used, indicates the color of the tracer. For example, a wire designated as WHT/BLK would have a white base color with a black tracer. Most manufacturers also number connectors and terminals for identification.

Standardized Wiring Designations

The Society of Automotive Engineers (SAE) is attempting to standardize the circuit diagrams used by various manufacturers. The system that is being developed may be similar to the Deutches Institut für Normung (DIN) used by import manufacturers. DIN assigns certain color codes to a particular circuit as follows:

❑ Red: For direct battery-powered circuits and also for ignition-powered circuits
❑ Black: For powered circuits controlled by switches or relays
❑ Brown: For grounds
❑ Green: For ignition primary circuits

A combination of wire colors is used to identify subcircuits. The base color still identifies the circuit's basic purpose. In addition to standardized color coding, DIN is attempting to standardize terminal identification and circuit numbering.

BLOWER MOTOR CIRCUIT

The blower motor is used to move air inside the vehicle for air conditioning, heating, defrosting, and ventilation. The motor is usually a permanent magnet-type located in the air-conditioning-heater system case/duct assembly. A blower motor switch mounted on the dash controls the speed by directing current flow to a resistor block that is wired in series between the switch and the blower motor, Figure 13–72.

The blower motor circuit includes the control assembly, blower switch, resistor block, the blower motor, and wiring harness. This system, using an insulated side switch and a grounded motor, supplies battery voltage to the control head when the ignition switch is in the RUN or ACC position. The current can flow from the control head to the blower switch and resistor block in any control head position except OFF.

When the blower switch is in the LOW position, the blower switch wiper opens the circuit. Current can only flow to the resistor block directly through the control head. Current must pass through all of the resistors before reaching the motor. With the voltage dropped over the resistors, the motor speed is slowed.

When the blower switch is placed in the MED 1, MED 2, or MED 3 position, current flows through the blower switch to the resistor block. Current must pass through three, two, or one of the resistors, depending on the speed selected. With more applied voltage to the motor, the fan speed is increased as the amount of resistance decreases. In high speed, full battery voltage is supplied to the motor through a high-speed relay circuit.

Current through the circuit will remain constant; varying the amount of resistance changes the voltage applied to the motor. Because the motor is a single-speed motor, it obtains its fast rotational speed with full battery voltage. The resistors drop the amount of voltage to the motor, resulting in slower speeds.

ELECTRIC DEFOGGER

Heat is generated when electrons are forced to flow through a resistance. Rear window defoggers use this principle of controlled resistance to heat the glass. The resistance is through a **grid** that is baked on the inside of the glass. The terminals are soldered to the vertical bus bars. One terminal supplies the current from the switch; the other provides the ground, Figure 13–73.

The system usually incorporates a timer circuit that controls the relay. The timer is used because of the high amount of current required to operate the system

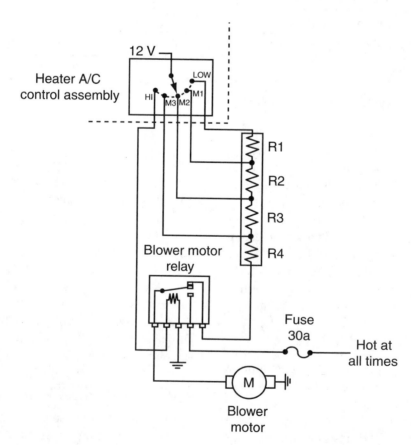

FIGURE 13–72 Typical blower motor circuit.

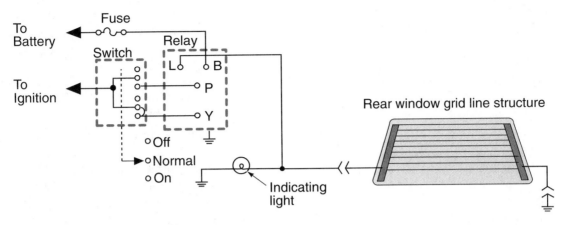

FIGURE 13–73 Rear window defogger circuit schematic.

(approximately 30 amperes). If this drain were allowed to continue for extended periods, battery and charging system failure could result. Because of the high-current draw, most vehicles equipped with a rear window defogger have a high-output AC generator.

Generally, the timer is initially activated for 10 minutes. If the switch is activated again at the completion of the timed cycle, the timer will energize the relay for an additional 5 minutes. The timer sequence is canceled by moving the switch to the OFF position or by turning the ignition switch off.

Because the ambient temperature has an effect on electrical resistance, the amount of current flow through the grid depends on the temperature of the grid. As the ambient temperature decreases, the resistance value of the grid also decreases. A decrease in resistance increases the current flow and results in quick warming of the window. The defogger system tends to be self-regulated to match the requirements for defogging.

REVIEW

Select the correct answer from the choices given.

1. The flow of electricity is due to the flow of:
   a. electrons.
   b. protons.
   c. Both A and B.
   d. Neither A nor B.

2. All of the following are good insulators, EXCEPT:
   a. steel.
   b. air.
   c. fiberglass.
   d. Teflon.

3. Which of the following is the *best* conductor?
   a. Copper
   b. Aluminum
   c. Silver
   d. Gold

4. Technician A says that the outer orbit of an atom is called the valence ring. Technician B says a good insulator atom will have five to eight electrons in its outer ring. Who is right?
   a. A only
   b. B only
   c. Both A and B
   d. Neither A nor B

5. Another term for electrical pressure is:
   a. resistance.
   b. voltage.
   c. amperage.
   d. all of the above.

6. The movement of electrons through a conductor is:
   a. known as current.
   b. measured in amperes.
   c. Both A and B.
   d. Neither A nor B.

7. All of the following are load devices, EXCEPT:
   a. a motor.
   b. a light-emitting diode.
   c. a resistor.
   d. an SPST switch.

8. Technician A says that a capacitor will charge to one-half the battery voltage. Technician B says that a capacitor is generally placed in series in a circuit. Who is right?
   a. A only
   b. B only
   c. Both A and B
   d. Neither A nor B

9. Technician A says a high-voltage spike occurs whenever the blower motor switch is opened. Technician B says a high-voltage spike may cause damage to delicate electronic circuits. Who is right?
   a. A only
   b. B only
   c. Both A and B
   d. Neither A nor B

10. All of the following are part of an electrical circuit, EXCEPT:
    a. the conductor
    b. the load device
    c. the power source
    d. the insulator

11. What type of wire is used in vehicle wiring harnesses?
    a. Solid
    b. Stranded
    c. Woven
    d. All of the above

12. When a second lamp is added to a circuit in parallel, the first lamp will be:
    a. dimmer.
    b. burned out.
    c. brighter.
    d. of the same brightness.

13. Technician A says that an insulator may be used to deflect magnetic lines of force. Technician B says magnetic lines of force extend from pole to pole in a magnet. Who is right?
    a. A only
    b. B only
    c. Both A and B
    d. Neither A nor B

14. A fusible link is generally ____ wire sizes smaller than the wire size of the circuit it protects.
    a. two
    b. three
    c. four
    d. five

15. The simplest switch is known as a(n):
    a. SPST.
    b. DPDT.
    c. SPDT.
    d. DPST.

16. Technician A says that a solenoid is often used to do mechanical work. Technician B says that a relay is often used to control high-current devices. Who is right?
    a. A only
    b. B only
    c. Both A and B
    d. Neither A nor B

17. All of the following may be electrical circuit defects, EXCEPT:
    a. a shorted circuit.
    b. an open circuit.
    c. excessive resistance.
    d. a grounded circuit.

18. Technician A says that most vehicles have over a mile of wire. Technician B says that most of the wires are contained in a single wiring harness. Who is right?
    a. A only
    b. B only
    c. Both A and B
    d. Neither A nor B

19. Technician A says the smaller the AWG number, the larger the wire. Technician B says the smaller the number in the metric system, the smaller the wire. Who is right?
   a. A only
   b. B only
   c. Both A and B
   d. Neither A nor B

20. The electric defogger is being discussed. Technician A says that the resistance of the electrical grid changes with a change in ambient temperature. Technician B says that the warming of the window is directly related to the temperature of the grid. Who is right?
   a. A only
   b. B only
   c. Both A and B
   d. Neither A nor B

TERMS

Write a brief description of the following terms:

1. capacitor
2. clamping diode
3. conductor
4. current
5. DC
6. grid
7. ground
8. insulators
9. involuntary
10. Ohm's law
11. open
12. orbits
13. parallel circuit
14. resistance
15. semiconductors
16. series circuit
17. short
18. spikes
19. valence ring
20. watt

CHAPTER 14

ELECTRICAL CIRCUITS

OBJECTIVES

On completion and review of this chapter, you should be able to:

- ❑ Understand the necessity for fuses and circuit breakers for electrical circuit protection.
- ❑ Recognize the electrical components of the heater and air-conditioning systems.
- ❑ Discuss the operation and function of climate control devices.
- ❑ Explain the function and purpose of the electromagnetic clutch.
- ❑ Compare the differences between controlling vehicle interior temperature by pressure- and temperature-actuated devices.
- ❑ Understand and explain the differences between low- and high-pressure control devices.
- ❑ Understand and demonstrate how to trace and troubleshoot an electrical circuit using a schematic.
- ❑ Identify and service the different types of blower motors and motor speed controllers.
- ❑ Perform troubleshooting and repair procedures to various electrical circuits and components.

INTRODUCTION

The basic motor vehicle air-conditioning (MVAC) electrical circuit, such as that found on an aftermarket system, is a simple one, as shown in Figure 14–1. Generally, it consists of a **fuse** or **circuit breaker**, master on/off blower speed control, control thermostat, blower motor, clutch coil, and simple wiring harness.

Note that only the positive (+) wire is shown from the battery to the air-conditioning electrical system. The other side, or negative (–) side, of the battery as well as the blower motor and clutch coil, terminate to a **common ground.** The symbol used to indicate a ground connection is shown in Figure 14–2.

The vehicle chassis, body, and all metal parts are common (ground) in a 12-volt, direct-current (DC) vehicle electrical system. A separate ground wire is not required unless the vehicle has fiberglass or other nonconducting body components.

The electrical **schematics** for factory-installed heater and air-conditioner systems are somewhat more complex, as shown in Figure 14–3. This schematic has been condensed so that it fits on one page. The actual schematic may require several pages in the manufacturer's service manual.

In some applications the heater system shares the same fuse or circuit breaker with the air-conditioning system in factory-installed units. The two systems most always share the same blower motor and motor speed controller. Additional electrical circuits associated with the cooling system are those used to warn of engine coolant heat conditions. Such a warning device may be a telltale dash light or a dash gauge. Each device has a sending unit located in the engine coolant system.

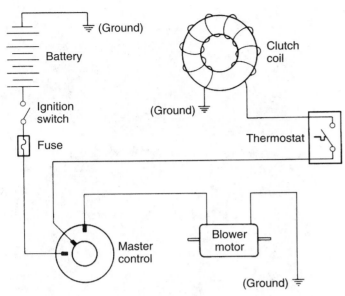

FIGURE 14–1 Typical wiring schematic.

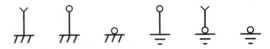

FIGURE 14–2 Electrical symbols used to identify a ground connection.

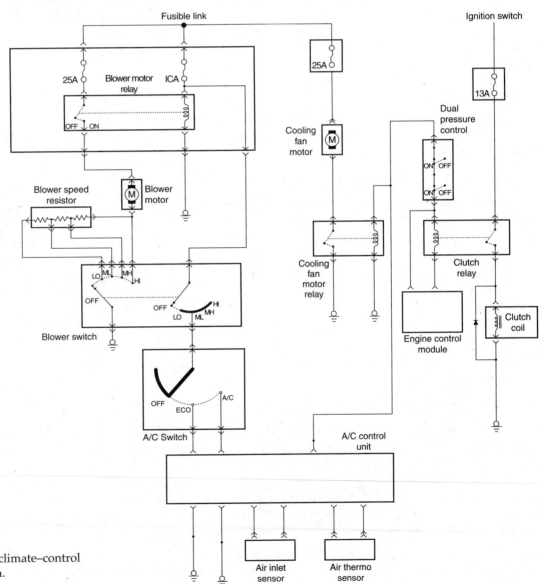

FIGURE 14–3 A typical climate–control system electrical diagram.

FUSES AND CIRCUIT BREAKERS

A fuse or circuit breaker is used to protect the air-conditioning and heating system components and wiring circuits. These devices are usually rated at 20 to 30 amperes, depending on the electrical system design. If found to be defective, a circuit protective device should be replaced with one of the same rating. Using a higher-rated device will not provide circuit protection, and a lower rated device will not carry the required electrical **load.**

Fuses

Three types of fuses, as shown in Figure 14–4, are currently found in vehicles One type consists of a thin ribbon of flat wire enclosed in a clear glass tube with metal ends. Another type, also consisting of a ribbon of flat wire, is enclosed in a ceramic tube with metal ends. The third type, again consisting of a ribbon of flat wire, is enclosed in a plastic case and has metal ends. One type must not replace another because of the difference in the metal ends that are used to secure the fuse in the fuse holder. The fuses are usually located in a main fuse block, as shown in Figure 14–5. Occasionally, a fuse may be found in an in-line fuse holder, like that shown in Figure 14–6. The in-line fuse holder is generally found on aftermarket **accessory** equipment, such as an underdash air-conditioning system.

Excessive current, such as that caused by a defective component, "burns out" or "blows" the fuse, interrupting the flow of electrical current to that component. Occasionally, fuses, like light bulbs, fail due to age only. If, however, the replacement fuse blows soon after it is replaced, the cause of the problem must be located and corrected.

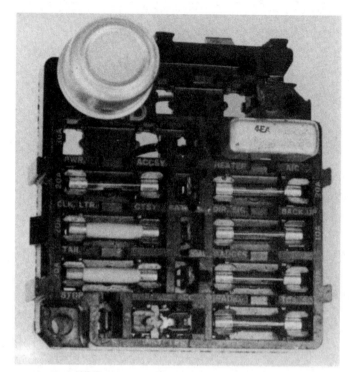

FIGURE 14–5 Typical fuse block. *(Courtesy of BET, Inc.)*

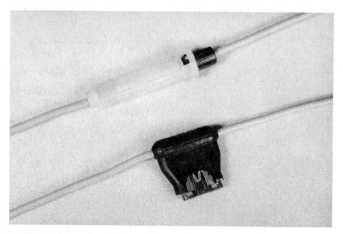

FIGURE 14–6 In–line fuse holders.

FIGURE 14–4 Types of fuses used to protect vehicle electrical circuits. *(Courtesy of BET, Inc.)*

Circuit breakers

Circuit breakers, like the one shown in Figure 14–7, are constructed of a **bimetallic** strip and a set of points **(contacts).** Excessive heat, such as that caused by excessive current of a defective component, causes the bimetallic strip to bend. When the strip bends, the electrical contacts open and current to the defective component is interrupted. When there is no current flow, the bimetallic strip cools, and the electrical contacts automatically reset (close). This opening and closing action will continue until the cause of the problem is located and corrected or until the circuit breaker fails due to fatigue.

FIGURE 14–7 Typical circuit breaker with lugs for fuse block mounting. *(Courtesy of BET, Inc.)*

TESTING FUSES AND CIRCUIT BREAKERS

Several methods may be used to check a fuse or circuit breaker: the in-vehicle testing with a voltmeter or non-powered test **lamp** (Part 1) and the out-of-vehicle testing with an **ohmmeter** or powered test lamp (Part 2).

Part 1

To test a fuse or circuit breaker in the vehicle, use a voltmeter or test lamp as follows:

1. Connect one lead of the test lamp or voltmeter to the body ground (–).
2. Touch the other lead to the hot side of the fuse or circuit breaker in the fuse block or holder, Figure 14–8. If the lamp does not light or if voltage is not indicated, power is not available and the problem is elsewhere. However, power is available if the lamp lights or if voltage is indicated. Proceed with step 3.
3. Touch the lead to the other side of the fuse or circuit breaker, Figure 14–9. If the lamp does not light or if voltage is not indicated, the fuse is **blown** or the circuit breaker is defective. Proceed with step 4. However, if the lamp lights or voltage is indicated, the problem is elsewhere and further testing is indicated.
4. Test protected components for shorts or overloads, then replace the fuse or circuit breaker.

Part 2

To test a fuse or circuit breaker that has been removed from the vehicle, follow this procedure:

1. Set the ohmmeter in the 1X scale; touch the leads together, and set the meter to zero or make sure the test lamp and battery are not defective.
2. Touch the two leads of the ohmmeter or test lamp to either side of the fuse or circuit breaker, Figure 14–10.

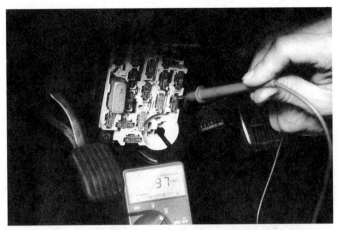

FIGURE 14–8 Touch the other lead to the hot side of the fuse block or fuse holder.

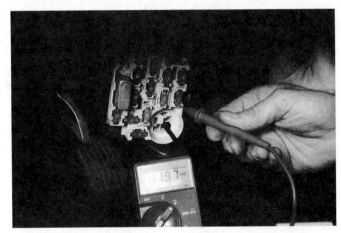

FIGURE 14–9 Touch the lead to the other side of the fuse.

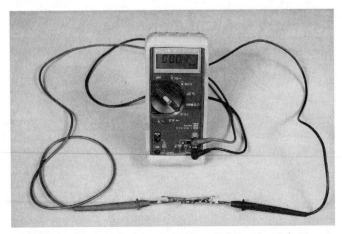

FIGURE 14–10 Touch the two leads to either side of the fuse.

a. If the ohmmeter indicates a low resistance or if the test lamp lights, the fuse or circuit breaker is good.

b. If there is no resistance indicated on the ohmmeter or if the test lamp does not light, the fuse is blown or the circuit breaker is defective.

MASTER CONTROL

The master control generally includes the blower motor speed control provisions. The **variable** (infinite) speed control, also known as a *rheostat*, shown in Figure 14–11, is popular with aftermarket systems. Two-, three-, four-,

and five-position blower speed controls, such as that shown in Figure 14–12, are generally found on factory-installed climate control systems.

The four-speed blower motor control shown in Figure 14–13 uses three internal **resistors** to provide motor speed control. There is no OFF position; whenever the master switch is closed, the blower motor will operate at its selected speed. Position 1 supplies current to the motor through all three resistors to provide low-speed operation. Position 2 supplies current to the motor through two resistors, providing medium-low speed. Current is supplied to the motor through only one resistor in Position 3 to provide medium-high blower speed.

FIGURE 14–11 Rheostat, a variable resistor used to provide infinite fan/blower speed control. *(Courtesy of BET, Inc.)*

FIGURE 14–12 Multiposition blower motor speed switch with dropping resistors (Nichrome wire). *(Courtesy of BET, Inc.)*

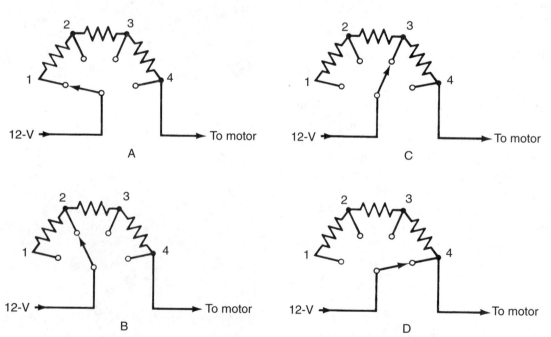

FIGURE 14–13 Resistance in the fan/blower circuit controls speed. Greatest resistance in circuit "A" provides for low speed. Less resistance in circuits "B" and "C" provides for medium speeds. No resistance in circuit "D" provides for high speed.

Position 4 bypasses the resistors and supplies full **battery voltage** to the blower motor. With the resistors out of the electrical circuit, full battery voltage provides high-speed operation of the blower motor.

Blower motor speed resistors, Figure 14–14, may be remotely mounted but are usually inserted into the case/duct system where circulated air provides cooling. Their operation is essentially the same as previously described. Regardless of where they are located, motor speed resistors are usually constructed of Nichrome™ (Ni/Cr) wire. They are generally placed in the air-conditioning or heater system air stream.

Some three-, four-, and five-position blower motor speed control switches, such as that illustrated in Figure 14–15, are not equipped with resistors. This type of switch provides full battery voltage to any of several **windings** in the blower motor. See the topic on multiwound blower motors later in this chapter.

The blower motor speed control may be the master ON and OFF control as well. When in either of the ON positions, the blower motor switch will provide full battery voltage to the temperature control thermostat. In this arrangement, the compressor clutch coil will not be energized if the blower motor is not running.

TEMPERATURE CONTROL

An electromagnetic clutch is used on the compressor of all motor vehicle air-conditioning (MVAC) systems to turn the compressor on when cooling is desired or off when it is not. The clutch is also used on most aftermarket and many factory-installed air conditioners to provide a means of temperature control. To accomplish this, the clutch is controlled by a temperature-sensitive switch known as a thermostat, Figure 14–16, or a pressure-sensitive switch, known as a low pressure control, Figure 14–17.

FIGURE 14–14 Remote duct–mounted blower motor speed dropping resistors (Nichrome wire). *(Courtesy of BET, Inc.)*

FIGURE 14–16 Thermostat or cold control. *(Photo by Bill Johnson)*

FIGURE 14–15 Multiposition switch used for fan or blower speed control with multitap motor or remote resistors. *(Courtesy of BET, Inc.)*

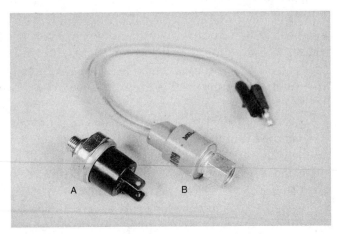

FIGURE 14–17 Low–pressure cutoff switch: (A) male thread; (B) female thread.

Thermostat

The thermostat or **remote bulb,** is located in the evaporator, where it senses the temperature of the air being delivered into the interior of the vehicle. The thermostat is initially set by the driver to a predetermined temperature. The thermostat cycles the clutch on and off at the selected setting to control the average in-vehicle temperature.

The thermostat is an electrical switch that is activated by a change in temperature. It senses either evaporator core air temperature or the temperature of the refrigerant as it enters or leaves the evaporator (depending on design). A temperature above the preselected one closes a thermostatic switch, and an electrical signal is sent to the clutch. The clutch is energized, and the air conditioner will operate. Similarly, a temperature at or below the preselected one will open the thermostatic switch to interrupt the electrical signal to the clutch. The clutch becomes de-energized, and compressor operation ceases.

Most thermostats have an OFF position so that the clutch can be turned off regardless of the temperature. In this way, the air conditioner or heater blower can be used to circulate air without a refrigerating effect.

Two basic types of thermostats are available for the control of the clutch: the **bellows** type and the bimetallic type, Figure 14–18. Both types of thermostats are temperature actuated. Although the principle of operation is different for each type of thermostat, they serve the same purpose in that they both control the evaporator temperature by cycling the compressor on and off through the clutch.

Bellows-Type Thermostat.

A diagram of the construction details of the bellows-type thermostat is shown in Figure 14–19. A capillary tube connected to the thermostat is filled with a temperature-sensitive fluid or vapor. The capillary is attached to a bellows within the thermostat. This bellows, in turn, is attached to a swinging frame assembly. Two electrical contact points are provided. One contact is fastened to the swinging frame through an insulator and the other electrical contact is fastened to the body of the unit, again through an insulator.

Operation of the Bellows-Type Thermostat.

When the gases inside the capillary tube expand, pressure is exerted on the bellows. As a result of this pressure, the bellows closes the electrical contacts at a preselected temperature, as illustrated in Figure 14–20. Manual temperature control is provided by a shaft connected to the swinging frame and an external control knob. When the knob is turned in a clockwise direction, the spring tension is increased against

REMOTE SENSING THERMOSTAT
(BELLOWS TYPE)

BIMETALLIC THERMOSTAT

FIGURE 14–18 Thermostats. *(Courtesy of BET, Inc.)*

the bellows. More pressure is then required to overcome the increased spring tension. The requirement for more pressure means that more heat is necessary. Because heat is being removed from the evaporator, a lower temperature is required to *open* the points. When the temperature rises, the heat again exerts pressure on the bellows to *close* the points and allow for cooling.

Another spring within the thermostat regulates the temperature interval through which the points open. This interval is usually a temperature rise of about 12°F (6.6°C) and gives sufficient time for the evaporator to defrost. This rise is known as the delta T (Δ_T) of the thermostat.

Most factory air-conditioning systems have a means to automatically cycle the clutch off at about 32°F (0°C). Some General Motors cycling clutch orifice tube (CCOT) systems use a preset thermostat to sense the temperature at the evaporator inlet. Some Chrysler H-valve-equipped systems have a preset thermostat to sense the temperature at the evaporator outlet.

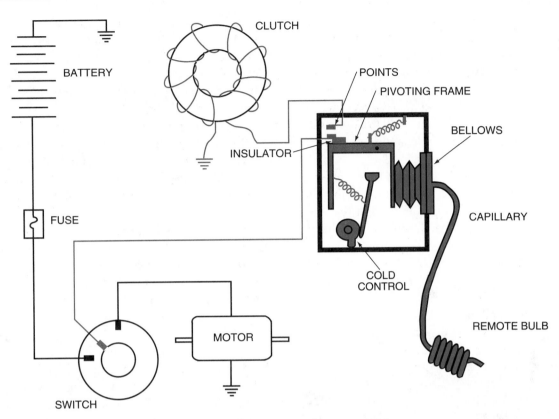

FIGURE 14–19 Temperature–controlled bellows–type thermostat in the open position.

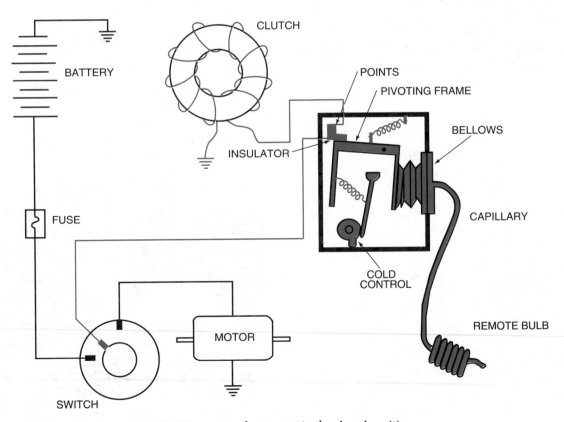

FIGURE 14–20 Temperature–controlled bellows–type thermostat in the closed position.

Many General Motors cycling clutch orifice tube (CCOT) and fixed orifice tube cycling clutch (FOTCC) systems use a pressure switch instead of a thermostat. The pressure switch, which is mounted on the accumulator, Figure 14–21, senses low-side pressure to cycle the clutch off at about 30 psig (207 kPa). This pressure corresponds to a temperature of about 32°F (0°C). Pressure switches are covered later in this chapter as well as in other chapters of this text.

Care must be exercised when handling a thermostat with a capillary tube. There should be no sharp bends or kinks in the capillary tube. A bend that must be made in the capillary tube, should be no sharper than one that can be formed around the end of the thumb, as depicted in Figure 14–22.

FIGURE 14–21 The low-pressure switch is usually found on the accumulator.

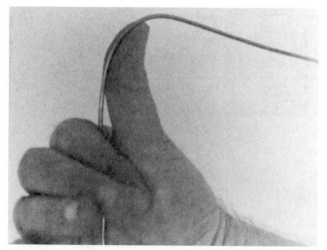

FIGURE 14–22 The bend of a capillary tube should be no sharper than one that can be formed around one's thumb. (*Courtesy of BET, Inc.*)

For best results, the end of the capillary tube should be inserted into the evaporator core between the **fins** to a depth of about 1 inch (25.4 mm), Figure 14–23. The capillary tube should not be inserted all the way through the fins because it may interfere with the blowers, which are often mounted behind the core.

If, for any reason, the capillary tube is damaged and has lost its charge of inert gas, the thermostat must be replaced. When there is no fluid in the capillary tube, the unit has no ON cycle. The capillary tube cannot be recharged using standard equipment.

Bimetallic-Type Thermostat. The bimetallic-type thermostat is often desirable as a replacement part for aftermarket hang-on-type air conditioners because of its lower cost. This thermostat does not have a capillary tube, and therefore depends on air passing over its bimetallic strips to maintain proper operation.

Operation of the Bimetallic-Type Thermostat. Manual temperature control with the bimetallic thermostat is achieved in the same manner as that described for the bellows-type thermostat. Cold air passing over the bimetallic leaf in the rear of the thermostat causes it to retract. By retracting, it bows enough to open a set of points. As the temperature increases, the other leaf of the bimetallic element reacts to the heat and pulls the points back together. The OFF cycle range (Δ_T) of this thermostat is also about 12°F (6.6°C) to allow for a sufficient defrost period.

The bimetallic thermostat is limited in application because it must be mounted inside the evaporator. In many instances, the bellows-type thermostat must be used, because its long capillary tube allows the thermostat to be placed some distance from the evaporator core.

Many thermostats are adjustable. In addition, a means is provided for regulating the range between the

FIGURE 14–23 Thermostat remote bulb inserted into the evaporator core.

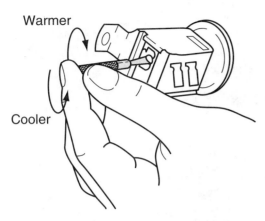

FIGURE 14–24 Typical thermostat adjustments.

opening and closing of the points. The adjustment in some thermostats is located under the control knob in the shaft; in other thermostats, the adjustment is located under a fiber cover on the body of the unit, Figure 14–24.

A thermostat that lacks a **setscrew** can be considered to be a nonadjustable type. Malfunction of this type of thermostat requires replacement of the complete unit.

TROUBLESHOOTING THE THERMOSTAT

There are basically two types of thermostats: adjustable and fixed. Testing either type is a simple matter if the thermostat has been removed from the vehicle. Troubleshooting the variable- or adjustable-type thermostat is covered in Part 1, and the fixed-type thermostat is covered in Part 2.

Part 1 Variable Type

1. Connect an ohmmeter (X1 scale) or powered test lamp to the two terminals of the thermostat, Figure 14–25.

2. While observing the ohmmeter or test lamp, rotate the thermostat shaft from fully clockwise (cw) to fully counterclockwise (ccw).
   a. If a low resistance was noted or if the test lamp was lit, the thermostat is probably good.
   b. If no resistance was noted or if the test lamp did not light, the thermostat is probably defective.
3. Repeat step 2 several times.
   a. The test lamp should go on and off consistently, or the meter should show high and low resistances.

Part 2 Fixed Type

1. Connect the ohmmeter or test lamp in the same manner as for the adjustable thermostat test.
2. Is low resistance noted or is the lamp lit? Generally, at ambient temperature, the thermostat will be closed.
   a. If yes, proceed with step 3.
   b. If no, proceed with step 5.
3. Immerse the capillary tube end or remote bulb into the ice bath, Figure 14–26.
4. Did the resistance increase or did the lamp go out? A reduction in temperature below the set point should open the thermostat contacts.
   a. If yes, proceed with step 5.
   b. If no, the contacts are stuck closed and the thermostat is defective.
5. Immerse the capillary tube in the hot bath.
6. Did the resistance decrease or did the lamp light?
   a. If yes, the thermostat is probably all right.
   b. If no, the thermostat is probably defective with its contacts stuck open.

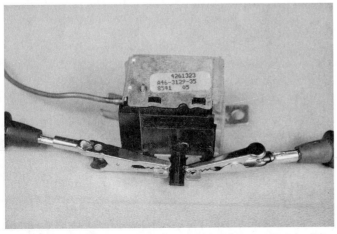

FIGURE 14–25 Connect an ohmmeter to the two terminals of the thermostat.

FIGURE 14–26 Immerse the capillary tube into an ice bath.

Low-Pressure Control

Many air-conditioning systems have a low-pressure cut-off switch in the clutch circuit to maintain in-vehicle temperature conditions. These normally closed (NC) switches are sensitive to system pressure and open in the event of a predetermined low system pressure. This, in turn, interrupts electrical current to the clutch coil to stop compressor action. A schematic symbol of a low-pressure control is shown in Figure 14–27.

The low-pressure cutoff switch is found in the system anywhere between the evaporator inlet and the compressor inlet. It is most often found on the accumulator-drier in a fixed orifice tube (FOT) system. In the event of a predetermined low system pressure, the switch will open to stop the compressor. When the air-conditioning system pressure rises to a predetermined high, the contacts again close and compressor action is restored.

In-vehicle temperature conditions are maintained in the same manner as with a thermostat. The only difference is that the control device is pressure sensitive instead of temperature sensitive. For a better understanding, recall the discussion of temperature and pressure relationships in Chapter 5.

BLOWER MOTOR

There are many styles and types of blower motors available, some of which are depicted in Figure 14–28. Blower motors may have a single- or double-shaft, depending on their design and application. Blower motors may be flange mounted and may have provisions for internal cooling. Regardless of the style or type, the blower motor drives one or two squirrel-cage blowers to move air across the evaporator and/or heater core, as shown in Figure 14–29.

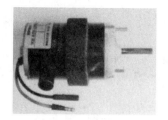
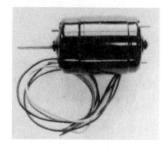

FIGURE 14–28 Styles and types of blower motors depend on their application. Generally, they are not interchangeable from one year or model vehicle to another. *(Courtesy of BET, Inc.)*

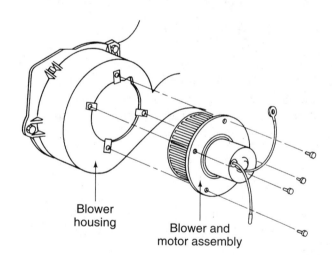

Blower housing

Blower and motor assembly

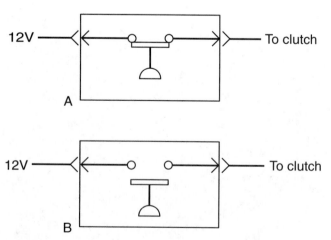

A

B

FIGURE 14–27 Low–pressure switch symbol: (A) closed; (B) open.

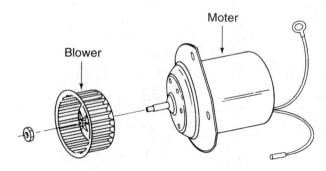

Blower

Moter

FIGURE 14–29 Squirrel–cage blower, which is attached to the blower motor, is found inside the blower housing. The direction of rotation of the motor is important, depending on blower housing design, to ensure airflow.

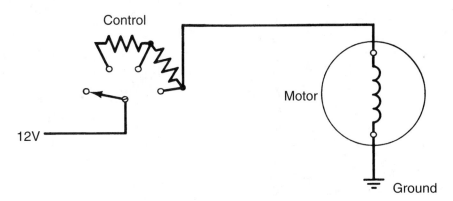

FIGURE 14-30 Resistors in the switch provide for low, medium, and high fan/blower speed. The switch is shown in the OFF position.

If the blower motor speed control is provided by resistors, the motor will have only one winding, as shown in Figure 14–30. Motor speed control may also be provided by a multiwound motor, like that shown in Figure 14–31, as well. Resistance for speed control is provided by the motor windings and no external resistors are required.

Blower motors generally are not repairable. They must be replaced if found to be defective. The common causes of failure include worn bushings and brushes or a defective internal wiring. Before replacing a motor thought to be defective, however, always check to ensure that the ground wire is secure, because most blower housings are constructed of nonconductive materials.

It must be noted that most replacement motors are nonreversible. It is important to note whether the defective motor turned clockwise (cw) or counterclockwise (ccw) when facing the shaft end of the motor. The replacement motor selected must turn in the same direction to ensure air delivery to the conditioned space. If, after motor replacement, there is little or no air delivery from the vents, suspect that the motor is turning in the wrong direction.

TROUBLESHOOTING THE BLOWER MOTOR

The blower motor often receives its power through a low blower relay or a high blower relay. To determine whether a blower motor is inoperative, it is necessary to bypass the other components in the control circuit. Follow the schematic in Figure 14–32 to troubleshoot a blower motor, as follows:

1. First, make certain that the ground wire (1) has not been disconnected.
2. Disconnect the hot lead wire (2) at the blower motor.
3. Connect a jumper wire from the battery positive (+) terminal (3) to the blower motor (2).
a. If the motor runs, it is good and the problem is elsewhere.
b. If the motor does not run, it is probably defective and must be replaced.

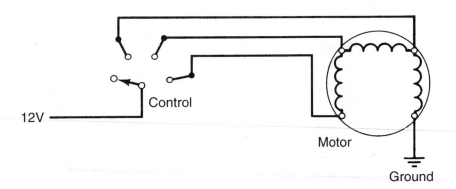

FIGURE 14-31 A multiwound motor provides the internal resistance required for low, medium, and high fan/blower speed. Switch is shown in the OFF position.

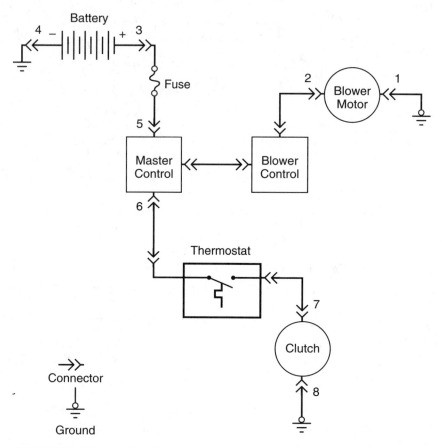

FIGURE 14–32 A typical blower motor schematic.

ELECTROMAGNETIC CLUTCH

Motor vehicle air-conditioning (MVAC) manufacturers use an electromagnetic clutch as a means of engaging and disengaging the compressor. For example, the compressor is engaged when cooling is required and is disengaged when the desired temperature is reached in the vehicle.

Basically, all clutches operate on the principle of magnetic attraction, hence the term *electromagnetic*. There have been two basic types of clutches used: those with a rotating field and those with a stationary field. Because the rotating field clutch has not been used since the early 1980s, they are only covered briefly.

Rotating Field Clutch

The rotating field clutch is only found on some vintage vehicles. It was last used on some Ford car lines in 1980 and on some Jeeps in 1981. The field is a part of, and turns with, the rotor. Current is applied to the field by means of brushes that are mounted on the compressor.

Stationary Field Clutch

The stationary field clutch is found on all vehicles since the

early 1980s. It is more desirable than the rotating field clutch because it is much lighter and has fewer parts to wear out.

The field coil is mounted to the compressor by mechanical means, depending on the type of compressor. The rotor is held on the armature by means of a bearing and snap rings. The rotor shown in Figure 14–33 is held on

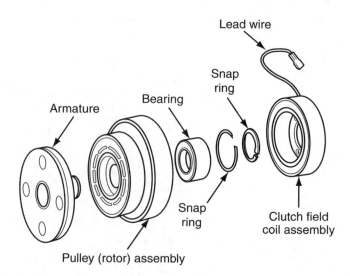

FIGURE 14–33 Typical electromagnetic clutch details.

the front head of the compressor by means of a bearing and snap rings. In either application, the armature is mounted on the compressor crankshaft.

When there is no current applied to the field coil, only a slight residual magnetic force is applied to the clutch. The rotor is free to turn and the armature remains stationary on the crankshaft.

When the temperature or **pressure control** switch is closed, current is applied to the field coil. Magnetic lines of force are established between the field coil and the armature. As a result, the armature is pulled into the rotor. When the armature becomes engaged with the rotor, the complete unit turns while the field remains stationary. The compressor crankshaft then begins to turn and the refrigeration cycle is initiated.

When the control device is opened, current to the field coil is interrupted and the magnetic field collapses. The armature disengages from the rotor and stops while the rotor continues to turn. Because the armature is attached to the compressor crankshaft, the pumping action of the compressor is stopped. The refrigeration cycle is interrupted until current is again applied to the clutch field coil.

CLUTCH DETAILS

Slots are machined in the armature and rotor of a clutch, Figure 14–34, to aid in concentrating the magnetic field, thereby increasing the attraction between them.

The clutch engages and disengages at high speeds

to maintain the desired preselected in-vehicle temperature conditions as often as required. Due to this action considerable scoring will occur on the armature and rotor surfaces of the clutch. Such scoring is expected and should not be a cause for concern. On the other hand, if the surfaces are burned or scorched, the problem should be corrected.

The spacing between the field coil and the rotor (pulley) is most important. The rotor should be as close as possible to the field coil, without touching, to achieve optimum magnetic flux travel. However, the pulley should not be so close to the rotor that it drags on the coil housing. This space is fixed and is not adjustable. Worn bearings, however, may allow the rotor to rub against the field coil. When this happens, replace the bearing and/or rotor.

The spacing between the rotor and the armature is also very important. If this spacing is too close, the armature will drag on the rotor when the unit is turned off. If the rotor and armature are too far apart, there will be poor contact between the armature and rotor when energized. Either one of these situations may result in serious clutch malfunction or damage.

The spacing of the rotor and armature should be such that when the clutch is off, there is no drag. Also, when the clutch is turned on, the proper spacing ensures that no slippage will occur (except for the moment when the clutch is first engaged). A nonmagnetic feeler gauge is used to measure clutch gap clearance, Figure 14–35.

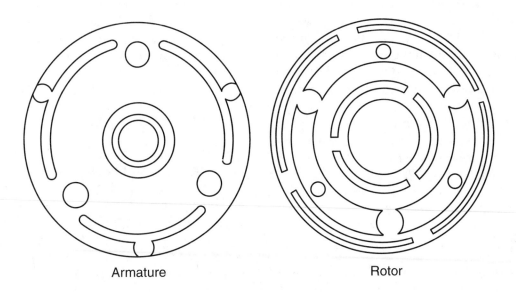

Armature   Rotor

FIGURE 14–34 Slots in the armature and rotor aid in magnetic attraction.

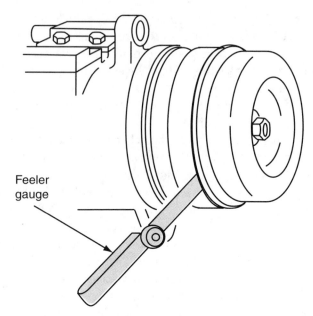

Feeler
gauge

FIGURE 14–35 Check the air gap between the armature and the rotor.

PRESSURE CUTOFF SWITCH

Many systems have a low- or high-pressure cutoff switch connected electrically in series in the clutch circuit. These normally closed (NC) switches are sensitive to system pressure and open in the event of abnormal pressure. This action interrupts electrical current to the clutch coil to stop compressor action.

Low-Pressure Cutoff Switch

The low-pressure cutoff switch is found in the system anywhere between the evaporator inlet and the compressor inlet. It is often found on the accumulator-drier in a fixed orifice tube (FOT) system. In the event of an abnormally low pressure of 10 psi (68.9 kPa), for example, the switch will open to stop the compressor. This action prevents further reduction of system pressure and protects the system from the possible entrance of air and/or moisture, as would be the case with a low-side leak.

A low-pressure cutoff switch may also be found in the high side of some systems to prevent clutch engagement if the system pressure is between 18 and 31 psig (124.1 and 213.7 kPa). This would be the case if most of the refrigerant had leaked out. The actual pressure rating of the control depends on system application.

High-Pressure Cutoff Switch

The high-pressure cutoff switch is found on some General Motors, Chrysler, Subaru, and other vehicles. This switch is found in the system anywhere between the compressor outlet and the evaporator inlet. It is often found on the receiver-drier on thermostatic expansion valve (TXV) systems.

In case of an abnormally high pressure of 300–500 psi (2,068.5–3,447.5 kPa), for example, this switch will open to stop compressor action. This prevents system component damage and/or hose rupture that may be caused by a further increase in pressure.

COMPRESSOR DISCHARGE PRESSURE SWITCH

Many factory systems use a compressor discharge pressure switch to disengage the compressor clutch electrical circuit. The action of this switch stops the compressor if the refrigerant charge is not adequate, for example, due to a leak, to provide sufficient circulation within the system. The compressor discharge pressure switch is also called a no-charge switch and ambient low-temperature switch, or a low-pressure cutoff switch. The switch is designed to open electrically to shut off the compressor when the high-side system pressure drops below 37 psig (255 kPa). This switch also performs the secondary function of an outside ambient air temperature sensor. When outside ambient air temperature falls below 25°F (–3.9°C), the reduced corresponding refrigerant pressure keeps the switch open.

The compressor discharge pressure switch, located in the rear head of the compressor, high-pressure discharge line from the compressor, or receiver-drier, cannot be repaired. It must be replaced with a new unit if it fails. Its function is to protect the compressor, and should not be defeated by bypassing it with a jumper wire.

Superheat Switch

The *superheat switch* was used on systems equipped with a thermal limiter fuse. These include some early models of Audi, General Motors, and Jaguar car lines. The switch, located in the rear head of the compressor, was used to protect the system against excessive superheat by "blowing" the thermal limiter fuse in the event of excess heat.

FACTORY-INSTALLED WIRING

The electrical schematic in Figure 14–36 is only one of hundreds showing the wiring of factory-installed heating, ventilating and air-conditioning (HVAC) systems. When servicing these systems, it is generally necessary to consult the appropriate manufacturer's service manual for specific information and schematics. As noted earlier, various methods of temperature control are used. Some may use a thermostat, whereas others may use a pressure control. Some early systems used an evaporator pressure control, whereas later systems use a blend of heated and cooled air to achieve

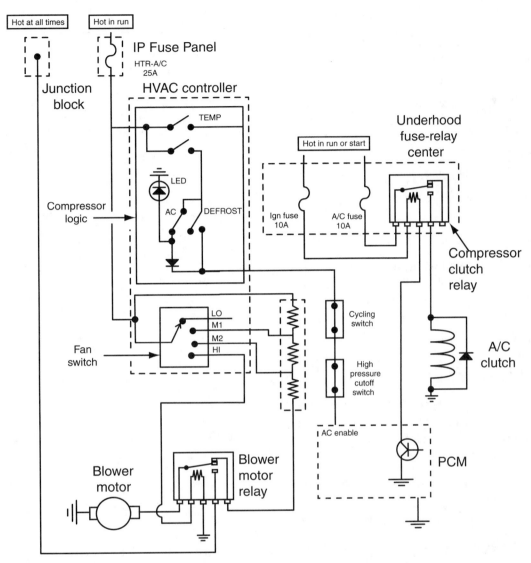

FIGURE 14–36 A typical electronic climate control (ECC) system schematic.

the desired temperature. For further information, refer to chapter 15 for the appropriate topic in this text.

COOLANT TEMPERATURE WARNING SYSTEM

The motor vehicle air-conditioning (MVAC) system, when operating, places a high demand on the engine cooling system. The condenser, which is mounted in front of the radiator, imposes an extra heat load. The ambient air that is made available to remove unwanted heat from the engine coolant in the radiator is heat laden from having first passed through the air-conditioning system condenser.

Often, a malfunctioning air-conditioning system will affect the engine coolant temperature. Conversely, an overheated engine will affect air-conditioning system performance. A dash light or a dash gauge is used to monitor engine coolant condition.

Lamp

The engine coolant lamp system is often called an *idiot light*. That is probably because the problem has already occurred by the time the light comes on. The system, shown in Figure 14–37, warns that the engine has overheated and immediate attention is required.

The contacts of the engine sending unit will close when engine coolant temperature reaches about 250°F (121°C). The actual temperature at which this switch closes depends on the engine-operating temperature design.

Pressurized engine cooling systems are covered in Chapter 17 of this text. The disadvantage of the telltale light system is obvious: the hot lamp generally does not light until after the problem has already occurred.

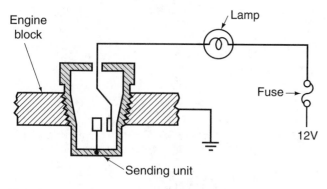

FIGURE 14–37 A one–lamp coolant "HOT" warning system.

Gauge

The coolant temperature gauge system, illustrated in Figure 14–38, consists of two parts: the dash (gauge) unit and the engine (sending) unit. The sending unit contains a sintered material that can change its resistance in relation to its temperature. This material is sealed in a metal bulb and screwed into a coolant passage of the engine. It has a high resistance when cold, and a low resistance when hot.

The varying resistance of the sending unit regulates the amount of current passing through the coil of the dash gauge which, in turn, moves the pointer accordingly.

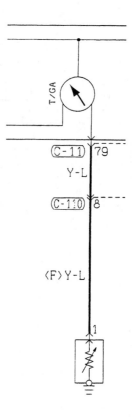

FIGURE 14–38 Engine coolant temperature gauge schematic.

REVIEW

Select the correct answer from the choices given.

1. Technician A says the purpose of a rheostat is to ensure that full battery voltage reaches the blower motor. Technician B says that a resistor block is used to provide a means of varying blower motor speed. Who is right?
   a. A only
   b. B only
   c. Both A and B
   d. Neither A nor B

2. Technician A says that a bellows-type thermostat has a capillary tube. Technician B says that a bimetallic-type thermostat must be placed in the air stream. Who is right?
   a. A only
   b. B only
   c. Both A and B
   d. Neither A nor B

3. Technician A says that a control thermostat is used on some motor vehicle air-conditioning (MVAC) systems as a means of temperature control. Technician B says that a low-pressure control is used on some motor vehicle air-conditioning (MVAC) systems as a means of temperature control. Who is right?
   a. A only
   b. B only
   c. Both A and B
   d. Neither A nor B

4. All of the following may be attached electrically to a common ground, EXCEPT:
   a. the control thermostat.
   b. the blower motor.
   c. the electromagnetic clutch coil.
   d. the dash temperature gauge.

5. All of the following are commonly shared by the heating and air-conditioning systems, EXCEPT:
   a. the case/duct system.
   b. the blower motor.
   c. the blower speed control.
   d. the fuse or circuit breaker.

6. Technician A says that the different types of fuses are not interchangeable. Technician B says a circuit breaker may be used to replace a fuse that blows frequently. Who is right?
   a. A only
   b. B only
   c. Both A and B
   d. Neither A nor B

7. Technician A says that a fuse that blows frequently always indicates that there is a problem in the circuit. Technician B says that an undersized fuse may blow frequently with no problem in the circuit. Who is right?
   a. A only
   b. B only
   c. Both A and B
   d. Neither A nor B

8. An electromagnetic clutch is slipping. A test lamp that is connected to the clutch lead wire, does not burn. Technician A says the voltage at the clutch lead is low. Technician B says that the test light bulb is burned out. Who is right?
   a. A only
   b. B only
   c. Both A and B
   d. Neither A nor B

9. How many blower motor speeds will a three-resistor speed reducer provide?
   a. Five
   b. Four
   c. Three
   d. Two

10. Typically, a control thermostat:
    a. controls in-vehicle temperature conditions.
    b. cycles the electromagnetic clutch on and off .
    c. can perform both A and B.
    d. cannot perform A or B.

11. The capillary tube of _____ thermostat(s) must be inserted into the condenser coil.
    a. the bimetallic-type
    b. the bellows-type
    c. both A and B
    d. neither A nor B

12. Technician A says that the operating ranges of all thermostats are adjustable. Technician B says the opening and closing ranges of all thermostats are adjustable. Who is right?
    a. A only
    b. B only
    c. Both A and B
    d. Neither A nor B

13. A high-pressure switch may be found anywhere in an air-conditioning system, EXCEPT:
    a. in the condenser outlet.
    b. in the evaporator outlet.
    c. in the compressor outlet.
    d. in the receiver-drier.

14. Technician A says a low-pressure control may be found in the high side of an air-conditioning system. Technician B says a high-pressure control may be found in the low-side of an air conditioning system. Who is right?
    a. A only
    b. B only
    c. Both A and B
    d. Neither A nor B

15. Technician A says that the air gap between the rotor and armature may be excessive if a clutch slips. Technician B says that the air gap between the rotor and armature may not be sufficient if a clutch slips. Who is right?
    a. A only
    b. B only
    c. Both A and B
    d. Neither A nor B

16. Which part of the clutch is attached directly to the compressor crankshaft?.
    a. Rotor
    b. Armature
    c. Both A and B
    d. Neither A nor B

17. What is used to measure clutch air gap?
    a. A dime
    b. A feeler gauge
    c. A micrometer
    d. Any of the above

18. A compressor discharge pressure switch:
    a. is a normally closed switch.
    b. opens at about 35–40 psig (241–276 kPa).
    c. prevents ambient air from entering the system.
    d. can perform all of the above.

19. Technician A says that a dash gauge indicating HOT may be caused by a grounded sending unit wire. Technician B says that a dash warning light indicating HOT may be caused by a grounded sending unit wire. Who is right?
    a. A only
    b. B only
    c. Both A and B
    d. Neither A nor B

20. Technician A says a temperature lamp is preferred because the operator is more likely to notice a dash light than a gauge. Technician B says a temperature gauge is preferred because an overheating problem may be noted earlier. Who is right?
    a. A only
    b. B only
    c. Both A and B
    d. Neither A nor B

TERMS

Write a brief description of the following terms:

1. accessory
2. battery voltage
3. bellows
4. bimetallic
5. blown
6. circuit breaker
7. common ground
8. contacts
9. fins
10. fuse
11. lamp
12. load
13. ohmmeter
14. pressure control
15. remote bulb
16. resistors
17. schematics
18. setscrew
19. variable
20. windings

CHAPTER 15

SYSTEM CONTROL DEVICES

OBJECTIVES

On completion and review of this chapter, you should be able to:

- ❑ Understand and discuss the operation of the manual, semiautomatic, and automatic functions of a motor vehicle air-conditioning (MVAC) system.
- ❑ Understand and explain the purposes and functions of various sensors, modules, actuators, and switches used in climate control systems.
- ❑ Discuss the operation of vacuum motors, switches, relays, and related components.
- ❑ Recognize vacuum system components and understand their function.
- ❑ Recognize automatic temperature control (ATC) components and understand their function.

INTRODUCTION

Many controls and control systems used in motor vehicle air-conditioning (MVAC) systems are pneumatically actuated with either negative (vacuum) or positive (pressure). Other types of manual, **semiautomatic,** and automatic controls are electrically or electronically actuated.

VACUUM CIRCUITS

To understand vacuum circuits, it is essential first to define and understand the term *vacuum.* One definition, given in a leading encyclopedia describes a vacuum as a "portion of space that is *entirely* devoid of matter"—literally, a space that contains nothing. Because all things in nature contain matter in some form, it would seem that there is no such thing as a vacuum. For all practical purposes, however, a vacuum must be thought of as a portion of space that is *partially* devoid of matter. For a better understanding, consider that a vacuum is a space in which pressure is *below* atmospheric pressure.

A good example of a vacuum is of a person drinking through a straw in Figure 15–1. A slight vacuum is created in the straw as the person sucks on it. Atmospheric pressure, which is greater than the vacuum pressure in the straw, is exerted against the surface of the liquid. This difference in pressure, known as delta P (Δ_P), forces the liquid up the straw.

Atmospheric pressure at sea level is 14.696 psia (101.328 kPa absolute). For all practical purposes, this value is usually rounded off to 14.7 psia (101.4 kPa absolute) or 15 psia (103.4 kPa absolute). At sea level, then, a pressure of 14 psia (96.5 kPa absolute) is a vacuum. Traditionally, English system vacuum pressure values are usually given in *inches of mercury* (in. Hg), as covered in other chapters of this text.

Most vehicle manufacturers' manuals only use the term *inches* when giving vacuum value requirements and specifications. In this text, any references to vacuum values are given in the English and metric absolute scales of pressure. A conversion chart is given in Figure 15–2 and may be used as a reference for comparing *inches to psia* and *kPa absolute.*

FIGURE 15–1 Sipping liquid through a straw produces a lower–than–atmospheric pressure (vacuum) at "A." The higher (atmospheric) pressure, "B," forces the liquid through the straw due to a difference in pressure.

VACUUM-OPERATED DEVICES

Most vacuum-operated devices, such as *heater coolant valves* and **mode doors,** are activated with a *vacuum pot,* also called a **vacuum motor** or *vacuum power unit.* The exertion (force) of atmospheric pressure on one side of a diaphragm causes it to move toward the lower (vacuum) pressure side, as shown in Figure 15–3. This moves the device to be controlled through a lever, arm, or rod linkage.

Dual-chamber vacuum pots (motors), like that shown in Figure 15–4, operate below atmospheric pressure, based on a pressure differential (Δ_P) from one side

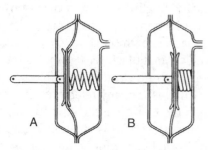

FIGURE 15–3 Typical vacuum pot (motor) operation: (A) no vacuum applied and (B) full vacuum applied.

Inches of Mercury (in. Hg)	Pounds Per Square Inch Absolute (psia)	Kilopascals Absolute (kPa absolute)
28.98	0.5	3.45
27.96	1.0	6.89
26.94	1.5	10.34
25.92	2.0	13.79
24.90	2.5	17.24
23.88	3.0	20.68
22.86	3.5	24.13
21.83	4.0	27.58
20.81	4.5	30.03
19.79	5.0	34.47
18.77	5.5	37.92
17.75	6.0	41.37
16.73	6.5	44.82
15.71	7.0	48.26
14.69	7.5	51.71
13.67	8.0	55.16
12.65	8.5	58.61
11.63	9.0	62.05
10.61	9.5	65.50
9.59	10.0	68.95
8.57	10.5	72.40
7.54	11.0	75.84
6.52	11.5	79.29
5.50	12.0	82.74
4.48	12.5	86.19
3.46	13.0	89.63
2.44	13.5	93.08
1.42	14.0	96.53
0.40	14.5	99.98
	15.0	103.42

FIGURE 15–2 Comparison of inches of mercury (in. Hg) to pounds per square inch absolute (psia) and kilopascals absolute (kPa absolute).

to the other. A higher pressure on either side will move the diaphragm to the side with a lower pressure. This provides a *push* or *pull effect* of the vacuum pot.

VACUUM SOURCE

The vehicle engine, when running, provides a ready source of vacuum. This source is usually taken off the intake manifold and is routed to the various components through small-diameter synthetic rubber, plastic, or nylon hoses. Engine vacuum supply source can vary from 0.1 in. Hg (14.6 psia or 101.12 kPa absolute) to 20 in Hg (4.89 psia or 33.7 kPa absolute) or more, depending on certain engine operating conditions. The reason for this vacuum variation is not important in this discussion. It is important, however, to be aware that engine vacuum does vary.

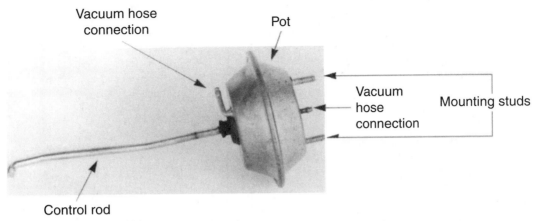

Vacuum hose connection

Pot

Vacuum hose connection

Mounting studs

Control rod

FIGURE 15–4 A dual–chamber vacuum motor. *(Courtesy of BET, Inc.)*

This vacuum variation necessitates using a **reserve tank** and *check valve* as shown in Figure 15–5. This combination of devices provides the means of maintaining maximum vacuum values required to properly operate heating, ventilating, and air-conditioning **(HVAC)** system vacuum controls during all engine operating conditions. More than one reserve tank or check valve may be found in the vacuum circuit.

CHECK VALVES AND RELAYS

Vacuum systems normally have a vacuum check valve or **check relay** to prevent vacuum loss during those periods when the engine manifold vacuum is less than the value required to operate a vacuum-actuated component. In addition, most vacuum systems contain a vacuum reserve tank. The check valve or check relay is usually located in the vacuum line between the reserve tank and the vacuum source.

Check Relay

The vacuum check relay serves two purposes: It prevents a vacuum loss during low manifold vacuum conditions, and it prevents system mode operation during these periods. Figure 15–6 is a typical schematic of a vacuum check relay.

Check Valve

The check valve, shown in Figure 15–7, opens whenever the manifold vacuum is greater than the reserve vacuum. In other words, the check valve is opened by normal engine vacuum. In this position, the check valve connects the source to the tank. The normal engine vacuum also opens the diaphragm and allows vacuum from the control to reach the vacuum motor. The check valve closes whenever the manifold vacuum drops below the value of the reserve pressure. When the valve closes, the diaphragm also closes and blocks the passage from the control to the motor. As a result, the reserve vacuum is not lost because it is not allowed to bleed back through the manifold.

The manifold vacuum drops during periods of acceleration and when the engine is stopped. The vacuum reserve is used to operate the air-conditioning system vacuum components as well as other accessory equipment in the vehicle, such as the door locks.

Many types and styles of check valves are used in the air-conditioning and heating vacuum circuit. Essentially, a check valve, illustrated in Figure 15–8, allows the flow of a fluid or vapor in one direction and blocks the flow in the opposite direction. A check valve is easily tested, as follows.

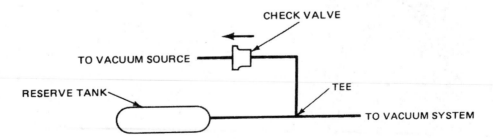

CHECK VALVE

TO VACUUM SOURCE

RESERVE TANK

TEE

TO VACUUM SYSTEM

FIGURE 15–5 A reserve tank provides vacuum source during periods of low- or no-engine vacuum. A check valve prevents vacuum loss back to the source during this period.

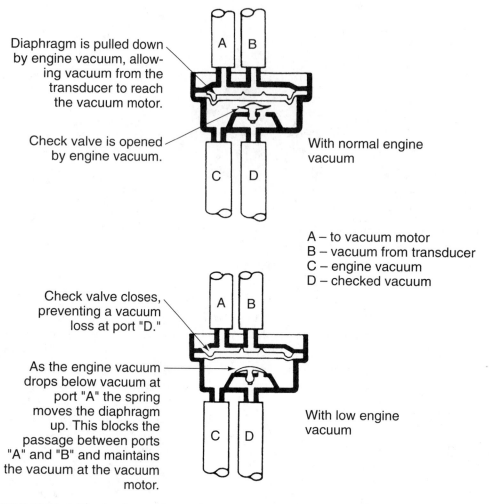

Diaphragm is pulled down by engine vacuum, allowing vacuum from the transducer to reach the vacuum motor.

Check valve is opened by engine vacuum.

With normal engine vacuum

A – to vacuum motor
B – vacuum from transducer
C – engine vacuum
D – checked vacuum

Check valve closes, preventing a vacuum loss at port "D."

As the engine vacuum drops below vacuum at port "A" the spring moves the diaphragm up. This blocks the passage between ports "A" and "B" and maintains the vacuum at the vacuum motor.

With low engine vacuum

FIGURE 15–6 Check relay used to maintain vacuum at the vacuum motor.

FIGURE 15–7 Typical vacuum check valve. *(Courtesy of BET, Inc.)*

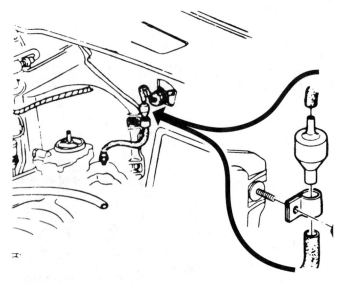

FIGURE 15–8 Check valve.

1. Remove the suspected valve from the vehicle and connect it, in series, to a vacuum pump and vacuum or compound gauge, as shown in Figure 15–9.

   NOTE: The direction of flow should be *away* from the pump. If a vacuum shows on the gauge when the pump is running, the valve is closed; if no vacuum shows, then valve is open and therefore defective.

2. Next turn the check valve so the direction of flow is *toward* the vacuum pump.

   NOTE: If a vacuum now shows on the gauge when the pump is running, the valve has a restriction and is considered to be defective. If no vacuum shows, then the valve is not defective—provided it passed previous tests. A defective check valve must be replaced because repairs are not usually possible or practical.

Reserve Tank

Reserve vacuum tanks are manufactured in a variety of sizes and shapes. Early tanks, like that shown in Figure 15–10, are made of metal and resemble a large juice can. Later tanks, like that in shown Figure 15–11, are made of plastic and resemble a sphere. Vacuum reserve tanks generally do not require maintenance but sometimes develop pinhole-size leaks due to exposure to the elements. When a tank is suspected of leaking, it may be removed from the vehicle and pressurized to about 5 psig (34.4

kPa) with shop air pressure. It is then leak tested with a soap solution or by immersing it in a tank of water. After releasing the pressure, the hole may be repaired by first cleaning the area with a wire brush or sandpaper, and then applying a two-part epoxy or fiberglass material.

Restrictor

Some vacuum systems have a **restrictor,** Figure 15–12, to delay or to slow the operation of a device. Restrictors have a small orifice that sometimes becomes clogged with lint or other airborne debris. Attempts to clean a restrictor usually prove unsuccessful; replacement is suggested. To test a restrictor, simply use a vacuum pump and gauge setup as previously described.

FIGURE 15–10 Vacuum reserve tank. *(Courtesy of BET, Inc.)*

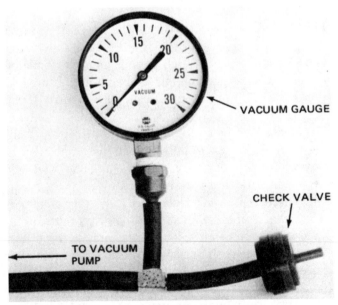

FIGURE 15–9 Typical setup to test a check valve. *(Courtesy of BET, Inc.)*

FIGURE 15–11 A typical vacuum reserve tank. *(Courtesy of BET, Inc.)*

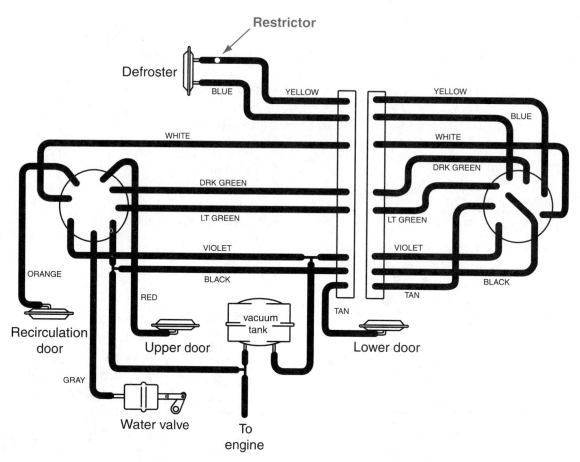

FIGURE 15–12 A typical vacuum system diagram showing the location of the restrictor.

The gauge should show a vacuum when the pump is running, with the restrictor connected in *either* or *both* directions. When the vacuum pump is stopped, the gauge should drift slowly to zero (atmospheric) pressure. If not, the restrictor is clogged and must be replaced.

VACUUM SYSTEM DIAGRAMS

The vacuum system in Figure 15–13 must be considered typical because there are hundreds of variations of these systems. Consequently, the vehicle manufacturer's vacuum system **diagrams** for a specific year or model vehicle must be followed. Basically, the vacuum system is used to open, close, or position the heater coolant valve and mode doors to achieve a desired in-vehicle preselected temperature and humidity level.

A description of the various vacuum-operated devices associated with the air-conditioning and heating systems is given later in this chapter. It should be noted, however, that not all air-conditioning systems have all of the devices discussed in this text.

PRESSURE CONTROLS

Figure 15–14 shows that within the range of about 20 psig (137.9 kPa) to 80 psig (551.6 kPa), the temperature and pressure of CFC-12 refrigerants are closely related. For HFC-134a refrigerant, this temperature-pressure relationship exists from about 10 psig (68.9 kPa) to 70 psig (482.6 kPa).

Liquid refrigerant is metered into the evaporator, Figure 15–15, by the metering device, thermostatic expansion valve, or orifice tube. The amount of refrigerant required is determined by the heat load on the evaporator. There is a corresponding decrease in the amount of refrigerant that is metered into the evaporator by the metering device as the heat load decreases. In this discussion of pressure controls, the technician should recall that water droplets that accumulate on the surface of the evaporator freeze when the temperature drops below 32°F (0°C). The temperature of the evaporator, then, must be kept just above the freezing point.

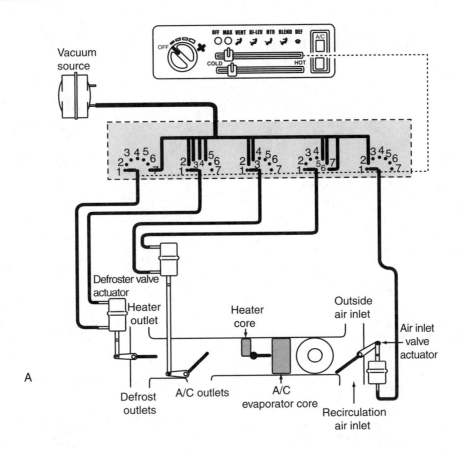

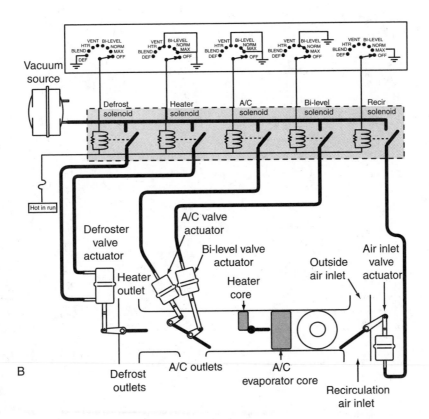

FIGURE 15–13 Typical (A) rotary vacuum valve control system, and (B) vacuum solenoid control system.

Temp. °F	Press. psig	Temp. °F	Press. psig	Temp. °F	Press. psig	Temp. °F	Press. psig	Temp. °F	Press. psig
0	9.1	35	32.5	60	57.7	85	91.7	110	136.0
2	10.1	36	33.4	61	58.9	86	93.2	111	138.0
4	11.2	37	34.3	62	60.0	87	94.8	112	140.1
6	12.3	38	35.1	63	61.3	88	96.4	113	142.1
8	13.4	39	36.0	64	62.5	89	98.0	114	144.2
10	14.6	40	36.9	65	63.7	90	99.6	115	146.3
12	15.8	41	37.9	66	64.9	91	101.3	116	148.4
14	17.1	42	38.8	67	66.2	92	103.0	117	151.2
16	18.3	43	39.7	68	67.5	93	104.6	118	152.7
18	19.7	44	40.7	69	68.8	94	106.3	119	154.9
20	21.0	45	41.7	70	70.1	95	108.1	120	157.1
21	12.7	46	42.6	71	71.4	96	109.8	121	159.3
22	22.4	47	43.6	72	72.8	97	111.5	122	161.5
23	23.1	48	44.6	73	74.2	98	113.3	123	163.8
24	23.8	49	45.6	74	75.5	99	115.1	124	166.1

A: English Temp/Press Chart for CFC–12

Temperature °F	Pressure psig	Temperature °F	Pressure psig
−5	4.1	39.0	34.1
0	6.5	40.0	35.0
5.0	9.1	45.0	40.0
10.0	12.0	50.0	45.4
15.0	15.1	55.0	51.2
20.0	18.4	60.0	57.4
21.0	19.1	65.0	64.0
22.0	19.9	70.0	71.1
23.0	20.6	75.0	78.6
24.0	21.4	80.0	86.7
25.0	22.1	85.0	95.2
26.0	22.9	90.0	104.3
27.0	23.7	95.0	113.9
28.0	24.5	100.0	124.1
29.0	25.3	105.0	134.9
30.0	25.3	110.0	146.3
31.0	27.0	115.0	158.4
32.0	27.8	120.0	171.1
33.0	28.7	125.0	184.5
34.0	29.5	130.0	198.7
35.0	30.4	135.0	213.5

B: English Temp/Pres Chart for HFC–134a

Evaporator Temperature °C	Evaporator Pressure Gauge Reading kiloPascal (Gauge)	Evaporator Pressure Gauge Reading kiloPascal (Absolute)	Ambient Temperature °C	High Pressure Gauge Reading kiloPascal (Gauge)
−16	73.4	174.7	16	737.7
−15	81.0	182.3	17	759.8
−14	87.8	189.1	18	784.6
−13	94.8	196.1	19	810.2
−12	100.6	201.9	20	841.2
−11	108.9	210.2	21	868.7
−10	117.9	219.2	22	901.8
−9	124.5	225.8	23	932.2
−8	133.9	235.2	24	970.8
−7	140.3	241.6	25	1 020.5
−6	149.6	250.9	26	1 075.6
−5	159.2	260.5	27	1 111.5
−4	167.4	268.7	28	1 143.2
−3	183.2	268.7	29	1 174.9
−2	186.9	288.2	30	1 206.6
−1	195.8	288.2	31	1 241.1
0	206.8	308.1	32	1 267.3
1	218.5	319.8	33	1 294.8
2	227.8	329.1	34	1 319.7
3	238.7	340.0	35	1 344.5
4	249.4	350.7	36	1 413.5
5	261.3	362.6	37	1 468.6
6	273.7	375.0	38	1 527.9
7	287.5	388.8	39	1 577.5
8	296.6	397.9	40	1 627.2
9	303.3	404.6	42	1 737.5
10	321.5	422.8	45	1 854.7

C: Metric Temp/Press Chart for CFC–12

Temperature °C	Pressure kPa	Temperature °C	Pressure kPa
−15.0	63	5.0	247
−12.5	83	7.5	280
−10.0	103	10.0	313
−7.5	122	12.5	345
−5.0	142	15.0	381
−4.5	147	17.5	422
−4.0	152	20.0	465
−3.5	157	22.5	510
−3.0	162	25.0	560
−2.5	167	27.5	616
−2.0	172	30.0	670
−1.5	177	32.5	726
−1.0	182	35.0	785
−0.5	187	37.5	849
0.0	192	40.0	916
0.5	198	42.5	990
1.0	203	45.0	1066
1.5	209	47.5	1146
2.0	214	50.0	1230
2.5	220	52.5	1315
3.0	225	55.0	1385
3.5	231	57.5	1480
4.0	236	60.0	1580
4.5	242	65.0	1795

D: Metric Temp/Pres Chart for HFC–134a

FIGURE 15–14 Typical temperature–pressure charts.

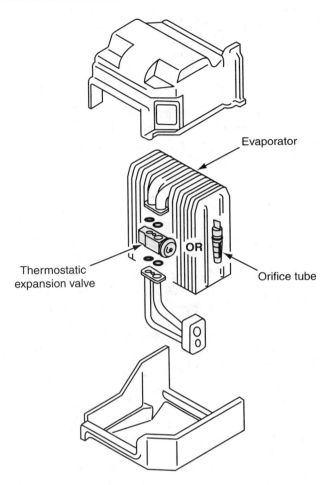

FIGURE 15–15 A typical evaporation core.

PRESSURE REGULATORS

Several types of evaporator pressure regulators were used through the early 1980s to control the pressure of the refrigerant in the evaporator. Pressure regulators were designed to prevent the pressure from falling below a predetermined range, usually 22 psig (151.6 kPa) to 30 psig (206.8 kPa), depending on system design.

The suction pressure regulator, as a part in a device called valves-in-receiver (VIR), was last used by Audi in 1983. Another device, called a suction throttling valve (STV), was last used on some Ford and Mercury car lines in 1981.

CONTROL SYSTEMS

Many different types of manual, semiautomatic, and automatic temperature control systems are in use today. In fact, there are so many that it is impossible to cover each system individually in this text. Systems are modified or changed from year to year and from vehicle model to vehicle model.

Diagnostic testing procedures differ, so there are no recommended typical testing procedures. For example, Test Point 12 of one vehicle may be Test Point 10 or 14 of another vehicle. The *shortening* of the wrong test point could destroy an expensive component.

Some solid-state components are so sensitive that even the 1.5-volt AAA battery used in an analog ohmmeter, Figure 15–16, may destroy them. A digital ohmmeter, Figure 15–17 must then be used whenever a manufacturer's specifications suggest that component resistance measurements be taken. Some components and circuits are so

FIGURE 15–16 Typical VOM with analog readout. *(Courtesy of Wavetek)*

FIGURE 15–17 Typical VOM with digital readout. *(Courtesy of Wavetek)*

sensitive to outside influence, however, that some schematics are labeled "do not measure resistance." Heed this caution to avoid damage to delicate electronic components.

Although temperature control systems differ in many respects, all are designed to maintain in-vehicle temperature and humidity conditions at a certain level (within system limitations), regardless of the temperature conditions outside the vehicle. The temperature control also holds the relative humidity within the vehicle to a healthful level and prevents window fogging. For example, if the desired temperature is 75°F (23.89°C), the automatic control system will maintain an in-vehicle environment of 75°F (23.89°C) at 45 to 55 percent humidity, regardless of the outside weather conditions.

In even the hottest weather, a properly operating system can rapidly cool the vehicle interior to the predetermined temperature (75°F or 23.89°C). The degree of cooling then cycles to maintain the desired temperature level. In mild weather conditions, the passenger compartment can be held to this same predetermined temperature (75°F or 23.89°C), without resetting or changing the control.

During cold weather, the system rapidly heats the passenger compartment to the predetermined 75°F (23.9°C) temperature level, and then automatically maintains it.

The intent of this test is to provide basic overview of the components of various systems, not to cover any particular system in detail. These components include, but are not limited to, coolant temperature **sensor,** in-vehicle temperature sensor, outside temperature sensor, high-side temperature switch, low-side temperature switch, low-pressure switch, high-pressure switch, vehicle speed sensor, **throttle** position sensor, **sunload** sensor, and power steering cutout switch.

Many automotive electronic temperature control systems have self-diagnostic test provisions, whereby an onboard microprocessor-controlled subsystem will display a code. This code (number, letter, or alphanumeric) is displayed to tell the technician the cause of the malfunction. Some systems also display a code to indicate which computer detected the malfunction. A typical trouble code chart is provided in Figure 15–18. However, manufacturers' specifications must always be followed to identify the malfunction display codes, because they differ from vehicle to vehicle. For example, in some General Motors vehicles ".7.0" will be displayed to indicate no malfunction if "no trouble" codes are stored in the computer. On some Ford vehicles, the "no trouble" code is "888."

It is also possible for the air-conditioning system to malfunction even though the self-check testing indicates

BCM DIAGNOSTIC TROUBLE CODES

CODE	NOTES	PROBLEM
F10	1	OUTSIDE TEMPERATURE SENSOR CIRCUIT
F11	1 - 2	A/C HIGH SIDE TEMPERATURE SENSOR CIRCUIT
F12	1 - 3	A/C LOW SIDE TEMPERATURE SENSOR CIRCUIT
F13	1	IN-CAR TEMPERATURE SENSOR CIRCUIT
F30	1	CCP TO BCM DATA CIRCUIT
F31	1	FDC TO BCM DATA CIRCUIT
F32	1 - 4	ECM TO BCM DATA CIRCUIT
F40	1	AIR MIX DOOR PROBLEM
F43	1	HEATED WINDSHIELD PROBLEM
F46	2	LOW REFRIGERANT WARNING
F47	2 - 5	LOW REFRIGERANT PROBLEM
F48	2 - 5	LOW REFRIGERANT PRESSURE
F49	1	HIGH TEMPERATURE CLUTCH DISENGAGE
F51	1	BCM PROM ERROR

NOTES:		
	1	Does not turn on any light
	2	Turns on SERVICE AIR COND light
	3	Disengages A/C clutch
	4	Turns on cooling fans
	5	Switches from AUTO to ECON

FIGURE 15–18 Body Control Module (BCM) diagnostic trouble codes lead the technician to the problem. *(Courtesy of BET, Inc.)*

that there are no problems. It is then necessary to follow the manufacturer's step-by-step procedure to troubleshoot and check the system. Again, typical diagnostic procedures are not practical because of the many different types of systems found in the modern vehicle. For example, the flow chart in Figure 15–19 illustrates diagnostic procedures for "no clutch operation" for a particular year or model vehicle. This procedure may not be applicable for that same vehicle in another model year or for other similar vehicles in that same model year.

SENSORS

Although they may vary in physical appearance, all sensors have the same general operating characteristics; that is, they are extremely sensitive to slight changes in temperature. The change in resistance value of each sensor is inversely proportional to a temperature change often referred to as **delta T** (Δ_T). For example, when the temperature decreases, the resistance of the sensor increases;

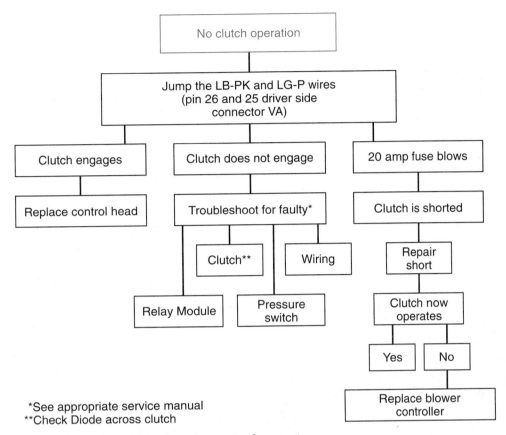

FIGURE 15–19 Typical flow chart for no clutch operation.

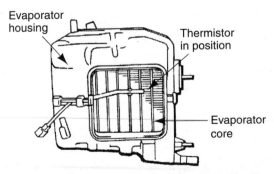

and when the temperature increases, the sensor resistance decreases.

The sensor is actually a resistor whose resistance value is determined by its temperature. This type of resistor is called a **thermistor,** shown in Figure 15–20. Although the theory of thermistor operation is not covered in this text, the student should be able to gain a good understanding of thermistor operation from the following description and Figure 15–21.

In Figure 15–21A, one thermistor is installed in a duct. With air at a temperature of 60°F (15.56°C) passing through the duct, the resistance value of the thermistor is 94 ohms. Refer to the thermistor value chart given in Figure 15–22. If the temperature in the duct is 90°F (32.22°C), Figure 15–21B, then the resistance of the thermistor decreases to about 45 ohms. If, however, the temperature is decreased to 40°F (4.44°C), the thermistor resistance is increased to 160 ohms, Figure 15–21C.

Figure 15–22 is a graph of individual sensor values at various temperatures. Compare the chart with the examples given up to this point. Note that each sensor has a different value for a particular temperature.

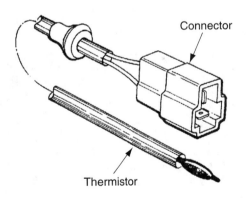

FIGURE 15–20 A thermistor (A); in position (B). (*From Green/Green and Dwiggins, Australian Automotive Air Conditioning,* © *Nelson, a division of Thomson Learning*)

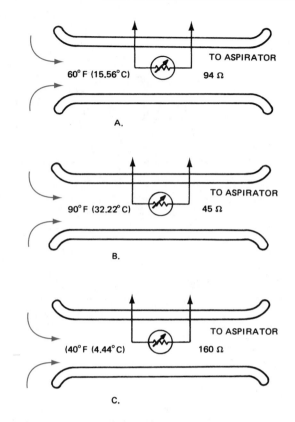

FIGURE 15–21 The resistance of a thermistor changes as the temperature changes.

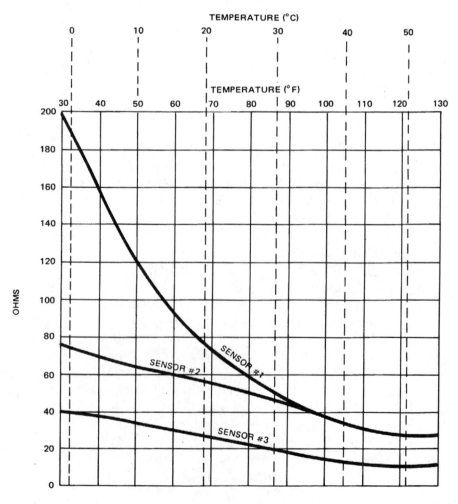

FIGURE 15–22 Typical resistance values of thermistors (sensor) at various temperatures.

ELECTRONIC TEMPERATURE CONTROL SYSTEMS

Many types of electronic temperature control systems are in use. The flow charts of Figure 15–23 and Figure 15–24 illustrate two typical systems. The following information relates to many of the components found in an electronic temperature control system. Not all components, however, are found in all systems.

Control Panel

The control panel is found in the instrument panel at a convenient location for both driver and front seat passenger access. Two types of control panels are shown in Figure 15–25. Some control panels have features that other panels do not have, such as provisions to display both inside and outside air temperatures in English or

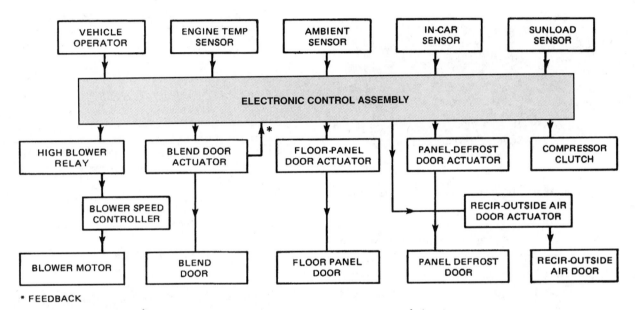

* FEEDBACK

FIGURE 15–23 Typical flow chart for electronic temperature control system with five inputs.

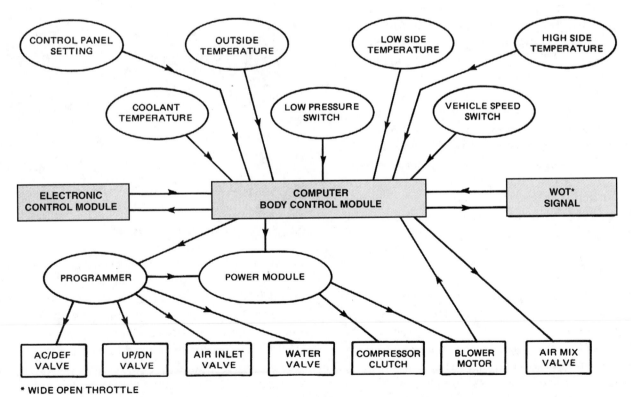

* WIDE OPEN THROTTLE

FIGURE 15–24 Typical flow chart for electronic temperature control system with nine inputs.

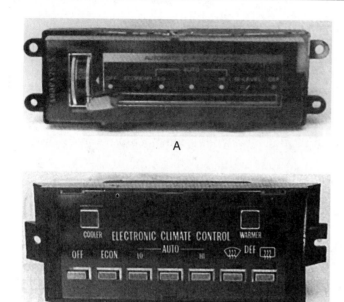

FIGURE 15–25 Control panels: (A) manual, and (B) pushbutton.

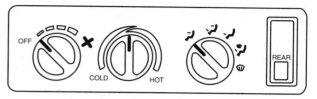

Heater control

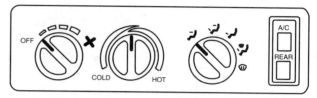

Manual A/C control

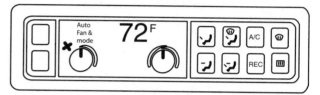

ATC control module

FIGURE 15–26 Typical master control heads.

metric units, Figure 15–26. A rear air-conditioning system control panel is shown in Figure 15–27. All serve the same purpose, that is, to provide operator input control for the heating, ventilating, and air-conditioning (HVAC) system.

Provisions are made on the control panel for the operator to select any temperature, generally between 65°F (18.3°C) and 85°F (29.4°C) in 1-degree increments. Some have an override feature that provides for a setting of either 60°F (15.6°C) or 90°F (32.2°C). Either one of these two settings will override all temperature control circuits to provide maximum cooling or heating conditions.

Usually, a microprocessor is located in the control head to input data to the **programmer,** based on operator-selected conditions. When the ignition switch is turned off, a memory circuit will remember the previous setting. These conditions are restored each time the ignition switch is turned on. If the battery is disconnected, however, the memory circuit clears and must be reprogrammed.

Programmer

The programmer, illustrated in Figure 15–28, receives electrical input signals from sensors and the main control panel. Based on all the inputs, the programmer provides output signals to turn the compressor clutch on and off, open and close the heater water valve, determine blower speed, and position all MIX/BLEND and FRESH/RECIR mode doors.

FIGURE 15–27 A typical rear control panel.

Blower and Clutch Control

The blower clutch control, shown in Figure 15–29, converts low-current signals from the control panel to high-current feed to the blower motor. Blower speeds with this control are infinitely variable. The speed is controlled through a resistor strip on the temperature door actuator. The resistor strip, then, functions the same as a rheostat does to input data to the control panel. The control panel, in turn, inputs the blower speed signal to the blower control.

FIGURE 15–28 A programmer. *(Courtesy of BET, Inc.)*

FIGURE 15–29 A typical blower clutch control. *(Courtesy of BET, Inc.)*

A power transistor circuit is included in the blower control, whose function is to engage the compressor clutch circuit. The metal strip on which the transistor is mounted serves as a heat sink. This assembly is located in the blower air stream to aid in heat dissipation.

Power Module

The **power module,** illustrated in Figure 15–30, controls the operation of the blower motor. The power module amplifies the blower drive signal from the programmer; its output signal is proportional to its input signals. This provides variable blower speeds as determined by in-vehicle conditions. If the in-vehicle temperature is considerably higher than the selected temperature, as in air conditioning, the blower will start at high speed and decrease to low speed as the in-vehicle temperature lowers.

FIGURE 15–30 A typical power module. *(Courtesy of BET, Inc.)*

Conversely, if the in-vehicle temperature is considerably lower than the selected temperature, as in heating, the blower will start at high speed and decrease to low speed as the in-vehicle temperature rises.

Clutch Diode

The clutch coil is an electromagnet that has a strong magnetic field when current is applied. This magnetic field is constant as long as power is applied to the coil. When power is removed, the magnetic field collapses and creates high-voltage *spikes.* These are harmful to the delicate electronic circuits of the computer and must be prevented.

A diode placed across the clutch coil, illustrated in Figure 15–31, provides a path to ground, holding the spikes to a safe level. This diode is usually taped inside the clutch coil connector, across the 12-volt lead and ground lead. A diode is checked with an analog ohmmeter according to the following steps:

1. Carefully cut the tape to expose the diode leads.
2. Unplug the connector from the compressor clutch coil.
3. Disconnect the ground wire. Isolate this wire so it does not touch ground.
4. Touch the ohmmeter leads to the diode leads. Observe the ohmmeter, Figure 15–32.
5. Reverse the ohmmeter leads and repeat step 4.

NOTE: In steps 4 and 5, the ohmmeter should indicate very high resistance in one step and little or no resistance in the other step. If the ohmmeter indicates very high resistance or little or no resistance in *both* steps, then the diode is probably defective and should be replaced.

6. Replace the diode, if indicated to be defective.
7. Retape the diode and reconnect the ground wire.
8. Plug the connector into the clutch coil.

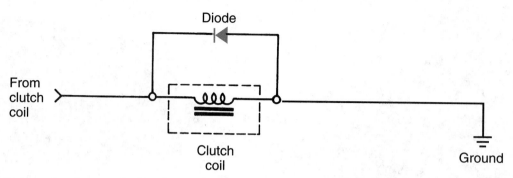

FIGURE 15–31 A diode is placed across the clutch coil to reduce spikes as the clutch is cycled on and off.

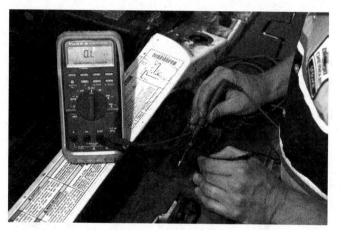

FIGURE 15–32 Connect an ohmmeter to the diode.

High-Side Temperature Switch

In Figure 15–33, the high-side temperature switch is located in the air-conditioning system liquid line between the condenser outlet and the orifice tube inlet. Although it is a temperature-sensing device, it provides

air-conditioner system pressure data to the processor. System temperature is determined by system pressure based on the temperature-pressure relationship of the refrigerant.

Low-Side Temperature Switch

The low-side temperature switch, Figure 15–33, is located in the air-conditioning system line between the orifice tube outlet and the evaporator inlet. Its purpose is to sense low-side refrigerant pressure and to provide this information to the microprocessor.

Evaporator Thermistor

The evaporator thermistor, Figure 15–34, is used on some systems to control evaporator temperature. It is a variable resistor that electrically connects to the compressor clutch microprocessor circuit to turn the compressor off when the evaporator temperature drops to 34°F (1.1°C). This prevents the formation of frost and ice on the fins of the evaporator.

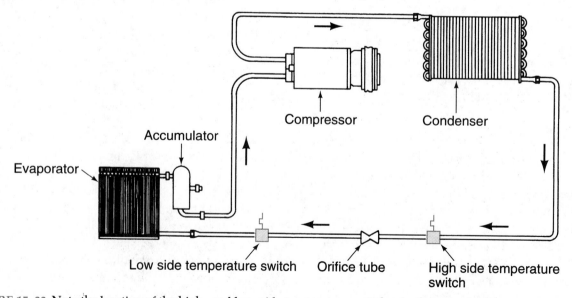

FIGURE 15–33 Note the location of the high– and low–side temperature switches in the system.

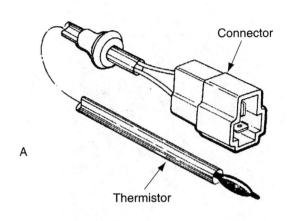

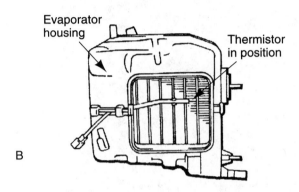

FIGURE 15–34 A thermistor (A); in position (B). *(From Green/Green and Dwiggins, Australian Automotive Air Conditioning, © Nelson, a division of Thomson Learning)*

High-Pressure Switch

The high-pressure switch is normally closed (NC) and opens if air-conditioning system pressure exceeds 425–435 psig (2,930–2,953 kPa). It closes when the system pressure drops below 200 psig (1,379 kPa). This swich provides for system safety if, for any reason, pressures exceed safe limits. Unlike the low-pressure switch, the high-pressure switch does not provide data to the microprocessor. This switch is usually in series with the compressor clutch circuit.

Low-Pressure Switch

The low-pressure switch is located in the low side of the air-conditioning system, usually on the accumulator, as shown in Figure 15–35. This normally closed (NC) switch opens when system low-side pressure drops below 2–8 psig (13.8–55.2 kPa). An open low-pressure switch signals the microprocessor to disengage the compressor clutch circuit to prevent compressor operation during low-pressure conditions. Low-pressure conditions may result from loss of refrigerant or a clogged metering device.

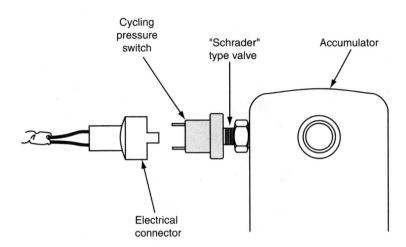

FIGURE 15–35 Typical low-pressure switch found on the accumulator.

Pressure Cycling Switch

A pressure cycling switch is found on some systems. It is used as a means of temperature control by opening and closing the electrical circuit to the compressor clutch coil. On cycling clutch systems, this switch usually opens at a low pressure of 25–26 psig (172.4–179.3 kPa) and closes at a high pressure of 46–48 psig (317.2–331 kPa). On some systems, this switch may be in line with the compressor clutch coil. On other systems, it may send data to the microprocessor to turn the compressor on and off.

Sunload Sensor

The sunload sensor, illustrated in Figure 15–36, is usually found atop the dashboard, adjacent to one of the radio speaker grilles. It is a photovoltaic diode that sends an appropriate signal to the microprocessor to aid in regulating the in-vehicle temperature.

Outside Temperature Sensor

The outside temperature sensor (OTS), Figure 15–37, also called an ambient temperature sensor (ATS), is usually

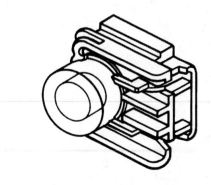

FIGURE 15–36 A typical sunload sensor.

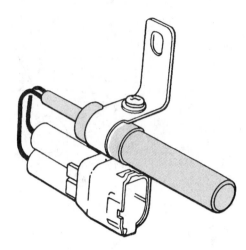

FIGURE 15–37 The outside temperature sensor is located behind the grille. *(From Green/Green and Dwiggins, Australian Automotive Air Conditioning, © Nelson, a division of Thomson Learning)*

FIGURE 15–38 A typical in–car temperature sensor. *(Courtesy of BET, Inc.)*

located just behind the radiator grille and in front of the condenser. It senses outside ambient temperature conditions to provide data to the microprocessor.

This sensor circuit has several programmed memory features to prevent false ambient temperature data input during periods of low-speed driving or when stopped close behind another vehicle, such as when waiting for a traffic signal.

Testing the Outside Temperature Sensor

The ohmmeter battery current through the sensor as well as heat from one's hand affect the ohmmeter reading. Do not handle or leave the ohmmeter connected for more that 5 seconds.

To test the ambient temperature sensor, carefully remove it from its socket first. Using a digital ohmmeter, quickly measure its resistance. At an ambient temperature of 70°F–80°F (21°C–27°C), the resistance should be 225Ω–235Ω. If it is not within this range, replace the sensor.

In-car Temperature Sensor

The in-car temperature sensor, shown in Figure 15–38 is also called an in-vehicle sensor. It is located in a tubular device called an *aspirator*. A small amount of inside air is drawn through the aspirator across the in-car sensor to provide average inside temperature data to the microprocessor.

Testing the In-Car Temperature Sensor

The ohmmeter battery current through the sensor affects the ohmmeter reading. Do not leave the ohmmeter connected for more that 5 seconds.

1. Disconnect the sensor.
2. Start the vehicle engine and set the air-conditioning system controls to COOL and blower motor speed to MED.
3. Place a thermometer in the air outlet nearest the sensor.
4. Depress, then pull out the NM-A/C button to turn off the compressor and close the coolant valve.
5. Operate the blower while quickly noting the sensor resistance.

   NOTE: Resistance should be 1,000Ω–1,500Ω at an ambient temperature of 70°F–80°F (21°C–27°C). If it is not within this value, replace the sensor.

Aspirator

The aspirator is a small duct system, illustrated in Figure 15–39, designed to cause a small amount of inside air to pass through it. The main air stream causes low pressure (suction) at the inlet end of the aspirator. This causes inside air to be drawn into the in-car sensor plenum. The in-car sensor,

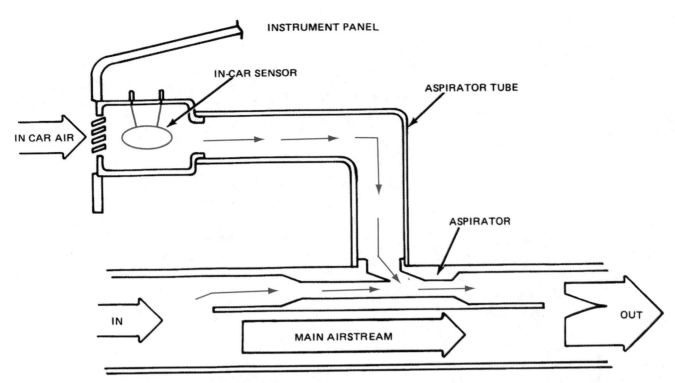

INSTRUMENT PANEL

IN-CAR SENSOR

ASPIRATOR TUBE

IN CAR AIR

ASPIRATOR

IN

OUT

MAIN AIRSTREAM

FIGURE 15–39 A typical aspirator. Note the location of the in–car temperature sensor.

which is located in the plenum, is continuously exposed to average inside air to monitor the inside air temperature.

Coolant Temperature Sensor

The coolant temperature sensor, not to be confused with a coolant temperature sending unit, is a thermistor that provides engine coolant temperature information to the microprocessor. This sensor also provides input information to other onboard computers to provide data for fuel enrichment, ignition timing, exhaust gas recirculate operation, canister purge control, idle speed control, and closed loop fuel control.

A defective coolant temperature sensor will cause poor engine performance, which will probably be evident before poor air-conditioning performance is noticed.

Vehicle Speed Sensor

The vehicle speed sensor is a pulse generator usually located at the transmission output shaft. Its purpose is to provide actual vehicle speed data to the microprocessor as well as other subsystems, such as the electronic control module (ECM).

Throttle Position Sensor

The throttle position sensor is actually a three-wire resistor-type electrical sensor, referred to as a potentiometer,

that has a voltage input from the processor or electrical control unit (ECU). The ECU determines the throttle position based on the value of the return voltage signal.

At the wide-open throttle (WOT) position, the electrical power to the clutch coil is interrupted and the compressor clutch is disengaged. Removing the compressor "load" from the engine provides maximum power during acceleration. This device, often called the WOT sensor or WOT switch, is most often found on vehicles that have diesel engines.

Heater Turn-on Switch

The heater turn-on switch is usually a bimetallic snap-action switch found in the coolant stream of the engine. If heat is selected, this switch prevents blower operation when engine coolant temperature is below 118°F-122°F (48.9°C-50°C). If cooling is selected, the programmer will override this switch to provide immediate blower operation, regardless of engine coolant temperature.

Brake Booster Vacuum Switch

The brake booster vacuum switch is found on some cars. It disengages the air-conditioning compressor whenever braking requires maximum effort. It is usually in series with the compressor clutch electrical circuit and does not provide data to the microprocessor.

Power Steering Cutoff Switch

The power steering **cutoff switch** found on some vehicles, is used to disengage the air-conditioning compressor whenever power steering requires maximum effort. On some vehicles, this switch is in series with the compressor control relay and does not provide data to the programmer. On other applications, this switch is in the electronic control module and provides feedback data to the microprocessor.

Mode Actuator

Two types of mode actuators are used to position the mode doors. Vacuum- and electrically operated actuators, often called "pots" or "motors," are covered in another chapter of this text.

Electric mode actuators, illustrated in Figure 15–40, have both drive and feedback circuitries. These provide the means for the mode actuators to be stopped at any specified position through 360 degrees of travel. Mode actuators are not reversible and travel in one direction only. The feedback circuit provides constant data to the control panel relative to the position in which it is stopped. If the feedback signal is not received by the control panel within 20 seconds, power to the actuator will be turned off. If this occurs, an LED will flash on the control panel on many systems to warn the operator that there is a problem with the system.

Water Valve Actuator

Two types of water valve actuators are available: vacuum and electric. The vacuum-operated actuator is covered in another chapter of this text. The electric water valve actuator, shown in Figure 15–41, operates in a manner similar to the mode actuator. It differs only in that it is reversible and travels only 90 degrees. It turns the water valve on and off. The water valve provides water to the heater core, which is located in the air-conditioning and heating duct system. Duct systems are covered in Chapter 16 of this text.

Ported Vacuum Switch

The ported vacuum switch was used in the 1980s by some General Motors and Chrysler car lines. It was used on some Jeep and Ford car lines through the 1991 model year.

The ported vacuum switch guards against engine overheating, when the engine is idling for long periods of time in hot weather. It has a heat-sensitive switch that shifts vacuum ports at a predetermined engine coolant temperature to advance engine timing. This, in turn, increases engine speed, increasing coolant flow and fan airflow through the radiator.

CONTROL DEVICES

Automatic and semiautomatic control devices are actuated directly or indirectly by lever, cable, or pressure (vacuum). Many controls are actuated electrically. Often a combination control is used, such as a lever (dash) and an electrical switch (on the device), with a cable linkage between the lever and the switch. In the modern vehicle, control devices are used to actuate door locks, deck lids, headlamp covers, windows, seats, antennas, and so on. This text, however, is concerned only with those devices used to control air-conditioning and heating system functions.

A   B

FIGURE 15–40 A typical electric mode actuator (A), with the cover removed (B). *(Courtesy of BET, Inc.)*

FIGURE 15–41 A typical electric water valve actuator. *(Courtesy of BET, Inc.)*

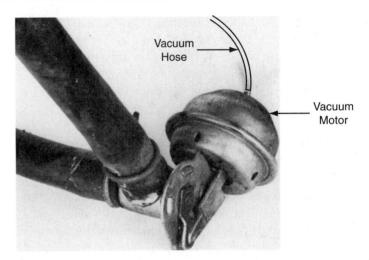

FIGURE 15–42 A vacuum–operated heater control valve. *(Courtesy of BET, Inc.)*

HEATER CONTROL

The heater control valve may be located on the engine, on the fender well, near the heater core, or inside the heater case. In some systems, the control is actuated by a cable. However, for the automatic temperature system, the control is governed by varying vacuum levels. A typical vacuum-operated hot water heater control valve is shown in Figure 15–42. A typical cable-controlled water valve is shown in Figure 15–43.

The cutaway view of a vacuum-operated water valve in Figure 15–44A shows that when no vacuum is applied, the control valve is closed. The combined water pressure and spring pressure help to keep the valve in the closed position. Most valves have some provision to prevent them from being installed backward. On some installations, however, the hoses can be accidentally reversed. Figure 15–44B shows the effect that reversing the hoses has on the water circulation.

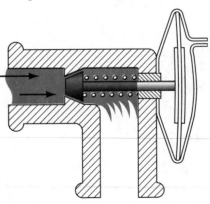

FIGURE 15–43 A cable–operated heater control valve. *(Courtesy of BET, Inc.)*

If it is assumed that no vacuum is applied to the control valve in Figure 15–44B, then the pressure of the water affects the pressure of the spring, which causes the valve to open. As a result, hot water is allowed to flow in the

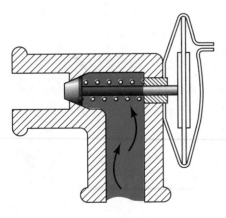

A Normal   B Reversed

FIGURE 15–44 Water valve with (A) normal coolant flow and (B) reversed coolant flow.

heater core. Actually, a pulsating effect is more likely to occur at speeds in excess of 50 mph (80.5 km/h), when water pressure is high. This condition has a marked effect on the operation of the temperature control.

In normal operation, the valve can be opened by varying degrees to control the water flow. Figure 15–45 shows the valve position for no vacuum (A), a partial vacuum (B), and a full vacuum (C). When referring to a vacuum schematic, a no vacuum condition is usually designated "nv," a partial vacuum "pv," and a full vacuum "fv."

Because one of the primary functions of an automatic temperature control system is to provide a vehicle relative humidity of 45–50 percent, the hot-water valve is often opened slightly (cracked) to allow a small amount of engine coolant to enter the system. The heated air from the hot coolant provides increased humidity. Thus, this hot air is mixed with the cooler air from the passenger compartment, and through the air-conditioner evaporator, to maintain the desired humidity level.

MODE DOORS

Vacuum-operated mode doors include the temperature deflector, diverter, defroster, outside-inside inlet, and heater and air-conditioner outlet doors.

There are a number of combinations of door positions for various conditions. For example, there are six different arrangements for the mode doors in the air-conditioning cycle alone. There is a different arrangement for the normal and defog settings for each of three operating conditions: full outside, full recirculate, and modulated air conditioning.

Mode doors are used to divert the movement of air from one passage to another, as shown in Figure 15–46. In Figure 15–46A, the air is deflected into the air-conditioner core. In Figure 15–46B, the air is diverted into the heater core.

Temperature Door

As indicated in the previous section, the temperature door regulates the air mixture. The position of the temperature mode door determines the temperature of the duct air in an automatic temperature control system. The temperature door is regulated by a vacuum motor, a temperature door actuator, or a servo.

VACUUM MOTORS

To bring about a change in conditions, the mode doors must be operated either manually or remotely. A cable is often used to control the door. However, for automatic temperature controls, a vacuum motor is used. This device is not a motor in the usual sense, but in the sense that it imparts motion. The vacuum motor is also known as a vacuum pot.

Figure 15–47 shows how a vacuum motor is used to operate the mode doors. In Figure 15–47A, the device is shown in the relaxed position. In Figure 15–47B, it is in the applied position. In the relaxed position, the spring keeps the arm extended. In the applied position, the vacuum overcomes the spring pressure and the arm is pulled to the IN position. The normal, or OFF, position of the vacuum motor is the relaxed position, as shown in Figure 15–47A.

Most vacuum motors are of the type shown. Some units have two vacuum hose attachments. These attachments serve as tee fittings for another vacuum motor, or they may be **double-action** fittings, such as those used for temperature control on automatic temperature units. This type of control is known as a *double-action vacuum motor*. Fittings on both sides indicate a double-action motor but

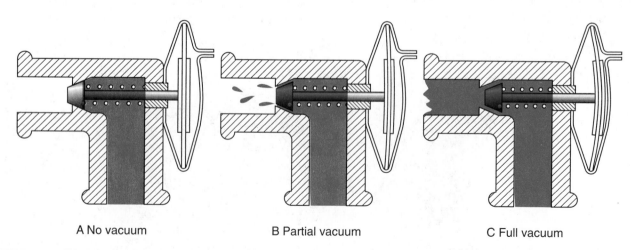

A No vacuum   B Partial vacuum   C Full vacuum

FIGURE 15–45 Water valve flow control.

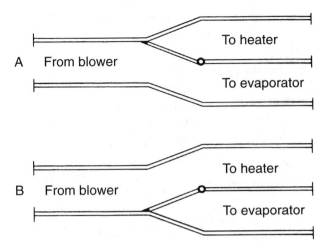

FIGURE 15–46 Mode door operation.

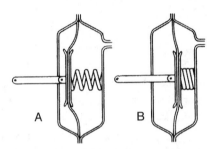

FIGURE 15–47 Movement of a single–chamber vacuum motor: (A) no vacuum applied; (B) full vacuum applied.

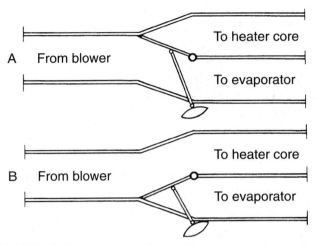

FIGURE 15–48 Air diverter door with vacuum motor.

Double-Action Vacuum Motors

The double-action vacuum motor, illustrated in Figure 15–49, has a double-action diaphragm that allows vacuum to be applied to either side of the motor. The vacuum causes the control arm attached to the rubber diaphragm to extend.

Some automatic temperature control units use a double-action vacuum motor known as a *temperature door actuator* (TDA).

BLOWER CONTROL

The blower control permits the driver to select a high or low volume of airflow. Although the airflow setting results in less air noise, it is less likely to maintain comfortable conditions within the passenger area of the vehicle. The high airflow setting is preferred for maximum benefit. In either range, however, the airflow varies automatically

not if both vacuum hose ports are on the same side of the vacuum motor.

Figure 15–48 shows the same type of duct arrangements as that given in Figure 15–46, but with the addition of vacuum motors.

Air-conditioner and heater duct systems with various mode door positions are covered in detail in Chapter 16.

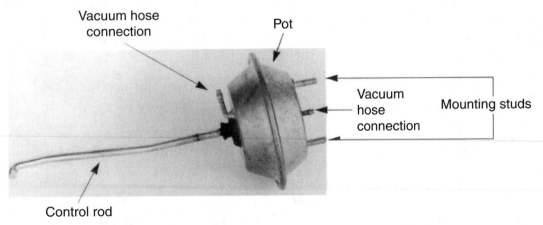

FIGURE 15–49 Double–action vacuum motor.

with the demand placed on the system by varying weather conditions. The high blower speed is obtained only when the servo is in the maximum air-conditioning position. This part of the operation of the system is electrically, not vacuum, controlled. The vacuum-controlled servo has electrical contacts (which complete various circuits) as does the vacuum switch section of the servo.

TIME-DELAY RELAY

The time-delay relay control is designed to prevent the heat cycle from coming on in the automatic system until after the engine coolant has reached a temperature of approximately 110°F (43.34°C). This control consists of two resistors, capacitors, and transistors. Figure 15–50 shows the time-day circuit of the wiring diagram.

ELECTROVACUUM RELAY

The **electrovacuum relay** (EVR) contains a normally closed (NC) vacuum solenoid valve and a normally closed (NC) electrical relay. The purpose of the EVR is to prevent blower operation when the system is in the "heat" mode until engine coolant temperature reaches about 115°F–120°F (46.1°C–48.8°C). When the coolant temperature is below this value, electrical contacts in the *engine temperature (sending) switch* (ETS) are closed. This grounds and completes the EVR circuit to open the ETR relay contacts. The ETR relay contacts, in series with the blower motor, interrupt current to the motor to open this circuit.

AMBIENT SWITCH

The **ambient switch,** shown in Figure 15–51, is an electrical switch actuated by changing ambient temperature. The ambient switch should not be confused with the ambient sensor. It is used in many custom and automatic systems.

The ambient switch is located outside the engine area, where it can sense the ambient temperature only. The actual switch location depends on its design. The switch is never mounted in a location where it is possible to sense engine heat.

If the master switch is pressed, the ambient switch will turn the air-conditioning compressor on at 35°F (1.67°C). The switch turns the compressor off if the ambient temperature falls to 25°F (−3.89°C). Whenever the ambient temperature is in the range of 64°F (17.78°C) and 55°F (12.78°C), the ambient switch bypasses the master control and time-delay relay, allowing the blower motor to run regardless of the engine coolant temperature.

When the air-conditioning compressor or blower is operated at low ambient temperatures, the humidity of the incoming air is reduced by condensing the moisture from it. In this way, window fogging is prevented when a vehicle is being operated during rainy, damp, or cool weather conditions.

THERMOSTATIC VACUUM VALVE

The thermostatic vacuum valve (TVV), depicted in Figure 15–52, is a vacuum-control valve that is sensitive to temperature. It is used to sense coolant temperature, such as on the side of the heater core. The TVV consists of a power element cylinder with a piston, vacuum parts, and spring. The power element is filled with a temperature-sensitive compound so that when the engine is cold and the coolant is not warm, the inlet part of the TVV is blocked and the outlet part is vented. When the coolant temperature reaches a specified range, usually 100°F–125°F (37.78°C–51.67°C), the compound in the cylinder expands and moves the piston until the vacuum flow starts.

In the automatic temperature control system, the vacuum flow proceeds from the selector vacuum disc switch to the program vacuum disc switch, master switch, vacuum diaphragm, and the outside-recirculate air-cooled diaphragm. On cold days, the TVV serves only as a **time delay.**

SUPERHEAT SWITCH

The superheat switch, located in the rear head of six-cylinder Harrison compressors, was used by General Motors through 1981 and on the Audi 5000 through 1983.

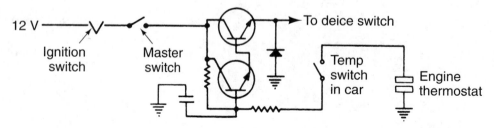

FIGURE 15–50 Time–delay relay schematic.

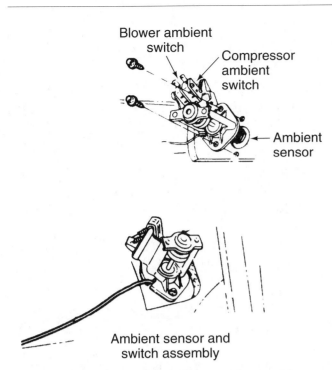

FIGURE 15–51 Ambient sensor and switch assembly.

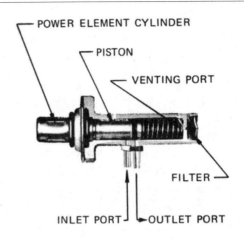

FIGURE 15–52 Thermostatic vacuum valve (TVV).

It is also used on later-model trucks equipped with Harrison six-cylinder compressors.

The superheat switch was replaced on many car lines in 1981 with a switch that is pressure actuated only. This switch was used on some General Motors car lines through 1987. The two switches are not interchangeable.

Using the wrong switch can reduce voltage to the clutch, causing erratic operation which can lead to a loss of compressor oil and result in a seized compressor.

SUMMARY

Many of the components of an automatic temperature control system are covered in this chapter. Others are covered throughout this text. Because of the complexity of the automatic system and its number of variations, it is essential that the manufacturers' specifications, manuals, and schematics be consulted for any specific year or model vehicle that is to be serviced.

REVIEW

Select the correct answer from the choices given.

1. The trouble code 888 on a certain Ford vehicle is being discussed. Although both technicians agree that this is a "no trouble" code on this particular vehicle, Technician A says that the customer's complaint is unfounded. Technician B says there may be a problem outside the microprocessor function. Who is right?
   a. A only
   b. B only
   c. Both A and B
   d. Neither A nor B

2. If the heater coolant control valve is installed backward, the *most likely* problem is that the control valve:
   a. will not operate.
   b. cannot be closed.
   c. cannot be opened.
   d. will pulsate.

3. Technician A says that 16 psia is a vacuum pressure. Technician B says that 179 kPa (absolute) is a vacuum pressure. Who is right?
   a. A only
   b. B only
   c. Both A and B
   d. Neither A nor B

4. Technician A says that in-vehicle humidity is controlled by tempering cooled air with heated air in the plenum/duct system. Technician B says that humidity is not important in a vehicle so long as the proper temperature is maintained. Who is right?
   a. A only
   b. B only
   c. Both A and B
   d. Neither A nor B

5. The compressor clutch coil diode:
   a. prevents unwanted spikes in the electrical system.
   b. ensures that 12 volts are supplied to the clutch coil.
   c. prevents reverse polarity in the clutch coil circuit.
   d. eliminates the magnetic field around the clutch coil.

6. Technician A says that an air-conditioning system control device may be actuated by a lever and cable. Technician B says an air-conditioning system control device may be actuated by a vacuum. Who is right?
   a. A only
   b. B only
   c. Both A and B
   d. Neither A nor B

7. Refer to Figure 15–2 and give the kiloPascal conversion nearest to 19 inches of mercury (19 in. Hg).
   a. 34.47 kPa absolute
   b. 37.92 kPa absolute
   c. Both A and B
   d. Neither A nor B

8. A vacuum motor that has a vacuum port on both sides is known as a _____ vacuum motor.
   a. dual-range
   b. double-action
   c. universal
   d. heat/cool door

9. The high-pressure switch is:
   a. normally open (NO); it closes on a pressure increase.
   b. normally open (NO); it closes on a decrease in pressure.
   c. normally closed (NC); it opens on a pressure increase.
   d. normally closed (NC); it opens on a decrease in pressure.

10. Technician A says a single-chamber vacuum motor is not affected by atmospheric pressure. Technician B says that a dual-chamber vacuum motor is not affected by atmospheric pressure. Who is right?
    a. A only
    b. B only
    c. Both A and B
    d. Neither A nor B

11. A time-delay relay:
    a. prevents the heating cycle from starting until the engine coolant reaches a predetermined temperature.

b. prevents the cooling cycle from starting until the engine coolant reaches a predetermined temperature.

c. delays the program advance on automatic temperature control (ATC) systems.

d. delays the "turn on" of the heating or cooling systems until the vehicle engine speed is adequate.

12. Vacuum will flow in either direction in the following, EXCEPT:
a. the check valve
b. the restrictor
c. the hose
d. the reserve tank

13. A mode door is used to:
a. provide heated air to the passenger compartment.
b. provide cool air to the passenger compartment.
c. provide both A and B.
d. provide neither A nor B.

14. Technician A says that any electronic control circuit may be measured using an analog volt-ohmmeter. Technician B says that all electronic control circuits may be measured using a digital volt-ohmmeter (DVOM). Who is right?
a. A only
b. B only
c. Both A and B
d. Neither A nor B

15. All of the following *may* be vacuum controlled, EXCEPT:
a. the heater coolant flow control valve.
b. the fresh/recirculate mode door actuator.

c. the compressor cycling clutch temperature control.
d. the bilevel or Hi/Lo mode door actuator.

16. A vacuum reserve tank leak may be repaired using:
a. two-part epoxy.
b. fiberglass.
c. Both A and B.
d. Neither A nor B.

17. A restrictor is found in the:
a. aspirator.
b. heater core.
c. in-car sensor.
d. vacuum system.

18. The evaporator core surface temperature should never be allowed to go below:
a. 36°F (2.2°C).
b. 34°F (1.1°C).
c. 32°F (0°C).
d. 30°F (–1.1°C).

19. All of the following are currently used for temperature control, EXCEPT:
a. the evaporator pressure regulator.
b. the control thermostat switch.
c. the variable displacement compressor.
d. the low-pressure control switch.

20. All sensors share the following traits, EXCEPT:
a. their resistances change with temperature changes.
b. they are a type of resistor.
c. they are sensitive to temperature changes.
d. they are interchangeable.

TERMS

Write a brief description of the following terms:

1. ambient switch
2. check relay
3. cutoff switch
4. delta T
5. diagrams
6. double action
7. electrovacuum relay
8. HVAC
9. mode door
10. power module

11. programmer
12. reserve tank
13. restrictor
14. semiautomatic
15. sensor
16. sunload
17. thermistor
18. throttle
19. time delay
20. vacuum motor

CHAPTER 16

CASE/DUCT SYSTEMS

OBJECTIVES

On completion and review of this chapter, you should be able to:

❏ Identify the types of case/duct systems.

❏ Discuss the air distribution through the case/duct system.

❏ Understand the airflow pattern through the case/duct system for defrost, heat, and cool modes.

❏ Troubleshoot, service, and adjust the operation of in-vehicle airflow through the case/duct system.

❏ Remove, adjust or repair, and replace case/duct system components, such as the evaporator or heater core, blower, linkage, and controls.

INTRODUCTION

This chapter provides a basic understanding of the motor vehicle heating, ventilating, and air-conditioning (HVAC) case/duct system for factory-installed climate control systems. The systems described in this chapter should be considered **typical,** but not representative, of any particular automotive case/duct system. An average motor vehicle heating and air-conditioning case/duct system is shown in Figure 16–1. At first glance, this system may seem to be a complicated maze of passages and doors. It is actually much simpler than it first appears.

The purpose of the case/duct system is twofold. First, it is used to house both the heater core and the air-conditioner evaporator core. Second, it directs conditioned air through selected components into the passenger compartment of the vehicle via selected **outlet** provisions. The supply air selected may be either fresh (outside) or **recirculated** (in-vehicle), depending on the system mode. After the air is heated or cooled, it is delivered to the floor outlet, dash (**panel**) outlets, and/or the **defrost** (windshield) outlets.

Two basic types of case assemblies are used to house the heater core and air-conditioner evaporator core: the independent case and the split case. The independent case, which is used on compact and small vehicles, may have an upstream blower, as shown in Figure 16–2, or a **downstream** blower, as shown in Figure 16–3. An upstream integral blower, illustrated in Figure 16–4, or an independent blower, shown in Figure 16–5, is used on split case systems. The split case system, which is used on larger vehicles, is located on both sides of the engine fire wall. The independent case system is usually located under the dash, on the inside of the fire wall.

For simplicity of understanding, a typical hybrid case/duct system is illustrated here. This system is divided into three sections, as shown in Figure 16–6. These sections are the *air intake* section, the *heater core and air-conditioner evaporator (plenum)* section, and the *air distribution* section.

Each of these sections is studied individually first and then as a complete system. Remember, however, that this is a discussion of a factory-installed or original equipment manufacturer (OEM) installation.

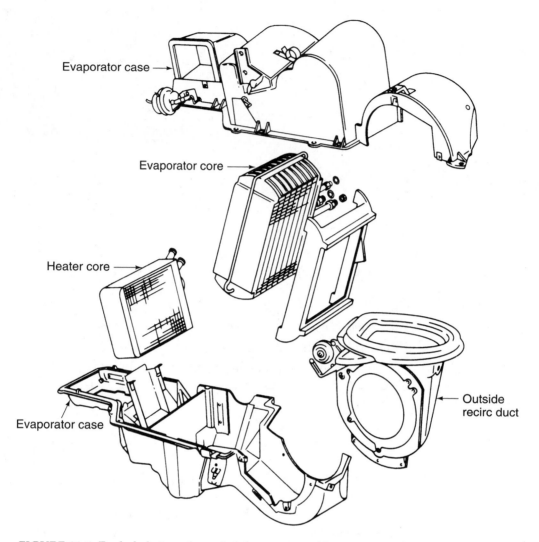

FIGURE 16–1 Exploded view of a typical duct system. (*Courtesy of Ford Motor Company*)

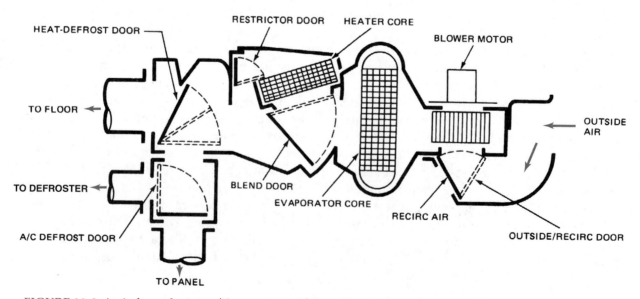

FIGURE 16–2 An independent case/duct system with an upstream blower.

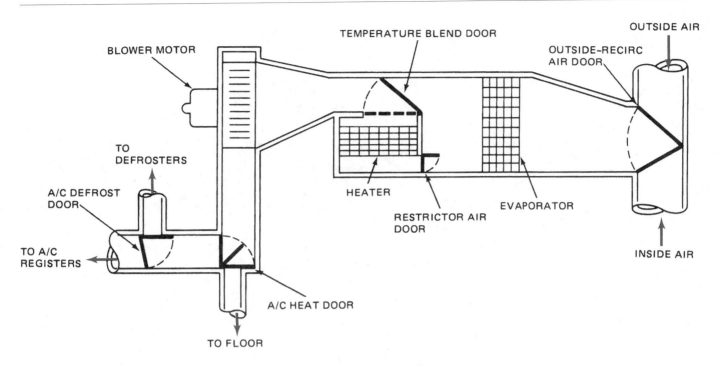

FIGURE 16–3 An independent case/duct system with a downstream blower.

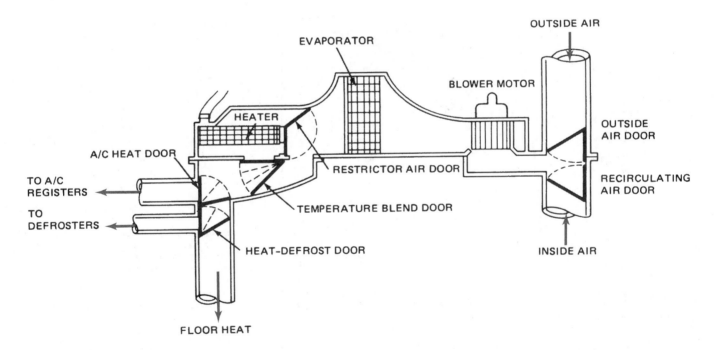

FIGURE 16–4 Split case system with upstream blower.

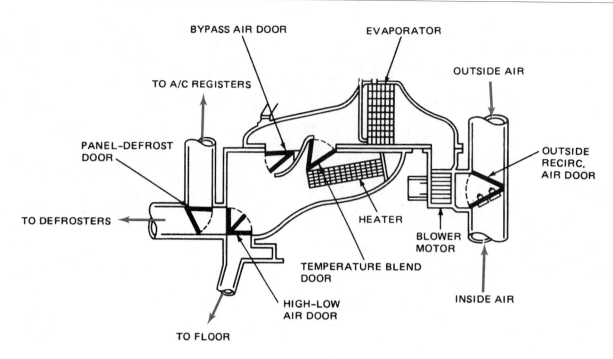

FIGURE 16–5 Case/duct system with independent (downstream) blower.

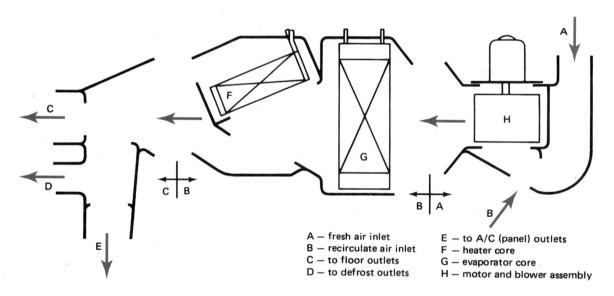

A — fresh air inlet
B — recirculate air inlet
C — to floor outlets
D — to defrost outlets

E — to A/C (panel) outlets
F — heater core
G — evaporator core
H — motor and blower assembly

FIGURE 16–6 Case/duct system split into three sections for illustration. Section C is air distribution, Section B is the heater core and evaporator plenum section, and Section A is the air intake.

FACTORY OR AFTERMARKET

There are **"custom"** aftermarket installations that have the appearance of factory-installed systems. This type of system has no provisions for outside air and is not connected to the heater duct system. If in doubt, use the following simple test to determine which type of system is being serviced. This test assumes that the system is in good working condition.

1. Note the in-vehicle ambient temperature.
2. Start the engine.
3. Turn the air-conditioning system on.
4. Move the temperature control to HOT.
5. Measure the temperature of the air coming out of the dash registers.

If the air temperature does not exceed ambient temperature, the system is a custom installation and has no connection with the heater circuit. If this condition is noted, the heater circuit is a separate and distinct unit. The explanation in this chapter only applies to the heater section of the case/duct system.

AIR INTAKE

The air intake or inlet section shown in Figure 16–7 consists of a fresh (outside) air inlet, a recirculate (inside) air inlet, a fresh-recirculate air door, a blower with motor, and an air outlet. The fresh air inlet provides the system with a fresh outside air supply; the recirculate air inlet provides a recirculated in-vehicle supply.

The position of the vacuum-motor-operated fresh-recirculate door depends on the system mode. For illustration purposes, 100 percent fresh air is provided in Figure 16–8A; and 100 percent recirculated air is provided in Figure 16–8B. Actually, in all modes except maximum cooling (**MAX** A/C), the air supply is from the outside. In

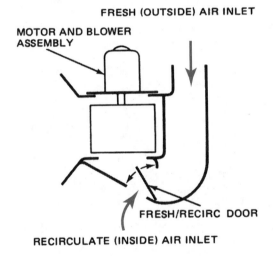

FIGURE 16–7 Air inlet (intake) section.

MAX A/C, the air supply is from the inside (recirculated). But even in this mode, some systems provide up to 20 percent fresh air, which results in a slightly **positive** in-vehicle pressure.

> CAUTION
> A SLIGHTLY POSITIVE PRESSURE INSIDE THE VEHICLE PREVENTS THE ENTRANCE OF DANGEROUS EXHAUST GASES THAT COULD PRODUCE A VERY HAZARDOUS IN-VEHICLE ATMOSPHERE WHEN ALL THE WINDOWS ARE TIGHTLY CLOSED.

CORE SECTION

The core section in Figure 16–9, more appropriately called the plenum section, is the center section of the duct system. It consists of the heater core, the air-conditioner evaporator, and a **blend door.** Air flows from right to left in the illustration. The blend door, usually Bowden **cable** operated, provides full-range control of airflow either through, or bypassing, the heater core. All air passes through the air-conditioner evaporator. It is in this section that full-range temperature conditions are provided for in-vehicle comfort. The following sections describe how this is accomplished.

Heating

The heater coolant valve is open to allow hot engine coolant to flow through the heater core. Cool outside air is heated as it passes through the heater core. The air conditioner is not operational; therefore, it has no effect on the air temperature as the air first passes through the evaporator. The desired temperature level is achieved by the position

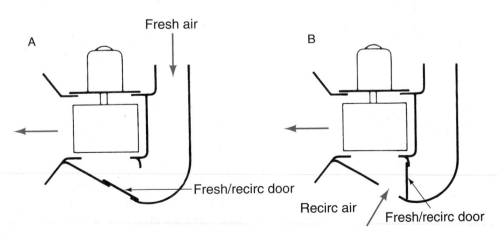

FIGURE 16–8 Providing the system with (A) 100 percent fresh air, and (B) 100 percent recirculated air. Note the position of the FRESH/RECIRC door.

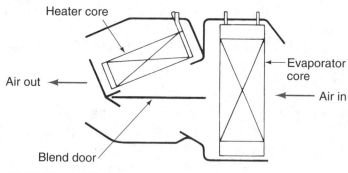

FIGURE 16–9 Core section.

of the blend door. This allows a **percentage** of the cool outside air to bypass the heater core to **temper** the heated air. The heated air and cool air are then blended in the plenum to provide the desired temperature and humidity level before passing on to the air distribution section.

Cooling

In maximum cooling (MAX A/C), recirculated air passes through the air-conditioner evaporator then is directed back into the vehicle. In other than MAX A/C, fresh outside air passes through the air-conditioner evaporator and is cooled before delivery into the vehicle. The desired temperature level is achieved by the position of the blend door. The blend door allows a percentage of cooled air to pass through the heater core to be reheated. The cooled air passing through the evaporator and the reheated air passing through the heater core are blended in the plenum to provide the desired temperature level. This tempered air is then directed to the air distribution section.

DISTRIBUTION SECTION

The air distribution section, illustrated in Figure 16–10, directs conditioned air to be discharged to the floor outlets, defrost outlets, or the dash panel outlets. Also, depending

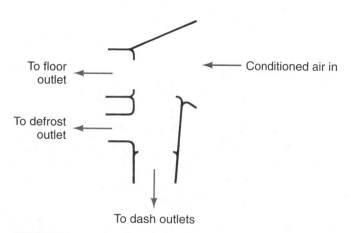

FIGURE 16–10 Air distribution section.

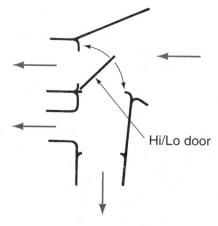

FIGURE 16–11 Air distribution section illustrating the HI/LO diverter door.

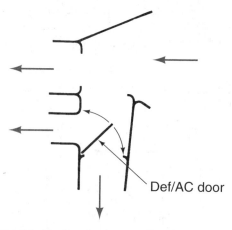

FIGURE 16–12 Air distribution section illustrating the DEF/AC diverter door.

on the position of the mode doors, conditioned air may be delivered to any combination of outlets. There are two mode (blend) doors in the air distribution section: the HI/LO door and the DEF/AC door. The HI/LO door, shown in Figure 16–11, provides 0–100 percent full-range conditioned air outlet control to the HI (dash) and LO (floor) outlets. The DEF/AC door, shown in Figure 16–12, provides conditioned air outlet control either to the defrost (windshield) outlets or to the dash panel outlets.

COMBINED CASE

The combined case/duct system provides full-range control of air circulation through the heater core and air-conditioner evaporator. Figure 16–13 illustrates 100 percent recirculated air flowing through the air-conditioner evaporator and out through the panel outlets. This may typically represent mode and blend door positions when maximum cooling (MAX A/C) is selected during high in-vehicle ambient temperature conditions.

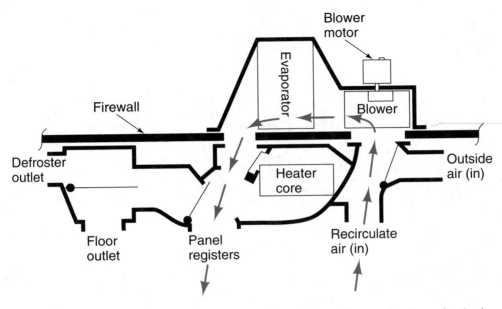

FIGURE 16–13 All recirculated air flowing through the evaporator and out the panel registers (outlets).

Figure 16–14 illustrates 100 percent fresh air circulation through the heater core and out through the floor outlets. This may typically represent the mode and blend door positions when heat is selected during low in-vehicle ambient temperature conditions. A variation, shown in Figure 16–15, shows some of the heated air **diverted** to the defrost outlets. This would be the typical application to clear the windshield of fog or light icing conditions.

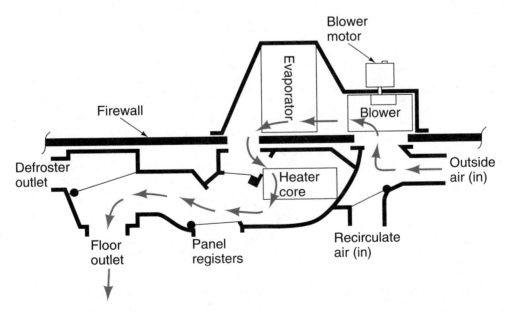

FIGURE 16–14 All fresh air flowing through the heater core and out the floor outlets. Although air flows through the evaporator, the compressor is not running and there is no cooling effect.

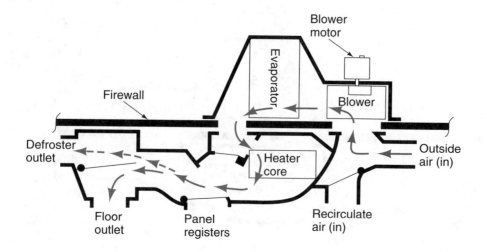

FIGURE 16–15 The same condition as illustrated in Figure 16–14 but with some air diverted to the defrost outlets.

AIR DELIVERY

In addition to the OFF position, there are six air selections for the passenger compartment of the vehicle. Some passengers may require recirculated air and other passengers may require fresh air. Although the select conditions may differ slightly from one vehicle model to another, typically, they are:

1. MAX
2. NORM
3. BI-LEVEL (MIX)
4. VENT
5. HEAT/COOL
6. DEFROST

The following are some of the typical duct door routing of conditioned air for the various selections available at the driver **control panel.**

Max

In MAX (maximum) cooling, shown in Figure 16–16, the compressor is running and the outside/recirculate air door is closed to ambient air. Flow is from in-vehicle air, through the evaporator, and out through the panel registers. **Bi-level,** as shown in Figure 16–17, may be selected, which will provide some air to the floor outlet.

In MAX heating, shown in Figure 16–18, the compressor is not running and the heater coolant valve is open. Airflow is from in-vehicle, through the evaporator and

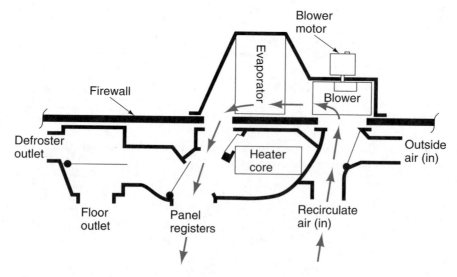

FIGURE 16–16 In MAX cooling, air flows from in–vehicle, through the evaporator, and out the panel outlets.

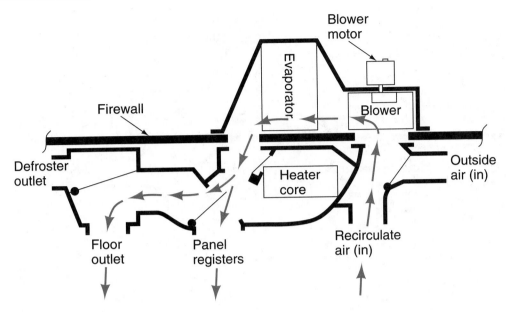

FIGURE 16–17 MAX cooling with BI-LEVEL selected.

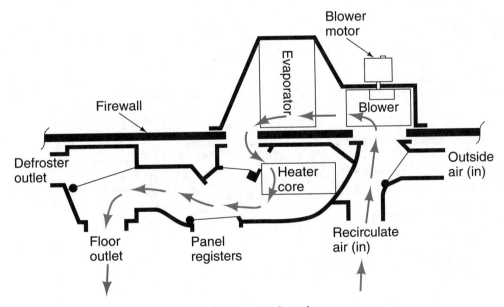

FIGURE 16–18 Airflow when MAX heating is selected.

heater cores, and out the floor outlet. If bi-level is selected, as shown in Figure 16–19, some air is directed to the panel registers. In either one of these conditions, a small amount of air is directed to the windshield to prevent **fogging.**

Norm

If normal air conditioning is selected, then the compressor is running. Air flows in from the outside (ambient),

through the evaporator, and out the panel registers, as illustrated in Figure 16–20. For humidity control, some air may be directed through the heater core, shown in Figure 16–21, as well. Bi-level air can also be selected, providing some conditioned air to the floor outlet.

If normal heating is selected, then the compressor is not running and the coolant control valve is open. Air flows in from the outside (ambient), through the heater

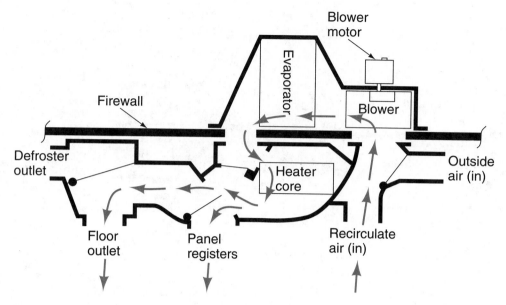

FIGURE 16–19 MAX heating with BI-LEVEL selected.

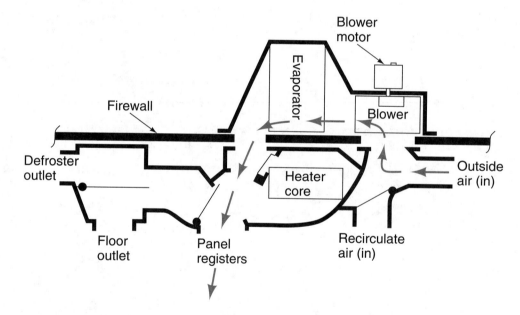

FIGURE 16–20 Normal cooling (air conditioning) is selected.

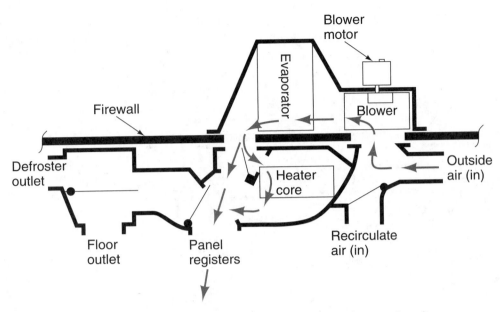

FIGURE 16–21 Airflow when humidity control is required with normal cooling.

core, and out the floor outlet, as shown in Figure 16–22. Bi-level may also be selected to provide some air through the panel registers, as shown in Figure 16–23. In either one of these conditions, a small amount of air is directed to the defrost outlet to prevent windshield fogging.

Bi-level

The BI-LEVEL condition, which is described for the MAX and NORM operations, is simply a means to provide conditioned air at two outlets: the panel and the floor. In

some systems this is referred to as HI-LO. It is similar in operation to MIX (discussed later).

Vent

The VENT setting brings in unconditioned, ambient air when neither heating nor cooling is desired. The compressor is not running and the heater coolant valve is not open. Air passage is from ambient air through the heater or evaporator core to the selected floor outlets and/or panel registers. Figure 16–24 shows the VENT selected with BI-LEVEL air delivery.

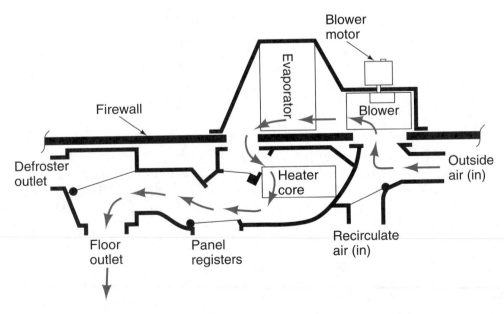

FIGURE 16–22 Airflow when normal heating is selected.

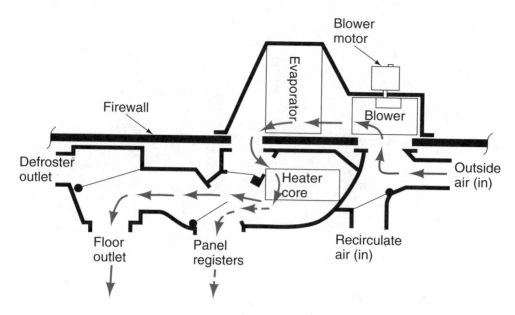

FIGURE 16–23 Airflow in the heating mode when BI-LEVEL is selected.

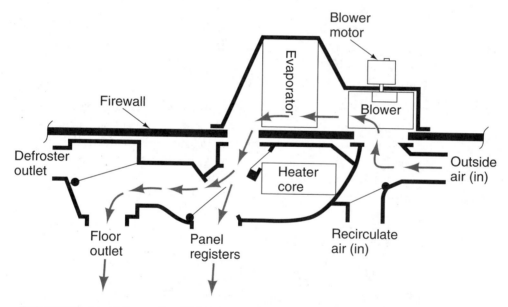

FIGURE 16–24 Airflow when VENT is selected in the bi-level condition.

Heat/Cool

A temperature control is generally provided to select in-vehicle temperature. There are two methods: manual/semiautomatic and automatic, which are illustrated in Figure 16–25. Temperature and mode are manually selected in the manual/semiautomatic temperature control (**SATC**) system. In the automatic temperature control (ATC) system, the selected temperature and mode are fully automatic functions of a digital microprocessor. The microprocessor compares data from designated sensors to maintain the desired in-vehicle temperature.

Defrost

In the DEFROST position, shown in Figure 16–26, outside ambient air passes through the heater core and is directed to the defroster outlets. A slight amount of heated air is directed to the floor outlets. If the outside ambient air temperature is below 50°F (10°C), then the compressor may operate to temper the heated air for humidity control.

Mix

MIX may be selected on some models. Generally, those with a MIX selection do not have a HI-LO select provi-

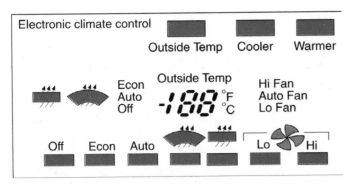

FIGURE 16–25 Climate control panel. *(Courtesy of General Motors Corporation, Service Operations)*

sion. When in the MIX position, the heater/defroster door opens halfway and the panel air door is closed. The defrost air door is opened, as shown in Figure 16–27. In this selection, conditioned air is delivered to the floor and defrost outlets. In-vehicle temperature is controlled by adjusting the temperature control lever. The compressor will operate if warrented by in-vehicle temperature conditions. The compressor will also operate if outside ambient air temperature is below 50°F (10°C), as an aid for in-vehicle humidity control.

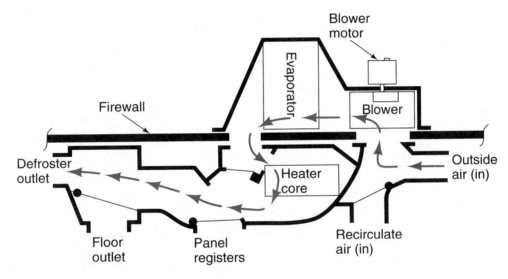

FIGURE 16–26 Airflow when DEFROST is selected.

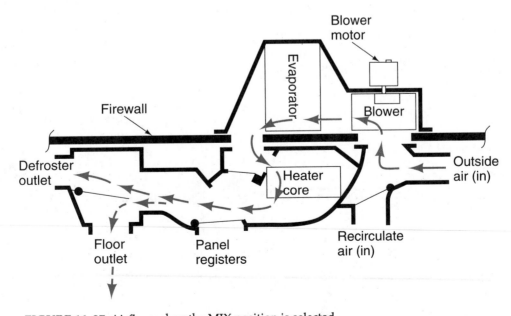

FIGURE 16–27 Airflow when the MIX position is selected.

DUAL-ZONE DUCT SYSTEM

The dual-zone duct system found in some cars has separate driver- and passenger-side duct systems, Figure 16–28. Both sides have a defrost/air-conditioning outlet door and a heater/floor outlet door that operate together. The passenger has control of the passenger-side temperature door only.

The dual-zone duct air distribution system can be controlled by either a manual or automatic climate control system and has a separate temperature control for the passenger. The passenger may adjust the temperature of the air at the outlet vents on the passenger side in an

automatic climate control system only within the limits set by the driver—generally up to 30°F (16.7°C) cooler or warmer than that selected by the driver. Passenger control on a manual climate control system is not usually restricted by the driver's temperature selection.

The passenger manually controls the position of the passenger-side temperature air door, controlling the discharge temperature of the passenger-side air outlets between max hot, Figure 16–29 and max cold, Figure 16–30. The actual passenger-side temperature depends on the general operation of the system. The passenger controls do not engage or disengage the compressor, change blower speed, or reposition the passenger mode door.

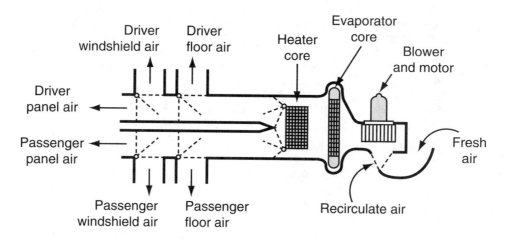

FIGURE 16–28 Typical dual–duct system.

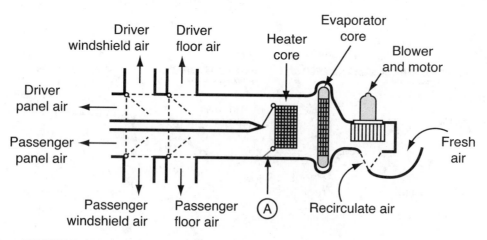

FIGURE 16–29 Typical dual–duct system with passenger full hot selected.

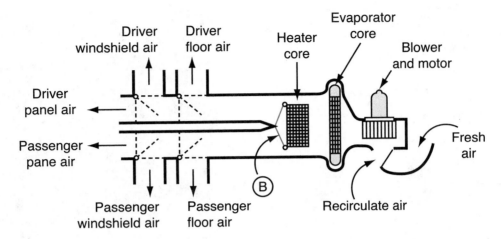

FIGURE 16–30 Typical dual–duct system with passenger full cold selected.

REAR HEAT/COOL SYSTEM

Some trucks and vans may be equipped with a rear air distribution system to provide rear heating or cooling, or a combination of both. The rear air distribution system is often referred to as an **auxiliary** air-conditioning system, Figure 16–31.

Depending on design, the rear air distribution system may have the following dual components: blower and motor, temperature door, evaporator core with metering device, heater core with flow control, outlet mode door, control panel(s), and controller.

The rear auxiliary system that provides only heating or cooling does not require an outlet or temperature mode door. The heat-only control panel has a blower speed control accessible to rear-seat passengers. The rear blower **master control** for the cooling-only system is gen-

erally in the front control panel. The switch in the REAR position permits the rear blower switch to select the speed of the rear blower.

Systems that provide both heating and cooling have an outlet mode door to direct outlet air to the upper or lower vents, Figure 16–32 and Figure 16–33. Some systems may have a temperature door controlling outlet air temperature, whereas others are controlled by the front master control.

The heat/cool system generally has both front and rear control panels for controlling the rear air distribution system. The control panels allow selection of the blower speed, the mode door, and, in some systems, the temperature door position.

Because the rear system is connected in parallel to the front system, the rear controls cannot override the

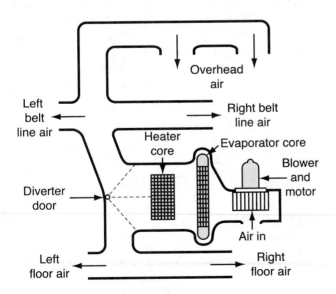

FIGURE 16–31 Typical rear (auxiliary) Heat–A/C system.

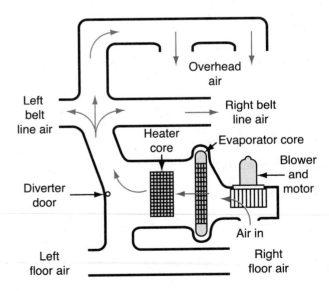

FIGURE 16–32 Air diverted to upper vents.

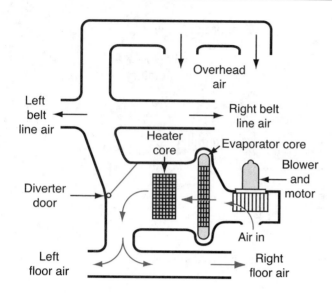

FIGURE 16–33 Air diverted to lower (floor) vents.

master controls, such as to energize or de-energize the compressor or heater control valve. The rear system can only provide cooling or heating when cooling or heating is selected in the front system.

EVAPORATOR DRAIN

Moisture extracted from the air in the evaporator is readily expelled from the evaporator case through a drain tube that extends through the floorboard of the vehicle. To prevent insects from entering it, the drain tube is a molded, accordian shape. The weight of moisture, as it collects, overcomes the rigidness of the tube closure and allows the water to pass.

Over time, however, airborne debris and dust may restrict the tube to the point that it causes water to back up inside the case. This is sometimes evidenced by water droplets exiting the dash outlets with the air or dripping onto the floor mats in the passenger compartment. There have even been reports of drivers getting a cold, wet right foot when making a hard right turn.

If this becomes a problem, it is necessary to clean out the drain tube to ensure that it will open and allow water to pass. This is best done from under the vehicle, Figure 16–34. For obvious reasons, do not stand directly under the tube while cleaning it.

VISUAL INSPECTION

When addressing a customer complaint for poor or insufficient heating or cooling, the first step is to make a visual inspection. The following should be included in the inspection:

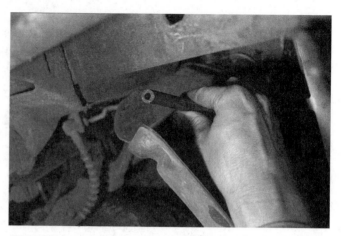

FIGURE 16–34 Evaporator drain hose.

- ❏ Is the case and ductwork sound? Check for cracks, broken, or disconnected ducts.
- ❏ Are the vacuum hoses sound? Check for disconnected, split, damaged, or kinked vacuum hoses.
- ❏ Is airflow restricted? Check to ensure that mode doors are opening. Check for leaves or other debris that may block airflow, such as at the fresh air inlet screen located at the base of the windshield.
- ❏ Are the cables secure? Check for loose, broken, or binding mode door control cables.

MODE DOOR ADJUSTMENT

A cable or vacuum actuator is used to position one or more of the mode doors in the duct system. The cable-operated system, Figure 16–35, consists of a steel cable encased in a plastic, nylon, or steel housing. It is used to connect the mode door to the control panel. Adjustments are made on the mode door end of the cable by repositioning the cable housing in its mounting bracket. The cable is usually held in place with a clip or a retainer secured in place with a hex-head screw.

The only adjustment possible for a vacuum actuator, Figure 16–36, is in the linkage, if there are adjustment pro-

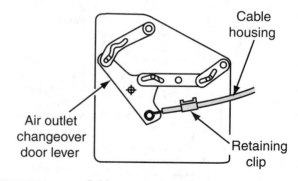

FIGURE 16–35 Cable-operated mode doors.

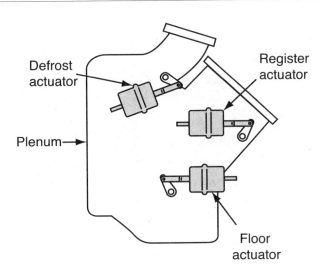

FIGURE 16–36 Vacuum-operated mode doors.

visions. If the problem proves to be a defective vacuum motor, however, it must be replaced. First, ensure that there is a vacuum signal at the vacuum motor indicating that the vacuum system is sound. More information on trou-

bleshooting and servicing of the vacuum and electrically operated actuators can be found in this text.

TEMPERATURE DOOR CABLE ADJUSTMENT

Remove access panels or components to gain access to the temperature door, Figure 16–37, and proceed as follows:

1. Loosen the cable-attaching fastener at the heater case assembly.
2. Make sure the cable is properly installed and routed to prevent binding and to ensure freedom of movement.
3. Place the temperature control lever in the full-cold position and hold it in place.
4. Tighten the cable fastener that was loosened in step 1.
5. Move the temperature control lever from full-cold to full-hot to full-cold positions.
6. Repeat step 5 several times and check for freedom of movement.
7. Recheck the position of the door. If it is loose or out of position, repeat steps 2 through 7. If it is still in position and secure, replace the access panels and covers.

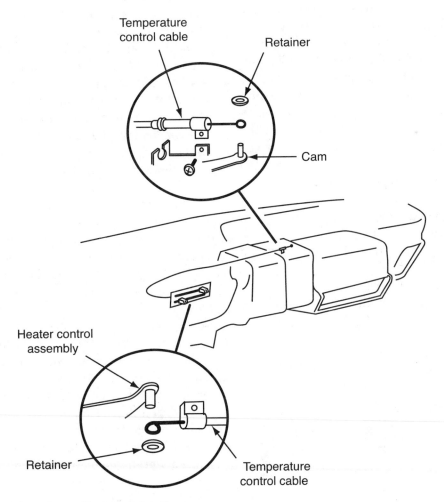

FIGURE 16–37 Temperature door adjustment details.

AIR FILTER

The introduction of air filters in the automotive air-conditioning systems of domestic vehicles has been slow. The first was found in the 1938 Nash; the next was not until over thirty-five years later, in the 1974 Oldsmobile Toronado and 88. It would be another twenty years later before electrostatically charged air filters appeared in some Ford Contour and Mercury Mystique car lines. Since 1988, however, many European vehicles have included some type of air-conditioning system air filter, Figure 16–38.

Referred to by many as a "cabin air filter," the electrostatically charged filters are designed to remove particles as small as 0.02 micron—less than one-millionth of an inch in diameter. Electrostatic filters eliminate such unwanted particles as pollen, mold spores, road dust, bacteria, and tobacco smoke. The filter ensures that the air inside the passenger compartment is clean during either the recirculate or fresh air modes. The filter, usually located in the case/duct system, should be serviced annually or every 15,000 miles (24,000 kilometers). A clogged filter can create an air pressure drop, placing a greater demand on the blower motor and perhaps leading to an early failure. Because it restricts airflow, a clogged filter will also affect the air-conditioning and heating systems, and defroster performance.

The procedure for cleaning or replacing the filter varies from vehicle to vehicle, so it is important to follow the manufacturer's recommended procedures. If the vehicle is equipped with an air filter, instructions may be found in the owner's manual or on a label inside the glove box. The following is a typical procedure:

1. Remove the dash undercover.
2. Remove the glove box.
3. Remove the instrument reinforcement from the instrument panel.
4. Remove the filter retaining clip.
5. Remove the air filter from the case/duct, Figure 16–39.

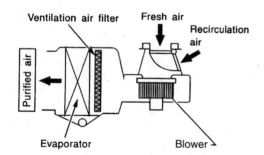

FIGURE 16–38 Some systems have an air filter.

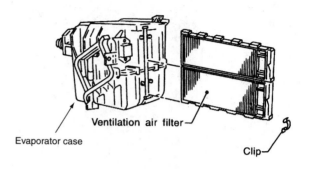

FIGURE 16–39 Remove the filter from the case.

6. Clean and/or install a new filter.
7. Replace the components in reverse order of removal.

SUMMARY

There are many variations of mode and blend door positions as well as many case/duct system designs. Doors may be operated electrically, by vacuum, or by cable. Some doors are either fully opened or fully closed; others are infinitely variable. Because of the many different applications and methods of control, it is necessary to consult the particular manufacturer's manual for specifications and testing procedures. The information given in this chapter must be considered typical, and not as representative of any particular case/duct system.

Block or ladder diagrams, covering several pages, are often used in manufacturers' service manuals to troubleshoot problems with motor vehicle air-conditioning (MVAC) case/duct systems. It is therefore recommended that the appropriate service manuals be consulted for specific troubleshooting procedures.

Always follow a manufacturer's recommended procedures and heed its cautions when troubleshooting any control system. The unintentional grounding of some circuits can cause immediate and permanent damage to delicate electronic components. The use of a test light, powered or nonpowered, is not recommended for underdash service. The battery in a powered test light or the added resistance on a nonpowered test lamp may be sufficient to cause failure to the delicate balance of solid-state electronic circuits.

Be especially cautious when servicing circuits that have a warning symbol, such as that shown in Figure 16–40. These circuits are susceptible to damage by electrostatic discharge (ED) by merely touching them.

NOTICE

CONTENTS SENSITIVE
TO
STATIC ELECTRICITY

FIGURE 16–40 Typical warning symbol.

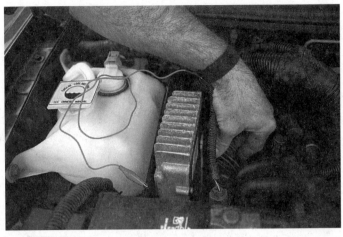

FIGURE 16–41 A typical grounding bracelet.

Electrostatic discharge is a result of static electricity, which "charges" a person simply by sliding across a seat, for example. To provide an extra margin of safety, the technician should wear a grounding bracelet (Figure 16–41), an electrical conducting device that surrounds the wrist and attaches to a known ground source. This device ensures that the body does not store damaging static electricity by providing a path to ground for it to be discharged.

REVIEW

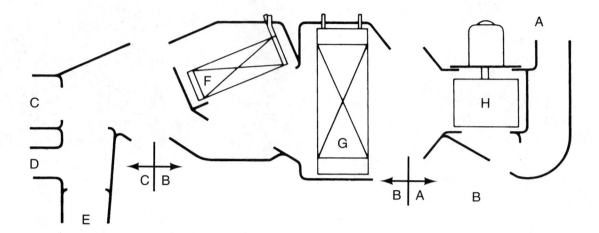

FIGURE 16–42 Typical case/duct system layout.

Select the correct answer from the choices given.
Refer to Figure 16–42 to answer questions 1 through 4.

1. What component is shown as "H"?
   a. Heat exchanger
   b. Blower motor
   c. Evaporator core
   d. Heater core

2. The airflow through the center section is:
   a. left to right only.
   b. right to left only.
   c. left to right, then left
   d. right to left, then right.

3. The fresh/recirculate air door is found in _____ section.
   a. the left
   b. the center
   c. the right
   d. either a, b, or c

4. The illustration depicts:
   a. a single in-dash air-conditioning system.
   b. a dual split front/rear air-conditioning system.
   c. a single-rear air-conditioning system.
   d. a dual-zone in-dash air-conditioning system.

5. Fresh air may be supplied in all modes, EXCEPT:
   a. MAX A/C
   b. BI-LEVEL (MIX)
   c. MAX HEAT
   d. DEFROST

6. Blend (mode) doors are found in _____ of the case/duct system.
   a. the first section
   b. the second section
   c. the third section
   d. all three sections

7. The airflow pattern of an air-conditioning case/duct system is being discussed. Technician A says HI-LO is similar to BI-LEVEL. Technician B says MIX is similar to HI-LO. Who is right?
   a. A only
   b. B only
   c. Both A and B
   d. Neither A nor B

8. BI-LEVEL will provide conditioned air at the:
   a. panel and floor registers.
   b. panel and defrost registers.
   d. defrost and floor registers.
   c. panel, floor, and defrost registers.

9. Heating is selected but no warm air flows from the panel registers. Technician A says the heat/cool mode door may be defective. Technician B says the air-conditioning system may be an aftermarket unit. Who is right?
   a. A only
   b. B only
   c. Both A and B
   d. Neither A nor B

10. Technician A says an independent case/duct system may have an upstream blower assembly. Technician B says an independent case/duct system may have a downstream blower assembly. Who is right?
    a. A only
    b. B only
    c. Both A and B
    d. Neither A nor B

11. Technician A says the vacuum reserve tank provides a steady vacuum source for the air-conditioning system controls while the engine is running. Technician B says the vacuum reserve tank is equipped with a check valve that preserves the vacuum source when the engine is not running. Who is right?
    a. A only
    b. B only
    c. Both A and B
    d. Neither A nor B

12. Technician A says that a defective diaphragm in a vacuum motor will not affect its operation. Technician B says that not all vacuum motors have a diaphragm. Who is right?
    a. A only
    b. B only
    c. Both A and B
    d. Neither A nor B

13. Technician A says that a dual-action vacuum motor has two diaphragms. Technician B says a dual-action vacuum motor has two vacuum ports. Who is right?
    a. A only
    b. B only
    c. Both A and B
    d. Neither A nor B

14. Air-conditioning system service is being discussed. Technician A says that it is always necessary to disconnect the battery before service. Technician B says that it is always necessary to disconnect the air bag restraint system before service. Who is right?

    a. A only
    b. B only
    c. Both A and B
    d. Neither A nor B

15. Technician A says that if refrigerant has been recovered, it may be reused. Technician B says that if coolant has been recovered, it must be properly disposed of. Who is right?
    a. A only
    b. B only
    c. Both A and B
    d. Neither A nor B

16. Safety is being discussed. Technician A says one can be burned by touching the blower motor speed resistor. Technician B says this danger can be avoided by first disconnecting the battery cable. Who is right?
    a. A only
    b. B only
    c. Both A and B
    d. Neither A nor B

17. A mode door may be operated by all of the following, EXCEPT:
    a. the cable.
    b. the solenoid.
    c. the electric motor.
    d. the vacuum motor.

18. Technician A says that the refrigerant must be recovered before replacing a heater core. Technician B says the coolant must be recovered before replacing an evaporator core. Who is right?
    a. A only
    b. B only
    c. Both A and B
    d. Neither A nor B

19. A musty odor from the case/duct system may be caused by all of the following, EXCEPT:
    a. mold.
    b. mildew.
    c. coolant leak.
    d. refrigerant leak.

20. The *least likely* cause of an inoperative vacuum motor is:
    a. a split or otherwise damaged vacuum hose.
    b. a defective check valve.
    c. a defective vacuum switch.
    d. a kinked or otherwise restricted vacuum hose.

TERMS

Write a brief description of the following terms:

1. auxiliary
2. bi-level
3. blend door
4. cable
5. control panel
6. custom
7. defrost
8. diverted
9. downstream
10. fogging

11. master control
12. MAX
13. outlet
14. panel
15. percentage
16. positive
17. recirculated
18. SATC
19. temper
20. typical

CHAPTER 17

ENGINE COOLING AND COMFORT HEATING

OBJECTIVES

On completion and review of this chapter, you should be able to:

❏ Recognize and identify the major components of the engine cooling system.
❏ Compare the different types of radiators.
❏ Discuss the function and purpose of the coolant (water) pump.
❏ Explain the need for a pressurized cooling system.
❏ Understand and describe the purpose of a coolant thermostat in the cooling system.
❏ Understand and practice the procedures used for testing the various components of the cooling system.
❏ Recognize the safety hazards associated with cooling system service.
❏ Understand and practice cooling system troubleshooting techniques and procedures.

INTRODUCTION

Excessive automotive engine heat, a product of combustion, is transferred to, and dissipated in the radiator. This is accomplished by two heat transfer principles known as *conduction* and *convection, which are* covered in Chapter 4. The cooling system, when operating properly, maintains an operational design temperature for the engine and automatic transmission.

The cooling system, as illustrated in Figure 17–1, functions by circulating a liquid coolant through the engine and the radiator. Engine heat is picked up by the coolant by conduction and is given up to the less-hot outside air in the radiator by convection. Unlike air-conditioning systems, however, the coolant in the cooling system does not change in state during this process. It remains a liquid throughout the cooling system.

The semiclosed cooling system, shown in Figure 17–2, consists of four main parts: *radiator, water pump, pressure (radiator) cap,* and *thermostat.* Other essential parts include the water-pump pulley and belt(s), fan, hoses, hose *clamps,* and *engine water passages.* Some late-model vehicles include a *thermostatic vacuum switch* or *thermostatic vacuum valve* (TVV).

Actually, the cooling system is responsible for removing only about 35 percent of the total engine heat. With an exhaust valve temperature as high as 4,500°F (2,482°C), the remaining unwanted heat is dissipated by the engine walls, heads, and pistons. A large percentage of heat also passes with exhaust gases through the exhaust system to the atmosphere.

TEST THE COOLING SYSTEM

1. Remove the radiator cap; make sure the engine is not running.
2. Ensure that the engine coolant is at the proper level in the radiator.
3. Wipe out the inside of the **filler neck** and examine the lower inside sealing seat for nicks, debris,

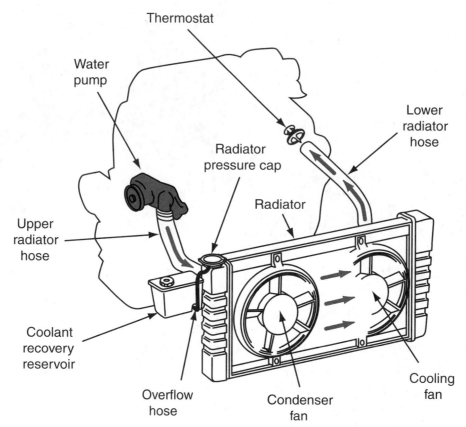

FIGURE 17–1 Major components of a typical liquid-cooling system. Arrows indicate the coolant flow.

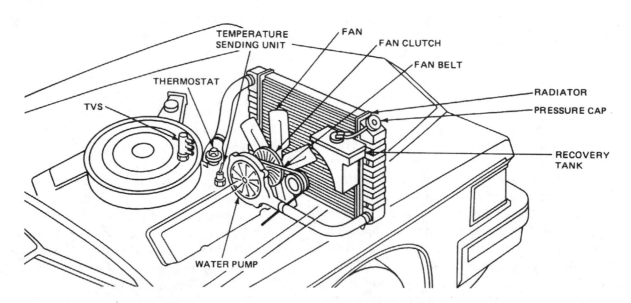

FIGURE 17–2 Components of a semiclosed cooling system. The heater core, control valve, and hoses are not shown.

and/or solder bumps.

4. Inspect the overflow tube to ensure that it is not restricted.

5. Inspect the cams on the outside of the filler neck. Carefully re-form cams that are bent. Take care not to break the solder joint between the radiator tank and the filler neck.

6. Attach the pressure tester to the radiator filler neck, as shown in Figure 17–3. Carefully press down on it, rotating it clockwise (cw) until the lock ears are fully engaged to the stop lugs of the filler neck. Do not force.

FIGURE 17–3 Attach a pressure tester to the radiator neck.

NOTE: A spacer washer may have to be used on some 1-inch (25.4 mm) deep necks, such as those found on many import vehicles.

7. For the following procedures, refer to current published specifications to determine the proper cooling system operating pressures for the vehicle being tested. These specifications may also be found in the **owner's manual** for the vehicle.
8. Operate the pump until the needle on the gauge reaches the arrow just beyond the end of the color band for the specific system pressure range, as illustrated in Figure 17–4.
9. Observe the gauge needle; proceed as required:
    a. If the needle is steady, follow steps 10 and 11. If the needle holds steady for 2 minutes, then there are no serious leaks.
10. Using a flashlight, examine all gaskets and hose connections for seepage.
11. If any leaks are found, then correct as necessary.
    b. If the needle drops slowly, follow steps 12 through 14. If the needle drops slowly, then a slight leak or seepage is indicated.
12. Using a flashlight, examine the radiator, heater core, gaskets, and hose connections.
13. Correct any leak(s) found, as necessary.
14. Recheck the system. Repeat this procedure, starting with step 12.
    c. If the needle drops rapidly, follow steps 15 through 17. This is an indication of a serious leak.
15. Examine the radiator, heater core, gaskets, and hose connections.
16. Correct any leak(s) found, as necessary.

NOTE: If no leaks are found externally, then the cooling system may be leaking into the

FIGURE 17–4 Operate the pump until the needle reaches the arrow just beyond the end of the color band. *(Courtesy of BET, Inc.)*

transmission cooler or into the engine. Follow the procedures outlined in the appropriate manufacturer's service manuals for further leak testing.

17. Recheck the system. Repeat this procedure starting with step 15.
18. Release the tester from the radiator filler neck with a counterclockwise rotation, as illustrated in Figure 17–5, to release system pressure.

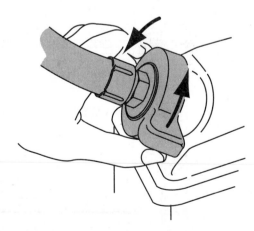

FIGURE 17–5 Use a counterclockwise action to release the tester. *(Courtesy of Stant Manufacturing Inc.)*

RADIATORS

Radiators may be constructed of copper or aluminum. The receiver and collector tanks of some late-model radiators are constructed of a special high-temperature plastic. These tanks are held in place with clips onto the headers. A special gasket prevents leakage. There are two basic types of radiators: *downflow* and *crossflow*. In the downflow type, illustrated in Figure 17–6, coolant flows from the top tank to the bottom tank. In the crossflow type, shown in Figure 17–7, coolant flows from one side to the other.

The first tank is referred to as the *collector* because it collects coolant from the engine. The second tank is referred to as the *receiver* because it receives coolant after passing through the many tubes of the radiator *core*.

The collector usually contains a baffle plate to aid in even distribution of coolant through the core. The receiver, on automatic transmission equipped cars, contains a *transmission oil cooler*, shown in Figure 17–8. Some cars with a trailer-towing package also have an external transmission cooler. In either case, the transmission cooler has little to do with the engine cooling system, except for adding a minor heat load to the coolant. Occasionally, the internal cooler develops a leak, which would be noticed by the presence of transmission fluid (oil) in the coolant. There could also be coolant leakage to the transmission, depending on pressure differential.

Two types of cores are found in automotive radiators: *cellular* (often called *honeycomb*) and *tubular*. The cellular core in Figure 17–9 is fabricated by soldering together thin, **preformed** sheets of metal, usually brass or copper. The tubular core in Figure 17–10 is constructed of smal,l round, or narrow oblong tubes that are soldered to the *headers* of the collector and receiver.

Radiator cores are very fragile. The tube walls of the core are only a few thousandths of an inch (millimeter) thick. Care must be taken when handling radiators to avoid costly damage.

Radiators develop leaks and/or become clogged. Radiators in this condition must be repaired. Repairs should be attempted only by those with the proper tools, equipment, and knowledge to do so.

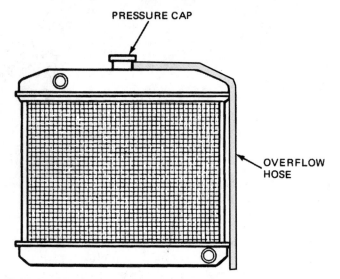

FIGURE 17–6 Details of a typical downflow radiator.

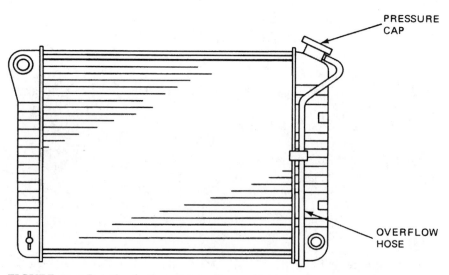

FIGURE 17–7 Details of a typical crossflow radiator.

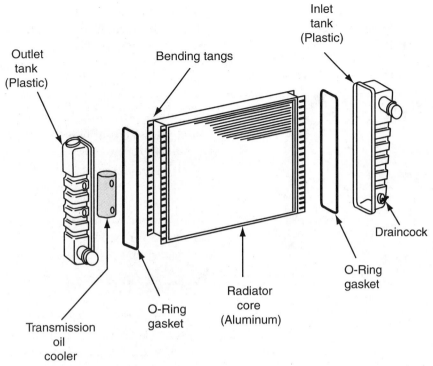

FIGURE 17–8 Details of a crossflow radiator showing the location of the transmission oil cooler.

FIGURE 17–9 Typical cellular core radiator details.

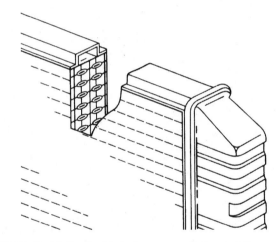

FIGURE 17–10 Typical tubular core radiator details.

WATER PUMP

A centrifugal-type water pump, like that shown in Figure 17–11, driven by a belt off the engine crankshaft, is used to pump coolant through the cooling system. At road speed, coolant may be circulated at a rate as great as 160–170 gallons (605-643 L) per minute, with the water pump turning as fast as 4,500 to 5,000 revolutions per minute (rpm).

The pump inlet is connected to a *neck* on the radiator receiver with a rubber hose and two clamps. This

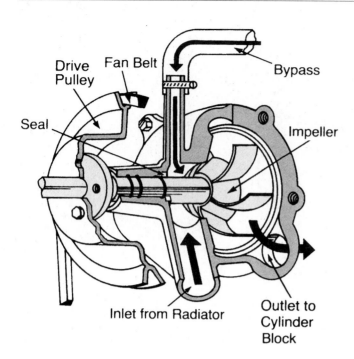

FIGURE 17-11 Parts of a water pump. (*Courtesy of The Gates Rubber Company*)

hose often has a wire insert to prevent it from collapsing, due to the suction of the pump impeller. The hose may also be preformed to fit a particular year or model engine. Engine information is generally given in cubic-inch displacement (CID) or liters (L).

The internal pump impeller is attached to the external pulley provisions by means of a steel shaft running through the pump housing. This shaft is also equipped with bearings and seals. The pump outlet, then, is connected directly from the back side of the impeller into the engine block coolant passages.

The most common causes of water pump failure are leaks and/or worn bearings. Although kits are generally available for making repairs, it is advisable to replace defective pumps with new or professionally rebuilt units.

PRESSURE CAP

Water (H_2O) boils at 212°F (100°C) at sea level atmospheric pressure (14.69 psi or 101.3 kPa). The boiling point of water (H_2O), or any other coolant, is raised by increasing its pressure, as illustrated by the graph in Figure 17-12. The boil-

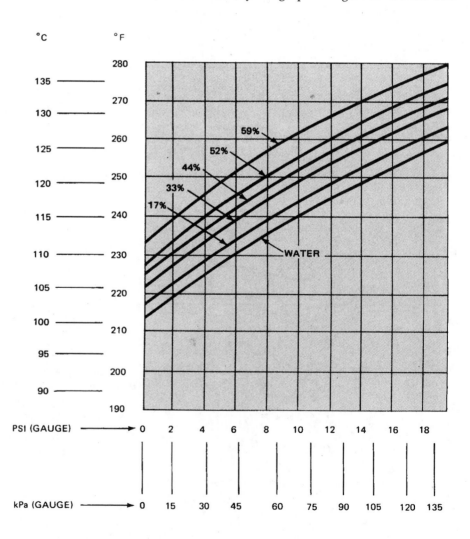

FIGURE 17-12 Boiling point of water and water/ethylene glycol solution at various pressures.

ing point is further increased if mixed with an ethylene glycol-type antifreeze, covered later in this chapter.

Radiator pressure caps, depicted in Figure 17–13, rated from 8 psi (55.1 kPa) to 17 psi (117.2 kPa), are used to increase the boiling point of cooling systems in order to maintain desired engine-operating temperatures. Radiator caps are equipped with a pressure valve and a vacuum valve, shown in Figure 17–14. At the predetermined pressure rating of the cap, the pressure valve opens, under certain conditions, to expel excess coolant vapor and pressure into the atmosphere. This action prevents above-maximum cooling system pressure buildup.

When the engine is stopped for any length of time, the cooling system will go from a positive to a negative pressure (vacuum). The vacuum valve will open as this occurs to admit enough ambient air to raise cooling system pressure to atmospheric pressure.

A defective pressure valve may cause one of the cooling system components, such as a hose or radiator tank, to rupture due to excessive pressure buildup. A defective vacuum valve will result in below-atmospheric pressure in the cooling system (when cool) and may be noted by a collapse of the upper radiator hose.

If the rubber gasket sealing surfaces of the pressure cap are distorted or damaged, then the system will not become pressurized. Also, the surfaces of the radiator that these gaskets seat against must be clean and free of dirt or other foreign matter. If there is not a good seal between the two surfaces, then overheating will occur and result in a loss of coolant.

As a rule of thumb, a pressure cap should be replaced with one of equal pressure rating when preventive **maintenance** is performed. It is important to note that a special aluminum cap is required on all aluminum radiators.

The effects of boiling points in relation to pressure are covered in Chapter 5. The effects of boiling points with antifreeze-type solution and operation of pressure caps in a recovery system are covered later in this chapter.

TESTING THE PRESSURE CAP

1. To avoid injury, make sure the engine is cool and the cooling system is not pressurized. Carefully remove the radiator pressure cap.
2. Inspect the mating surfaces. Inspect the rubber gaskets of the pressure cap.
3. Install the pressure cap adapter on the pressure tester, as shown in Figure 17–15.
4. Ensure that all seating surfaces of the adapter and pressure cap are clean. The lever should be in the closed position of safety-type pressure caps.
5. Wet the rubber gasket in water (H_2O) and install the pressure cap on the tester, as shown in Figure 17–16. Make sure the locking ears of the cap stop on the adapter cams.
6. Hold the tester with the gauge facing you and operate the pump until the needle reaches its highest point.
7. Stop pumping and observe the needle.
   a. If the needle is below the rated range of the pressure cap, replace the cap.

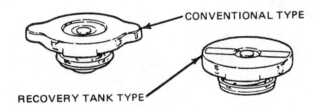

FIGURE 17–13 A typical radiator pressure cap.

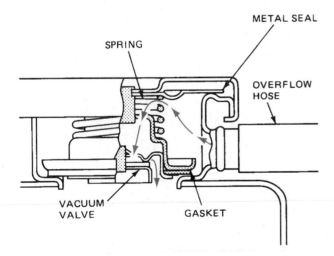

FIGURE 17–14 Pressure and vacuum valve details of a pressure cap.

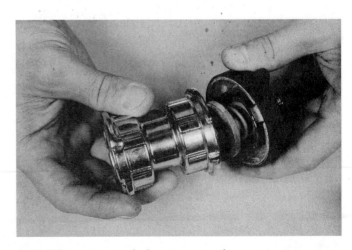

FIGURE 17–15 Attach the adapter to the pressure tester.

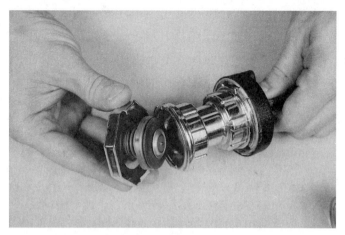

FIGURE 17–16   Install the pressure cap.

b. If the needle is above the rated range of the pressure cap, replace the cap.

c. If the needle falls rapidly, replace the pressure cap.

NOTE: The needle should remain in the proper range for at least 30 seconds. If in doubt, replace the cap.

8. Remove the cap and the adapter from the tester.

THERMOSTAT

The thermostat, illustrated in Figure 17–17, is located at the outlet of the engine coolant passage, usually under the return hose flange or *thermostat housing.* Engine coolant temperature, sensed by a sensing element, causes the normally closed (NC) thermostat to open at a predetermined temperature from 140°F (60°C) to 180°F (82.2°C).

Typically, a thermostat rated at 170°F (76.6°C) will start to open at its rated temperature and will be fully open at 195°F (90.5°C). A thermostat's opening and closing, illustrated in Figure 17–18, is gradual as the temperature of the coolant increases or decreases.

Two important facts regarding thermostats should be noted:

❑ Thermostats are a design component of the engine cooling system and should not be omitted.

❑ A thermostat will not affect maximum engine temperature, unless it is defective.

A thermostat ensures *minimum* engine-operating temperature. This is particularly important if the vehicle is often driven only for short distances. For example, if a thermostat rated at 180°F (82.2°C) is used, coolant will not circulate through the radiator until the coolant in the engine

FIGURE 17–17   A typical cooling system thermostat.

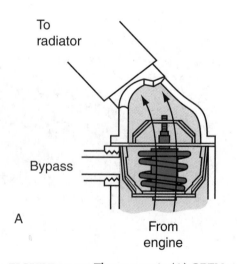

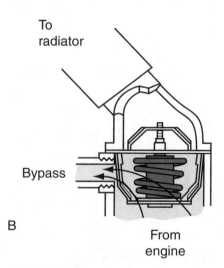

FIGURE 17–18   Thermostat in (A) OPEN position, and (B) CLOSED position.

has reached this design temperature. The purpose of a thermostat, then, is to protect against engine *overcooling*.

A thermostat, unless defective, will not cause overheating. A thermostat rated at 180°F (82.2°C) is wide open at 205°F (96.1°C), and full coolant flow is provided through the cooling system. A thermostat rated at 160°F (71.1°C), wide open at 185°F (85°C), would provide no more or no less coolant flow.

Thermostats fail, however. If failure occurs while the thermostat is in the *closed* position, then severe engine overheating will result. This may be noted by an extremely hot engine and a cool-to-warm radiator. If failure occurs while the thermostat is in the *open* position, a longer-than-normal warmup period may be noted by the "cold" dash lamp or temperature gauge.

Thermostat Testing

This procedure presumes that the thermostat has been removed from the vehicle cooling system.

1. Note the condition of the thermostat.
2. Is the thermostat corroded or open?
   a. If no, proceed with step 3.
   b. If yes, replace the thermostat.
3. Note the temperature range of the thermostat.
4. Suspend the thermostat in a heatproof glass container filled with water (H_2O).
5. Suspend a thermometer in the container. Neither the thermostat nor the thermometer should touch the container or touch each other.
6. Place the container and contents on a stove burner and turn on the burner, Figure 17–19.
7. Observe the thermometer. Does the thermostat start to open at about 200°F (110°C) below its rating?
   a. If yes, proceed with step 8.
   b. If no, replace the thermostat.
8. Observe the thermometer. Is the thermostat fully open at its rated temperature?
   a. If yes, the thermostat is good.
   b. If no, replace the thermostat.

PULLEY AND BELT

Two types of belt systems are used to drive the air-conditioning compressor and other accessories: the *V belt*, shown in Figure 17–20, and the *serpentine belt* (also referred to as the *V-groove belt*). Figure 17–21 illustrates the serpentine belt found on most late-model cars. This is a single-belt system, whereby one belt drives all of the accessories.

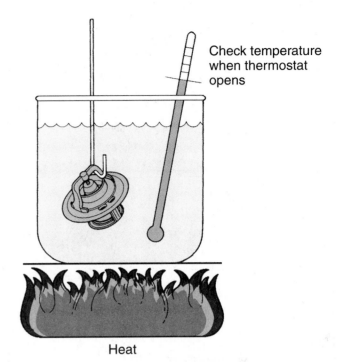

Check temperature when thermostat opens

Heat

FIGURE 17–19 Checking thermostat operation. (*Courtesy of Ford Motor Company*)

The pitch and width of belts in the V-belt system are important in that they must match those of the drive (engine crankshaft) and driven (engine accessories) pulleys, as illustrated in Figure 17–22. If the compressor and/or alternator are driven with two V-belts, then they should be replaced with a matched set. Replace the pair even when just one appears to be damaged. V-belts are tensioned in foot-pounds (N•m), according to manufacturers' specifications. It is generally necessary to recheck the V-belt after a "run-in" period of a few hundred miles (kilometers) to ensure proper belt tensioning.

The serpentine belt must be replaced with an exact duplicate. Its length, width, and V-groove characteristics are important to ensure proper fit. This system has a spring-loaded belt tensioner (idler pulley). Therefore, it is not necessary to manually tension this type of belt. If the belt will not remain tight, then the tensioner must be replaced.

Note that the water pump pulley shown in Figure 17–20 turns in the opposite direction from that shown in Figure 17–21. It is therefore possible that the same engine in two different cars, one with a V-belt drive and the other with a serpentine belt drive, will require two different water pumps and two different engine coolant fans. Always follow manufacturers' specifications when replacing these or any other components.

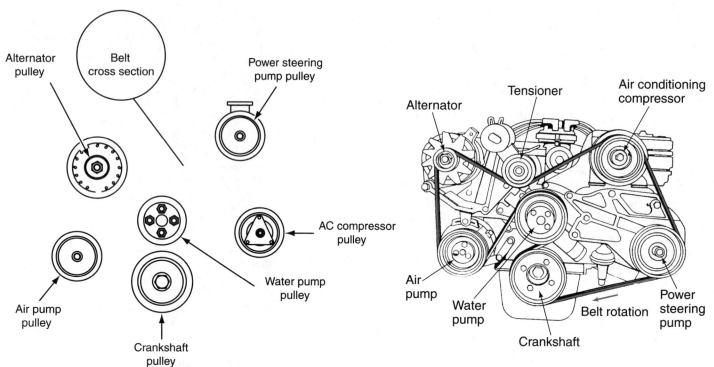

FIGURE 17–20 A typical V–belt drive system.

FIGURE 17–21 A serpentine V–ribbed belt drive. (*Courtesy of The Gates Rubber Company*)

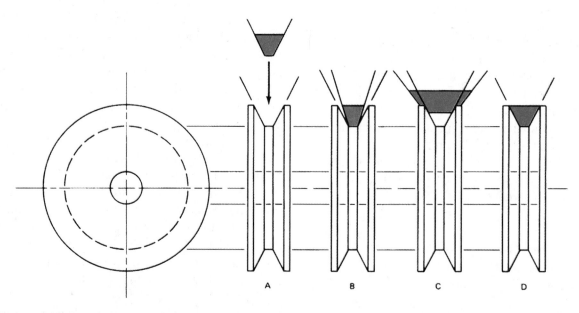

FIGURE 17–22 The belt should fit the pulley snugly as shown in A and D. The belt in pulley B is too narrow and has an improper pitch. The belt in pulley C (exaggerated) is too wide and has an improper pitch.

AUTOMATIC BELT TENSIONER

The drive belt(s) on most late-model vehicle engines is (are) equipped with a spring-loaded automatic tensioner. An automatic belt tensioner may be used with all belt configurations, such as with or without power steering and/or air conditioning.

Belt-driven engine accessories are often replaced due to noise or other problems only to learn that the automatic tensioner was at fault.

Replacing Belt Tensioner

The following procedure is typical for replacing an automatic belt tensioner. Always follow the particular manufacturer's recommended procedures for each vehicle.

1. Attach a socket wrench to the mounting bolt of the automatic tensioner pulley bolt.
2. Rotate the tensioner assembly clockwise (cw) until the belt tension has been relieved, Figure 17–23.
3. Remove the belt from the idler pulley first, then remove the belt from the other pulleys.
4. Disconnect and remove and/or set aside any components that are hindering tensioner removal.
5. Remove the tensioner assembly from the mounting bracket.

WARNING

BECAUSE OF HIGH SPRING PRESSURE, DO NOT DISASSEMBLE THE AUTOMATIC TENSIONER.

6. Remove the pulley bolt and remove the pulley from the tensioner.
7. Install the pulley and the pulley bolt to the tensioner. Tighten the bolt to 45 ft.-lb. (61 N•m).
8. Install the tensioner assembly to the mounting bracket. An indexing tab, Figure 17–24, is generally

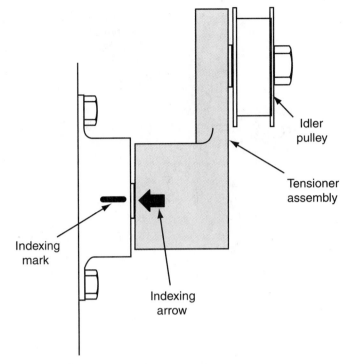

FIGURE 17–24 Tensioner indexing tab.

located on the back of the tensioner to align with the slot in the mounting bracket.
9. Tighten the nut to 50 ft.-lb. (67 N•m).
10. Replace any components that were moved in step 4.
11. Position the drive belt over all the pulleys, except for the idler pulley.

FANS

The engine-driven fan is mounted onto the water pump shaft in front of the pulley. Five- or six-blade fans are usually found on air-conditioned cars, whereas a four-blade fan is standard on non-air-conditioned cars. A six-blade fan, such as that shown in Figure 17–25, may be made of steel, nylon, or fiberglass. They are precisely balanced to prevent water pump bearing and/or seal damage.

It should be noted that some engine-driven fans are designed to turn clockwise (cw), and some are designed to turn counterclockwise (ccw). For proper airflow, it is important that the replacement fan be suitable for the design. An improper fan will result in poor (or no) air circulation and will cause engine overheating or failure.

Fans are especially required for idle- and low-speed driving to pull sufficient air through the radiator and across the engine for proper cooling. At road speeds, *ram air* would be sufficient for this purpose. To satisfy the needs for low-speed cooling and to reduce engine load at high speed, a *fan clutch, flexible fan,* or an *electric fan* is often used. The electric fan is found on most late-model applications.

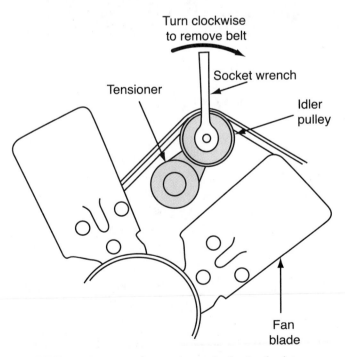

FIGURE 17–23 Rotate the tensioner clockwise (cw) to loosen the belt.

FIGURE 17–25 Six–blade, heavy–duty engine cooling fan. (*Courtesy of BET, Inc.*)

Fan Clutch

A fan clutch is installed between the water pump pulley and fan of an engine-driven fan. The fan clutch is sensitive to engine speed as well as underhood temperature. One type of fan clutch, shown in Figure 17–26, uses a temperature-sensitive silicone fluid in lands and grooves. When the underhood temperature is below about 160°F (71.1°C), the fan clutch allows the fan to turn at water pump speed up to about 800 rpm. When the temperature is above this, the fan will turn at about the same speed as the water pump, up to a maximum speed of about 2,600 rpm. It will not turn faster than 2,600 rpm, regardless of how fast the water pump turns above that speed.

Note that fan clutches are not *omni-directional*, which means they will not function properly in both directions. The replacement fan clutch selected must meet the rotation requirements of the application, either clockwise (cw) or counterclockwise (ccw).

A defective fan clutch must be replaced because no provisions are made for its repair. Failure is usually due to lockup, which is indicated by excessive noise, or a fluid leak, which is indicated by a tacky fluid at the clutch bearing area.

Flexible Fans

Flexible or *flex* fans, illustrated in Figure 17–27, have blades made of a material (metal, plastic, nylon, or fiberglass) that will flex or change pitch, based on engine speed. The fan blade pitch decreases, as engine speed increases. The extreme pitch at slower engine speeds, as shown in Figure 17–28A, provides maximum airflow to cool the engine and coolant.

FIGURE 17–26 Typical fan clutch. (*Courtesy of BET, Inc.*)

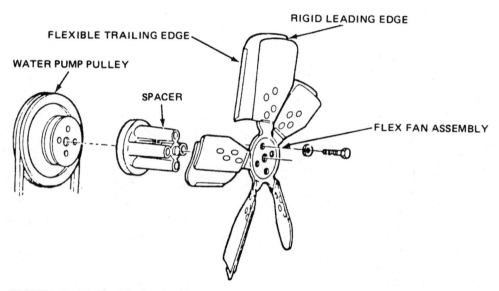

FIGURE 17–27 Flexible fan details.

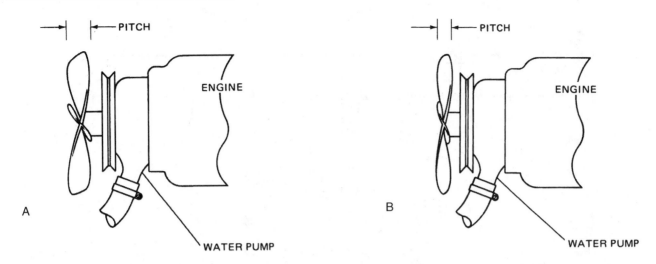

FIGURE 17–28 Extreme pitch at low speed (A); reduced pitch at high speed (B).

The need for fan-forced air is reduced at higher engine speeds. Engine and coolant air is provided by ram air, which is produced by the forward motion of the car. The flex blades feather, reducing pitch, as shown in Figure 17–28B, which, in turn, saves engine power and reduces the noise level.

Electric Fan

In late-model applications, the belt-driven fan has been replaced with an electrically driven fan, like that shown in Figure 17–29, to save power and reduce noise. This fan and motor are mounted to the radiator shroud and are not con-

nected mechanically or physically to the engine coolant (water) pump. The 12-volt, motor-driven fan is electrically controlled by either an engine coolant temperature switch (thermostat) and/or the air-conditioner select switch. In some applications, an air-conditioner-equipped vehicle would have two electric fans working independently of each other, depending on temperature conditions.

Following the schematic of Figure 17–30, the cooling fan motor is connected to the 12-volt battery supply through a normally open (NO) set of contacts (points) in the cooling fan relay. Protection for this circuit is provided by a fusible link (F/L). During normal operation,

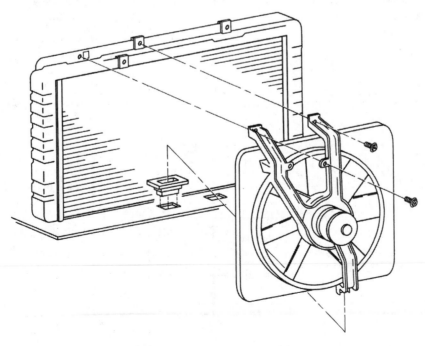

FIGURE 17–29 Exploded view of an electric engine cooling fan assembly.

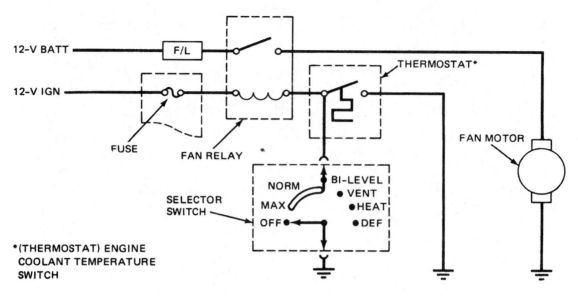

FIGURE 17–30 An electrical schematic of an engine cooling fan system.

with the air conditioner off and the engine coolant below a predetermined temperature of approximately 215°F (102°C), the relay contacts are open and the fan motor does not operate.

Should the engine coolant temperature exceed approximately 230°F (110°C), the engine coolant temperature switch will close, as shown in Figure 17–31, to energize the fan relay coil, which, in turn, will close the relay contacts—assuming the ignition switch is in the RUN position.

The 12-volt supply for the relay coil circuit is independent of the 12-volt supply for the fan motor circuit. The coil circuit is from the RUN terminal of the ignition switch, through a fuse in the fuse panel, and to ground through the relay coil and thermostat.

Should the air-conditioner select switch be turned to any cool position, regardless of engine temperature, a circuit will be completed through the relay coil to ground through the select switch. This action will close the relay contacts to provide 12 volts to the fan motor. The fan will then operate as long as the air-conditioner and ignition switches are on.

There are many variations to operate the electric cooling fan. Some systems provide a *cool-down period*, whereby the fan continues to operate after the engine has been stopped and the ignition switch has been turned off. The fan stops only when the engine coolant falls to a predetermined safe temperature, usually about 210°F (99°C).

In some systems, the fan does not start when the air-conditioner select switch is turned on unless the

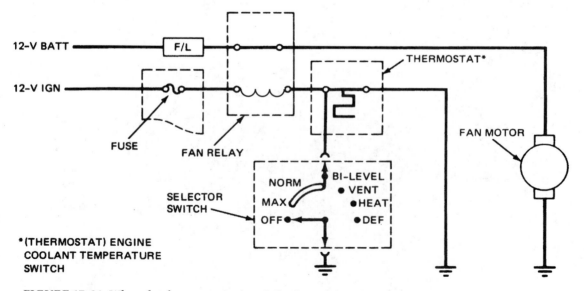

FIGURE 17–31 When the thermostat is closed, the fan motor circuit is complete.

system high side is above a predetermined high pressure. The fan does not run if the high side is below a predetermined pressure unless the engine coolant is above a predetermined safe temperature.

CAUTION
SOME FAN MOTORS WILL START AND RUN WITHOUT WARNING, EVEN THOUGH THE IGNITION SWITCH MAY BE IN THE OFF POSITION. THEREFORE, EXTREME CAUTION SHOULD BE EXERCISED WHEN WORKING UNDER THE HOOD OF A VEHICLE EQUIPPED WITH AN ELECTRIC COOLING FAN.

Because of the variations of electric fan systems, manufacturers' schematics and specifications must be consulted for troubleshooting and repairing any particular year or model car.

Follow the schematic in Figure 17–30 for testing and troubleshooting an engine cooling fan system. The following procedure is typical.

1. Start the engine and bring the coolant up to operating temperature.
2. Turn on the air conditioner.
3. Disconnect the cooling fan motor electrical lead **connector,** Figure 17–32.
4. Make sure the ground wire is not disturbed. If the ground wire is a part of the electrical connector, establish a ground connection with a jumper wire, Figure 17–33.
5. Connect a test lamp from ground to the hot wire

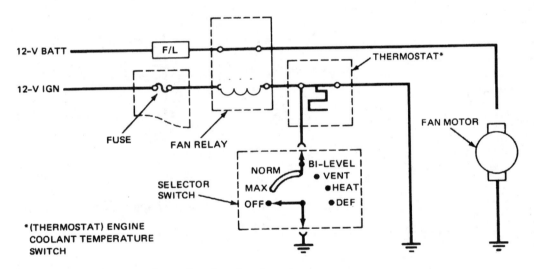

FIGURE 17–32 Disconnect the cooling fan electric motor.

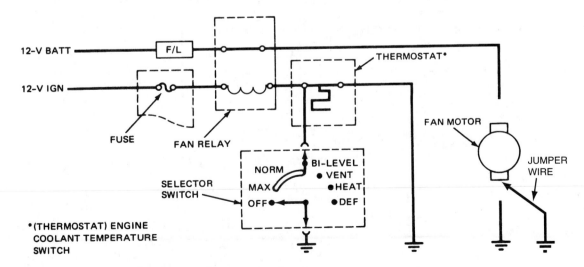

FIGURE 17–33 Establish ground with a jumper wire.

of the connector, Figure 17–34. Make sure the lamp is good.

6. Did the lamp light? If yes, proceed with step 7. If no, check for a defective fan relay or temperature switch. It is also possible that the engine is not up to sufficient temperature to initiate cooling fan action.

7. Connect a fused jumper wire from the battery positive (+) terminal to the cooling fan connector,

Figure 17–35. Make sure the fuse is good.

8. Did the fan start and/or run? If yes, the fan is all right. The system is apparently not warm enough to initiate fan action. If no, proceed with step 9.

9. Again check the fuse in the jumper wire. Is it blown? If yes, the motor is shorted and must be replaced. If no, the motor is open and must be replaced.

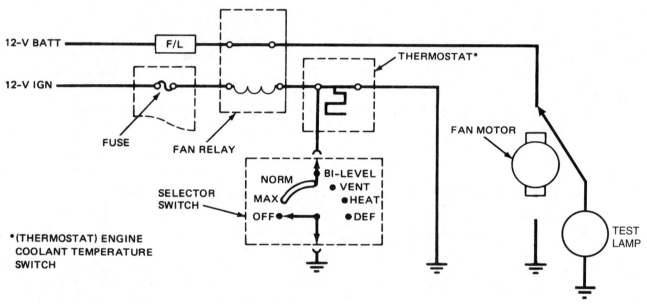

FIGURE 17–34 Connect a test lamp from ground to the hot wire.

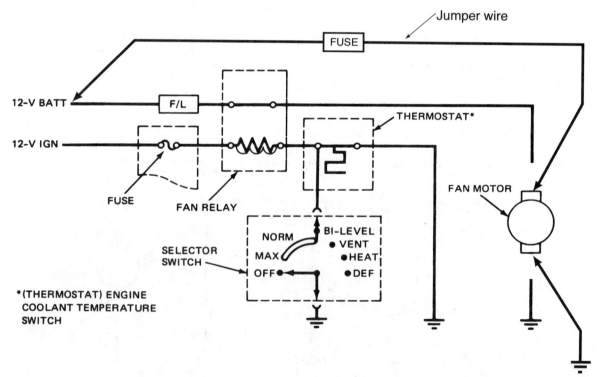

FIGURE 17–35 Connect a fused jumper wire from the battery positive (+) terminal to the cooling fan connector.

HOSES AND CLAMPS

Radiators usually have two hoses: an upper hose and a lower hose. These are constructed of ozone- and oil-resistant reinforced **synthetic** rubber. Preformed lower hoses, like that shown in Figure 17–36A, have a spiral tempered steel wire molded into them to prevent collapse due to suction of the water pump impeller. Upper hoses, not subject to this condition, usually do not have the wire. Unless the coolant has a rust retarder, such as ethylene glycol, and the cooling system is sound (free of leaks), this wire will rust away in a short period of time. (See the discussion of coolants later in this chapter.)

Universal flexible hoses with wire inserts, shown in Figure 17–36B, are available. These hoses are usually used when preformed hoses are not available. Because of body and engine parts, hoses must often be critically **routed.** The flexible hose, shown in Figure 17–36C, is not always so easily routed.

Many types of hose clamps are used. One popular type of replacement is the high-torque, worm-gear clamp with a carbon steel screw and a stainless steel band, shown in Figure 17–37. It is important that the clamp be properly positioned at 90 degrees to the hose and that it not be overtightened. Overtightening and/or mispositioning a hose clamp causes it to cut into the hose, often breaking the reinforcing fabric.

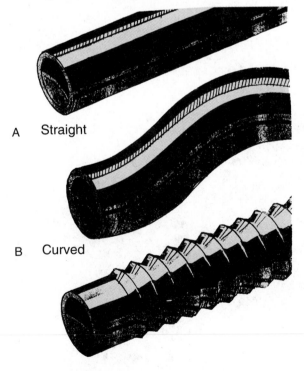

A Straight

B Curved

C Flex or Universal

FIGURE 17–36 Three kinds of hoses: (A) straight; (B) curved: (C) flex. (*Courtesy of The Gates Rubber Company*)

COOLANT RECOVERY TANK

A *coolant recovery tank* is used to **capture** venting coolant and vapor from the radiator during the time the pressure valve of the radiator cap is open. When the vacuum valve of the cap opens, this same vented coolant is metered back into the cooling system.

The nonpressurized recovery tank, as shown in Figure 17–38, has a capacity of about 0.5 gallon (1.8 L) and contains approximately one pint (0.47 L) of coolant when the cooling system is cold. The coolant level is checked, not by removing the radiator cap, but by noting the coolant level in the recovery tank. It should be filled to the perimeters marked *cold, normal,* or *hot* on the side of the tank.

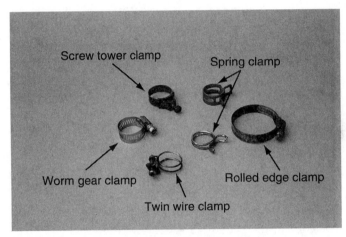

FIGURE 17–37 Several types of hose clamps.

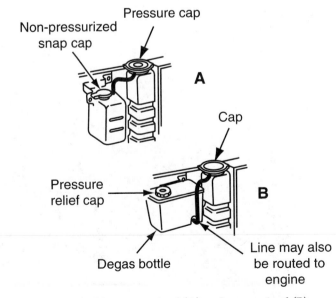

FIGURE 17–38 Nonpressurized (A) and pressurized (B) coolant recovery tanks.

HEATER SYSTEM

The automotive heater system consists of two parts in addition to hoses and clamps. They are the *heater core* and the **coolant flow valve.** The heater housing and duct are part of the air-conditioning duct system, which is covered in Chapter 16.

Heater Core

The heater core, shown in Figure 17–39, resembles a small radiator without a pressure cap. When the control valve is open, a small percentage of heated engine coolant is circulated through the heater core instead of through the radiator core. This provides hot air inside the passenger compartment, when desired.

Leaks are common causes of heater problems. Leaks in a heater core are detected by an obvious loss of engine coolant and a wet front floor mat, usually on the passenger's side. These leaks are repaired by those qualified in radiator repair and service.

Coolant Flow Valve

The heater control valve may be either cable operated, as shown in Figure 17–40, or vacuum operated, as shown in Figure 17–41. When heating, the amount of heated coolant allowed to enter the heater core—from off to full flow—is metered by the control valve. The amount of coolant that flows depends on the control selector cable position or the amount of vacuum signal applied.

When cooling in a climate control system, hot coolant is allowed to flow through the heater core to temper the cool air flowing from the air-conditioner evaporator to maintain in-vehicle relative humidity.

Hoses and Clamps

Like radiator hoses, heater hoses are constructed of oil- and ozone-resistant synthetic rubber. However, these hoses are smaller and somewhat more flexible than radiator hoses. Three popular sizes of heater hose are 1/2-inch, 5/8-inch, and 3/4-inch inside diameter (ID).

Although more may be found on some applications, three hoses are usually associated with the heater system:

❏ Engine outlet to the control valve
❏ Control valve to the heater core
❏ Heater core to the water pump inlet

The hose clamps used on heater hoses (both shown in Figure 17–42) are similar to those previously described

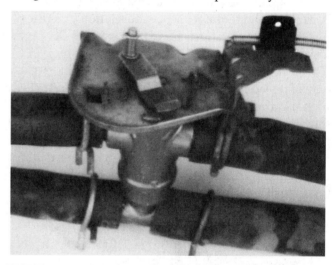

FIGURE 17–40 Cable–operated heater control valve. *(Courtesy of BET, Inc.)*

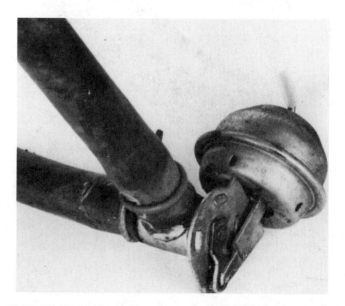

FIGURE 17–41 Typical vacuum–controlled heater (coolant) shutoff valve. *(Courtesy of BET, Inc.)*

FIGURE 17–39 Typical heater core. *(Courtesy of BET, Inc.)*

for use on radiator hoses. The only difference is that the heater hose clamps are smaller.

ADDITIVES

Many **additives,** inhibitors, and "remedies" are available for use in the automotive cooling system. These include, but are not limited to, stop-leak, water pump lubricant, engine **flush,** and acid neutralizers. Extreme caution should be exercised when using any additive in the cooling system. Read the directions and precautions on the label to *know* in advance the end results of any additive used. For example, caustic solutions must never be used in aluminum radiators; alcohol-based remedies should never be used in any cooling system; and so on.

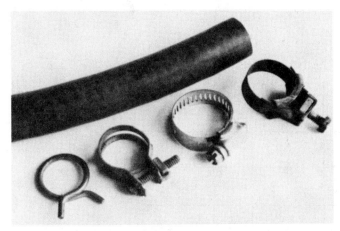

FIGURE 17–42 Heater hose and common types of heater hose clamps. *(Courtesy of BET, Inc.)*

If a cooling system is maintained in good order by a program of preventive maintenance, the use of additives and inhibitors should never be necessary. Only car line manufacturer-recommended ethylene glycol based antifreeze-type solution should be added.

Antifreeze

Most manufacturers of automotive cooling systems recommend and specify ethylene glycol based antifreeze mixed with water (H_2O) for cooling system protection. A 50-50 (percent) mixture is recommended for year-round protection. Most manufacturers warn against the use of alcohol-based antifreeze solutions. The ethylene glycol and water (H_2O) coolant mixture is essential for four important reasons:

- ❏ It lowers the freezing temperature point of the coolant.
- ❏ It raises the boiling temperature point of the coolant.
- ❏ It provides water pump lubrication.
- ❏ It inhibits rust and corrosion.

The freezing temperature of water (H_2O) coolant, at ambient sea level pressure (14.69 psi or 101.3 kPa), is 32°F (0°C). If one-third (33.3 percent) of the coolant is ethylene glycol and two-thirds (66.6 percent) is water (H_2O), the **freezing point** is reduced to 0°F (–17.7°C). A half-and-half (50 percent) solution protects the cooling system to about –32°F (–35.6°C). Typical antifreeze and **antiboil** products are shown in Figure 17–43.

FIGURE 17–43 EG– (A), and PG–based (B), antifreeze.

The coolant should contain no less than 30 percent ethylene glycol, giving protection to about 5°F (–15°C), and no more than 60 percent, giving protection to about –62°F (–52.2°C). After 60 percent, protection is actually reduced. For example, 100 percent ethylene glycol will freeze at about –2°F (–18.8°C), as shown in the graph in Figure 17–44. Protection should, of course, be to the lowest temperature expected. In warmer climate zones, protection is required for the benefits of antirust and anticorrosion inhibitors, as well as for water pump lubrication.

Ethylene glycol also increases the boiling point of the coolant. In a 15 psi (103.4 kPa) pressurized system, water (H_2O) boils at 250°F (121.1°C). A 50-50 mixture of coolant in the same system boils at about 265°F (129.4°C), which would be a 15°F (8.3°C) advantage.

Most *permanent* antifreeze solutions are formulated to withstand 2 years of *normal* operation. A driver rarely experiences only normal operation in 2 years. Some manufacturers recommend changing the solution at intervals of 24 months or 24,000 miles (38 616 km); others recommend doing this every 12 months or 12,000 miles (19 308 km). Always follow the directions on the label for adding or changing coolant fluid.

FLUSH THE COOLING SYSTEM

If an existing antifreeze/coolant with an extended-life antifreeze/coolant is being replaced, the cooling system must first be completely drained and flushed. This is necessary to gain the full benefits of the longer-lasting for-

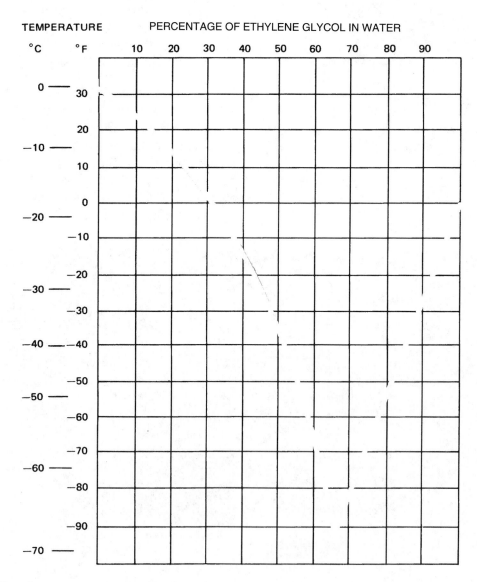

FIGURE 17–44 Freezing temperature of coolant, based on percentage of ethylene glycol solution in water.

mula. For example, if an extended-life antifreeze/coolant is currently being used in a vehicle and a regular-type antifreeze is added to the cooling system, the extended life protection will be lost. When adding antifreeze, always add the same type as that previously used in the cooling system. If the type is unknown, it is generally recommended to drain, flush, and refill the cooling system. The procedure that follows is typical:

> NOTE: Often, the owner's manual includes instructions to open an air bleed valve on the engine or to remove a heater hose to purge air that may have entered the engine during draining. Always follow the manufacturer's instructions.

1. Remove the radiator cap.
2. Place a drain pan under the vehicle.
3. Open the **drain cock** valve to drain the cooling system, Figure 17–45.
4. Close the drain valve.
5. Recycle or dispose of the used coolant according to local laws and regulations, Figure 17–46.

> NOTE: Label the container clearly as used antifreeze. Do not use a beverage container to store antifreeze, new or used. If stored, keep the containers away from children and animals. If the antifreeze is to be disposed of, do so promptly and properly.

6. Flush the cooling system to clean the engine block of any **scale,** rust, or other debris before refilling with new antifreeze/coolant.

FIGURE 17–46 Remove the pans and dispose of the coolant in a manner consistent with local regulations.

> NOTE: A cooling system flush may be used to remove stubborn rust, grease, and sediment that may not be removed by plain water alone.

7. Fill the radiator with cleaner and/or water.
8. Start and run the engine with the heater on HI and the temperature gauge reading normal-operating temperature for the time recommended on the flush product label.

> NOTE: An infrared (IR) thermometer is a handy tool to use for determining when "normal" engine-operating temperature has been reached.

9. Stop the engine and allow it to cool.
10. Again, open the drain valve and drain the cooling system.
11. Close the drain valve and refill the radiator with plain water.
12. Start and run the engine for about 15 minutes at normal engine-operating temperature.
13. Stop the engine and allow it to cool; open the drain valve and drain the cooling system.
14. Close the drain valve and refill the cooling system.

> NOTE: Refer to the owner's manual for the cooling system capacity. Follow all special service instructions to ensure a proper refill.

15. Once the radiator is filled, run the engine at normal-operating temperature with the heater on HI for 15 minutes to mix and disperse the coolant fully throughout the cooling system.
16. Shut off the engine and allow the cooling system to cool.

FIGURE 17–45 Open the radiator drain cock and allow the radiator to drain until the flow stops.

17. Check the coolant level and concentration of antifreeze. Adjust, if necessary.
18. After a few days of driving, recheck the cooling system, Figure 17–47.

NOTE: A **hydrometer** may be used for EG testing, and test strips may be used for PG testing.

PREVENTATIVE MAINTENANCE

Preventative maintenance (PM) should be performed on the cooling system every 12,000–15,000 miles (20,000–35,000 kilometers).

Replace any belts that are frayed, **glazed,** or obviously damaged. Replace any hoses that are found to be brittle, soft, or otherwise deteriorated, Figure 17–48.

The design consideration of a cooling system is to provide for a minimum sustained road speed operation of 90 mph (145 **km/h**) at an ambient temperature of 125°F (52°C). Another criterion is for 30 minutes of driving in congested stop-and-go traffic at an ambient temperature of 115°F (46°C) without experiencing any overheating problems. These design considerations exceed the conditions that one is likely to encounter in regular day-to-day driving.

If the engine overheats, the problem should be found and corrected. The life of a vehicle engine or transmission that is habitually allowed to overheat is greatly reduced. The high-limit properties of lubricating oil require adequate and proper heat removal to preserve formulated lubricating characteristics.

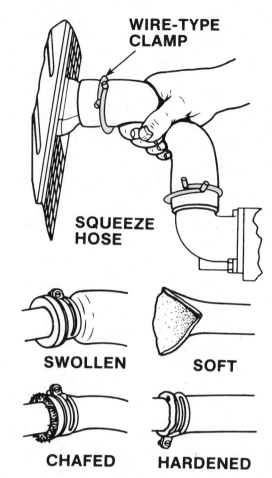

FIGURE 17–48 Typical hose defects.

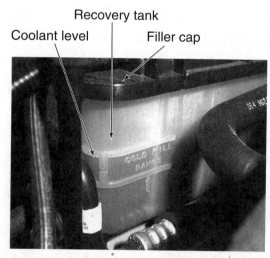

FIGURE 17–47 Check the coolant level in the recovery reservoir.

A preventative maintenance program should include the following:

❏ Test and/or replace the thermostat.
❏ Test and/or replace the pressure cap.
❏ Inspect and/or replace the radiator hose(s).
❏ Inspect and/or replace the heater hoses.
❏ Pressure test the cooling system.
❏ Test and/or replace the antifreeze solution.
❏ Visually inspect the coolant pump, heater, control valve, and belt(s).

SUMMARY

The cooling system, which is often neglected, is one of the most important systems of a vehicle. If well maintained, the cooling system should give years of trouble-free service. The cost of maintenance every 12,000–15,000 miles (19 308–24 135 km), is more than offset by the cost of initial breakdown and consequent repairs. These repairs, incidentally, often result in expensive engine service.

Design considerations of a typical cooling system provide for minimum sustained road speed operation of 90 mph (144.8 km/h) with an ambient temperature as high as 125°F (51.6°C), or 30 minutes of driving in congested stop-and-go traffic with an ambient temperature as high as 115°F (46.1°C) without overheating problems. These design considerations exceed the conditions one is likely to encounter in day-to-day driving. So, if the engine over-heats, the problem must be found and corrected.

The life of an engine and a transmission that are habitually allowed to overheat is greatly reduced. The high-limit properties of lubricating oil in the engine and transmission require adequate and proper heat removal to preserve formulated lubricating characteristics.

In a word, repair; do not patch.

REVIEW

Select the correct answer from the choices given.

1. A thermostat is used in the cooling system to prevent engine:
   a. overheating.
   b. overcooling.
   c. Both A and B.
   d. Neither A nor B.

2. Generally, the maximum speed of an engine-mounted declutch fan is ____ rpm.
   a. 1,200
   b. 2,600
   c. 3,400
   d. 4,800

3. A typical cooling system removes about ____ percent of the heat generated by the engine.
   a. 25
   b. 35
   c. 45
   d. 75

4. By design, the coolant flow in a radiator is:
   a. side to side.
   b. top to bottom.
   c. either A or B.
   d. neither A nor B.

5. A typical coolant thermostat will be fully open:
   a. a few degrees below its temperature rating.
   b. at its temperature rating.
   c. a few degrees above its temperature rating.
   d. coolant thermostats do not fully open.

6. The recommended concentration of antifreeze is being discussed. Technician A says 50 percent EG-type antifreeze provides maximum protection. Technician B says only 40 percent PG-type antifreeze provides the same protection. Who is right?
   a. A only
   b. B only
   c. Both A and B
   d. Neither A nor B

7. Technician A says that an overheating problem can often be corrected by replacing the pressure cap with one that has a higher rating. Technician B says that most overheating problems appear because the coolant thermostat has been removed. Who is right?
   a. A only
   b. B only
   c. Both A and B
   d. Neither A nor B

8. Technician A says that antifreeze lowers the boiling point of the coolant. Technician B says that antifreeze lowers the freezing point of the coolant. Who is right?
   a. A only
   b. B only
   c. Both A and B
   d. Neither A nor B

9. A leaking external transmission oil cooler is being discussed. Technician A says that transmission oil may be noted in the coolant. Technician B says that coolant may be noted in the transmission. Who is right?
   a. A only
   b. B only
   c. Both A and B
   d. Neither A nor B

10. The coolant recovery system is being discussed. Technician A says the pressure cap allows excess coolant to be released to the recovery tank. Technician B says the pressure cap allows coolant to return to the radiator from the recovery tank. Who is right?
    a. A only
    b. B only
    c. Both A and B
    d. Neither A nor B

11. Technician A says a serpentine belt must be retensioned after a few hours' run-in. Technician B says V-belts are pretensioned and do not require a run-in. Who is right?
    a. A only
    b. B only
    c. Both A and B
    d. Neither A nor B

12. All of the following may be considered a part of the cooling system, EXCEPT the:
    a. accumulator.
    b. heater core.
    c. thermostat.
    d. pressure cap.

13. Engine overheating is being discussed. Technician A says that a defective electric cooling fan relay or switch may be the cause. Technician B says that a defective coolant thermostat may be the cause. Who is right?
    a. A only
    b. B only
    c. Both A and B
    d. Neither A nor B

14. Although the dash temperature gauge indicates HOT, the coolant is in the normal temperature range. Technician A says the wire from the sending unit to the dash gauge may be grounded. Technician B says the dash gauge may have an open circuit. Who is right?
    a. A only
    b. B only

15. Damaged fan blades are being discussed. Technician A says a loose blade should be secured by welding. Technician B says a bent blade should be straightened. Who is right?
    a. A only
    b. B only
    c. Both A and B
    d. Neither A nor B

16. Electric engine cooling fans are being discussed. Technician A says they are generally independent of the ignition switch and may start and run without notice. Technician B says that most electric engine cooling fans are protected by a heavy shroud and pose no personal safety hazard. Who is right?
    a. A only
    b. B only
    c. Both A and B
    d. Neither A nor B

17. Radiators are constructed of all the following materials, EXCEPT:
    a. magnesium.
    b. aluminum.
    c. copper.
    d. plastic.

18. The *least likely* problem associated with a cooling system that has had the thermostat removed is:
    a. poor heater performance.
    b. erratic computer engine control.
    c. lower than normal operating temperature.
    d. loss of coolant.

19. The *most likely* use for an engine coolant temperature switch is to electrically energize the:
    a. compressor clutch.
    b. cooling system fan motor.
    c. blower motor.
    d. coolant hot warning light.

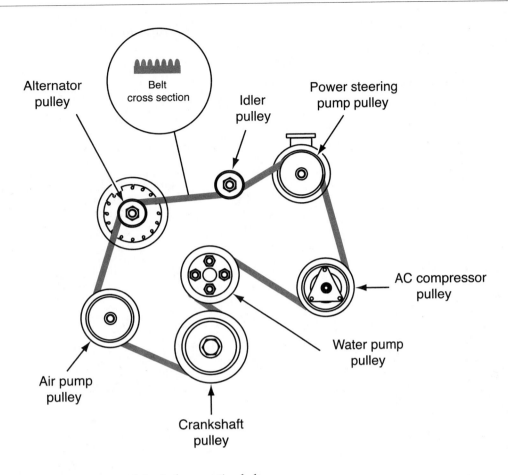

FIGURE 17–49 A typical serpentine belt.

20. If the crankshaft pulley turns clockwise (cw), all of the following statements about the illustration in Figure 17–49 are true, EXCEPT:

   a. the compressor pulley will turn clockwise (cw).
   b. the idler pulley will turn counterclockwise (ccw).
   c. the water pump pulley will turn clockwise (cw).
   d. the alternator pulley will turn clockwise (cw).

TERMS

Write a brief description of the following terms.:

1. additives
2. antiboil
3. capture
4. caution
5. clamps
6. connector
7. coolant flow valve
8. drain cock
9. filler neck
10. flush

11. freezing point
12. glazed
13. hydrometer
14. km/h
15. maintenance
16. owner's manual
17. preformed
18. routed
19. scale
20. synthetic

CHAPTER 18

TROUBLESHOOTING AND REPAIRS

OBJECTIVES

On completion and review of this chapter, you should be able to:

- Diagnose and correct the causes of noise in air-conditioning systems.
- Determine the causes of compressor noise and possible remedies.
- Troubleshoot air-conditioning systems for problems relating to intermittent, insufficient, or no cooling conditions.
- Diagnose and correct conditions that may cause engine overcooling or overheating.
- Troubleshoot an engine cooling system for problems and solutions relating to loss of coolant.
- Troubleshoot and diagnose air-conditioning system functions and malfunctions based on gauge pressures.

INTRODUCTION

An accurate **diagnosis** and determination of automotive air-conditioning system function and, more importantly, **malfunction,** depend largely on the ability of the technician to interpret gauge pressure readings. The use of a refrigeration technician's manifold and gauge set may be compared with that of a physician's stethoscope.

For the following exercises in temperature and pressure relationships, it will be necessary to refer to the appropriate temperature-pressure chart, Figure 18–11 through Figure 18–14, found on pages 404 and 405. A self-stick note tab on the appropriate page will make it easier to reference these pages while performing these exercises.

TEMPERATURE-PRESSURE RELATIONSHIPS

The following ten practice exercises are provided as an aid to become familiar with temperature and pressure similarities and the relationships of CFC-12 and HFC-134a refrigerants. Consider that all **conditions** are normal for an average-operating cycling clutch or non-cycling clutch air-conditioning system for these exercises.

TEMPERATURE–PRESSURE RELATIONSHIP 1

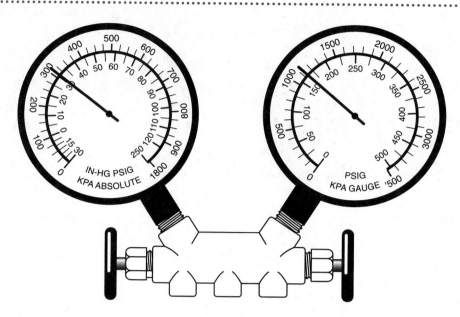

FIGURE 18–1 Temperature–pressure relationship–1.

1. What approximate pressure is indicated on the low-side gauge? _____ psig

    _____ kPa

2. This pressure corresponds to a CFC-12 evaporator temperature _____ °F
    of approximately
    _____ °C

3. This pressure corresponds to an HFC-134a evaporator temperature _____ °F
    of approximately
    _____ °C

4. What approximate pressure is indicated on the high-side gauge? _____ psig

    _____ kPa

5. This CFC-12 high-side pressure corresponds to an ambient _____ °F
    temperature of approximately
    _____ °C

6. This HFC-134a high-side pressure corresponds to an ambient _____ °F
    temperature of approximately
    _____ °C

TEMPERATURE–PRESSURE RELATIONSHIP 2

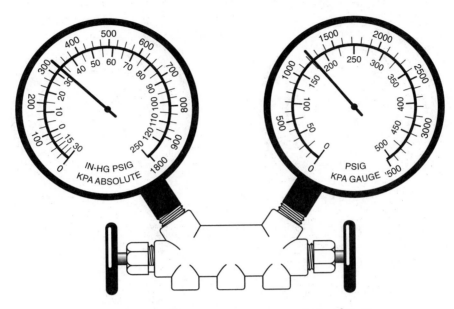

FIGURE 18–2 Temperature–pressure relationship–2.

1. What approximate pressure is indicated on the low-side gauge? _____ psig

   _____ kPa

2. This pressure corresponds to a CFC-12 evaporator temperature of _____ °F

   approximately _____ °C

3. This pressure corresponds to an HFC-134a evaporator temperature _____ °F

   of approximately _____ °C

4. What approximate pressure is indicated on the high-side gauge? _____ psig

   _____ kPa

5. This CFC-12 pressure corresponds to an ambient temperature of _____ °F

   approximately _____ °C

6. This HFC-134a pressure corresponds to an ambient temperature _____ °F

   of approximately _____ °C

TEMPERATURE–PRESSURE RELATIONSHIP 3

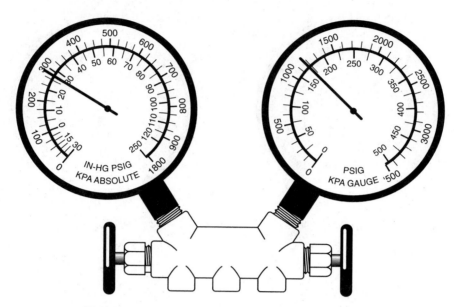

FIGURE 18–3 Temperature–pressure relationship–3.

1. What approximate pressure is indicated on the low-side gauge? _____ psig

   _____ kPa

2. This pressure corresponds to a CFC-12 evaporator temperature of approximately _____ °F

   _____ °C

3. This pressure corresponds to an HFC-134a evaporator temperature of approximately _____ °F

   _____ °C

4. What approximate pressure is indicated on the high-side gauge? _____ psig

   _____ kPa

5. This CFC-12 pressure corresponds to an ambient temperature of approximately _____ °F

   _____ °C

6. This HFC-134a pressure corresponds to an ambient temperature of approximately _____ °F

   _____ °C

TEMPERATURE–PRESSURE RELATIONSHIP 4

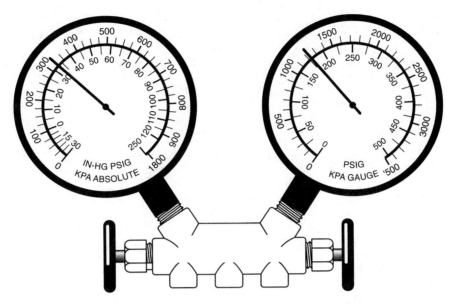

FIGURE 18–4 Temperature–pressure relationship–4.

1. What approximate pressure is indicated on the low-side gauge?

   _____ psig
   _____ kPa

2. This pressure corresponds to a CFC-12 evaporator temperature of approximately

   _____ °F
   _____ °C

3. This pressure corresponds to an HFC-134a evaporator temperature of approximately

   _____ °F
   _____ °C

4. What approximate pressure is indicated on the high-side gauge?

   _____ psig
   _____ kPa

5. This CFC-12 pressure corresponds to an ambient temperature of approximately

   _____ °F
   _____ °C

6. This HFC-134a pressure corresponds to an ambient temperature of approximately

   _____ °F
   _____ °C

TEMPERATURE–PRESSURE RELATIONSHIP 5

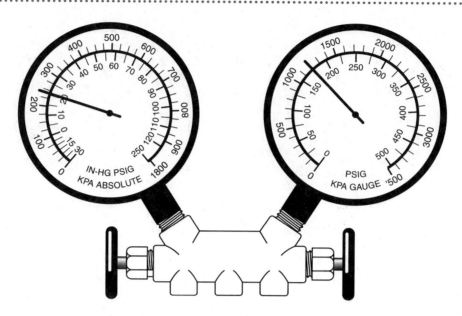

FIGURE 18–5 Temperature–pressure relationship–5.

1. What approximate pressure is indicated on the low-side gauge?

_____ psig

_____ kPa

2. This pressure corresponds to a CFC-12 evaporator temperature of approximately

_____ °F

_____ °C

3. This pressure corresponds to an HFC-134a evaporator temperature of approximately

_____ °F

_____ °C

4. What approximate pressure is indicated on the high-side gauge?

_____ psig

_____ kPa

5. This CFC-12 pressure corresponds to an ambient temperature of approximately

_____ °F

_____ °C

6. This HFC-134a pressure corresponds to an ambient temperature of approximately

_____ °F

_____ °C

TEMPERATURE–PRESSURE RELATIONSHIP 6

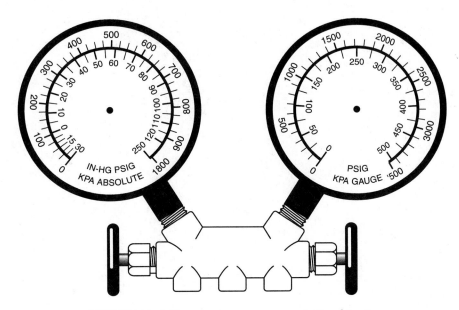

FIGURE 18–6   Temperature–pressure relationship–6.

CONDITIONS

The ambient temperature is 95°F (35°C).
The evaporator temperature is 33°F (0.5 °C).

1. What is the normal head pressure for a CFC-12 system? _____ psig

   _____ kPa

   a. Indicate this pressure on the gauge in the
      diagram, using a red pen.

2. What is the normal head pressure for an _____ psig

   HFC-134a system? _____ kPa

   a. Indicate this pressure on the gauge in the
      diagram, using a blue pen.

3. What is the normal suction pressure for a _____ psig

   CFC-12 system? _____ kPa

   a. Indicate the CFC-12 pressure on the gauge
      in the diagram, using a red pen.

4. What is the normal suction pressure for an _____ psig

   HFC-134a system? _____ kPa

   a. Indicate the HFC-134a pressure on the gauge
      in the diagram, using a blue pen.

TEMPERATURE–PRESSURE RELATIONSHIP 7

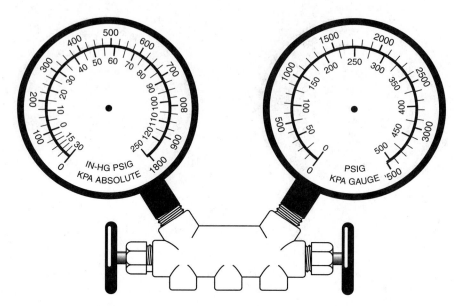

FIGURE 18–7 Temperature–pressure relationship–7.

1. A high-side (head) pressure of 185 psig (1,276 kPa) is normal for:

   a. A CFC-12 system at what ambient temperature?   _____ °F

   _____ °C

   b. An HFC-134a system at what ambient temperature?   _____ °F

   _____ °C

2. Show this pressure reading on the high-side gauge in the diagram.

   NOTE: Assume that the evaporator temperature is 31°F (–0.5°C):

3. What is the low-side gauge reading for a CFC-12 system?   _____ psig

   _____ kPa

   a. Show this pressure reading on the low-side gauge in the
   diagram, using a red pen.

4. What is the low-side gauge reading for an HFC-134a system?   _____ psig

   a. Show this pressure reading on the low-side gauge in the   _____ kPa
   diagram, using a blue pen.

TEMPERATURE–PRESSURE RELATIONSHIP 8

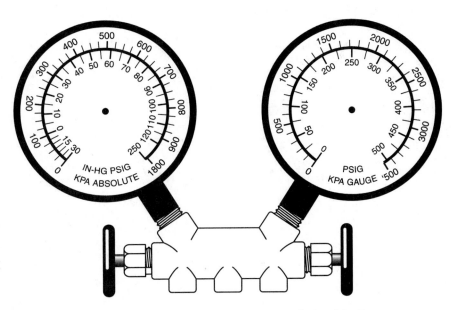

FIGURE 18–8 Temperature–pressure relationship–8.

CONDITIONS

The ambient temperature is 100°F (37.8°C).
The low-side gauge indicates 26 psig (179 kPa).

1. What is the normal high-side pressure for a
    CFC-12 system?
       _____ psig
       _____ kPa

    a. Show the pressure reading on the gauge in the
        diagram, using a red pen.

2. What is the normal high-side pressure for an
    HFC-134a system?
       _____ psig
       _____ kPa

    a. Show the pressure reading on the gauge in the
        diagram, using a blue pen.

3. What is the temperature of the CFC-12
    refrigerant in the evaporator?
       _____ °F
       _____ °C

    a. Show the CFC-12 pressure reading on the gauge in
        the diagram, using a red pen.

4. What is the temperature of the HFC-134a
    refrigerant in the evaporator?
       _____ °F
       _____ °C

    a. Show the HFC-134a pressure reading on
        the gauge in the diagram, using a blue pen.

TEMPERATURE–PRESSURE RELATIONSHIP 9

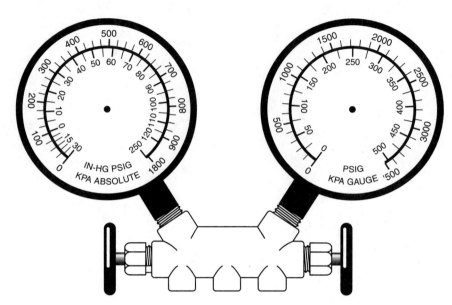

FIGURE 18–9 Temperature–pressure relationship–9.

CONDITIONS

The high-side gauge reads 195 psi (1,345 kPa).
The low-side gauge indicates 30 psig (207 kPa).

1. Show the **high side** reading on the high-side gauge.

2. Assuming a CFC-12 system, what is the   _____ °F
   ambient temperature in this exercise?   _____ °C

3. Assuming an HFC-134a system, what is the   _____ °F
   ambient temperature in this exercise?   _____ °C

4. Show the low-side reading on the low-side gauge.

5. What is the CFC-12 evaporator   _____ °F
   temperature in this example?   _____ °C

6. What is the HFC-134a evaporator   _____ °F
   temperature in this example?   _____ °C

TEMPERATURE–PRESSURE RELATIONSHIP 10

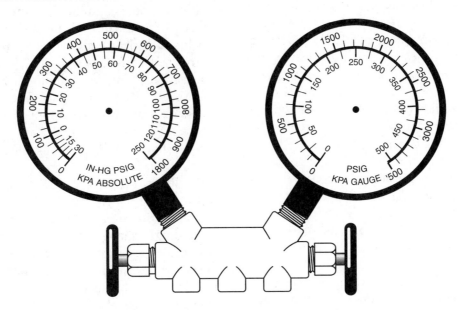

FIGURE 18–10 Temperature–pressure relationship–10.

CONDITIONS

The high-side gauge reads 205 psi (1,413 kPa).
The low-side gauge indicates 32 psig (221 kPa).

1. Show the high-side reading on the high-side gauge.

2. Assuming a CFC-12 system, what is the ambient
   temperature in this exercise?

   _____ °F
   _____ °C

3. Assuming an HFC-134a system, what is the ambient
   temperature in this exercise?

   _____ °F
   _____ °C

4. Show the low-side reading on the low-side gauge.

5. What is the CFC-12 evaporator temperature in
   this example?

   _____ °F
   _____ °C

6. What is the HFC-134a evaporator temperature in
   this example?

   _____ °F
   _____ °C

SYSTEM DIAGNOSTICS BY GAUGE PRESSURE

A gauge reading will relate to a specific air-conditioning system problem. More than one problem may be associated with a particular gauge reading, however. A system that is operating normally should have a low-side gauge pressure reading that corresponds with the temperature of the liquid refrigerant as it becomes a vapor while removing heat in the evaporator. The high-side gauge pressure reading should correspond with the temperature of the refrigerant vapor as it becomes a liquid while giving up its heat in the condenser.

Any deviation from normal gauge readings, other than a slight one, indicates a malfunction of the air-conditioning system. This malfunction may be caused by a faulty control device, a restriction, or a **defective** component. An **improper** installation of a component may also affect air-conditioning system performance.

An overheated engine cooling system or an improperly tuned engine may affect air-conditioning system performance as well, and will be noted by abnormal gauge readings.

Diagnosis of system malfunction is made easier with the knowledge that the low-side temperature, in degrees Fahrenheit, and pressure, in pounds per square inch gauge (psig), are closely **related.** This is true of refrigerant CFC-12 as well as of refrigerant HFC-134a. It is easily determined in the English system of **measure** between 20 psig and 60 psig. The English temperature-pressure relationship chart for CFC-12, Figure 18–11, and the temperature-pressure chart for HFC-134a, Figure 18–12, show that there are only slight variations between the temperature and pressure within this range.

It may, therefore, be correct to assume that for every pound (psig) of pressure change in the low side of the air-conditioning system, the temperature will correspondingly change by 1 degree Fahrenheit (1°F). For example, a pressure of 33.1 psig for CFC-12 indicates (on the chart) an evaporating temperature of 36°F. A pressure increase of 2 pounds (2 psig) to 35.1 psig will result in a temperature increase of 2°F to 38°F.

ENGLISH TEMPERATURE-PRESSURE CHART

Low-Side Pressure psi		Temperature °F	High Side Pressure psi		Temperature °F
Absolute	Gauge		Absolute	Gauge	
25.9	11.2	4	120	105	60
27	12.3	6	124	109	62
28.1	13.4	8	128	113	64
29.3	14.6	10	132	117	66
30.5	15.8	12	137	122	68
31.8	17.1	14	141	126	70
33	18.3	16	147	132	72
34.4	19.7	18	152	137	74
35.7	21	20	159	144	76
37.1	22.4	22	167	152	78
38.5	23.8	24	175	160	80
40	25.3	26	180	165	82
41.5	26.8	28	185	170	84
43.1	28.4	30	190	175	86
44.7	30	32	195	180	88
46.4	31.7	34	200	185	90
47.8	33.1	36	204	189	92
49.8	35.1	38	208	193	94
51.6	36.9	40	215	200	96
53.5	38.8	42	225	210	98
55.4	40.7	44	235	220	100
57.3	42.6	46	243	228	102
59.3	44.6	48	251	236	104

FIGURE 18–11 English temperature–pressure chart for CFC–12.

Temperature °F	Pressure psig	Temperature °F	Pressure psig
–5	4.1	39.0	34.1
0	6.5	40.0	35.0
5.0	9.1	45.0	40.0
10.0	12.0	50.0	45.4
15.0	15.1	55.0	51.2
20.0	18.4	60.0	57.4
21.0	19.1	65.0	64.0
22.0	19.9	70.0	71.1
23.0	20.6	75.0	78.6
24.0	21.4	80.0	86.7
25.0	22.1	85.0	95.2
26.0	22.9	90.0	104.3
27.0	23.7	95.0	113.9
28.0	24.5	100.0	124.1
29.0	25.3	105.0	134.9
30.0	26.1	110.0	146.3
31.0	27.0	115.0	158.4
32.0	27.8	120.0	171.1
33.0	28.7	125.0	184.5
34.0	29.5	130.0	198.7
35.0	30.4	135.0	213.5
36.0	31.3	140.0	229.2
37.0	32.2	145.0	245.6
38.0	33.2	150.0	262.8

FIGURE 18–12 HFC–134a temperature-pressure chart (English).

If the refrigerant is HFC-134a, a pressure of 31 psig corresponds to a temperature of 36°F. An increase of 2 psig will result in a temperature increase of 2°F–38°F.

This handy information does not, however, hold true in the metric system of measure. Unfortunately, there is no direct correlation between kilopascal (kPa) pressure, absolute or gauge, and the Celsius (C) temperature scale. The metric temperature-pressure charts of Figure 18–13 and Figure 18–14 may be used when troubleshooting an air-conditioning system, using gauges that are **calibrated** in metric units.

In either case, the actual temperature of the air after immediately passing over the tubes and fins of the evaporator coil will be several degrees warmer than that shown in the temperature-pressure charts. This is because of a temperature difference, called delta T (Δ_T), due to the expected rise of temperature through the tubes and fins of the evaporator coil.

Other refrigerants have similar temperature-pressure relationships but are not covered in this text because they are not recommended for motor vehicle air-conditioning (MVAC) system service. The only refrigerants that an MVAC service technician should be concerned with are CFC-12 and the industry recognized and approved replacement, HFC-134a.

Several exercises in gauge interpretations in both the English and metric scales are provided in this chapter. These exercises serve as an aid in the understanding of gauge pressure readings relating to system function and malfunction. Color plates are also included in this chapter to serve as visual aids in determining system function and malfunction. These plates may be keyed to appropriate diagnostic gauge sets, if desired.

UNITS OF PRESSURE

The customary English pressures, pounds per square inch (psi), pounds per square inch gauge (psig), or pounds per square inch absolute (psia) all become kilopascal (kPa) in the metric system of pressure measurement. One psi (English) is equal to 6.895 kPa (metric). Conversely, one kPa (metric) is equal to 0.145 psi (English).

The English terms *psi* and *psig* reference zero pressure as atmospheric pressure. On gauges calibrated psi or psig, all pressures above atmospheric are scaled accordingly: psi or psig. All pressures below atmospheric are scaled in inches of mercury (in. Hg or "Hg).

METRIC TEMPERATURE-PRESSURE CHART					
Low-Side Pressure kPa		Temperature °C	High Side Pressure kPa		Temperature °C
Absolute	Gauge		Absolute	Gauge	
175	74	−15	910	809	19
195	94	−13	942	831	20
210	109	−11	970	869	21
230	129	−9	1 010	909	22
245	144	−7	1 030	929	23
260	159	−5	1 070	969	24
280	179	−3	1 122	1 021	25
295	194	−1	1 175	1 074	26
308	207	0	1 210	1 109	27
320	219	1	1 245	1 144	28
340	239	3	1 275	1 174	29
363	262	5	1 308	1 207	30
385	284	7	1 340	1 239	31
420	319	9	1 375	1 274	32
437	336	11	1 390	1 289	33
480	379	13	1 405	1 304	34
510	409	16	1 446	1 345	35
540	439	18	1 515	1 414	36

FIGURE 18–13 Metric temperature–pressure chart for CFC–12.

Temperature °C	Pressure kPa	Temperature °C	Pressure kPa
−15.0	63	5.0	247
−12.5	83	7.5	280
−10.0	103	10.0	313
−7.5	122	12.5	345
−5.0	142	15.0	381
−4.5	147	17.5	422
−4.0	152	20.0	465
−3.5	157	22.5	510
−3.0	162	25.0	560
−2.5	167	27.5	616
−2.0	172	30.0	670
−1.5	177	32.5	726
−1.0	182	35.0	785
−0.5	187	37.5	849
0.0	192	40.0	916
0.5	198	42.5	990
1.0	203	45.0	1066
1.5	209	47.5	1146
2.0	214	50.0	1230
2.5	220	52.5	1315
3.0	225	55.0	1385
3.5	231	57.5	1480
4.0	236	60.0	1580
4.5	242	65.0	1795

FIGURE 18–14 HFC–134a temperature-pressure chart (metric).

The English term *psia* references zero pressure as absolute zero. All pressures above and below atmospheric pressure are scaled in psia. The chart in figure 18–15 compares psig, psia, kPa gauge, and kPa absolute.

The metric term *kPa* may reference zero pressure either as absolute or as atmospheric (gauge). The term must be qualified, however. The term "absolute" must be used when applicable. If this term is not used, the reference to kPa is assumed to be "gauge." The kPa absolute scale references zero pressure as absolute zero. On this scale, sea level atmospheric pressure is 101.3 kPa absolute. This value is usually rounded off to 100 kPa. The kPa gauge scale, on the other hand, references atmospheric pressure as zero. For comparison, see the chart in Figure 18–15.

It must be noted, as discussed in Chapter 3, that the kilopascal (kPa) is not used in negative terms. Therefore, compound gauges in the metric system of measure must be calibrated in absolute values. Pressure gauges, on the other hand, may be calibrated in either scale: absolute or atmospheric.

To provide experience in both atmospheric (gauge) and absolute values, the gauges in this section are scaled as follows.

Compound Gauge

The compound gauge, Figure 18–16, includes both an English (30 in. Hg to 120 psig) atmospheric (gauge) scale and a metric (0 to 900 kPa) absolute scale, with retard to 250 psig and 1,800 kPa. The scale range from 120 psig to 250 psig is referred to as "retard." This means that pressures as high as 250 psig may be applied without affecting the accuracy of the gauge. English pressures below atmospheric are given in inches of mercury (in. Hg). Metric pressures below atmospheric are given in kilopascals absolute (kPa absolute). English pressures above atmospheric are given in pounds per square inch gauge (psig). Metric pressures above atmospheric may be given in kPa gauge or absolute.

The **English scale** below atmospheric pressure is calibrated in 5 in. Hg divisions with 15 in. Hg and 30 in. Hg major divisions. Above atmospheric pressure major divisions are 10 psig through 120 psig. Minor divisions of the English scale are 5 psig.

Major divisions of the metric scale are 100 kPa from absolute, with minor divisions of 25 kPa. The retard range of the metric scale is from 900 kPa to 1,800 kPa. There is no accuracy, however, in the retard range of the gauge.

Pressure Gauge

The pressure gauge, Figure 18–17, includes a psi or psig English scale and a kPa gauge metric scale. Both English and metric values are given, using atmospheric pressure as zero reference.

Pounds Per Square Inch		Kilopascal	
Gauge	Absolute*	Gauge	Absolute*
30"Hg	0	-	0
15"Hg	7	-	51
0	15	0	101
15	30	103	205
30	45	207	308
45	60	310	412
60	75	414	515
75	90	517	618
100	115	690	791
150	165	1 034	1 136
200	215	1 379	1 480
250	265	1 724	1 825
300	315	2 069	2 170
400	415	2 758	2 859
500	515	3 448	3 549

*To nearest whole number

FIGURE 18–15 Comparison of psi gauge and absolute with kilopascal gauge and absolute.

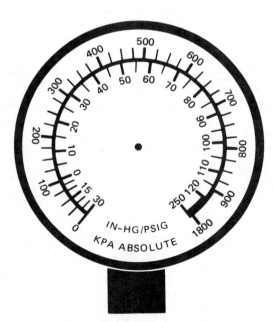

FIGURE 18–16 Comparison of psi gauge and absolute with kilopascal gauge and absolute.

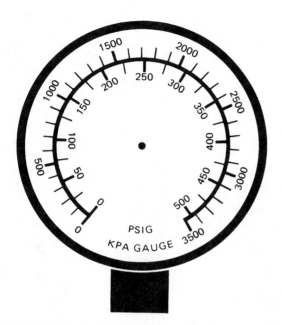

FIGURE 18–17 Typical high–side (pressure) gauge, with English (gauge) and metric (gauge) scales.

The English scale major divisions are 50 psi, with minor divisions of 25 psi. Major divisions of the metric scale are 500 kPa, with minor divisions of 100 kPa.

Conversions

To convert the English psig scale to psia, add 14.69, usually rounded off to 14.7 or 15, to the psig reading. To convert the metric gauge scale to absolute, add 101.3, usually rounded off to 101 or 100, to the kPa gauge reading. To convert psig to kPa gauge, multiply the psig reading by 6.895. The same **multiplier** may be used to convert psia to kPa absolute. To convert psig to kPa absolute, first convert psig, then apply the multiplier.

SYSTEM DIAGNOSIS

The following sixteen exercises in system diagnostics may be studied to become familiar with the troubleshooting techniques, using a manifold and gauge set to determine air-conditioning system problems.

For the following system diagnosis exercises, it will be necessary to refer to the appropriate temperature-pressure chart, Figure 18-11 through Figure 18-14, found on pages 404 and 405. A self-stick note tab on the appropriate page will make it easier to reference these figures while performing the exercises.

SYSTEM DIAGNOSIS 1
THE COMPRESSOR—TXV OR FOT SYSTEM

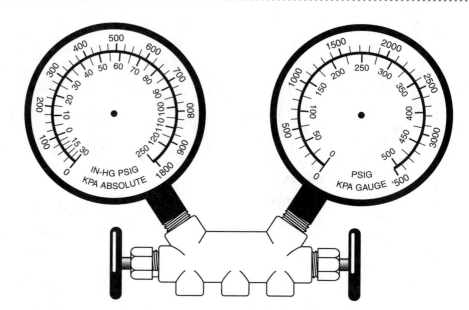

CONDITIONS

Ambient temperature: 90°F (32.2°C)
Low-side gauge: 50 psig (345 kPa)
High-side gauge: 120 psi (827 kPa)

FIGURE 18–18 System diagnosis 1: the compressor–cycling clutch system.

EXERCISE

1. Show the high- and low-side readings on the gauges in the diagram.

DIAGNOSIS

1. What should the normal high-side reading be:

    a. For CFC-12? _____ psig _____ kPa

    b. For HFC-134a? _____ psig _____ kPa

2. What is the evaporator temperature in this problem:

    a. For CFC- 12? _____ °F _____ °C

    b. For HFC-134a? _____ °F _____ °C

3. A low-side reading of 50 psig (345 kPa) is *high/low.*

4. A high-side reading of 120 psi (827 kPa) is *high/low.*

5. This condition results in *good/poor/no* cooling from the evaporator.

6. An internal _____ of the compressor is indicated by these conditions.

7. To correct this condition, a new _____ and/or _____ must be installed.

8. This condition is generally caused by excessive _____ .

SYSTEM DIAGNOSIS 2
THE CONDENSER—CYCLING CLUTCH TXV OR FOT SYSTEM

CONDITIONS

Ambient Temperature: 95°F (35°C)
Low-side gauge: 55 psig (379 kPa)
High-side gauge: 300 psi (2,069 kPa)

EXERCISE

1. Show the high- and low-side gauge readings on the gauges in the diagram.

DIAGNOSIS

1. What should the normal high-side gauge reading be:

   a. For a CFC-12 system?

   _____ psig _____ kPa

   b. For an HFC-134a system? _____ psig _____ kPa

2. What is the evaporator temperature in this problem:

   a. For a CFC-12 system? _____ °F _____ °C

   b. For an HFC-134a system? _____ °F _____ °C

3. A low-side reading of 55 psig (379 kPa) is *high/low.*

4. A high-side reading of 300 psi (2,069 kPa) is *high/low.*

5. This condition results in *good/poor/no* cooling from the evaporator.

6. Give two conditions outside the air-conditioning system that can cause this pressure.

   a. _____

   b. _____

7. Give two conditions inside the air-conditioning system that can cause this pressure.

   a. _____

   b. _____

8. Give one type of damage that can occur by operating the air conditioner with this pressure.

   a. _____

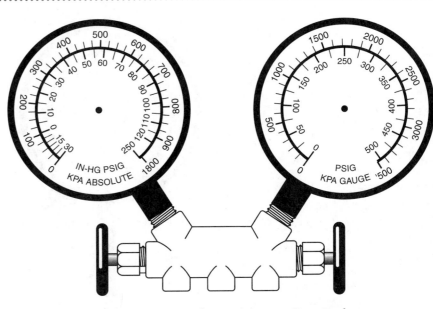

FIGURE 18–19 System diagnosis 2: the condenser–cycling clutch system.

SYSTEM DIAGNOSIS 3
THE RECEIVER-DRIER–CYCLING CLUTCH TXV SYSTEM

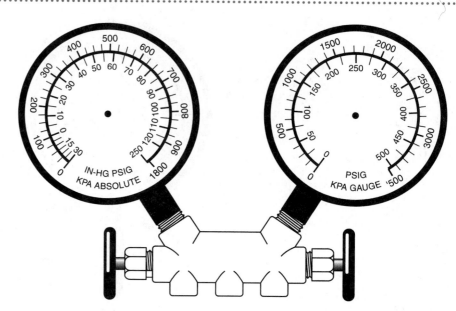

CONDITIONS

Ambient temperature: 100°F (37.8°C)
Low-side gauge: 5 psig (35 kPa)
High-side gauge: 305 psi (2,103 kPa)

EXERCISE

1. Show the high- and low-side gauge readings on the gauges in the diagram.

FIGURE 18–20 System diagnosis 3: the receiver-drier–cycling clutch TXV system.

DIAGNOSIS

1. What should the normal high-side reading be:

   a. For a CFC-12 system? _____ psig _____ kPa

   b. For an HFC-134a system? _____ psig _____ kPa

2. What is the evaporator temperature in this problem:

   a. For a CFC-12 system? _____ °F _____ °C

   b. For an HFC-134a system? _____ °F _____ °C

3. Explain: _____

4. A low-side reading of 5 psig (35 kPa) is *high/low.*

5. A high-side reading of 305 psi (2,103 kPa) is *high/low.*

6. This condition results is *good/poor/no* cooling from the evaporator.

7. A restriction at the _____ is indicated by these readings.

8. Frosting is likely to occur at the point of _____.

9. How can this system be repaired? _____

   _____

SYSTEM DIAGNOSIS 4
THE ACCUMULATOR-CYCLING CLUTCH FOT SYSTEM

CONDITIONS

Low-side service valve located on the
    accumulator
Ambient temperature: 95°F (35°C)
Low-side gauge: 45 psig (310 kPa)
High-side gauge: 165 psig (1,138 kPa)

EXERCISE

1. Show the high- and low-side
   gauge readings on the gauges
   in the diagram.

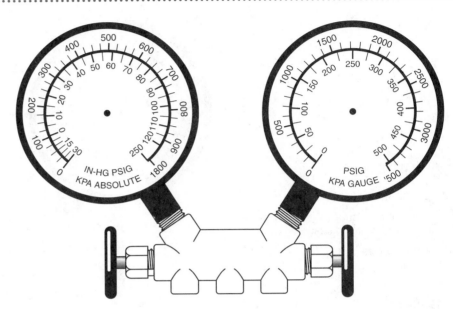

FIGURE 18–21 System diagnosis 4: the accumulator-drier–cycling clutch system.

DIAGNOSIS

1. What should the normal high-side gauge reading be:

   a. For a CFC-12 system? _____ psig _____ kPa

   b. For an HFC-134a system? _____ psig _____ kPa

2. What is the evaporator temperature in this problem:

   a. For a CFC-12 system? _____ °F _____ °C

   b. For an HFC-134a system? _____ °F _____ °C

3. Explain: _____

4. A low-side reading of 45 psig (310 kPa) is *high/normal/low.*

5. A high-side reading of 165 psig (1,138 kPa) is *high/normal/low.*

6. This condition results in *good/poor/no* cooling.

7. A restriction in the _____ is indicated by these readings.

8. Frosting is likely to occur at the point of_____.

9. How can this system be repaired? _____

   _____

SYSTEM DIAGNOSIS 5
THE ACCUMULATOR-CYCLING CLUTCH FOT SYSTEM

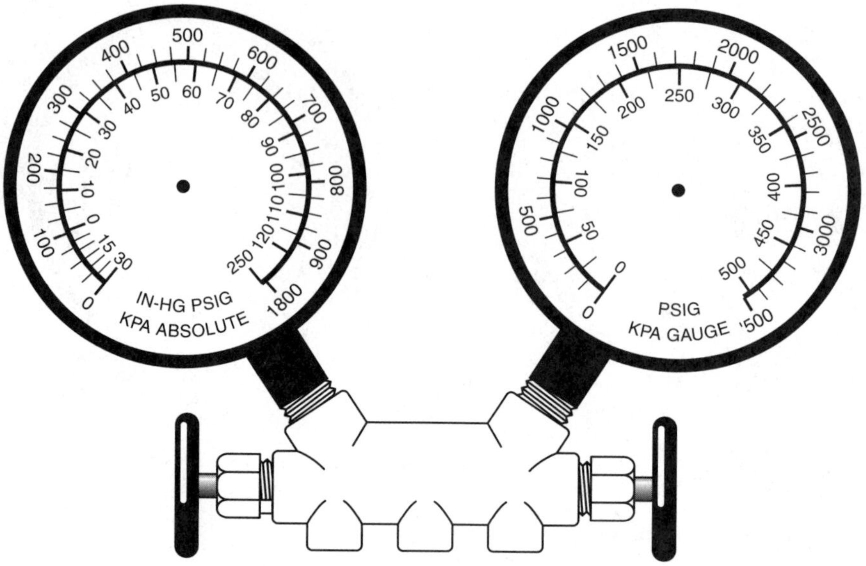

CONDITIONS

Low-side service valve located
    downstream of the accumulator
Ambient temperature: 95°F (35°C)
Low-side gauge: 5 psig (35 kPa)
High-side gauge: 165 psig (1,138 kPa)

EXERCISE

1. Show the high- and low-side
   gauge readings on the gages in
   the diagrams.

FIGURE 18–22 System diagnosis 5: the accumulator-drier–cycling clutch system.

DIAGNOSIS

1. What should the normal high-side reading be:

   a. For a CFC-12 system? _____ psig _____ kPa

   b. For an HFC-134a system? _____ psig _____ kPa

2. What is the evaporator temperature in this problem:

   a. For a CFC-12 system? _____ °F _____ °C

   b. For an HFC-134a system? _____ °F _____ °C

3. A low-side reading of 5 psig (35 kPa) is *high/normal/low.*

4. A high-side reading of 165 psig (1,138 kPa) is *high/normal/low.*

5. This condition results in *good/poor/no* cooling.

6. A restriction in the _____ is indicated by these readings.

7. Frosting is likely to occur at the point of _____.

8. How can this system be repaired? _____

   _____

SYSTEM DIAGNOSIS 6
THERMOSTATIC EXPANSION VALVE-CYCLING CLUTCH TXV SYSTEM

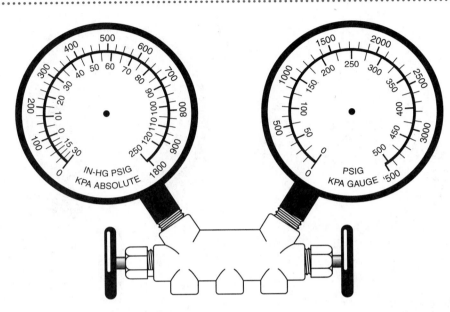

CONDITIONS

Ambient temperature: 95°F (35°C)
Low-side gauge: 2 psig (14 kPa)
High-side gauge: 170 psi (1,172 kPa)

EXERCISE

1. Show the high- and low-side manifold readings on the gauges in the diagram.

FIGURE 18–23 System diagnosis 6: the thermostatic expansion valve–cycling clutch system.

DIAGNOSIS

1. What should the normal high-side reading be:

   a. For a CFC-12 system? _____ psig _____ kPa

   b. For an HFC-134a system? _____ psig _____ kPa

2. What is the evaporator temperature in this problem:

   a. For a CFC-12 system? _____ °F _____ °C

   b. For an HFC-134a system? _____ °F _____ °C

3. Explain: _____

4. A low-side reading of 2 psig (14 kPa) is *high/normal/low.*

5. A high-side reading of 170 psi (1,172 kPa) is *high/normal/low.*

6. This condition results in *good/poor/no* cooling from the evaporator.

7. This condition indicates a *starved/flooded* evaporator due to a defective _____.

8. This condition is usually accompanied by frosting at the _____ of the _____ valve.

9. How can this condition be corrected? _____

   _____

SYSTEM DIAGNOSIS 7
THERMOSTATIC EXPANSION VALVE-CYCLING CLUTCH TXV SYSTEM

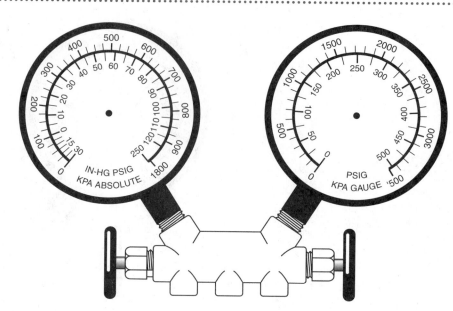

CONDITIONS

Ambient temperature: 95°F (35°C)
Low-side gauge: 55 psig (379 kPa)
High-side gauge: 160 psi (1,103 kPa)

EXERCISE

1. Show the high- and low-side
   manifold readings on the
   gauges in the diagram.

DIAGNOSIS

FIGURE 18–24 System diagnosis 7: the thermostatic expansion valve–cycling clutch system.

1. What should the normal
   high-side reading be:

   a. For a CFC-12 system? _____ psig _____ kPa

   b. For an HFC-134a system? _____ psig _____ kPa

2. What is the evaporator temperature in this problem:

   a. For a CFC-12 system? _____ °F _____ °C

   b. For an HFC-134a system? _____ °F _____ °C

3. A low-side reading of 55 psig (379 kPa) is *high/normal/low.*

4. A high-side reading of 160 psi (1,103 kPa) is *high/normal/low.*

5. This condition results in *good/poor/no* cooling from the evaporator.

6. This condition indicates a *starved/flooded* evaporator due to a malfunctioning _____.

7. Give two possible causes for this malfunction.

   a. _____

   b. _____

8. Can moisture in the system cause this malfunction? *Yes/No.*

9. Explain: _____

SYSTEM DIAGNOSIS 8

THE ORIFICE TUBE-CYCLING CLUTCH FOT SYSTEM

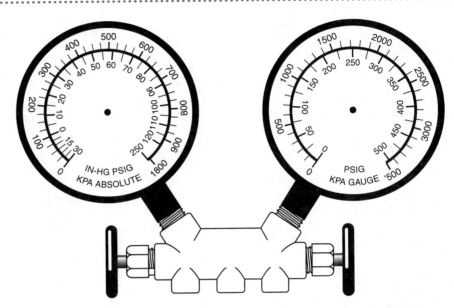

CONDITIONS

Ambient temperature: 95°F (35°C)
Low-side gauge: 3 psig (21 kPa)
High-side gauge: 172 psig (1,186 kPa)

EXERCISE

1. Show the high- and low-side gauge readings on the guages i n the diagram..

DIAGNOSIS

FIGURE 18–25 System diagnosis 8: the fixed orifice tube–cycling clutch system.

1. What should the normal high-side reading be:

   a. For a CFC-12 system? _____ psig _____ kPa

   b. For an HFC-134a system? _____ psig _____ kPa

2. What is the evaporator temperature:

   a. For a CFC-12 system? _____ °F _____ °C

   b. For an HFC-134a system? _____ °F _____ °C

3. Explain: _____

4. A low-side reading of 3 psig (21 kPa) is *high/normal/low*.

5. A high-side reading of 172 psig (1,186 kPa) is *high/normal/low*.

6. This condition results in *good/poor/no* cooling.

7. This condition indicates a *starved/flooded* evaporator and / or a clogged _____.

8. This condition is usually accompanied by frosting at the

   _____

9. How can this condition be corrected? _____

   _____

SYSTEM DIAGNOSIS 9
THE THERMOSTAT-CYCLING CLUTCH TXV OR FOT SYSTEM

CONDITIONS

Ambient temperature: 97°F (36°C)
Low-side gauge: 10 psig (69 kPa)
High-side gauge: 205 psi (1,413 kPa)

EXERCISE

1. Show the high- and low-side
   gauge readings on the gauges
   in the diagram.

DIAGNOSIS

1. A high-side pressure of 205
   psi (1413 kPa) is
   high/normal/low.

2. A low-side pressure of
   10 psig (69 kPa) is:
   high/normal/low.

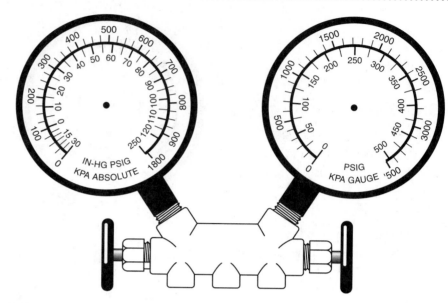

FIGURE 18–26 System diagnosis 9: the control thermostat–cycling clutch system.

3. What is the evaporator temperature in this problem:

   a. For a CFC-12 system? _____ °F _____ °C

   b. For an HFC-134a system? _____ °F _____ °C

4. This condition results in *good/poor/no* cooling from the evaporator.

5. This condition can be accompanied by frosting of the _____ , which blocks off airflow.

6. List two possible causes of a malfunctioning thermostat that can result in this problem.

   a. _____

   b. _____

7. Can the customer unintentionally cause this problem? *Yes/No*

   How/why? _____

8. List and describe two types of control thermostats.

   a. _____

   b. _____

9. Are all thermostats adjustable? *Yes/No*

   Explain: _____

SYSTEM DIAGNOSIS 10
THE THERMOSTAT-CYCLING CLUTCH TXV OR FOT SYSTEM

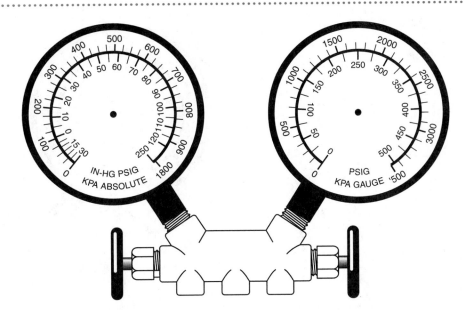

CONDITIONS

Ambient temperature: 98°F (37°C)
Low-side gauge: 60 psig (414 kPa)
High-side gauge: 210 psig (1,448 kPa)

EXERCISE

1. Show the high- and low-side gauge readings on the gauges in the diagram.

FIGURE 18–27 System diagnosis 10: the control thermostat–cycling clutch system.

DIAGNOSIS

1. A high-side pressure of 210 psig (1,448 kPa) is *high/normal/low*.

2. A low-side pressure of 60 psig (414 kPa) is *high/normal/low*.

3. What is the evaporator temperature:

    a. For a CFC-12 system? _____ °F _____ °C

    b. For an HFC-134a system? _____ °F _____ °C

4. This condition results in *good/poor/no* cooling from the evaporator.

5. Give two possible causes for this malfunction.

    a. _____

    b. _____

6. Can the customer unintentionally cause this problem? *Yes/No*

    How / why? _____

7. Are all thermostats adjustable? *Yes/No*

    Explain: _____

SYSTEM DIAGNOSIS 11
THE SYSTEM-CYCLING CLUTCH TXV OR FOT SYSTEM

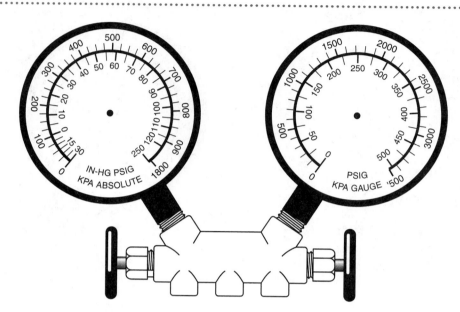

FIGURE 18–28 System diagnosis 11: the cycling clutch system.

CONDITIONS

Ambient temperature: 90°F (32°C)
Low-side gauge: 80 psig (552 kPa)
High-side gauge: 80 psi (552 kPa)

EXERCISE

1. Show the high- and low-side gauge readings on the gauges in the diagram.

DIAGNOSIS

1. What should the normal high-side gauge reading be:

   a. For a CFC-12 system? _____ psig _____ kPa

   b. For an HFC-134a system? _____ psig _____ kPa

2. The high-side gauge is *high/normal/low.*

3. The low-side gauge is *high/normal/low.*

4. List three possible problems with this system.

   a. _____

   b. _____

   c. _____

SYSTEM DIAGNOSIS 12
THE SYSTEM-CYCLING CLUTCH TXV OR FOT SYSTEM

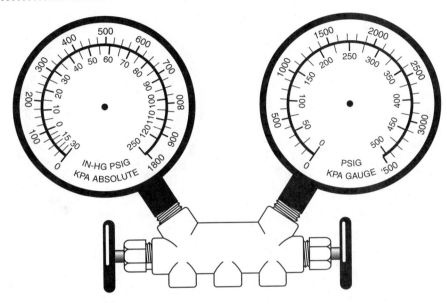

FIGURE 18–29 System diagnosis 12: the cycling clutch system.

CONDITIONS

Ambient temperature: 80°F (27°C)
Low-side gauge: 20 psig (138 kPa)
High-side gauge: 155 psig (1,069 kPa)

EXERCISE

1. Show the high- and low-side gauge readings on the gauge set in the diagram.

DIAGNOSIS

1. The low-side gauge is *low/normal/high*.

2. The high-side gauge is *low/normal/high*.

3. List four possible problems with this system.

   a. _____

   b. _____

   c. _____

   d. _____

SYSTEM DIAGNOSIS 13

THE SYSTEM-CYCLING CLUTCH TXV OR FOT SYSTEM

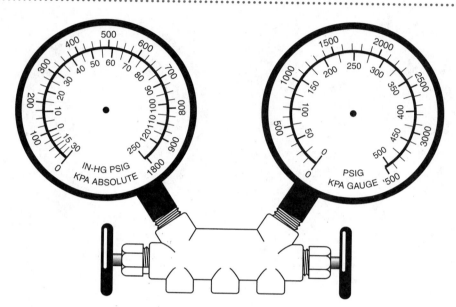

FIGURE 18–30 System diagnosis 13: the cycling clutch system.

CONDITIONS

Ambient temperature: 83°F (28°C)
Low-side gauge: 37 psig (255 kPa)
High-side gauge: 160 psi (1,103 kPa)

EXERCISE

1. Show the gauge readings on the gauge set in the diagram.

DIAGNOSIS

1. What is the evaporator temperature in this problem:

   a. For a CFC-12 system? _____ °F _____ °C

   b. For an HFC-134a system? _____ °F _____ °C

2. The evaporator temperature is *low/normal/high.*

3. List four possible causes for this malfunction.

   a. _____

   b. _____

   c. _____

   d. _____

SYSTEM DIAGNOSIS 14
THE SYSTEM-CYCLING CLUTCH TXV OR FOT SYSTEM

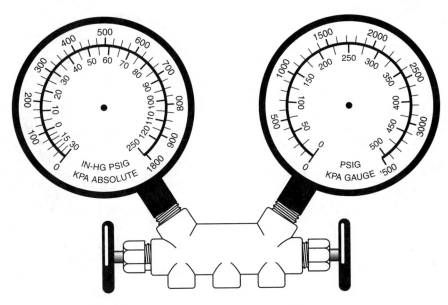

FIGURE 18–31 System diagnosis 14: the cycling clutch system.

CONDITIONS

Ambient temperature: 90°F(32°C)
Low-side gauge: 50 psig (345 kPa)
High-side gauge: 170 psi (1,172 kPa)

EXERCISE

1. Show the gauge readings on the gauge set in the diagram.

DIAGNOSIS

1. The high-side pressure is *high/normal/low.*

2. List four possible causes for this malfunction.

   a. _____

   b. _____

   c. _____

   d. _____

SYSTEM DIAGNOSIS 15
THE SYSTEM-CYCLING CLUTCH TXV OR FOT SYSTEM

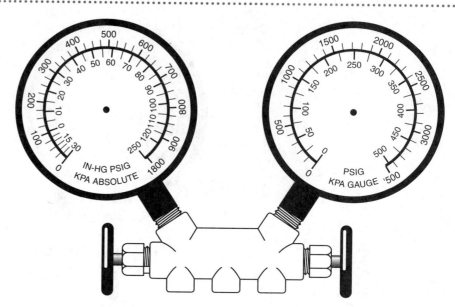

FIGURE 18–32 System diagnosis 15: the cycling clutch system.

CONDITIONS

Ambient temperature: 95°F (35°C)
Low-side gauge: 37 psig (255 kPa)
High-side gauge: 250 psi (1,724 kPa)

EXERCISE

1. Show the gauge readings on the gauge set in the diagram.

DIAGNOSIS

1. The low-side gauge is *high/normal/low.*

2. The high-side gauge is *high/normal/low.*

3. List two possible causes for this malfunction.

   a. _____

   b. _____

4. An *(undercharge/overcharge)* of refrigerant can cause this problem.

SYSTEM DIAGNOSIS 16
THE SYSTEM-CYCLING CLUTCH TXV OR FOT SYSTEM

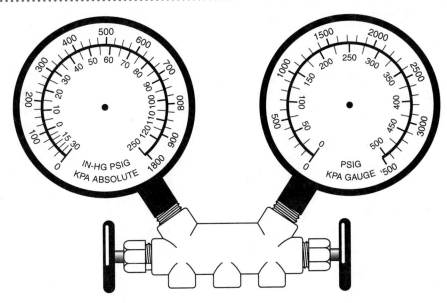

FIGURE 18–33 System diagnosis 16: system-cycling clutch TXV or FOT system.

CONDITIONS

Ambient temperature: 96°F (36°C)
Low-side gauge: 39 psig (269 kPa)
High-side gauge: 325 psi (2,241 kPa)

EXERCISE

1. Show the gauge readings on the gauge set in the diagram.

DIAGNOSIS

1. List six possible causes of excessive head pressure.

    a. _____

    b. _____

    c. _____

    d. _____

    e. _____

    f. _____

2. High head pressure is *always/not always* accompanied by high suction pressure.

    Explain: _____

COMPONENT REPLACEMENT

The following procedures are typical for step-by-step replacement of air-conditioning system components. For specific replacement details, refer to the shop service manual for the particular year or model of the unit.

Tools

• Manifold and gauge set • Service hose set • Recovery unit • Vacuum pump • Hand tools, as required

Materials

• Refrigerant • Refrigeration lubricant • Gaskets and/or O-rings • Components and parts, as required

PROCEDURE

The following procedures may be considered typical. Follow the instructions given in the appropriate manufacturer's service manual for specific applications.

Preparation

1. Recover all of the refrigerant from the air-conditioning system.
2. Disconnect the battery (ground cable), Figure 18–34.
3. Locate the component that is to be removed for repair or replacement.
4. Remove the necessary access panel(s) or other hardware to gain access to the component.

Remove the Component

1. For a thermostatic expansion valve:
   a. Remove the insulation and clamp to free the remote bulb.
   b. Disconnect the external equalizer, if so equipped.
   c. Remove the liquid line from the inlet of the TXV.
   d. Remove the evaporator inlet fitting from the outlet of the TXV.
   e. Remove the holding clamp (if provided on the TXV) and carefully lift the TXV from the evaporator. Do not damage the **remote bulb** or capillary tube.
2. For an expansion tube: see Chapter 12.
3. For an accumulator:
   a. Remove the accumulator inlet fitting.
   b. Remove the accumulator outlet fitting.
   c. Remove the bracket and/or attaching screws.
   d. Lift the accumulator from the vehicle, Figure 18–35.
4. For a compressor:
   a. Remove the inlet and outlet hoses or the service valves from the compressor.
   b. Remove the clutch lead wire(s).
   c. Loosen and remove the belt(s).
   d. Remove the mounting bolts from the compressor brackets and braces.
   e. Lift the compressor from the vehicle, Figure 18–36.
5. For a condenser:
   a. Remove the inlet and outlet hoses.
   b. Remove the mounting hardware.
   c. Lift the condenser from the vehicle.
6. For an evaporator:
   a. Aftermarket (add-on) unit:
      1. Remove the inlet hose from the TXV.
      2. Remove the suction line from the evaporator.
      3. Disconnect the electrical lead wire(s).
      4. Remove the mounting hardware.
      5. Lift the evaporator from the vehicle.
   b. Factory-installed unit:

FIGURE 18–34 Carefully disconnect the battery cable.

FIGURE 18–35 Remove the accumulator from the vehicle.

FIGURE 18–36 Remove the compressor from the vehicle.

1. Remove the inlet hose from the TXV, H-valve, or orifice tube, as required.
2. Remove the outlet hose from the evaporator or accumulator, as required.
3. Remove any mechanical linkage or vacuum line(s) from the evaporator controls.
4. Remove the mounting bolts and hardware from the evaporator housing, Figure 18–37.
5. Carefully lift the evaporator assembly from the vehicle. Do not force the assembly.
7. For a receiver-drier:
   a. Remove the low-pressure switch wire, if applicable.
   b. Remove the inlet and outlet hoses from the drier.
   c. Remove the mounting hardware.
   d. Lift the receiver-drier from the vehicle.
   e. Remove the low-pressure switch from the drier, if so equipped.

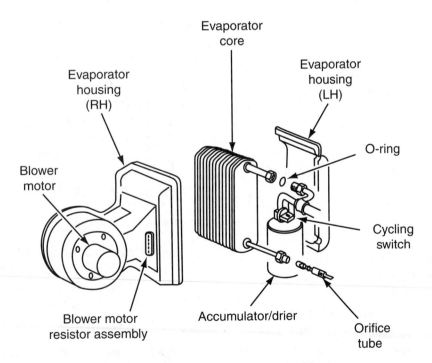

FIGURE 18–37 Exploded view of an evaporator assembly.

Component Replacement

1. Use only components designated for CFC-12 or HFC-134a, as applicable.
2. Use new gaskets and/or O-rings.
3. Coat all connections, gaskets, and O-rings liberally with clean mineral oil or O-ring lubricant before reassembly.
4. For reassembly, reverse the removal procedure.

Return the System to Service

1. Replace any access panels, clamps, or other hardware that were previously removed.
2. Carefully reconnect the battery ground cable.
3. Evacuate the air-conditioning system.
4. Charge the system with CFC-12 or HFC-134a, as applicable.
5. Conduct a performance test as required.

HEATING AND COOLING SYSTEM

The following procedure is typical for the replacement of engine cooling system and heater components. For specific details, refer to the manufacturer's shop service manuals for the particular year or model vehicle being serviced.

PROCEDURE

The following procedures are typical. Always refer to specific manufacturer's instructions when servicing a motor vehicle.

> NOTE: Antifreeze solution is considered to be a hazardous material. *Do not* dispose of antifreeze in any way other than in an environmentally safe manner. Refer to local ordinances and regulations.

Preparation

1. Ensure that the engine is cool and the cooling system is not under pressure.
2. Open the radiator drain provision, Figure 18–38, and drain the cooling system.

> NOTE: If the coolant is to be reused, drain it into a clean container, Figure 18–39. If, however, it is not to be reused, then it must be disposed of or recycled in a manner considered to be environmentally safe. Refer to local ordinances and regulations regarding the disposal of ethylene glycol type of antifreeze solutions.

FIGURE 18–38 Open the drain cock.

FIGURE 18–39 If coolant is to be reused, drain it into a clean container.

3. Locate the component to be removed.
4. Remove the necessary fasteners, panels, wiring, or hardware to gain access to the component.

> CAUTION
> DISCONNECT THE BATTERY GROUND CABLE IF YOU ARE WORKING UNDER THE DASH. ALSO, TAKE EXTREME CAUTION NOT TO INFLATE AN AIR BAG. REFER TO MANUFACTURER'S SPECIFICATIONS AND PRECAUTIONS *BEFORE* WORKING UNDER THE DASH.

Remove the Component

1. For a temperature sending unit:
   a. Disconnect the wire(s) from the sending unit.
   b. Remove the sending unit, using the proper wrench. This is usually a special wrench.

> NOTE: It is not always necessary to drain the cooling system to replace the temperature sending unit.

2. For a heater control valve:
   a. Remove the cable linkage or vacuum hose(s) from the control valve, Figure 18–40.
   b. Loosen the hose clamps and remove the hoses from the control valve.
   c. Remove the heater control.
   d. Inspect the hose ends. If they are hard or split, cut 0.5 inch (12.7 mm) from the damaged ends.
3. For heater hose(s):

   NOTE: It is good practice to replace all heater hoses if one is found to be defective.

   a. Remove the hose clamps from both ends of the hose.
   b. Remove the hose. Do not use unnecessary force when removing the hose end from the heater core.
4. For radiator hose(s):

   NOTE: It is good practice to replace all radiator hoses if one is found to be defective, Figure 18–41.

   a. Remove the hose clamps from both ends of the hose.
   b. Remove the hose. Do not use unnecessary force when removing the hose end from the radiator.
5. For a thermostat:
   a. Remove the bolts holding the thermostat housing onto the engine, Figure 18–42. It is not necessary to remove the radiator hose from the housing.
   b. Lift off the thermostat housing.

NOTE: Observe the pellet-side down position of the thermostat to ensure proper replacement. Do not reinstall the thermostat backward.

   c. Lift out the thermostat, Figure 18–43.
   d. Clean all of the old gasket material from the thermostat housing and engine mating surfaces.
6. For a radiator:
   a. If the fan is equipped with a shroud, remove the attachments and slide the shroud toward the engine.
   b. Carefully remove the upper and lower radiator hoses from the radiator, Figure 18–44.
   c. Remove the transmission cooler lines (if equipped with an automatic transmission) from the radiator. Plug the lines to prevent transmission fluid loss.

FIGURE 18–41 A hose that is defective on the inside. *(Courtesy of The Gates Rubber Company)*

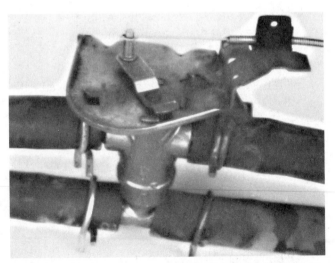

FIGURE 18–40 A cable–operated heater control valve. *(Courtesy of BET, Inc.)*

FIGURE 18–42 Remove the bolts holding the thermostat housing.

FIGURE 18–43 Lift out the thermostat.

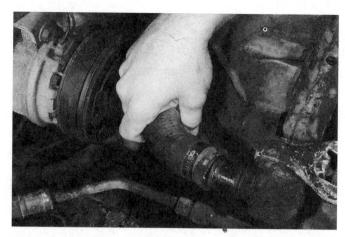

FIGURE 18–44 Remove the hoses.

d. Remove the radiator attaching bolts and brackets.
e. Carefully lift out the radiator.

NOTE: When removing the radiator, take care not to damage the delicate fins of the core.

7. For a heater core:

NOTE: It is often necessary to follow the manufacturer's service manual procedures for heater core service.

   a. Remove the access panel(s) or the split heater/air-conditioning case to gain access to the heater core.
   b. Remove the heater coolant hoses.
   c. Remove the cable and/or vacuum control lines (if so equipped).
   d. Remove the heater core securing brackets and/or clamps.
   e. Lift the core from the case, Figure 18–45. Do not use force. Take care not to damage the fins of the heater core.

8. For a water pump:

NOTE: It is often necessary to remove such accessories as the power-steering pump, the air-conditioning compressor, the alternator, the air pump, and so forth, to gain access to the water pump. If necessary, refer to the specific manufacturer's service manuals.

   a. Remove the radiator as outlined in step 6, if it is necessary to gain access to the water pump.
   b. Loosen and remove all of the belts.
   c. Remove the fan, fan/clutch assembly, and water pump pulley.

FIGURE 18–45 A typical heater core.

   d. Remove accessories as necessary to gain access to the water pump bolts.
   e. Remove the lower radiator hose from the water pump.
   f. Remove the bypass hose, if so equipped.
   g. Remove the bolts securing the water pump to the engine, Figure 18–46.

NOTE: Note the length of the bolts removed. All of the bolts may not be of the same length.

   h. Tap the water pump lightly to remove it from the engine, if necessary.
   i. Clean the old gasket material from all of the surfaces.

Component Replacement

1. Inspect all of the components removed, such as hoses, clamps, housings, and so forth. Replace any that are defective.

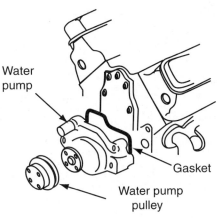

Water pump

Gasket

Water pump pulley

FIGURE 18–46 A typical coolant pump.

2. Use new gaskets where gaskets are required.
3. For reassembly, reverse the removal procedure.

Return to Service

1. Replace any access panels, clamps, or hardware that were previously removed.
2. Reconnect the battery ground cable, if it was previously disconnected.
3. Replace the coolant or install new coolant.
4. Check the cooling system for leaks.

TROUBLESHOOTING THE AIR-CONDITIONING SYSTEM

This procedure is a quick reference for the service technician who needs to isolate the conditions that can cause a noise or create a problem with the air-conditioning system, Figure 18–47. The four parts outline the typical customer complaints.

❑ Part 1 covers a **noisy** system. It suggests the causes of noise and how to correct the problem.
❑ Part 2 covers **intermittent** cooling, which is used as an aid to troubleshoot the customer complaint "Sometimes it works; sometimes it does not." Suggested causes and corrections are given.

❑ The common customer complaint "It just is not cool enough" is covered in Part 3. Causes of insufficient cooling and the suggested remedies are given.
❑ The complaint "It does not cool at all," which is the easiest to diagnose, is covered in Part 4. Possible causes and corrections are discussed.

It should be noted that air-conditioning problems may often be caused by, or may cause, cooling system problems. Conversely, cooling system problems may be caused by, or may cause, air-conditioning system problems. Therefore, it is also necessary to refer to "Engine Cooling System," for troubleshooting procedures.

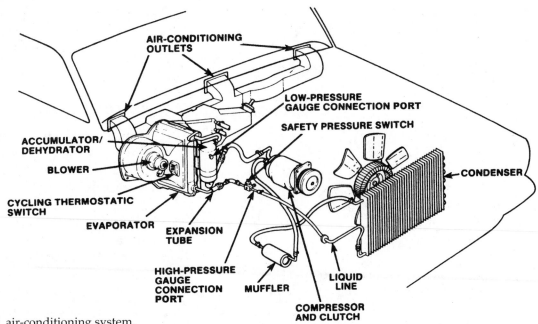

FIGURE 18–47 A typical air-conditioning system.

PROCEDURE

PART 1 — NOISY SYSTEM

Possible Causes

1. Loose electrical connection, causing "clutch chatter"
2. Defective clutch coil
3. Defective clutch
4. Defective clutch bearing(s)
5. Loose belt(s)
6. Broken belt (if two-belt drive)
7. Worn or frayed belt(s)
8. Loose compressor mount
9. Loose compressor brace(s)
10. Broken compressor mount
11. Broken compressor brace(s)
12. Blower fan is rubbing against the case
13. Defective blower motor
14. Defective idler pulley bearing(s)

Possible Corrections

1. Tighten or repair the connection, as necessary.
2. Replace the clutch coil.
3. Replace the clutch.
4. Replace the clutch bearing(s).
5. Tighten the belt(s). Do not overtighten.
6. Replace both belts as a pair.
7. Replace belt(s).
8. Tighten the compressor mount.
9. Tighten the compressor brace(s).
10. Repair/replace the compressor mount.
11. Repair/replace the brace(s).
12. Adjust or reposition the blower fan.
13. Replace the blower motor.
14. Replace the idler pulley assembly or bearing(s), as applicable.

NOISY COMPRESSOR

Possible Causes

1. Overcharge of the refrigerant
2. Undercharge of the refrigerant

3. Overcharge of the lubricant
4. Undercharge of the lubricant
5. Moisture in the system

6. Defective compressor

Possible Corrections

1. Adjust the refrigerant charge.
2. Locate and repair the leak. Evacuate and recharge the system.
3. Remove excess lubricant.
4. Locate and repair the leak Replace the lubricant.
5. Recover the refrigerant. Replace the receiver- or accumulator-drier, and evacuate and recharge the system.
6. Repair or replace the compressor.

PART 2 — SYSTEM COOLS INTERMITTENTLY

Possible Causes

1. Defective circuit breaker
2. Circuit breaker trips on overload
3. Loose wiring
4. Defective blower speed control
5. Defective blower speed resistors
6. Defective blower motor
7. Defective clutch coil
8. Loose belt(s)
9. Loose ground connection at the blower
10. Loose clutch coil ground at the connection
11. Clutch slipping; low voltage

Possible Corrections

1. Replace the circuit breaker.
2. Correct the cause of excessive current.
3. Repair or replace the wiring.
4. Replace the control (switch).
5. Replace the resistor block.
6. Replace the blower motor.
7. Replace the clutch coil.
8. Tighten the belt(s). Do not overtighten.
9. Tighten or repair the blower ground.
10. Tighten or repair the connection.
11. Determine the cause and correct.

12. Clutch slipping; excessive wear
13. Improperly adjusted thermostat
14. Defective thermostat
15. Defective low-pressure control
16. Defective high-pressure control
17. Moisture in the system

12. Replace worn clutch part(s).
13. Adjust the thermostat.
14. Replace the thermostat.
15. Replace the low-pressure control.
16. Replace the high-pressure control.
17. Recover the refrigerant. Replace the receiver- or accumulator-drier, and evacuate and recharge the system.

PART 3 — INSUFFICIENT COOLING

Possible Causes

1. "Sluggish" (runs slow) blower motor

2. Clutch slipping; low voltage
3. Clutch slipping; excessive wear
4. Clutch cycles too often

5. Defective thermostat
6. Defective low-pressure control
7. Insufficient airflow from the evaporator

8. Insufficient airflow over the condenser

9. Partially clogged screen in the receiver-drier or accumulator
10. Partially clogged screen in the expansion valve or orifice tube

11. Partially clogged screen in the compressor inlet

12. Loose thermostatic expansion valve remote bulb

13. No insulation on the TXV remote bulb
14. Moisture in the system

15. Excess refrigerant in the system
16. Excess lubricant in the system

17. Partially clogged receiver- or accumulator-drier
18. Defective thermostatic expansion valve
19. Undercharge of the refrigerant
20. Cooling system problem

Possible Corrections

1. Check for loose connections. If there is none, replace the motor.
2. Determine the cause and correct it.
3. Replace worn clutch part(s).
4. Adjust or replace the thermostat. Replace the low-pressure control.
5. Replace the thermostat.
6. Replace the low-pressure control.
7. Clean the evaporator and/or repair sticking or binding "blend" doors.
8. Clean the condenser and/or correct the problem (see Cooling System Problems).
9. Replace the receiver-drier or accumulator.

10. Clean the screen in the expansion valve or replace the orifice tube. Replace the receiver- or accumulator-drier.
11. Clean the screen, determine the cause, and correct as necessary.
12. Clean the contact area and tighten the remote bulb. Wrap it with cork tape.
13. Insulate the bulb with cork tape.
14. Recover the refrigerant. Replace the receiver- or accumulator-drier, and evacuate and recharge the system.
15. Adjust the refrigerant charge.
16. Drain the lubricant to the proper level or change the lubricant.
17. Replace the receiver- or accumulator-drier.
18. Replace the thermostatic expansion valve.
19. Repair the leak. Evacuate and recharge the system.
20. See cooling system remedies.

PART 4 — NO COOLING

Possible Causes

1. Blown fuse
2. Defective circuit breaker
3. Broken electrical wire
4. Disconnected electrical wire
5. Corroded electrical wire (high-resistance connection)
6. Defective clutch coil
7. Defective blower motor
8. Defective thermostat
9. Defective low-pressure control
10. Loose compressor drive belt(s)
11. Broken drive belt(s)
12. Defective compressor suction valve plate(s) and/or gasket(s)
13. Defective compressor discharge valve plate(s) and/or gasket(s)
14. Defective compressor
15. Undercharge or no refrigerant
    a. Leaking compressor shaft seal
    b. Defective hose (leaking)
    c. Leaking **fusible plug**
    d. Refrigerant leak in the system
16. Plugged (clogged) line or hose
17. Clogged inlet screen in the expansion valve
18. Defective thermostatic expansion valve
19. Clogged expansion tube
20. Clogged accumulator-drier
21. Clogged screen in the receiver-drier
22. Excessive moisture in the system

Possible Corrections

1. Correct the problem; replace the fuse.
2. Correct the problem; replace the circuit breaker.
3. Repair or replace the wire.
4. Reconnect the wire.
5. Disconnect, clean and reconnect, or replace the wire.
6. Replace the clutch coil.
7. Replace the blower motor.
8. Replace the thermostat.
9. Replace the low-pressure control.
10. Tighten the belt(s). Do not overtighten.
11. Replace the belt(s).
12. Replace defective parts or the compressor.

13. Replace defective parts or the compressor.

14. Repair or replace the compressor.
15. Locate and repair the leak.
    a. Replace the shaft seal.
    b. Replace the defective hose.
    c. Replace the fusible plug (do not repair).
    d. Locate and correct the leak.
16. Clean or replace the line or hose.
17. Clean the screen and replace the receiver-drier.
18. Replace the expansion valve.
19. Replace the expansion tube.
20. Replace the accumulator-drier.
21. Replace the receiver-drier.
22. Recover the refrigerant. Replace the receiver- or accumulator-drier, and evacuate and charge the system.

ENGINE COOLING SYSTEM

ENGINE COOLING SYSTEM

This procedure is a quick reference for the service technician who needs to isolate many of the conditions that can cause improper engine cooling system and/or heater operation, Figure 18–48. This procedure is given in three parts: Part 1, **engine overcooling;** Part 2, engine overheating; and Part 3, loss of coolant. The customer's probable complaint would be for an overheating condition. If the problem is due to a loss of coolant, the customer may complain that coolant or water (H2O) must be added frequently.

It should be noted that cooling system problems are often caused by, or may cause, air-conditioning system problems. Conversely, air-conditioning problems may be caused by, or may cause, cooling system problems. Therefore, it may be necessary to refer to "Troubleshooting the Air-Conditioning System," while using this procedure.

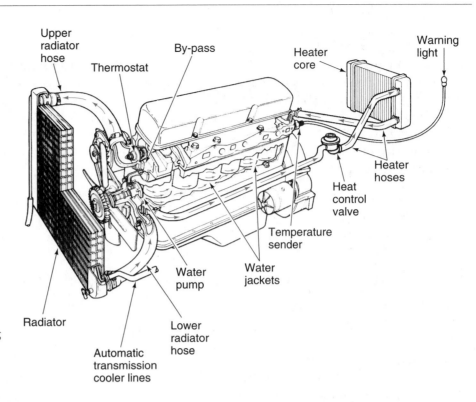

FIGURE 18–48 Typical engine cooling system. *(From Erjavec, Automotive Technology © Delmar, a division of Thomsom Learning)*

PART 1 — ENGINE OVERCOOLING

Possible Causes

1. Missing thermostat
2. Defective thermostat
3. *Defective temperature sending unit
4. *Defective dash gauge (if so equipped)
5. *Broken or disconnected wire (if equipped with a dash gauge unit)
6. *Grounded or shorted "cold" indicator wire (if so equipped with "cold" lamp)

Possible Corrections

1. Install a thermostat and replace the gasket.
2. Replace the thermostat and gasket.
3. Replace the sending unit.
4. Replace the dash gauge.
5. Repair or replace the wire.

6. Repair or replace the wire.

*Symptoms indicating overheating through engine temperature may be within safe limit.

PART 2 — ENGINE OVERHEATING

Possible Causes

1. Collapsed radiator hose
2. Coolant leak
3. Defective water pump
4. Loose fan belt(s)
5. Defective fan belt(s)
6. Broken belt (if two-belt drive)
7. Bent or damaged fan
8. Broken fan
9. Defective **fan clutch**
10. Exterior of radiator is dirty
11. Dirty **"bug" screen**

Possible Corrections

1. Replace the radiator hose.
2. See Part III, "Loss of Coolant."
3. Replace the water pump and gasket.
4. Tighten the fan belt(s). Do not overtighten.
5. Replace the fan belt(s).
6. Replace both belts as a pair.
7. Replace the fan. Do not straighten.
8. Replace the fan.
9. Replace the fan clutch.
10. Clean the radiator.
11. Clean or remove the screen.

12. Damaged radiator
13. Improperly timed engine
14. Out of tune engine
15. *Defective temperature sending unit
16. *Defective dash gauge (if so equipped)
17. *Grounded or shorted wire (if equipped with a dash gauge unit)
18. *Grounded or shorted "hot" indicator wire (if indicator lamp equipped)

12. Repair or replace the radiator.
13. Have the engine timed.
14. Have the engine tuned up.
15. Replace the sending unit.
16. Replace the dash gauge.
17. Repair or replace the wire.
18. Repair or replace the wire.

*Symptoms indicating overheating through engine temperature may be within safe limit.

PART 3 — LOSS OF COOLANT

Possible Cause

1. Ruptured (leaking) radiator hose
2. Ruptured (leaking) heater hose
3. Loose hose clamp
4. Leaking radiator (external)
5. Leaking transmission cooler (internal)
6. Leaking water pump shaft seal
7. Leaking gasket(s)
8. Leaking core plug(s)
9. Loose engine head(s)
10. Warped engine head(s)
11. Excessive coolant

12. Defective radiator pressure cap
13. Incorrect pressure cap
14. Defective thermostat
15. Incorrect thermostat
16. Rust in the cooling system
17. Clogged radiator
18. Leaking heater core
19. Leaking heater shut-off valve

Possible Corrections

1. Replace the radiator hose.
2. Replace the heater hose.
3. Replace and/or tighten the hose clamp.
4. Have the radiator repaired.
5. Have the radiator repaired.
6. Replace the water pump.
7. Replace the gasket(s).
8. Replace all of the core plugs.
9. Torque the head(s) to specifications.
10. Have the head(s) milled flat or replace the heads.
11. Adjust the coolant level. Coolant recovery tank may be installed.
12. Replace the cap.
13. Replace with correct cap.
14. Replace the thermostat.
15. Replace with proper thermostat.
16. Clean the cooling system
17. Have the radiator cleaned.
18. Repair or replace the heater core.
19. Replace the shutoff valve.

REVIEW

Select the correct answer from the choices given.

1. Technician A says that worn clutch parts may cause the compressor clutch to slip under a heavy cooling load. Technician B says that low voltage to the clutch coil may cause the compressor clutch to slip under any cooling load. Who is right?
   a. A only
   b. B only
   c. Both A and B
   d. Neither A nor B

2. All of the following may cause a loss of coolant, EXCEPT:
   a. a missing coolant thermostat.
   b. a defective fan clutch or fan.
   c. a defective radiator pressure cap.
   d. warped head(s) or blown head gasket(s).

3. Compressor noise is being discussed. Technician A says an overcharge of refrigerant will not cause a noisy compressor. Technician B says an overcharge of lubricant may cause a noisy compressor. Who is right?
   a. A only
   b. B only
   c. Both A and B
   d. Neither A nor B

4. Which of the following will *most likely* require replacement if found to be out of adjustment?
   a. Control thermostat
   b. Low pressure control
   c. Thermostatic expansion valve
   d. Compressor clutch assembly

5. Technician A says that an air-conditioning system problem may cause an engine cooling system problem. Technician B says that a cooling system problem may cause an air-conditioning system problem. Who is right?
   a. A only
   b. B only
   c. Both A and B
   d. Neither A nor B

6. How is the desiccant replaced in an orifice tube system?
   a. By replacing the accumulator
   b. By replacing the receiver
   c. By either A or B, as applicable
   d. By neither A nor B

7. Which of the following will *most likely* cause engine overheating or overcooling?
   a. Loose or worn belt(s)
   b. Incorrect engine timing
   c. Defective declutching fan
   d. Defective coolant thermostat

8. Which of the following is *most likely* to cause a noisy air-conditioning system as well as intermittent, insufficient, or no cooling?
   a. Blown fuse or defective circuit breaker
   b. An overcharge of refrigerant and/or lubricant
   c. A defective control thermostat or temperature control
   d. Loose or worn serpentine or V-belt

9. Air in an air-conditioning system may cause all of the following problems, EXCEPT:

a. an overheated engine.
b. a noisy air-conditioning compressor.
c. inadequate air-conditioning system performance.
d. a lower-than-normal low-side pressure.

10. All of the following are standard above atmospheric pressure terms, EXCEPT:
    a. kPa (absolute).
    b. kPa.
    c. psia.
    d. in. Hg.

11. Which of the following metric pressures is closest to 10 psig?
    a. 69 kPa
    b. 690 kPa
    c. 6.9 kPa
    d. 0.69 kPa

12. Which of the following is *most likely* to cause liquid refrigerant slugging in the compressor?
    a. Clogged screen in the receiver-drier
    b. Clogged screen in the accumulator-drier
    c. Misadjusted thermostatic expansion valve
    d. Clogged or restricted orifice tube

13. Which of the following temperatures of the Engl;ish scale is closest to 37°C ?
    a. 98°F
    b. 99°F
    c. 100°F
    d. 67°F

14. Which metric fastener can be used to replace an English 1/2-16 bolt?
    a. M18
    b. M20
    c. Either A or B
    d. Neither A nor B

15. Which English fastener can be used to replace a metric M7 bolt?
    a. 1/4-28
    b. 5/16-24
    c. Either A or B
    d. Neither A nor B

16. Which of the following is *most likely* to cause engine overcooling?
    a. Missing coolant thermostat
    b. Defective coolant thermostat
    c. Defective coolant (water) pump
    d. Loose fan belt(s)

17. Technician A says that intermittent cooling may be caused by moisture in the air-conditioning system. Technician B says that moisture in the air-conditioning system can cause insufficient cooling. Who is right?
    a. A only
    b. B only
    c. Both A and B
    d. Neither A nor B

18. Technician A says that an intermittent compressor operation may be caused by a loose clutch coil ground wire. Technician B says that an intermittent compressor operation may be caused by a loose clutch coil lead wire. Who is right?
    a. A only
    b. B only
    c. Both A and B
    d. Neither A nor B

19. Technician A says that as much as 20 percent of HFC-134a may be added to a CFC-12 air-conditioning system without harm. Technician B says that as much as 20 percent of CFC-12 may be added to an HFC-134a air-conditioning system without harm. Who is right?
    a. A only
    b. B only
    c. Both A and B
    d. Neither A nor B

20. None of the following components are omni-directional (works in either direction), EXCEPT:
    a. a cooling system termostat.
    b. a vacuum line check valve.
    c. a fixed orifice tube.
    d. a capillary tube.

TERMS

Write a brief description of the following terms:

1. bug screen
2. calibrated
3. clogged
4. conditions
5. cycling clutch
6. defective
7. diagnosis
8. English scale
9. fan clutch
10. fusible plug
11. high side
12. improper
13. intermittent
14. malfunction
15. measure
16. multiplier
17. noisy
18. overcooling
19. related
20. remote bul

COLOR PLATES

The following color plates may be used as visual aids to determine air-conditioning system conditions in various functions and malfunctions. These plates illustrate the state of the refrigerant in various parts of the air-conditioning system.

In a properly operating vehicle air-conditioning system, the four states of the refrigerant within the system are:

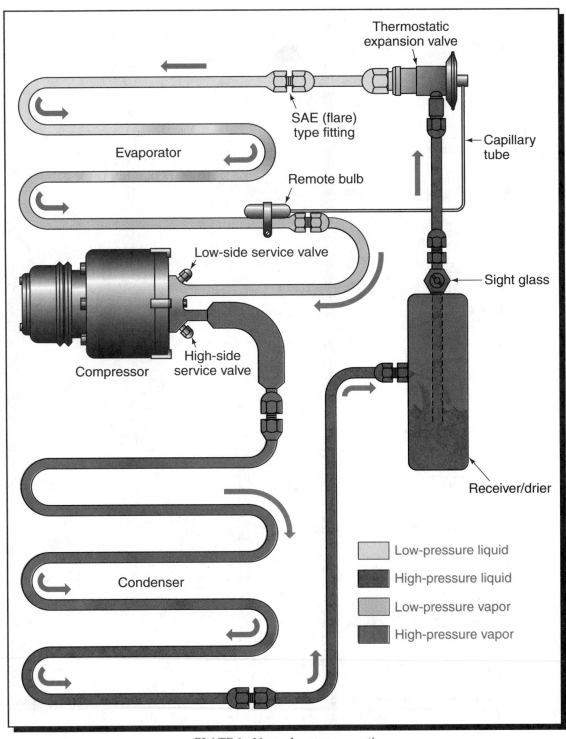

PLATE 1. Normal system operation

❏ Low-pressure liquid
❏ High-pressure liquid
❏ Low-pressure vapor
❏ High-pressure vapor

SYSTEM DIAGNOSIS

In each of the full color plates, the state or condition of the refrigerant may be studied to determine the effects of the particular malfunction. Additionally, if desired, the

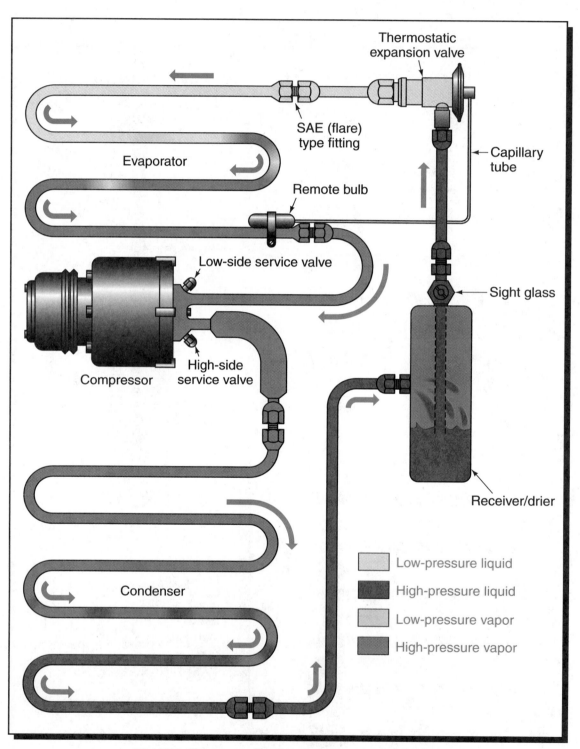

Low-pressure liquid
High-pressure liquid
Low-pressure vapor
High-pressure vapor

PLATE 2. Evaporator flooding defective expansion valve

plates can be keyed to the gauge pressure exercises of the System Diagnostics by Gauge Pressure assignments 1 through 15, as applicable.

Normal Operation

Plates 1 and 9 depict normal motor vehicle air-condi-

tioning (MVAC) system function. As noted in these plates, the refrigerant is a low-pressure liquid from the metering device into the evaporator. In the evaporator, the refrigerant changes to a low-pressure vapor while picking up heat. The a low-pressure vapor goes into the compressor, where it is changed into a high-

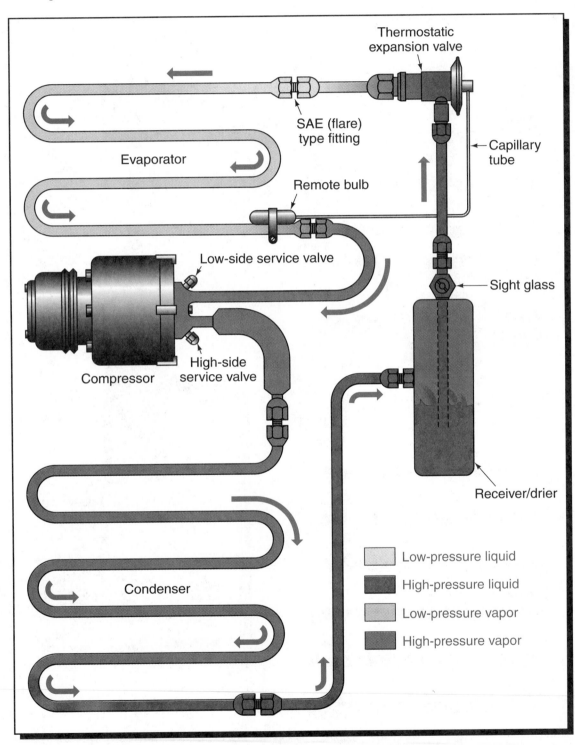

PLATE 3. Evaporator starving defective expansion valve

pressure vapor. This condition exists into the condenser, where it changes to a high-pressure liquid while giving up its heat. The high-pressure liquid is then stored in the receiver-drier where it is available to the metering device.

Abnormal Operation

Plates 2 and 8 depict the condition of a flooded evaporator, whereas plates 3 and 11 depict the condition of a starved evaporator. An undercharge of refrigerant is

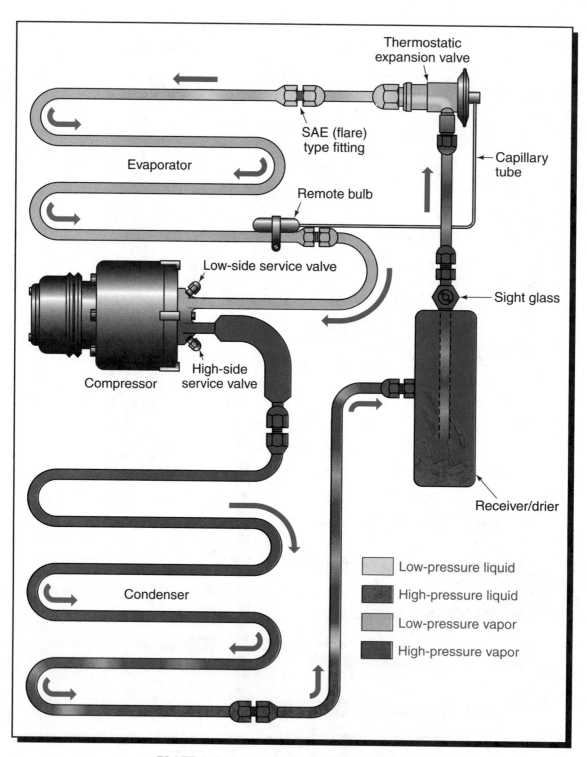

PLATE 4. System undercharged with refrigerant

shown in plates 4 and 12, while an overcharge of refrigerant is shown in plates 5 and 13. Plates 6 and 14 depict the effects of a defective compressor, such as bad valve plate(s). Plates 7 and 15 depict no compressor action, as would be the case with a defective com-

pressor, clutch, clutch coil, or broken drive belt. Finally, plate 10 shows the effect of a **clogged** or restricted accumulator. Plate 16 provides retrofit labels that may be reproduced on a color copier for use when retrofitting an MVAC system.

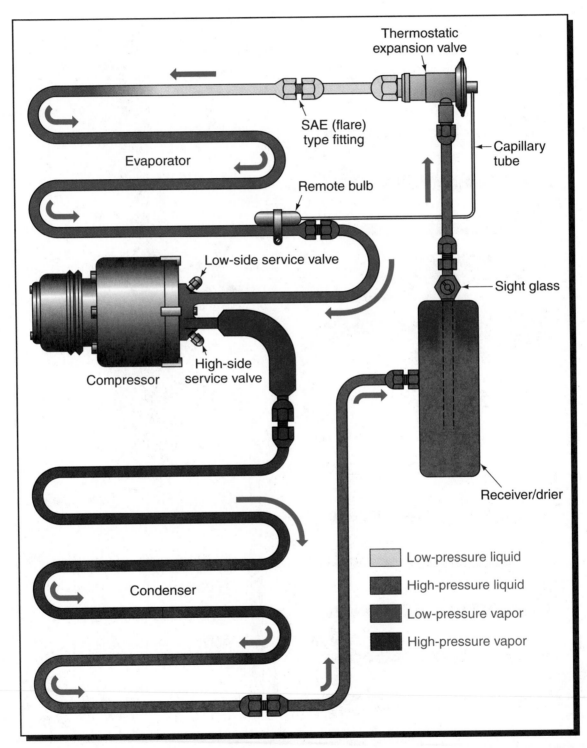

PLATE 5. Overcharge of refrigerant

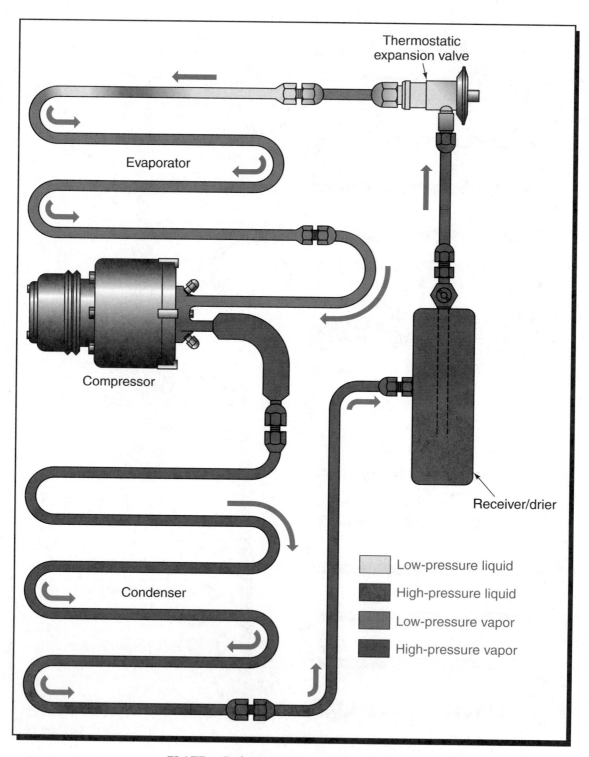

PLATE 6. Defective compressor (valve plate)

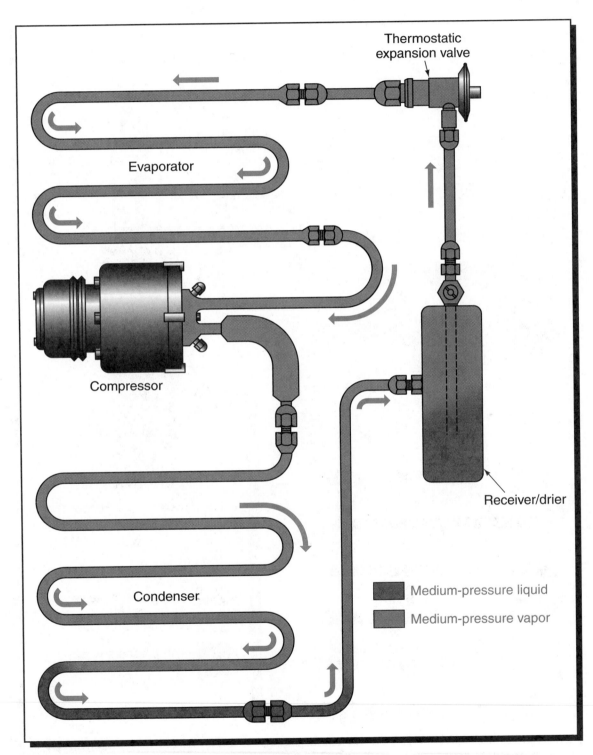

Thermostatic expansion valve

Evaporator

Compressor

Receiver/drier

Condenser

Medium-pressure liquid

Medium-pressure vapor

PLATE 7. No compressor action

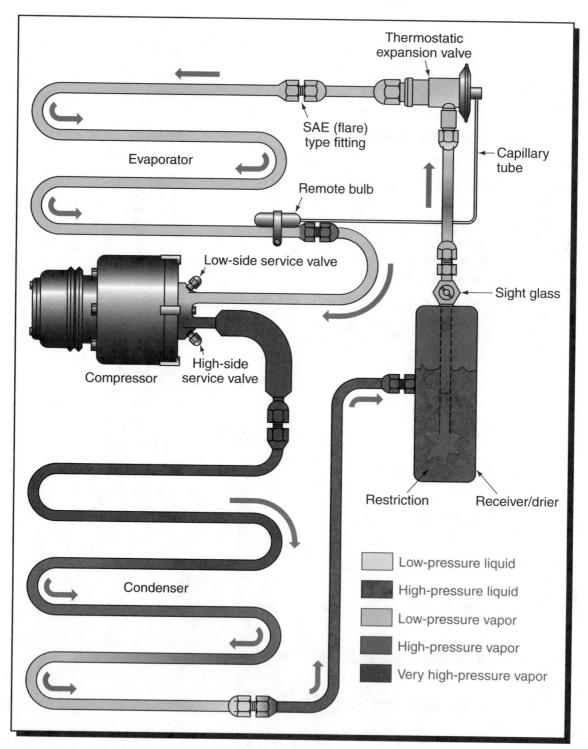

PLATE 8. Restriction in receiver-drier (at pickup tube inlet strainer)

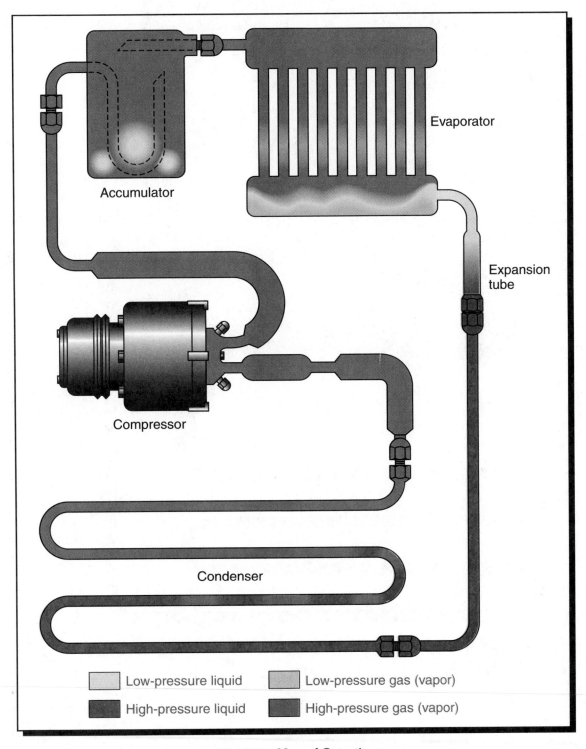

PLATE 9. Normal Operation

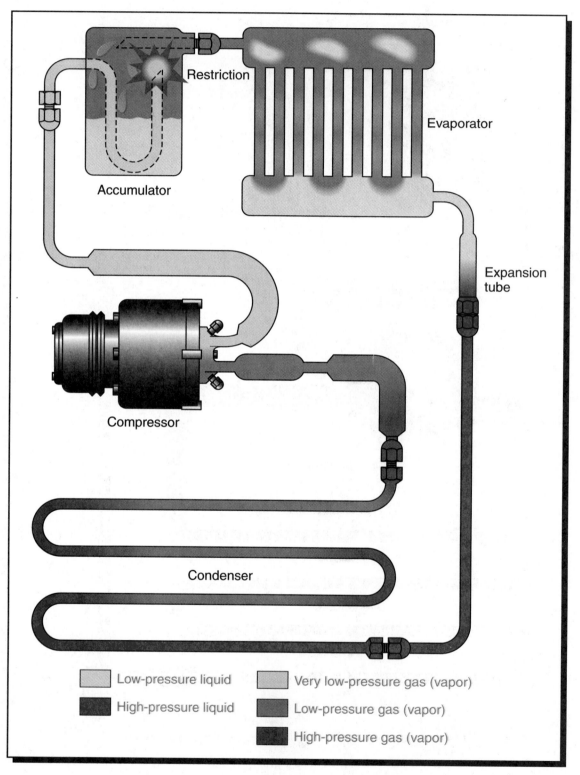

PLATE 10. Restriction in the accumulator

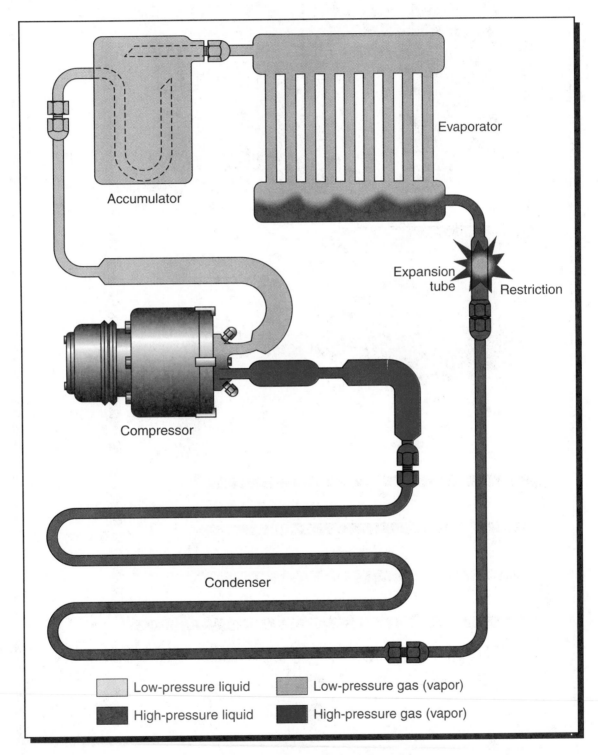

PLATE 11. Restriction in the expansion tube

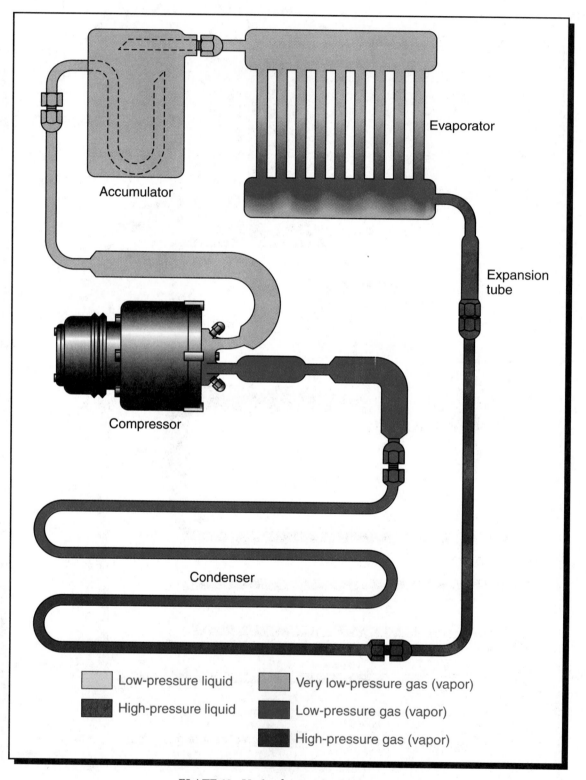

PLATE 12. Undercharge of refrigerant

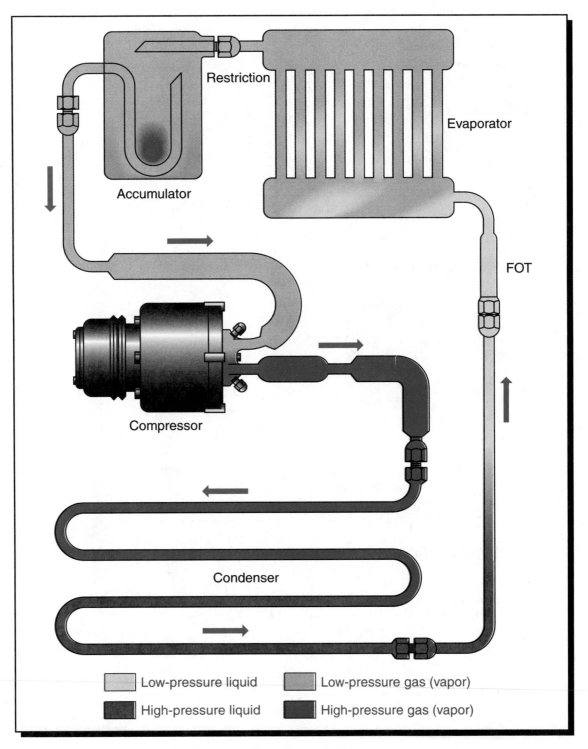

PLATE 13. Overcharge of refrigerant

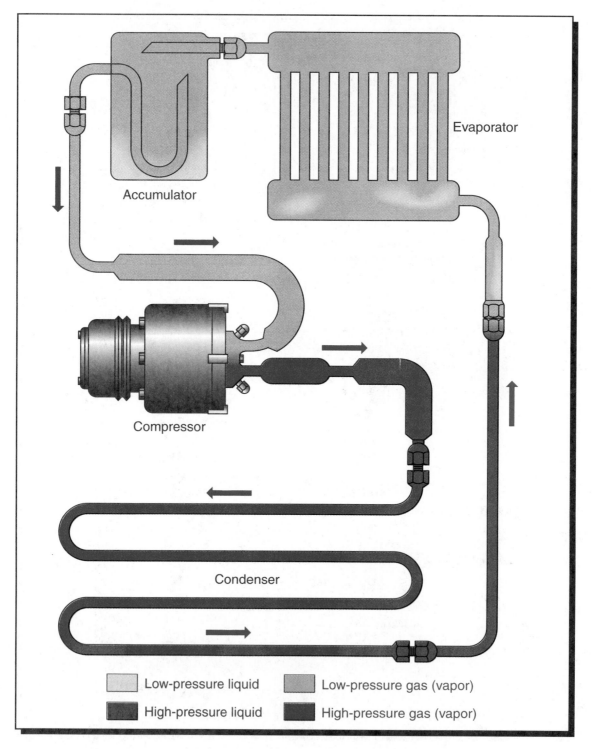

PLATE 14. Defective compressor (valve plate)

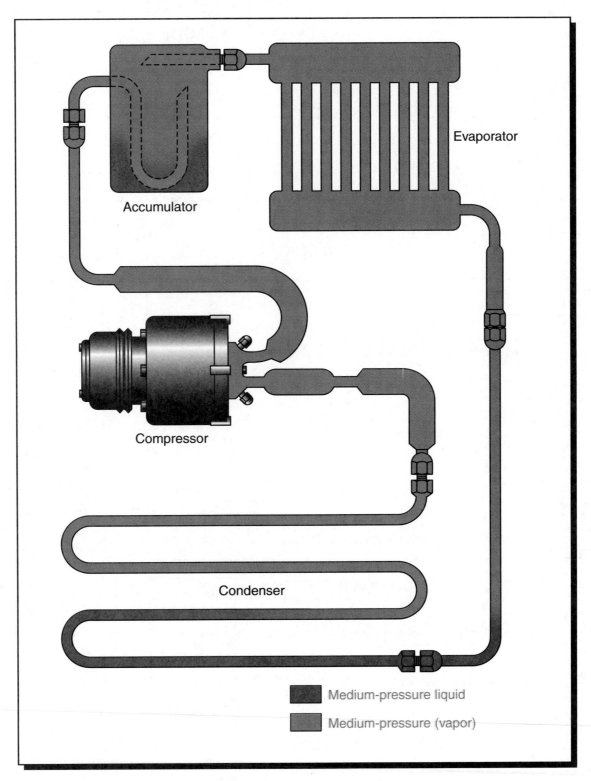

Evaporator

Accumulator

Compressor

Condenser

Medium-pressure liquid

Medium-pressure (vapor)

PLATE 15. No compressor action

CAUTION	NOTICE: RETROFITTED TO R-134a
⚠️ **R-134a** **REFRIGERANT** **UNDER** **HIGH PRESSURE** SYSTEM TO BE SERVICED BY QUALIFIED PERSONNEL ONLY. IMPROPER SERVICE METHODS MAY CAUSE PERSONAL INJURY ©DWIGGINS 2001	RETROFIT PROCEDURE PERFORMED TO SAE J1661 STANDARD USE ONLY R-134a REFRIGERANT AND SYNTHETIC OIL OR EQUIVALENT. OTHERWISE THE AIR-CONDITIONING SYSTEM WILL BE DAMAGED DATE _____ RETROFITTER NAME _____ ADDRESS _____ CITY/STATE _____ REFRIGERANT CHARGE/AMOUNT _____ LUBRICANT/TYPE _____ AMOUNT _____ AUXILIARY LABEL LOCATION _____

CAUTION	NOTICE: RETROFITTED TO R-134a		HFC-134a
⚠️ **R-134a** **REFRIGERANT** **UNDER** **HIGH PRESSURE** SYSTEM TO BE SERVICED BY QUALIFIED PERSONNEL ONLY. IMPROPER SERVICE METHODS MAY CAUSE PERSONAL INJURY ©DWIGGINS 2001	RETROFIT PROCEDURE PERFORMED TO SAE J1661 STANDARD USE ONLY R-134a REFRIGERANT AND SYNTHETIC OIL OR EQUIVALENT. OTHERWISE THE AIR-CONDITIONING SYSTEM WILL BE DAMAGED DATE _____ RETROFITTER NAME _____ ADDRESS _____ CITY/STATE _____ REFRIGERANT CHARGE/AMOUNT _____ LUBRICANT/TYPE _____ AMOUNT _____ AUXILIARY LABEL LOCATION _____		NON-CFC HFC-134a NON-CFC HFC-134a NON-CFC HFC-134a NON-CFC HFC-134a NON-CFC

CAUTION	NOTICE: RETROFITTED TO R-134a	HFC-134a	HFC-134a
⚠️ **R-134a** **REFRIGERANT** **UNDER** **HIGH PRESSURE** SYSTEM TO BE SERVICED BY QUALIFIED PERSONNEL ONLY. IMPROPER SERVICE METHODS MAY CAUSE PERSONAL INJURY ©DWIGGINS 2001	RETROFIT PROCEDURE PERFORMED TO SAE J1661 STANDARD USE ONLY R-134a REFRIGERANT AND SYNTHETIC OIL OR EQUIVALENT. OTHERWISE THE AIR-CONDITIONING SYSTEM WILL BE DAMAGED DATE _____ RETROFITTER NAME _____ ADDRESS _____ CITY/STATE _____ REFRIGERANT CHARGE/AMOUNT _____ LUBRICANT/TYPE _____ AMOUNT _____ AUXILIARY LABEL LOCATION _____	NON-CFC HFC-134a NON-CFC HFC-134a NON-CFC HFC-134a NON-CFC HFC-134a NON-CFC	NON-CFC HFC-134a NON-CFC HFC-134a NON-CFC HFC-134a NON-CFC

PLATE 16. Retrofit

RETROFIT CFC-12 TO HFC-134A

OBJECTIVES

On completion and review of this chapter, you should be able to:

- ❏ Determine the purity of refrigerants and understand the problems associated with contaminated refrigerants.
- ❏ Discuss the various refrigerants that are available; identify those that are approved and those that are not.
- ❏ Understand and discuss the necessity of using dedicated equipment for the recovery of refrigerant.
- ❏ Determine when all of the refrigerant has been removed from an air-conditioning system.
- ❏ Perform all of the procedures required for a successful retrofit as outlined in this text and manufacturers' service manuals.
- ❏ Compare the differences between components that are used for CFC-12 and HFC-134a refrigerant air-conditioning systems.

INTRODUCTION

The term *retrofit* generally applies to any modification accomplished to update or upgrade a system or a device. In this text, the term retrofit refers to modifying a motor vehicle air-conditioning (MVAC) system, specifically, to **replace** an ozone-depleting substance with one that does not harm the ozone layer. For MVAC service this means to convert the air-conditioning system from CFC-12 (R-12), a chlorofluorocarbon, to HFC-134a (R-134a), a hydrofluorocarbon refrigerant.

There are several refrigerant manufacturers and suppliers that claim to have a substance that may be used to replace CFC-12. Only HFC-134a has been **approved** or recommended by the motor vehicle industry as the "refrigerant of choice" for replacing CFC-12. HFC-134a has the approval of the Environmental Protection Agency (EPA), Society of Automotive Engineers (SAE), Mobile Air-Conditioning Society (MACS), International Mobile Air-Conditioning Association (IMACA), and all of the major vehicle manufacturers. Because no other refrigerant is endorsed or approved

at this time for MVAC service, other refrigerants will not be considered for such service in this text.

Refrigerant Sales

The Clean Air Act (CAA) prohibits the distribution and sale of small containers of CFC-12, Figure 19–1, to anyone who is not properly certified. Small containers of CFC-12 are generally available for sale at parts houses to the properly certified automotive technician. Also, one may occasionally find small containers of HCFC-22, Figure 19–2, and HFC-134a, Figure 19–3, in the automotive section of department, discount, or chain stores.

One must be cautioned that some available refrigerants are flammable under certain conditions. Others are simply not compatible and will contaminate the air-conditioning system. If the price of a **cylinder** of refrigerant sounds "too good to be true," it probably is. There are a lot of "black market" and "mixed refrigerant" offered for sale as CFC-12. One should only purchase refrigerant from a well-established reputable supply house.

FIGURE 19–1 Small containers of CFC–12 can no longer be sold to the general public. *(Courtesy of BET, Inc.)*

FIGURE 19–2 Small container of HCFC–22. *(Courtesy of BET, Inc.)*

FIGURE 19–3 Small container of HFC–134a. *(Courtesy of BET, Inc.)*

CONTAMINATED REFRIGERANT

By broad definition, if the refrigerant contains air, it is considered to be contaminated. To the MVAC technician, however, contaminated refrigerant usually refers to a mixture of two, or more, types of refrigerant in an air-conditioning system.

Contaminated refrigerant will not perform chemically and physically as intended in an air-conditioning system. The air-conditioning system will not function properly, if it functions at all. Contaminated refrigerant can cause serious damage to the air-conditioning system **components** and/or recovery/recycle equipment.

With the transition to CFC-free air-conditioning systems, the likelihood of cross-mixing refrigerants is a growing concern. Different refrigerants, as well as their lubricants, are not compatible and should not be mixed. It is possible, however, for the wrong refrigerant to be mistakenly charged into an air-conditioning system, or for refrigerants to be mixed in the same recovery tank. Also, because recovery/recycling equipment is generally designed for a particular refrigerant, inadvertent mixing can cause damage to the equipment.

If there is any doubt as to the purity of the refrigerant in the vehicle, the refrigerant should be checked before attempting repairs. "Recovery only" equipment, Figure 19–4, that meets the Society of Automotive Engineers' (SAE) J2209 standards and proper recovery cylinders that meet rigid Department of Transportation (DOT) specifications must be used for recovering conta-

minated refrigerant. The cylinders should be clearly marked "CONTAMINATED REFRIGERANT" for easy identification. Only contaminated refrigerant, should be recovered into this cylinder, Figure 19–5.

> **CAUTION**
>
> DISPOSABLE CYLINDERS, KNOWN AS DOT 39S, MUST NOT BE USED FOR RECLAIMED REFRIGERANT. IT IS A VIOLATION OF FEDERAL LAW TO REFILL THESE CYLINDERS.

Contaminated refrigerant may be reclaimed to Air Conditioning Refrigeration Institute (ARI) 700-88 standards, or it may be destroyed by fire. This is usually accomplished at an off-site reclamation facility that is equipped to handle contaminated refrigerants. It is the

FIGURE 19–4 A typical recovery only system. *(From Whitman/Johnson Refrigeration and Air Conditioning Technology, © Delmar, a division of Thomsom Learning)*

FIGURE 19–5 A typical DOT-approved recovery cylinder.

FIGURE 19–6 A pressure test gauge. *(Courtesy of BET, Inc.)*

technician's responsibility to ensure that contaminated refrigerant is properly disposed of.

Under the CAA, anyone performing repairs for consideration (pay) to a motor vehicle air conditioning (MVAC) system must be certified and must use approved recovery/recycle equipment.

Do not take any chances. If there is any doubt, either use a purity tester, turn the vehicle away or keep it overnight for a period of 12 hours or more to check for refrigerant purity before attempting repairs.

To determine the purity of the refrigerant in the vehicle, it is best to use a purity tester, often referred to as a refrigerant identifier. If a purity tester is not available, perform the following static purity test.

Static Purity Test

Although much less reliable than a purity tester, the static purity test may be performed if a tester is not available. The idea is that any test is better than no test.

Procedure

1. Park the vehicle inside the shop in an area relatively draft free, where the ambient temperature is not expected to go below 70°F (21°C) overnight.
2. Before closing shop for the day, perform steps 3 through 6.
3. Raise the hood.
4. Determine the type of refrigerant the air-conditioning system was designed to use: CFC-12 or HFC-134a.
5. Attach a gauge appropriate for the refrigerant being used, Figure 19–6.

6. Place a thermometer in the immediate area of the vehicle to measure ambient temperature, Figure 19–7.
7. First thing the next morning, follow steps 8 and 9.
8. Note and record the pressure reading shown on the gauge.
9. Compare the gauge reading with the appropriate table:

   ❏ Figure 19–8: (A) English or (B) metric for CFC-12
   ❏ Figure 19–9: (A) English or (B) metric for HFC-134a

Allow for reasonable inaccuracies of the gauge, the thermometer, and the reader. The pressure should nearly match that expected for any given temperature if the refrigerant is pure.

Other factors may give an inaccurate reading. For example, if there is air in the system an accurate reading may not be noted. The refrigerant should be treated as if it were contaminated if there is any doubt as to its purity.

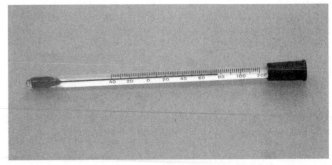

FIGURE 19–7 A spirit thermometer used to measure ambient temperature. *(Courtesy of BET, Inc.)*

Temperature Fahrenheit	Pressure PSIG	kPa	Temperature Fahrenheit	Pressure PSIG	kPa
70	80	551	86	103	710
71	82	565	87	105	724
72	83	572	88	107	738
73	84	579	89	108	745
74	86	593	90	110	758
75	87	600	91	111	765
76	88	607	92	113	779
77	90	621	93	115	793
78	92	634	94	116	800
79	94	648	95	118	814
80	96	662	96	120	827
81	98	676	97	122	841
82	99	683	98	124	855
83	100	690	99	125	862
84	101	696	100	127	876
85	102	703	101	129	889

A

Temperature Celsius	Pressure PSIG	kPa	Temperature Celsius	Pressure PSIG	kPa
21.1	551	80	30.0	710	103
21.7	565	82	30.5	724	105
22.2	572	83	31.1	738	107
22.8	579	84	31.7	745	108
23.3	593	86	32.2	758	110
23.9	600	87	32.8	765	111
24.4	607	88	33.3	779	113
25.0	621	90	33.9	793	115
25.6	634	92	34.4	800	116
26.1	648	94	35.0	814	118
26.7	662	96	35.6	827	120
27.2	676	98	36.1	841	122
27.8	683	99	36.7	855	124
28.3	690	100	37.2	862	125
28.9	696	101	37.8	876	127
29.4	703	102	38.3	889	129

B

FIGURE 19–8 Temperature/pressure chart for CFC–12: (A) English and (B) metric.

Temperature Fahrenheit	Pressure PSIG	kPa	Temperature Fahrenheit	Pressure PSIG	kPa
70	76	524	86	192	703
71	77	531	87	103	710
72	79	545	88	105	724
73	80	551	89	107	738
74	82	565	90	109	752
75	83	572	91	111	765
76	85	586	92	113	779
77	86	593	93	115	793
78	88	607	94	117	807
79	90	621	95	118	814
80	91	627	96	120	827
81	93	641	97	122	841
82	95	655	98	125	862
83	96	662	99	127	876
84	98	676	100	129	889
85	100	690	101	131	903

A

Temperature Celsius	Pressure PSIG	kPa	Temperature Celsius	Pressure PSIG	kPa
21.1	524	76	30.0	703	102
21.7	531	77	30.5	710	103
22.2	545	79	31.1	724	105
22.8	551	80	31.7	738	107
23.3	565	82	32.2	752	109
23.9	572	83	32.8	765	111
24.4	586	85	33.3	779	113
25.0	593	86	33.9	793	115
25.6	607	88	34.4	807	117
26.1	621	90	35.0	814	118
26.7	627	91	35.6	827	120
27.2	641	93	36.1	841	122
27.8	655	95	36.7	862	125
28.3	662	96	37.2	876	127
28.9	676	98	37.8	889	129
29.4	690	100	38.3	903	131

B

FIGURE 19–9 Temperature/pressure chart for HFC–134a: (A) English and (B) metric.

Using a Purity Tester

A refrigerant purity tester, also referred to as a refrigerant identifier, Figure 19–10, is far superior to pressure-temperature comparisons because, at certain temperatures, the pressures of CFC-12 and HFC-134a are too similar to differentiate with a standard gauge. This is easily noted

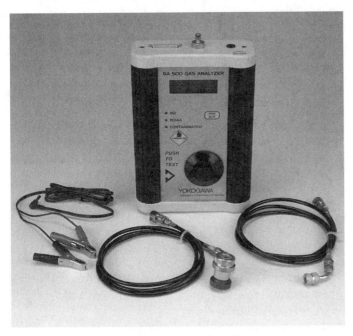

FIGURE 19–10 Typical refrigerant identifier.

in the chart shown in Figure 19–11. For example, at 90°F (32.2°C), both 95 percent CFC-12 and 95 percent HFC-134a have about the same pressure: 111 and 112 psig, respectively. Given that this chart is accurate to plus/minus 2 percent (±2%), there is really no way of determining which type of refrigerant is in the air-conditioning system or tank. Also, because other substitute refrigerants and blends may have been introduced into the motor vehicle air-conditioning (MVAC) system, they will contaminate a system or tank and may not be detected by the pressure/temperature method. A refrigerant identifier would conclude the refrigerant in our example to be UNKNOWN.

Use of a refrigerant identifier, often called a purity tester, should be the first step in servicing a motor vehicle air conditioning (MVAC) system. That way, the technician does not have to be concerned about customer dissatisfaction or damage to the vehicle that could occur if the wrong refrigerant is used. Further, testing refrigerant protects refrigerant supplies as well as recovery/recycling equipment. At today's prices, preventing just one tank of refrigerant from contamination can save several hundred dollars plus the high cost of disposing of the contaminated refrigerant.

Always follow the manufacturer's instructions for using any type of test equipment. The following procedure for using the Sentinel identifier is typical:

AMB TEMP		CFC-12/HFC-134a PERCENT BY WEIGHT										
°F	°C	100/0	98/2	95/5	90/10	75/25	50/50	25/75	10/90	5/95	2/98	0/100
65	18.3	64	67	71	74	83	84	78	73	70	67	64
70	21.1	70	74	79	82	90	92	87	81	77	74	71
75	23.9	77	81	85	91	99	101	96	89	85	83	79
80	26.7	84	88	93	99	107	110	105	98	95	92	87
85	29.4	92	96	101	108	116	120	114	106	103	100	95
90	32.2	100	105	111	116	125	130	125	116	112	109	104
95	35.0	108	114	119	126	135	140	135	126	122	119	114
100	37.8	117	123	127	135	145	151	145	136	133	130	124
105	40.6	127	132	138	146	158	164	159	149	144	141	135
110	43.3	136	142	147	156	170	176	173	164	157	152	146
115	46.1	147	152	159	166	183	192	184	175	168	163	158
120	48.9	158	164	170	177	195	205	196	187	181	176	171

CFC-12/HFC-134a Cross Contamination Chart. All pressures are given in psig. For kPa, multiply psig by 6.895. For example, 100% CFC-12 at 95°F (35°C) is 108 psig or 744.7 kPa.

FIGURE 19–11 Temperature/pressure chart of CFC–12 and HFC–134a mixed refrigerants.

Procedure

1. Turn on the main power switch; the unit automatically clears the last refrigerant sample and is made ready for a new sample.
2. When *READY* appears on the display, connect a service hose from the tester to the vehicle air-conditioning system or tank of refrigerant being tested.
3. The tester automatically pulls in a sample and begins processing it; *TESTING* shows on the display.
4. Within approximately one minute, the display will show *R-12*, *R-134a* or *UNKNOWN*. If *UNKNOWN* is displayed, the sample is a mixture or is some other type of refrigerant. In either case, it should not be added to previously recovered refrigerant. Also, it should not be recycled or reused.
5. Turn off the main power switch and disconnect the service hose.

RECOVER ONLY, AN ALTERNATE METHOD

Although the law requires that the service technician own recovery equipment, there are other methods that are effective. One such method, although slow, can be used for recovery only if the refrigerant is suspected to be contaminated. This procedure is based on the premise that refrigerant, at rest, will seek the coldest and lowest area of the system. The idea, then, is to provide a sub-cold area for the refrigerant to condense. This method is presented for information only and should be performed by, or under the direct supervision of, an experienced technician. The most important consideration is that the recovery cylinder will not be filled to more than 60 percent capacity when the temperature is increased to ambient.

Procedure

Refer to Figure 19–12 to follow these step-by-step instructions:

1. Place an identified recovery cylinder into a tub of ice on the floor near the vehicle.

   NOTE: The recovery cylinder should be below the level of the air-conditioning system.

2. Add water and rock (ice cream) salt to lower the temperature to about 0°F (–18°C).

   NOTE: A chemical that will lower the bath to about –15°F (–26°C) is available.

3. Connect a hose from the high-side fitting of the air-conditioning system to the vapor (gas) valve of the recovery cylinder.
4. Open all of the appropriate service valves.
5. Cover the recovery cylinder and tub with a blanket to insulate them from ambient air.

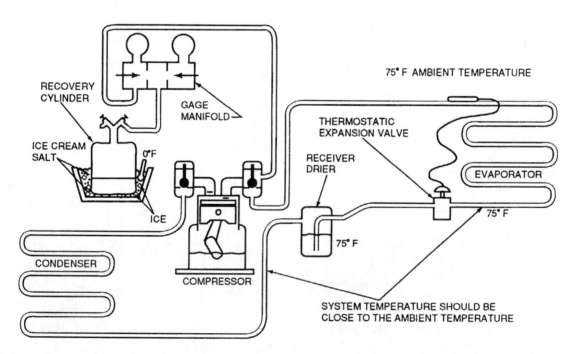

FIGURE 19–12 A typical ice–bath set–up for the refrigerant recovery cylinder. *(From Whitman/Johnson Refrigeration and Air Conditioning Technology, © Delmar, a division of Thomson Learning)*

6. Place shop light(s), or another heat source, near the accumulator or receiver.
7. Allow a minimum of two to three hours for recovery.

The actual time that is required will depend on the ambient temperature and the amount of refrigerant to be recovered. This method will not remove as much refrigerant as a quality recovery system may but it will remove enough to satisfy the requirements of the EPA: 80 percent to 90 percent of the refrigerant in the system.

State or local laws may, however, negate this procedure. If this procedure is used, it should be followed by a good pump down (evacuation) and a change of lubricant.

RETROFIT COMPONENTS

The following is an overview of some of the problems and conditions associated with components, listed in no particular order, when retrofitting a motor vehicle automotive air-conditioning (MVAC) system from CFC-12 to HFC-134a refrigerant.

Access Valves

There is a distinct difference in the access valves used on an HFC-134a system, Figure 19–13, when compared with those used on CFC-12 systems, Figure 19–14. Adapters, Figure 19–15, that are used on CFC-12 fittings during retrofit procedures to make them compatible with HFC-134a equipment are available. A special adapter, called a "Saddle Clamp Access" valve, Figure 19–16, is available for installation where space does not permit the HFC-134a adapter to convert the CFC-12 valve.

For installation of the saddle valve:

1. Make certain that the system is free of refrigerant.
2. Select the proper location for the valve.
   a. Will there be clearance for the hose access adapter?

b. Will there be **adequate** clearance to close the hood with the protective covers in place?
c. Will access to other critical components be restricted or blocked?

FIGURE 19–14 A typical CFC–12 access valve fitting. *(Courtesy of BET, Inc.)*

FIGURE 19–15 Access valve fitting adapters for retrofitting CFC–12 to HFC–134a. *(Courtesy of BET, Inc.)*

FIGURE 19–13 A typical HFC–134a access valve fitting. *(Courtesy of BET, Inc.)*

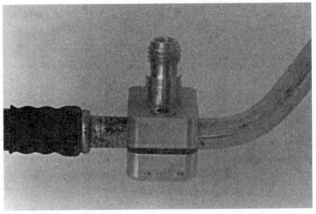
FIGURE 19–16 A saddle valve may be used. *(Courtesy of BET, Inc.)*

d. Is the tubing straight, clean, and sound?
3. Select the proper valve for the application.
   a. For low- or high-side use (the low-side valve is larger)
   b. The size of the tube the valve is to be installed on
4. Position both halves of the saddle valve on the tube, Figure 19–17.

NOTE: Make sure the O-ring is in position.

5. Place the screws (usually socket head) and tighten them evenly. Do not overtighten them; 20–30 in.-lb. (2–3 N•m) is usually recommended.

NOTE: A method other than that outlined in steps 5, 6, and 7 may be recommended. Follow the recommendations provided by the manufacturer of the saddle valve when they differ from those given here.

6. Insert the piercing pin in the head of the access port fitting, Figure 19–18.

7. Tighten the pin until the head touches the top of the access port, Figure 19–19.
8. Remove the piercing pin and replace it with the valve core, Figure 19–20.
9. Tighten the valve core securely, Figure 19–21.
10. Install the cap (or pressure switch) on the installed fitting.

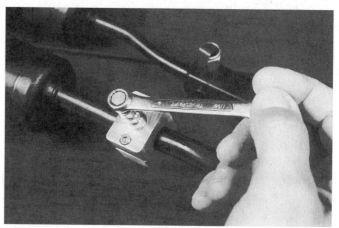

FIGURE 19–19 Tighten the pin.

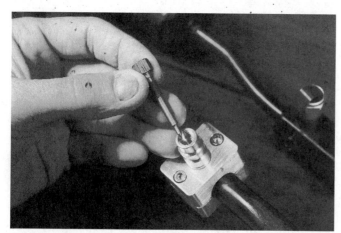

FIGURE 19–17 Position both halves of the valve on the tube.

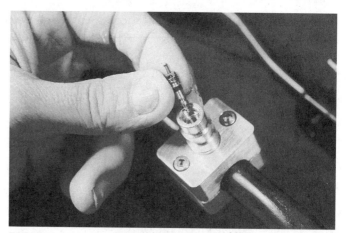

FIGURE 19–20 Replace the pin with the valve core.

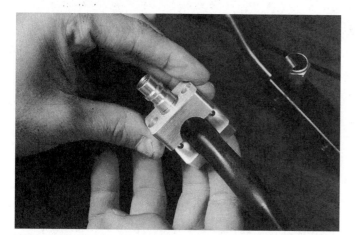

FIGURE 19–18 Insert the piercing pin.

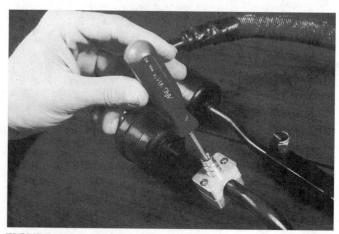

FIGURE 19–21 Tighten the valve core.

Accumulator

Accumulators, Figure 19–22, in CFC-12 systems typically have a desiccant designated as XH5. This desiccant is not compatible with HFC-134a refrigerant. The desiccant to be used in HFC-134a systems is designated XH7 or XH9. This desiccant is found in accumulators and receivers designated for HFC-134a service. General Motors and Ford do not recommend changing their accumulators because the desiccants used are compatible with HFC-134a. Both XH7 and XH9 desiccants are compatible with CFC-12 as well as HFC-134a.

If a clutch cycling pressure switch (CCPS) has to be

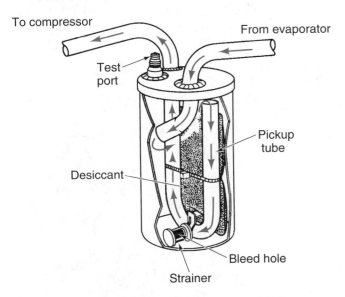

FIGURE 19–22 A typical suction line accumulator.

changed, however, the accumulator may have to be replaced to accommodate the metric threads found on the switch. In some retrofit packages, an adapter may be included for English/metric thread conversion.

Compressor

Compressors, Figure 19–23, are being redesigned to withstand the slight increase in pressures associated with HFC-134a. Most compressor rebuilders are also incorporating these design changes into their rebuilding procedures. When purchasing a new or rebuilt compressor for an HFC-134a system, make sure that it has been identified for that application.

It is not recommended that a compressor be replaced as a matter of course for retrofitting. The compressor should only be replaced if it is defective. One should not replace a compressor simply because the system is being retrofitted.

Condenser

To change any part of an original design is to change the performance of the equipment. This may be especially true for the condenser. The engine cooling system may also be affected by the slight increase in pressure (and temperature) of the condensing HFC-134a refrigerant. The following may be considered to reduce the problems.

Dams

A *dam*, loosely defined, is the sealing provision located between the radiator and condenser that helps to direct

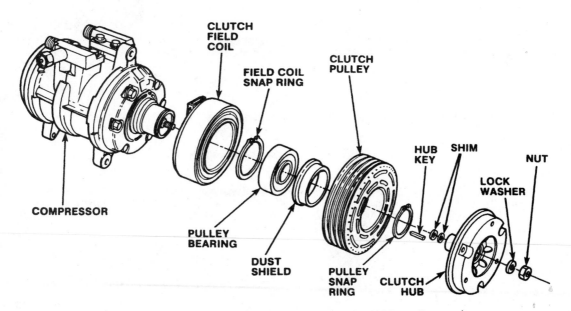

FIGURE 19–23 A typical compressor showing clutch details. *(Courtesy of Ford Motor Company)*

ambient and ram air through both components. It is critical that all condenser and radiator seals be in place. All holes, regardless of their size, that could allow air to bypass either component should be blocked off to ensure maximum airflow.

In some installations the condenser will be changed. Because the mounting space is limited, this usually means a condenser with more fin area, fins per inch (FPI). A higher rpm motor may be used to replace the original motor. In other cases a second motor and fan, a pusher-type, may be placed in front of the condenser. The idea is to improve or increase airflow to remove more heat.

If a fan and motor are added, a relay should also be added to ensure that the electrical system is not overloaded. The coil of the relay may be wired in with the compressor clutch circuit, Figure 19–24, to ensure that the fan is running when the air-conditioning system is turned on. An in-line fuse is included to protect the circuit.

Evaporator

The evaporator, Figure 19–25, is not replaced unless it is leaking. There have been no problems reported when using CFC-12 evaporators for HFC-134a retrofit.

Minor changes are made, however, for evaporators

designated for use with HFC-134a refrigerant to accommodate the slightly higher pressure that may be expected.

Hose

Generally, hoses used for automotive air-conditioning service on 1989 and later-year or model vehicles need not be replaced when retrofitting from CFC-12 to HFC-134a. The exception is if the hose, Figure 19–26, leaks during retrofit procedures.

O-Rings and Seals

Although O-rings made of epichlorohydrin and designated for CFC-12 service are not compatible with HFC-134a, it is not recommended that they be replaced when retro-

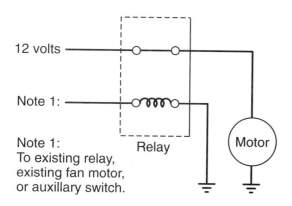

FIGURE 19–24 Electrical schematic for adding a condenser fan and motor.

FIGURE 19–25 Typical evaporator core.

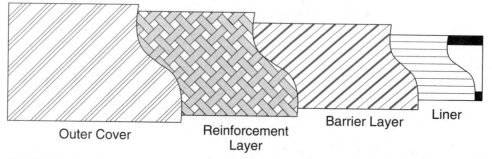

FIGURE 19–26 Construction details of a barrier hose. (*Courtesy of BET, Inc.*)

fitting an air-conditioning system. The exception is if the fitting is found to be leaking. In that case use only O-rings and seals, Figure 19–27, designated for the refrigerant being used in the system. Generally, O-rings and seals for CFC-12 systems are black. Unfortunately, some manufacturers prefer black HFC-134a O-rings and seals as well. Many, however, color code the O-rings and seals designated for HFC-134a service. When in doubt, use color-coded neoprene or HSN/HNBR O-rings or seals; they are also compatible with CFC-12.

Metering Device

Metering devices should not be changed as a matter of practice when retrofitting a system. There are two types of metering devices used in the modern automotive air-conditioning system. They are the thermostatic expansion valve (TXV) and the fixed orifice tube (FOT).

Thermostatic Expansion Valve. The thermostatic expansion valve (TXV), Figure 19–28, does not have to be replaced when retrofitting a system from CFC-12 to HFC-134a. If, however, a TXV is found to be defective, it should be replaced with a model designed for use with the system refrigerant. A CFC-12 TXV used in an HFC-134a system will result in higher superheat and improved overall evaporator temperature. An HFC-134a valve used in a CFC-12 system will have reduced superheat and will not perform as well. Because superheat has a direct effect on performance, it is not advisable that the superheat be allowed to increase more than 3°F (1.7°C)

over that of the operating CFC-12 system. Superheat is discussed in Chapters 9 and 15 of this text.

If a new TXV is required, use only those designed for the specific refrigerant in the system. As a rule of thumb, CFC-12 valves should not be used on originally equipped R-134a systems and, conversely, HFC-134a valves should not be used on CFC-12 systems.

Orifice Tube. With the exception of one vehicle manufacturer, it is not recommended that the orifice tube, Figure 19–29, be replaced when retrofitting an air-conditioning system. The manufacturer, Volvo, recommends changing the orifice tube to one that has a 0.002-inch (0.0508 mm) smaller orifice. If the orifice tube is changed, a slight increase in high-side pressure may be noted.

Pressure Switch

Either one of two switches, or both, may be recommended for change during some retrofit procedures. These switches are the clutch cycling pressure switch (CCPS) and the refrigerant containment device (RCD). A brief description follows.

Clutch Cycling Pressure Switch. The clutch cycling pressure switch (CCPS), Figure 19–30, may be changed for some HFC-134a retrofits. The difference is that the HFC-134a switches are calibrated for slightly higher clutch cycling pressures. Also, the mounting threads are metric to prevent the connection of an English-thread CFC-12 switch in an HFC-134a system.

FIGURE 19–27 Typical O–rings and seals.

FIGURE 19–28 A thermostatic expansion valve. *(Courtesy of Singer Controls)*

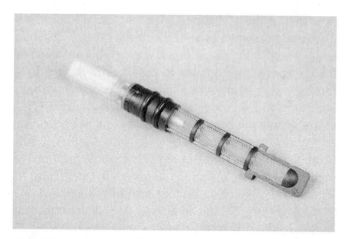

FIGURE 19–29 An orifice tube.

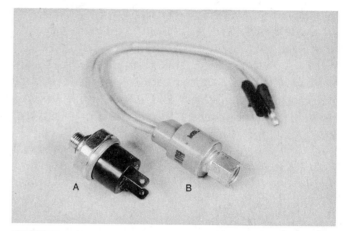

FIGURE 19–30 Typical pressure switch: (A) male thread and (B) female thread.

Refrigerant Containment Device. This device, new for 1994 and later-model year vehicles, may also be included in some retrofit kits for earlier model year vehicles. The refrigerant containment device (RCD) includes models for single- and dual-function refrigerant containment switches.

Each has its specific application. The single switch is, in general, used for controlling the compressor clutch; the dual switch also includes provisions to control the condenser fan. The transducer is used in some solid-state temperature control systems.

Receiver-Drier

The receiver-drier, Figure 19–31, used in CFC-12 systems typically has XH5 desiccant. This desiccant is not compatible with HFC-134a refrigerant. To be sure, the receiver-drier should be replaced during retrofit procedures with a unit designated for HFC-134a service and PAG or ESTER lubricants. This desiccant, designated XH7 or XH9, is also compatible with CFC-12 refrigerant and mineral oils.

RETROFIT

Specific procedures to retrofit any particular make or model vehicle are provided by the respective vehicle manufacturers. According to information released by automotive manufacturers, the procedure, methods, and materials vary considerably from vehicle to vehicle.

Some manufacturers require draining mineral oil, while others do not; some require flushing the system, others do not. Also, some require replacing components, such as the accumulator or receiver-drier and/or the condenser, while others do not.

FIGURE 19–31 Typical receiver-driers. (*Courtesy of BET, Inc.*)

The Society of Automotive Engineers (SAE), in mid-June 1993, issued their standard J1661 "Procedure for Retrofitting CFC-12 (R-12) Mobile air-conditioning systems to HFC-134a (R-134a)." This service procedure, considered typical, is therefore based on SAE's J1661.

Procedure

The following step-by-step procedures are to be considered typical for the retrofitting of any vehicle from refrigerant CFC-12 (R-12) to refrigerant HFC-134a (R-134a). For specific procedures, follow the manufacturer's instructions.

WARNING/CAUTION
DO NOT ATTEMPT TO USE ANY OTHER TYPE OF REFRIGERANTS.

Connect the Manifold and Gauge Set

Follow this procedure when connecting the CFC-12 manifold and gauge set into the system for service.

1. Place fender covers to avoid damage to the vehicle finish.
2. Remove the protective caps from the service valves, Figure 19–32. Some caps are made of light metal and can be removed by hand; others may require a wrench or pliers.

   WARNING
   REMOVE THE CAPS SLOWLY TO ENSURE THAT REFRIGERANT DOES NOT LEAK PAST THE SERVICE VALVE.

   NOTE: The service hoses must be equipped with a Schrader valve **depressing** pin, Figure 19–33. If the hoses are not so equipped, a suitable adapter, Figure 19–34, must be used.

3. Make sure the manifold hand shutoff valves, Figure 19–35, are closed.
4. Make sure the hose shutoff valves, Figure 19–36, are closed.
5. Connect the low-side manifold hose to the suction side of the system, finger tight.
6. Connect the high-side manifold hose to the discharge side of the system, finger tight.

   NOTE: The CFC-12 high-side fitting on most later-model car lines requires a special adapter, Figure 19–37, to be connected to the hose before being connected to the fitting.

7. If the valve is a hand shutoff type service valve, Figure 19–38, use a service valve wrench and **rotate** the valve stems two or three turns clockwise (cw).
8. Connect the service hose to the CFC-12 recovery system.
9. Continue with "Refrigerant Recovery."

Refrigerant Recovery

Until the early 1990s it was common practice for the service technician to vent refrigerant into the atmosphere. At that time refrigerant was inexpensive and the cost of recovery would have probably been greater than the cost of the refrigerant. The Clean Air Act (CAA) Amendments of 1990 changed that practice. The CAA enacted by the Environmental Protection Agency (EPA) states that, after July 1, 1992, no refrigerants may be intentionally vented into the atmosphere.

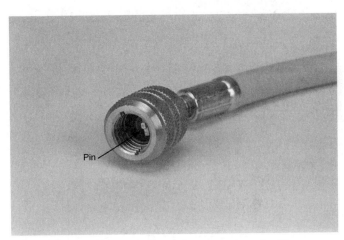

FIGURE 19–33 CFC–12 service hoses equipped with Schrader valve depressing pin.

FIGURE 19–32 Remove the protective caps from the service valves. (*Courtesy of BET, Inc.*)

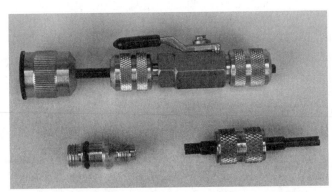

FIGURE 19–34 CFC–12 adapters for Schrader access valves.

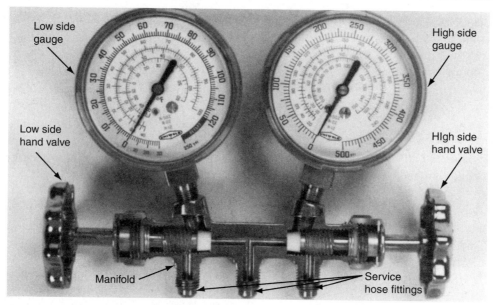

FIGURE 19–35 Typical manifold and gauge set. *(Courtesy of Uniweld Products)*

FIGURE 19–36 Typical hose shutoff valves. *(Courtesy of BET,*

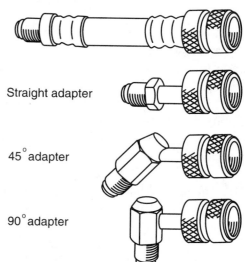

FIGURE 19–37 Special adapters.

CAUTION

THE UNINTENTIONAL VENTING OF REFRIGERANT COULD BE HAZARDOUS TO ONE'S HEALTH. ACCORDINGLY, ADEQUATE VENTILATION MUST BE MAINTAINED DURING THIS PROCEDURE. DO NOT DISCHARGE REFRIGERANT NEAR AN OPEN FLAME AS A HAZARDOUS TOXIC GAS MAY BE FORMED.

1. Start the engine and adjust its speed to 1,250–1,500 rpm.
2. Set all of the air-conditioning controls to the maximum cold position with the blower on high speed.

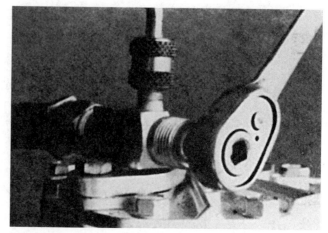

FIGURE 19–38 Use the appropriate wrench. *(Courtesy of BET, Inc.)*

NOTE: Certain system malfunctions, such as a defective compressor, may make this step impossible.

3. Operate for 10-15 minutes to stabilize the system.
4. Return the engine speed to normal idle to prevent dieseling.
5. Turn off all of the air-conditioning controls.
6. Shut off the engine.
7. If not **integrated** in the recovery system, use a service hose and connect the recovery system to an approved recovery cylinder.
8. Open all of the hose shutoff valves, Figure 19–39.
9. Open both low- and high-side manifold hand valves, Figure 19–40.
10. Open the recovery cylinder shutoff valves, as applicable.
11. Connect the recovery system into an approved electrical outlet, Figure 19–41, and turn the main power switch on.

WARNING

IF AN EXTENSION CORD IS USED, MAKE CERTAIN THAT IT HAS AN ELECTRICAL RATING EQUAL TO, OR GREATER THAN, THE RATING OF THE RECOVERY SYSTEM.

12. Turn the recovery system compressor switch on, Figure 19–42.
13. Operate until a vacuum pressure is indicated.
14. If not equipped with an automatic shutoff, turn the compressor switch off after achieving a vacuum (step 13).
15. Make sure the vacuum holds for a minimum of 5 minutes, Figure 19–43.
   a. If not, repeat the procedures starting with step 12 and continue until the system holds a stable vacuum for a minimum of 2 minutes.
   b. If the vacuum holds, proceed with step 16.
16. Close all of the valves.

FIGURE 19–41 Connect the refrigerant recovery system to an approved electrical power supply.

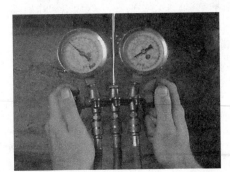

FIGURE 19–39 Open all hose shutoff valves.

FIGURE 19–42 Turn on the recovery (compressor) switch.

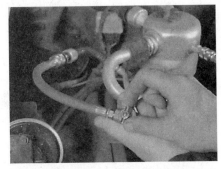

FIGURE 19–40 Open both the high– and low–side manifold hand valves.

FIGURE 19–43 Make sure the vacuum holds for a minimum of 5 minutes.

a. At the recovery cylinder
b. Recovery system
c. Service hoses
d. Manifold
e. Compressor

17. Disconnect all of the hoses that were previously connected.

CAUTION
SOME RECOVERY SYSTEMS HAVE AUTO-MATIC SHUTOFF VALVES. BE CERTAIN THEY ARE OPERATING PROPERLY BEFORE DISCONNECTING THE HOSES TO AVOID REFRIGERANT LOSS, WHICH COULD RESULT IN PERSONAL INJURY.

Repair/Replace Components

1. Determine what repairs, if any, are required.
2. If oil change is required, proceed with step 3; if not, proceed with step 4.
3. Remove the necessary components to drain the oil from the component, Figure 19–44.
4. **Flush** the individual components while they are removed from the vehicle.

NOTE: See Figure 19–45 for a typical setup for this procedure.

5. Replace the components, such as the accumulator, receiver-drier, and/or condenser, if required.
6. Add or replace the electrical fail-safe components, such as the refrigerant containment and high-pressure switch, if required.
7. Perform any other modifications and/or procedures required by the specific vehicle manufacturer.

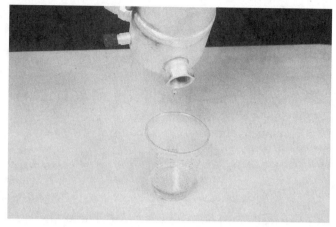

FIGURE 19–44 Draining oil from an accumulator.

8. Replace/reinstall all of the components serviced in steps 3 and 7.
9. If not accomplished by the requirements of steps 5, 6, or 7, repair any problems determined in step 1.

Prepare the System for HFC-134a

WARNING
IF THE SYSTEM WAS FLUSHED, CHARGE OIL DIRECTLY INTO THE COMPRESSOR TO PROVIDE LUBRICATION AT STARTUP.

1. Charge the system with the proper type and **quantity** of lubricant, as recommended by the vehicle manufacturer for HFC-134a refrigerant.
2. Change the service ports from CFC-12 to HFC-134a access type, Figure 19–46.
3. Check for leaks.
4. Affix a **decal** to identify the type of refrigerant for future service, Figure 19–47.

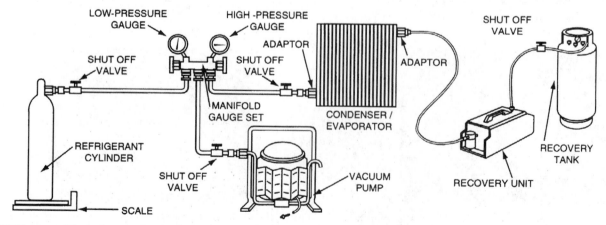

FIGURE 19–45 Typical setup for flushing a component.

FIGURE 19–46 Use adapters to change the service ports from CFC–12 (A) to HFC–134a (B) service. *(Courtesy of BET, Inc.)*

FIGURE 19–47 Apply the identification decal in a conspicuous place. *(Courtesy of BET, Inc.)*

Evacuating the System

The air-conditioning system must be evacuated whenever it is serviced to the extent that the refrigerant has been removed. Proper evacuation not only **eliminates** all traces of refrigerant, it also rids the system of unwanted air and moisture that may have entered during service.

SAE standard J1661 requires that a vacuum pump be capable of achieving a vacuum level of 29.2 in. Hg (2.7 kPa absolute) adjusted to the altitude. The boiling point of water (H_2O) at this vacuum level is 69°F (20.5°C), Figure 19–48.

Some refrigerant recovery systems may have an integral vacuum pump that may be used to evacuate an air-conditioning system. Whenever a recovery unit is

System Vacuum Inches of Mercury	Temperature °F Boiling Point
24.04	140
25.39	130
26.45	120
27.32	110
27.99	100
28.50	90
28.89	80
29.18	70
29.40	60
29.66	50
29.71	40
29.76	30
29.82	20
29.86	10
29.87	5
29.88	0
29.90	–10
29.91	–20

A

System Vacuum kilopascals absolute	Temperature °C Boiling Point
19.66	60.0
15.61	54.4
12.02	48.8
9.07	43.3
6.80	37.7
5.08	32.2
3.75	26.6
2.77	21.1
2.03	15.5
1.15	10.0
0.98	4.4
0.81	–1.1
0.60	–6.7
0.47	–12.2
0.44	–15.0
0.40	–17.8
0.33	–23.0
0.30	–28.8

B

FIGURE 19–48 Boiling point of water (H_2O) under a vacuum: (A) English; (B) metric.

used for this purpose, the manufacturer's **prescribed** procedures must be followed for best results.

This service procedure is to be used for the typical independent vacuum pump, Figure 19–49 or charging station, Figure 19–50.

Procedure

1. Make sure the high- and low-side manifold hand valves are in the closed position, Figure 19–51.
2. Make sure that the service hose shutoff valves are closed, Figure 19–52.
3. Remove the protective caps and covers from all service access fittings.
4. Connect the HFC-134a manifold and gauge set to the system in the same manner as previously outlined for the CFC-12 manifold and gauge set.
5. Place the high- and low-side compressor service valves, if so equipped, in the cracked position.

FIGURE 19–50 Typical charging station. (*Courtesy of Robinair, SPX Corp.*)

FIGURE 19–49 Typical high vacuum pump. (*Courtesy of Robinair, SPX Corp.*)

FIGURE 19–51 Make sure that manifold hand shutoff valves are closed. (*Courtesy of Uniweld Products*)

FIGURE 19–52 Make sure that the service hose shutoff valve is closed. *(Courtesy of Uniweld Products Inc.)*

6. Remove the protective caps from the inlet and exhaust of the vacuum pump.

WARNING
MAKE SURE THE PORT CAP IS REMOVED FROM THE EXHAUST PORT TO AVOID DAMAGE TO THE VACUUM PUMP.

7. Connect the center manifold hose to the inlet of the vacuum pump.
8. Open all service hose shutoff valves.
9. Start the vacuum pump.
10. Open the low-side manifold hand valve.
11. Observe the low-side (compound) gauge needle. The needle should indicate a slight vacuum.
12. After 5 minutes, the **compound gauge** should indicate 20 in. Hg (33.8 kPa absolute) or less, Figure 19–53.
13. The high-side (pressure) gauge needle should be slightly below the zero index of the gauge, Figure 19–54.

NOTE: If the high-side gauge does not drop below zero (unless restricted by a stop), a system blockage is indicated. If the system is blocked, discontinue the evacuation. Repair or remove the obstruction. If the system is clear, continue the evacuation with step 14.

14. Open the high-side manifold hand valve.
15. Operate the pump for 15 minutes and observe the gauges. The system should be at a vacuum of 24–26 in. Hg (20.3–13.5 kPa absolute) minimum if there is no leak.
16. If the vacuum is not down to 24–26 in. Hg (20.3–13.5 kPa absolute), close the low-side hand valve and observe the compound gauge.

NOTE: If the compound gauge needle rises, indicating a loss of vacuum, there is a leak that must be repaired before the evacuation is continued. Leak check the system. If no leak is evident, continue with the pump down, step 17.

17. Pump for a minimum of 30 minutes, as required by SAE J1661.

NOTE: A longer pump down is much better, if time permits. For maximum air-conditioning system performance, a triple pump down is recommended by many.

18. After pump down, close the high- and low-side manifold hand valves.
19. Shut off the vacuum pump.
20. Close all of the valves.
    a. Service hose
    b. Vacuum pump
    c. Compressor, if equipped
21. Disconnect the manifold hoses.
22. Replace any protective caps previously removed.

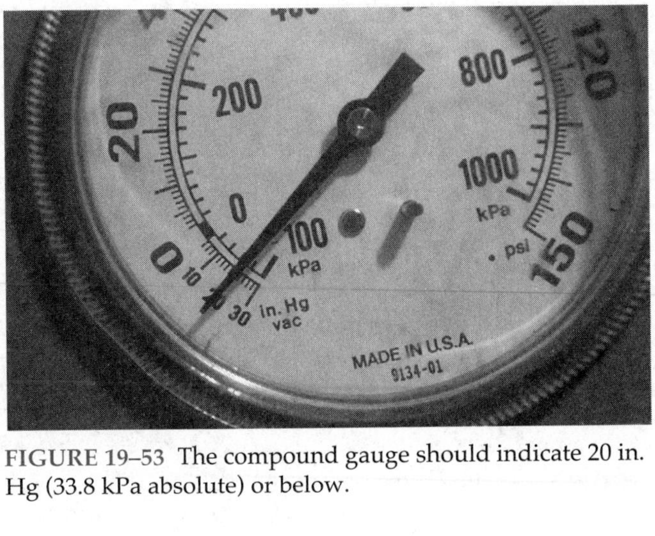

FIGURE 19–53 The compound gauge should indicate 20 in. Hg (33.8 kPa absolute) or below.

FIGURE 19–54 The high–side gauge should drop below zero.

Charging an HFC-134a Air-Conditioning System

It may be noted that HFC-134a is an "ozone friendly" refrigerant. Nonetheless, the Environmental Protection Agency (EPA) has required that this refrigerant also be recovered after mid-November 1995.

Procedure

1. Place the vehicle in a reasonably draft-free work area. This is an aid in detecting small leaks.
2. Make sure that all of the valves are closed.
   a. Service valves, if so equipped
   b. Manifold gauge
   c. Service hose shut-off valves
   d. Refrigerant cylinder or charging station shutoff valve
3. Connect the manifold and gauge set to the system, following the procedures previously outlined.
4. Connect the service hose to the refrigerant source.

NOTE: If a charging station is used, it is very important that specific instructions, provided by the manufacturer of the equipment, are followed.

WARNING

DO NOT OPEN THE MANIFOLD AND GAUGE SET HAND VALVES UNTIL INSTRUCTED TO DO SO. EARLY OPENING COULD CONTAMINATE THE SYSTEM WITH MOISTURE-LADEN AIR.

5. Open the service hose shutoff valves, Figure 19–55.
6. Open the system service valves, if so equipped.
7. Observe the gauges.
   a. Confirm that the system is in a vacuum. If so, proceed with step 8.
   b. If not, follow the procedure outlined for "Evacuating the System" before proceeding.
8. **Dispense** one "pound" can of HFC-134a refrigerant into the system.
   a. Attach a **can tap** valve, Figure 19–56.
   b. Invert the can for liquid dispensing, Figure 19–57.
   c. Open the high-side manifold hand valve.
   d. Empty the contents of the can into the system.
   e. Close the manifold high-side valve.
   f. Rotate the clutch armature several revolutions by hand to ensure that no liquid refrigerant is in the compressor.
9. Attach electronic thermometer **probes,** Figure 19–58, to the inlet and outlet of the evaporator, Figure 19–59.

NOTE: Be sure the end of the probe makes good contact with the metal tubes of the evaporator.

10. Open all of the windows.

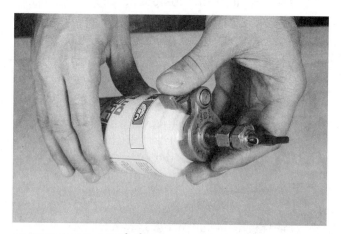

FIGURE 19–56 Attach the can tap valve.

FIGURE 19–55 Open the shutoff valve of the three service hoses.

FIGURE 19–57 Invert the can for liquid dispensing.

FIGURE 19–58 A typical two–probe electronic thermometer. *(Courtesy of BET, Inc.)*

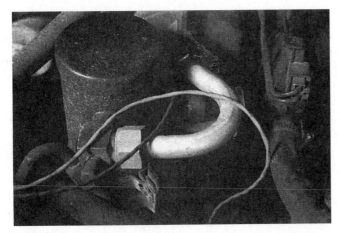

FIGURE 19–59 Attach the electronic thermometer probes to the inlet and outlet of the evaporator.

11. Place a **jumper** wire across the terminals of the temperature or pressure control, usually found on the accumulator.
12. Start the engine.
13. Set all of the air-conditioner controls to HIGH.
14. Allow the engine to reach normal-operating temperature.
15. Note and record the temperature of the two thermometer probes. Calculate the difference in temperature between the inlet and outlet tubes of the evaporator.
16. Wait a few minutes and record the temperatures again to confirm the readings.
17. Note and record the ambient temperature. Compare it with the chart in Figure 19–60, as applicable.
18. Follow the temperature differential chart (step 15) to determine how much refrigerant must be added to the system to ensure a proper charge.
19. Continue charging, as required.
    a. Tap a pound can of HFC-134a.
    b. With the can upright, open the manifold low-side valve, Figure 19–61.
    c. Dispense the contents of the can into the system.
    d. Close the low-side manifold valve.
    e. Repeat 19a, b, and c, as required.
20. Turn off the air conditioner.
21. Stop the engine.

AMBIENT TEMPERATURE (°F)						AMOUNT OF HFC-134a TO ADD (OUNCES)
60	70	80	90	100	110	
Evaporator Inlet To Outlet Temperature Difference						
–8	–8	–8	–8	–8	–8	0
–7	–7	–7	–7	–7	–7	2
–6	–6	–6	–6	–6	–6	4
–5	–5	–5	–5	–5	–5	6
+13	+13	+13	+17	+20	+25	8
+21	+25	+29	+33	+37	+42	12
+40	+45	+50	+55	+60	+65	14

A

AMBIENT TEMPERATURE (°C)						AMOUNT OF HFC-134a TO ADD (mL)
16	21	27	32	38	43	
Evaporator Inlet To Outlet Temperature Difference						
–5	–5	–5	–5	–5	–5	0
–4	–4	–4	–4	–4	–4	59
–3	–3	–3	–3	–3	–3	118
–3	–3	–2	–1	0	0	177
+7	+7	+7	+9	+11	+14	237
+10	+14	+16	+18	+21	+23	335
+22	+25	+28	+31	+33	+36	414

B

FIGURE 19–60 Compressor discharge pressure vs. ambient temperature chart: (A) English; (B) metric.

FIGURE 19–61 Dispense one can of refrigerant into the system.

22. Remove the jumper wire from the temperature-pressure switch (see step 11).
23. Close all of the valves.
    a. Manifold, Figure 19–62
    b. Hose shutoff, Figure 19–63
    c. Service, if so equipped
24. Disconnect all of the hoses from the system, Figure 19–64.
25. Replace all of the protective covers and caps, Figure 19–65.

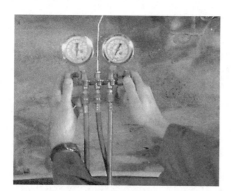

FIGURE 19–62 Make sure the high– and low–side manifold hand valves are in the closed position.

FIGURE 19–63 Close the hose shutoff valves.

FIGURE 19–64 Disconnect the hoses.

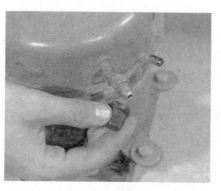

FIGURE 19–65 Replace protective caps and covers.

SUMMARY

After a retrofit procedure, appropriate decals must be **affixed** in a conspicuous place under the hood of the vehicle to identify the type of refrigerant found in the system. Decal masters, which are suitable for this purpose, are found in the color plate section of this textbook (Color Plate 16). For reproduction, take this textbook to a copy center and have a color copy of the page made. For convenience, there are several sizes of each decal. Decals may be laminated and/or affixed with rubber cement. They may also be protected from the elements with wide transparent carton sealing tape.

A light blue decal, Figure 19–66, should be placed over the current CFC-12 decal (if there is one). Although not required, decals may also be placed around the hoses at the service fittings, Figure 19–67.

According to a technical paper written by a staff engineer and a senior technician at Elf Atochem's fluorochemicals research and development center, retrofitting is not a difficult task. The paper was presented by Elf Atochem, a leading manufacturer of refrigerants, at an International CFC and Halon Alternatives Conference.

According to the report, a fleet of thirty-seven employee-volunteered vehicles were retrofitted to HFC-134a; seventeen with polyalkylene glycol (PAG) lubricant

NOTICE: RETROFITTED TO HFC-134a

RETROFIT PROCEEDURE PERFORMED TO SAE J1661
USE ONLY HFC-134a REFRIGERANT AND SYNTHETIC
OIL TYPE: _____1_____ PN: _____2_____OR
EQUIVALENT, OR A/C SYSTEM WILL BE DAMAGED

REFRIGERANT CHARGE/AMOUNT: _____3_____
LUBRICANT AMOUNT:_____4_____ PAG ☐ ESTER ☐5

RETROFITTER NAME: _____6_____ DATE: _____7_____
ADDRESS: _____8_____
CITY: _____9_____ STATE: ___10___ ZIP: ___11___

1 Type: manufacturer of oil
  (Saturn, GM, Union Carbide,
  etc.).

2 PN: Part number assigned
  by manufacturer.

3 Refrigerant charge / amount:
  Quantity of charge installed.

4 Lubricant amount: Quantity
  of oil installed (indicate
  ounces, cc. ml).

5 Kind of oil installed (check
  either PAG or ESTER).

6 Retrofitter name: Name of
  facility that performed
  the retrofit.

7 Date: Date retrofit
  is performed.

8 Address: Address of facility
  that performed the retrofit.

9 City: City in which the facility
  is located.

10 State: State in which the
   facility is located.

11 Zip: Zip code of the facility.

FIGURE 19–66 Retrofit label.

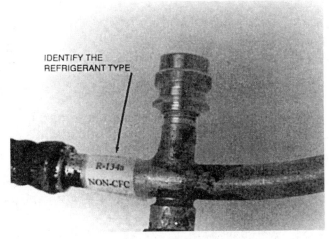

FIGURE 19–67 Service valve fitting identification. *(Courtesy of BET, Inc.)*

and twenty with polyol ester (**POE**) lubricant. Only the refrigerant and lubricant were replaced, not the system components. Some system components were power flushed to remove as much of the mineral oil as possible, others were simply drained and refilled. The study involved a random selection of both domestic and import cars and light trucks.

Regardless of the procedure, flush or no flush, there was little or no noticeable difference in the performance of any of the vehicles. There were only two reported failures; both lost their complete HFC-134a charge due to O-ring failure. Another vehicle lost 6 percent of its charge of HFC-134a due to a leak. The worst "leaker" in the study was actually a control vehicle that had not been retrofitted at all. This vehicle lost 38 percent of its CFC-12 charge due to a leak.

The paper concluded, "While retrofitting the existing auto CFC-12 fleet is not a trivial or cheap proposition, it is definitely not the disaster once thought."

Conclusion

The production of all CFC refrigerants ended on January 1, 1996 in the United States. CFC-12 cannot be imported into the United States from any other country except for medical use. Several attempts have been made to smuggle CFC-12 into this country, which have resulted in very heavy fines. This means that the only CFC-12 refrigerant that will be available for motor vehicle air-conditioning (MVAC) use will be that which has been recovered and recycled.

Some Final Points

❏ After a retrofit, ensure that proper labels are in place.
❏ Do not remove the HFC-134a fitting adapters from the CFC-12 fittings. Once installed they are to become a permanent part of the air-conditioning system.
❏ Ensure that the proper lubricant, Figure 19–68, has been added to the air-conditioning system for maximum system performance.
❏ Do not overcharge the air-conditioning system with refrigerant. The typical proper charge of HFC-134a for retrofit is about 90 percent of the original CFC-12 refrigerant charge. Refer to the chart in Figure 19–69 for the 90 percent rule.

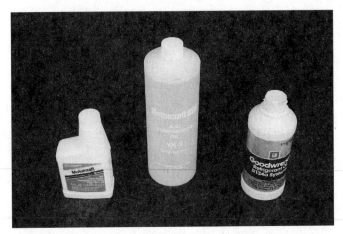

FIGURE 19–68 Refrigeration oil is packaged in several container sizes.

CFC-12		HFC-134A	
OUNCES	MILLILITERS	OUNCES	MILLILITERS
48	1420	43.2	1278
44	1302	39.6	1171
40	1183	36.0	1065
36	1065	32.4	958
32	947	28.8	852
30	887	27.0	799
28	828	25.2	745
26	769	23.4	692
24	710	21.6	639
22	651	19.8	586
20	592	18.0	532
18	532	16.2	479
16	473	14.4	426
14	414	12.6	373

FIGURE 19–69 The 90 percent rule for HFC–134a versus CFC–12 refrigerant charge.

REVIEW

Select the correct answer from the choices given.

1. Retrofit procedures are being discussed. Technician A says that it is always necessary to drain all mineral oil if PAG lubricant is to be used. Technician B says that it is always necessary to drain all mineral oil if POE lubricant is to be used. Who is right?
   a. A only
   b. B only
   c. Both A and B
   d. Neither A nor B

2. All of the following lubricants are approved for motor vehicle air-conditioning (MVAC) use, EXCEPT:
   a. Castor.
   b. Mineral.
   c. PAG.
   d. POE.

3. Which of the following refrigerant mixtures need *not* be considered contaminated for recovery purposes?

   a. CFC-12 with HCFC-22
   b. CFC-12 with HFC-134a
   c. CFC-12 with air
   d. All of the above

4. Technician A says a minimum pump down of 15 minutes is required to remove excess moisture. Technician B says a pump down is not required if a new receiver-drier or accumulator-drier is installed. Who is right?
   a. A only
   b. B only
   c. Both A and B
   d. Neither A nor B

5. Which of the following components is *least likely* to be replaced during retrofit procedures?
   a. Accumulator-drier
   b. Receiver-drier
   c. Evaporator
   d. Condenser

6. Which of the following components is *most likely* to be replaced during retrofit procedures?
   a. Nonbarrier hoses
   b. O-rings/gaskets
   c. Thermostatic expansion valve
   d. Orifice tube

7. The sealing provisions between the radiator and the condenser is referred to as a:
   a. barrier.
   b. dam.
   c. restrictor.
   d. shroud.

8. What is the color of the decal that is to be affixed to the vehicle to identify HFC-134a refrigerant?
   a. Yellow
   b. Blue
   c. Red
   d. Green

9. Which of the following is considered a contaminant in an HFC-134a air-conditioning system?
   a. CFC-12
   b. Air
   c. Both A and B
   d. Neither A nor B

10. Technician A says that DOT 39 cylinders may be used to store and transport contaminated refrigerant. Technician B says cylinders that contain contaminated refrigerant must be clearly identified. Who is right?
    a. A only
    b. B only
    c. Both A and B
    d. Neither A nor B

11. Technician A says that immersing a recovery cylinder in an ice bath will hasten refrigerant recovery. Technician B says that heating the receiver-drier or accumulator-drier will hasten refrigerant recovery. Who is right?
    a. A only
    b. B only

c. Both A and B
d. Neither A nor B

12. An empty DOT 39 cylinder:
    a. may be used as an air tank.
    b. may be used to store virgin refrigerant.
    c. may be used for contaminated refrigerant.
    d. must be properly disposed of.

13. Refrigerant HFC-134a cylinders are:
    a. white.
    b. blue.
    c. green.
    d. yellow.

14. Contaminated refrigerant:
    a. may be vented to the atmosphere.
    b. must be recovered for reuse.
    c. may be either A or B, depending on the contaminant.
    d. is neither A nor B.

15. Contaminated refrigerant can cause damage to:
    a. the air-conditioning system.
    b. recovery/recycle equipment.
    c. both A and B.
    d. neither A nor B.

16. Technician A says that federal laws require shops performing MVAC service to be equipped with certain certified equipment. Technician B says the technician may also be held accountable if certified equipment is not used. Who is right?
    a. A only
    b. B only
    c. Both A and B
    d. Neither A nor B

17. The color of a CFC-12 cylinder should be:
    a. white.
    b. yellow.
    c. green.
    d. blue.

NOTE: For questions 18 and 19, consider the following:

After evacuating an air-conditioning system for 30 minutes, the low-side gauge indicates less than 29 in. Hg. The manifold service valves are closed and the vacuum pump is turned off.

18. After 5 minutes the low-side gauge indicates 10 in. Hg. The most probable cause of this condition is:
    a. air in the air-conditioning system.
    b. a leak in the air-conditioning system.
    c. not all of the refrigerant was removed.
    d. all of the above.

19. After 5 minutes, the low-side gauge indicates 28 in. Hg. The most probable cause of this condition is:
    a. air in the air-conditioning system.
    b. a leak in the air-conditioning system.
    c. not all of the refrigerant was removed.
    d. all of the above.

20. After evacuating an air-conditioning system for 30 minutes, the low-side gauge only indicates about 15 in. Hg. The manifold service valves are closed and the vacuum pump is turned off. After about 5 minutes, the low-side gauge still indicates about 15 in. Hg. The most probable cause of this condition is:
    a. air in the air-conditioning system.
    b. a leak in the air-conditioning system.
    c. the service hose shutoff valve was not opened.
    d. an inefficient vacuum pump.

TERMS

Write a brief description of the following terms:

1. adequate
2. affixed
3. approved
4. can tap
5. components
6. compound gauge
7. cylinder
8. decal
9. depressing
10. dispense

11. eliminates
12. flush
13. integrated
14. jumper
15. POE
16. prescribed
17. probes
18. quantity
19. replace
20. rotate

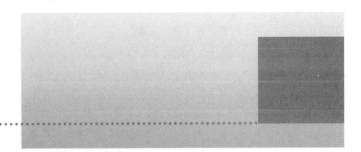

APPENDIX A

UNITED STATES ENVIRONMENTAL PROTECTION AGENCY SECTION 609 TECHNICIAN CERTIFICATION PROGRAMS

The following list is updated when other technician certification programs are approved. Section 609 covers technician certification in the motor vehicle sector only. Becoming certified allows the technician to:

- Purchase CFC-12 and ozone-depleting blend substitutes for CFC-12 refrigerant (at the present time, all blends are ozone depleting)
- Perform refrigerant servicing of vehicles with CFC-12, HFC-134a, or blend refrigerants

Although one must be certified to perform refrigerant servicing on vehicles equipped with an HFC-134a air-conditioning system, currently one does not have to be certified to purchase HFC-134a refrigerant. On June 11, 1988, however, the EPA issued a proposed regulation that would restrict the sale of HFC-134a refrigerant to certified technicians. If one becomes certified now, certification under Section 609 would allow the purchase of HFC-134a refrigerant when sales restriction is instituted.

Some programs offer home study and some offer training and testing on the World Wide Web (www). The following list, given in alphabetical order, was complete at the time of printing. For an up-to-date list, contact the EPA's Global Programs Division toll-free at 1 (800) 296-1996 or on the Internet at http://www.epa.gov/ozone/title6/609/609certs.html.

Air Conditioning Contractors of America/Ferris State University
1712 New Hampshire Avenue, NW
Washington, DC 20009
(202) 483-9370
http://www.acca.org

C.F.C. Reclamation and Recycling Service, Inc.
Post Office Box 560
Abilene, Texas 79604
(915) 675-5311

E F Technical Institute, Inc.
1860 Crown Drive
Suite 1400
Dallas, Texas 75234
(972) 831-8845

ESCO Institute
1350 West Northwest Highway
Suite 205
Mount Prospect, Illinois 60056
(800) 726-9696
http://www.escoinst.com/

The Greater Cleveland Automobile Dealers' Association
6100 Rockside Woods Boulevard, Suite 235
Independence, Ohio 44131
(216) 328-1500 or (888) 740-2886

International Mobile Air Conditioning Association (IMACA)
P.O. Box 9000
Fort Worth, Texas 76147-2000
(817) 732-4600
http://www.info@imaca.org

Mainstream Engineering Corporation
200 Yellow Place
Rockledge, Florida 32955
(407) 631-3550
http://www.epatest.com/
Mechanic's Education Association
1805 Springfield Avenue
Maplewood, New Jersey 07040-2910
(973) 763-0086

Mobile Air Conditioning Society Worldwide (MACS)
Post Office Box 88
225 S. Broad Street
Lansdale, Pennsylvania 19446
(215) 631-7020; fax: (215) 631-7017
http://www.info@macsw.org

National Institute for Automotive Service Excellence (ASE)
13505 Dulles Technology Drive
Herndon, Virginia 22071-3415
(703) 713-3800
talktoase@asecert.org

New York State Association of Service Stations and Repair Shops, Inc.
Automotive Technician Training Program
8 Elk Street
Albany, New York 12207
(518) 452-4367

New York State Department of Motor Vehicles,
Division of Vehicle Safety
Technical Training Unit
Empire State Plaza
Swan Street Building, Room 111
Albany, New York 12228
(518) 474-4049

Texas Engineering Extension Service
San Antonio Training Division
The Texas A & M University System
9350 South Presa
San Antonio, Texas 78223-4799
(210) 633-1000

Waco Chemicals, Inc.,
12306 Montague Street
Pacoima, California 91331
(818) 897-3018 or (800) 266-9226

Universal Technical Institute
3823 North 34th Avenue
Phoenix, Arizona 85017
(800) 859-7249

Vatterott College
10265 St. Charles Rock Road
St. Louis, Missouri 63074
(314) 843-4200
www.vatterot-college.com

The following programs are for their employees only.

Jiffy Lube International
P.O. Box 2967
Houston, Texas 77252-2967
(713) 546-4100

Los Angeles County Metropolitan Transportation Authority (MTA)
900 Lyon Street
Los Angeles, California 90012
(213) 922-5159

Potomac Electric Power Company
8400B Old Marlboro Pike
Upper Marlboro, Maryland 20772
(301) 967-5294

U.S. Army Ordnance Center and School
Aberdeen Proving Ground
Aberdeen, Maryland 21005-5201
(410) 278-2302 or 278-9774

Whayne Supply Company
P.O. Box 35900
Louisville, Kentucky 40323-5900
(502) 774-4441

Yellow Freight System, Inc.
10990 Roe Avenue
P.O. Box 7270
Overland Park, Kansas 66207
(913) 345-3000

The following no longer offer a certification program; however, certification cards and certificates issued by them are still valid:

Geneva Steel
K-Mart
Marine Safety Consultants/Tidewater School of
    Navigation
Minnesota Department of Transportation
Pense Auto Centers
Rancho Santiago College
Refrigerant Certification Services
Rider Trucks
Snap-On Tools

APPENDIX B

PRACTICE QUESTIONS FOR TECHNICIAN CERTIFICATION

1. Technician A says that an empty disposable refrigerant cylinder should be evacuated before being discarded. Technician B says that a disposable refrigerant cylinder should never be refilled to more than 60 percent. Who is right?
   a. A only
   b. B only
   c. Both A and B
   d. Neither A nor B

2. The ozone layer is in:
   a. the ionosphere.
   b. the stratosphere.
   c. the atmosphere.
   d. all of the above.

3. Refrigerant recovery/recycle equipment must be capable of cleaning refrigerant to what SAE standard?
   a. J1989
   b. J1990
   c. J1991
   d. J1992

4. What protects the earth from ultraviolet (UV) radiation?
   a. Ozone layer
   b. Cloud layer
   c. Atmospheric layer
   d. All of the above

5. When refilling an approved refrigerant cylinder with recovered refrigerant, what percentage of the cylinder's rated gross weight should not be exceeded?

   a. 60.
   b. 65.
   c. 70.
   d. 75.

6. Excessive ultraviolet (UV) radiation striking the earth causes:
   a. harm to plant and marine life.
   b. harm to animal and human life.
   c. both A and B.
   d. neither A nor B.

7. Technician A says that an approved refrigerant recovery cylinder should never be filled more than 70 percent with refrigerant. Technician B says that an approved refrigerant cylinder need not be evacuated if it is being refilled with the same type of refrigerant. Who is right?
   a. A only
   b. B only
   c. Both A and B
   d. Neither A nor B

8. Sunlight striking a CFC molecule in the atmosphere causes what to be released?
   a. Chlorine (Cl)
   b. Fluorine (F)
   c. Ozone (O_3)
   d. Hydrogen (H)

9. Recovered refrigerant should not be reused until it has been checked for:
   a. hydrocarbons.
   b. contamination.
   c. both A and B.
   d. neither A nor B.

10. The hole in the ozone was first discovered in:
    a. 1970.
    b. 1975.
    c. 1980.
    d. 1985.

11. All of the following statements are true, EXCEPT:
    a. never handle any chemical, including refrigerant, without first consulting a material safety data sheet (MSDS).
    b. never use an electrical outlet that is less than 12 inches (30 cm) above the floor.
    c. never fill a refrigerant storage tank more than 60 percent of its rated capacity.
    d. never transfer refrigerant into another tank unless it is DOT4BA or DOT4BW approved.

12. Containers for the storage of recycled refrigerant must meet what standard?
    a. DOT CFR Title 49
    b. SAE J1141
    c. ASHRAE 4BA
    d. All of the above

13. A hole in the ozone was first discovered over:
    a. Africa.
    b. North America.
    c. South America.
    d. Antarctica.

14. A static temperature-pressure test of a cylinder of refrigerant is being held. The ambient temperature is 72°F (54°C) and the tank pressure is 78 psig (528 kPa). Which of the following statements is most nearly correct?
    a. There is excessive air trapped in the refrigerant.
    b. Air, if any, in the refrigerant is not excessive.
    c. The refrigerant cylinder has been overfilled with refrigerant.
    d. The refrigerant contains an excessive amount of lubricant.

15. The use of recovery/recycle equipment when servicing mobile air-conditioning systems is established by the:
    a. Society of automotive Engineers (SAE).
    b. Motor Vehicle Manufacturers Association (MVMA).
    c. Air Conditioning and Refrigeration Institute (ARI).
    d. American Independent Air Conditioning Society (AIMCS).

16. All of the following statements are true, EXCEPT:
    a. closing the manifold and service valves prevents air from entering the system.
    b. closing the manifold and service valves prevents refrigerant from escaping the system.
    c. the burst pressure of service hoses is addressed in SAE Standard J1990.
    d. service hoses should be leak tested every 5 years.

17. How long should a mobile air conditioning system hold a stable vacuum to ensure that all of the refrigerant has been removed?
    a. 1 minute
    b. 5 minutes
    c. 15 minutes
    d. 1 hour

18. Technician A says that SAE Standard J1991 addresses the maximum allowable limit of moisture in recovered/recycled refrigerant. Technician B says SAE Standard J1991 addresses the maximum allowable noncondensible (air) gas in the recovered/recycled refrigerant. Who is right?
    a. A only
    b. B only
    c. Both A and B
    d. Neither A nor B

19. Which of the following statements is correct?
    a. Refrigerant recovery/recycle equipment approved by Underwriters Laboratories (UL) will clean refrigerant to the Society of Automotive Engineers (SAE) J1991 purity standards.
    b. The refrigerant recovery/recycle equipment being used must have a label stating that it is UL approved and certified to meet J1991 standards.
    c. Both A and B.
    d. Neither A nor B.

20. Refrigerant service hoses must have positive shut-off valves located:
    a. within 12 inches (30 cm) of each end.
    b. within 12 inches (30 cm) of the service end.
    c. at both ends
    d. at the service end.

21. When not in use, the hose shutoff valve(s) should be:
    a. closed.
    b. open.
    c. cracked.
    d. either of the above.

22. Technician A says the maximum allowable moisture content in recovered/recycled refrigerant is not to exceed 20 ppm. Technician B says the maximum allowable oil content of recovered/recycled refrigerant is not to exceed 400 ppm. Who is right?
    a. A only
    b. B only
    c. Both A and B
    d. Neither A nor B

23. Technician A says that, at the present time, a technician must be certified to service an air-conditioning system containing CFC-12. Technician B says that, at the present time, a technician does not have to be certified to service an air-conditioning system containing HFC-134a. Who is right?
    a. A only
    b. B only
    c. Both A and B
    d. Neither A nor B

24. All of the following statements are true, EXCEPT:
    a. Recovered refrigerant may be returned to the same vehicle without being processed.
    b. Refrigerant used for testing purposes, because it is such a small amount, can be released to the atmosphere.
    c. Recovered contaminated refrigerant should not be mixed with other refrigerant and thereby become contaminated.
    d. It is unlawful to vent any refrigerant, including CFCs or HFCs, to the atmosphere.

25. As the refrigerant condenses in the condenser, its pressure:
    a. is increased.
    b. is reduced.
    c. remains about the same.
    d. either of the above, depending on its temperature.

26. A "blend" refrigerant means that:
    a. it contains more than one component.
    b. it may be "blended" with another refrigerant.
    c. it is both A and B
    d. it is neither A nor B

27. When a refrigerant container is inverted, the air-conditioning system must be charged through the:
    a. high side with the compressor off.
    b. high side with the compressor running.
    c. low side with the compressor off.
    d. low side with the compressor running.

28. All of the following statements are true, EXCEPT: A vacuum pump is used to remove:
    a. air from an air-conditioning system.
    b. moisture from an-air conditioning system.
    c. trace refrigerant from an air-conditioning system.
    d. debris from an air-conditioning system.

29. How often must refrigerant recovery cylinders be inspected?
    a. Annually
    b. Every time they are refilled
    c. Every 5 years
    d. Whenever rust spots appear

30. The high-side gauge reads below 0 psig. This could be an indication that the gauge:
    a. is out of calibration and should be adjusted to zero.
    b. is hooked up to a system that is under a vacuum.
    c. is both A and B.
    d. is neither A nor B.

31. To pass a purity test, the refrigerant sample must be at least:
    a. 99 percent pure.
    b. 98 percent pure.
    c. 97 percent pure.
    d. 96 percent pure.

32. All of the following are required on a retrofit label, EXCEPT:
    a. the date of retrofit.
    b. the company or technician's certificate number.
    c. the type and quantity of refrigerant.
    d. the type and quantity of lubricant.

33. What is the minimum number of manifold and gauge sets required of a full-service repair facility to ensure compliance with federal regulations?
    a. One
    b. Two
    c. Four
    d. Six

34. A refrigerant purity analyzer will not identify CFC-12 or HFC-134a refrigerants if either one contains impurities greater than:
    a. 2 percent.
    b. 3 percent.
    c. 4 percent.
    d. 5 percent.

35. After a 30-minute pumpdown in a Miami service center, the low-side gauge indicates 28 in.-Hg. Five minutes after the vacuum pump is stopped, the low-side gauge reads 25 in.-Hg. Technician A says the problem is a leak in the air-conditioning system. Technician B says the vcuum pump may be defective. Who is right?
    a. A only
    b. B only
    c. Both A and B
    d. Neither A nor B

36. All of the following statements are true, EXCEPT:
    a. it is not always necessary to flush the air-conditioning system when retrofitting from CFC-12 to HFC-134A.
    b. the air-conditioning system capacity for HFC-134A refrigerant is about 90 percent of the CFC-12 capcity.
    c. a refrigerant identifier can be used to detect any hydrocarbon that may be present in a refrigerant.
    d. when used properly, a refirgerant purity tester will identify HFC-134a in any concentration.

37. What air-conditioning system pressure is required to conduct a leak test?
    a. A minimum of 40 psig (276 kPa)
    b. A maximum of 60 psig (414 kPa)
    c. Both A and B
    d. Neither A nor B

38. How often should refrigerant recovery cylinders be tested?
    a. Every 5 years
    b. Every 10 years
    c. Every year
    d. Every 2 years

39. Under the Clean Air Act (CAA) any person servicing a motor vehicle air-conditioning (MVAC) system:
    a. must be properly certified by an approved agency.
    b. must use properly certified recovery/recycle equipment.
    c. is both A and B.
    d. is neither A nor B.

40. Which of the following may be vented to the atmosphere?
    a. CFC-12
    b. HCFC-22
    c. Nitrogen
    d. None of the above.

41. Technician A says that PAG lubricant may be used in a CFC-12 air-conditioning system. Technician B says that mineral oil may be used in an HFV-134a air-conditioning system. Who is right?
    a. A only
    b. B only
    c. Both A and B
    d. Neither A nor B

42. All of the following alternate refrigerants are approved for automotive air-conditioning system service, EXCEPT:
    a. Freeze-12.
    b. FRIGC FR-12.
    c. Ikon-12.
    d. Duracool 12A.

43. Technician A says that DOT39 cylinders may be used for recoverying contaminated refrigerant. Technician B agrees but warns that the cylinder must be marked for identification—CONTAMINATED REFRIGERANT. Who is right?
    a. A only
    b. B only
    c. Both A and B
    d. Neither A nor B

44. Technician A says immersing a recovery cylinder in an ice bath will hasten refrigerant recovery. Technician B says heating a receiver or accumulator will hasten refrigerant recovery. Who is right?
    a. A only
    b. B only
    c. Both A and B
    d. Neither A nor B

45. Technician A says that a red label is used to identify HFC-134a in a retrofit CFC-12 air-conditioning system. Technician B says that a yellow labed is used to identify PAG lubricant in a retrofit air-conditioning system. Who is right?
    a. A only
    b. B only
    c. Both A and B
    d. Neither A nor B

46. Section 609 of the Clean Air Act (CAA) was signed into law by the president in:
    a. 1985.
    b. 1990.
    c. 1991.
    d. 1995.

47. An agreement, known as the Montreal Protocol, was signed by twenty-two countries in September:
    a. 1985.
    b. 1987.
    c. 1989.
    d. 1991.

48. Technician A says that unique fittings must be used on an air-conditioning system that is retrofit to HFC-134a. Technician B says that unique fittings must be used on an air-conditioning system retrofit to any other type of approved refrigerant. Who is right?
    a. A only
    b. B only
    c. Both A and B
    d. Neither A nor B

49. Technician A says that using a recovery cylinder beyond its reinspection date can result in heave penalties. Technician B agrees but adds that there is a 6-month grace period for reinspection. Who is right?
    a. A only
    b. B only
    c. Both A and B
    d. Neither A nor B

50. Technician A says that "subject to use conditions" means that certain system components must be replaced in some air-conditioning systems when using a particular refrigerant. Technician B says that "subject to use conditions" always means that a sticker and unique fittings are required when using any alternate refrigerant in an air-conditioning system. Who is right?
    a. A only
    b. B only
    c. Both A and B
    d. Neither A nor B

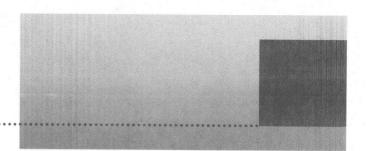

APPENDIX C

METRIC CONVERSIONS

	to convert these	to these,	multiply by:
TEMPERATURE	Celsius Degrees	Fahrenheit Degrees	1.8 then + 32
	Fahrenheit Degrees	Celsius Degrees	0.556 after - 32
LENGTH	Millimeters	Inches	0.03937
	Inches	Millimeters	25.4
	Meters	Feet	3.28084
	Feet	Meters	0.3048
	Kilometers	Miles	0.62137
	Miles	Kilometers	1.60935
AREA	Square Centimeters	Square Inches	0.155
	Square Inches	Square Centimeters	6.45159
VOLUME	Cubic Centimeters	Cubic Inches	0.06103
	Cubic Inches	Cubic Centimeters	16.38703
	Cubic Centimeters	Liters	0.001
	Liters	Cubic Centimeters	1,000
	Liters	Cubic Inches	61.025
	Cubic Inches	Liters	0.01639
	Liters	Quarts	1.05672
	Quarts	Liters	0.94633
	Liters	Pints	2.11344
	Pints	Liters	0.47317
	Liters	Ounces	33.81497
	Ounces	Liters	0.02957
	Milliliters	Ounces	0.3381497
	Ounces	Milliliters	29.57
WEIGHT	Grams	Ounces	0.03527
	Ounces	Grams	28.34953
	Kilograms	Pounds	2.20462
	Pounds	Kilograms	0.45359
WORK	Centimeter Kilograms	Inch-Pounds	0.8676
	Inch-Pounds	Centimeter-Kilograms	1.15262
	Meter Kilograms	Foot-Pounds	7.23301
	Foot-Pounds	Newton-Meters	1.3558
PRESSURE	Kilograms/Square Centimeter	Pounds/Square Inch	14.22334
	Pounds/Square Inch	Kilograms/Square Centimeter	0.07031
	Bar	Pounds/Square Inch	14.504
	Pounds/Square Inch	Bar	0.0689
	Pounds/Square Inch	Kilopascals	6.895
	Kilopascals	Pounds/Square Inch	0.145

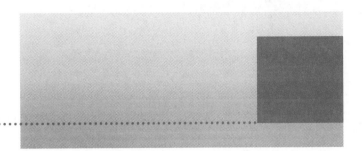

APPENDIX D

AUTOMOTIVE HEATING AND AIR CONDITIONING SPECIAL TOOL SUPPLIERS

Bright Solutions, Inc.
Troy, MI

Carrier Corporation
Syracuse, NY

Clardy Manufacturing Corporation
Fort Worth, TX

Classic Tool Design, Inc.
New Windsor, NY

Component Assemblies, Inc.
Bryan, OH

Corrosion Consultants, Inc.
Roseville, MI

CPS Products, Inc.
Hialeah, FL

Envirotech Systems, Inc.
Niles, MI

FJC, Inc.
Davidson, NC

Floro Tech Inc.
Pitman, NJ

Four Seasons
Division of Standard Motor Products, Inc.
Lewisville, TX 75057

Interdynamics, Inc.
Brooklyn, NY

K. D. Binnie Engineering Pty. Ltd.
Kirrawee, Australia NSW

KD Tools
Lancaster, PA

Kent Moore Division
SPX Corporation
Warren, MI

Lincor Distributors
N. Hollywood, CA

MAC Tools
Washington Courthouse, OH

Mastercool, Inc.
Rockaway, NJ

Neutronics, Inc.
Exton, PA

OTC Division
SPX Corporation
Owatonna, MN

Owens Research, Inc./Tubes 'N Hoses
Dallas, TX

P & F Technologies Ltd.
Mississauga, ONT, Canada

Ritchie Engineering Company Inc.
Garrett, IN

Robinair Division
SPX Corporation
Montpelier, OH

RTI Technologies, Inc.
York PA

The S. A. Day Manufacturing Company, Inc.
Buffalo, NY

Snap-On Tools Corporation
Kenosha, WI

Superior Manufacturing Company
Morrow, GA

Technical Chemicals Company
Dallas, TX

Thermolab, Inc.
Farmersville, TX

TIF Instruments, Inc.
Miami, FL

Tracer Products Division
Spectronics Corporation
Westbury, NY

Uniweld Products
Ft. Lauderdale, FL

Uview Ultraviolet
Mississauga, ONT, Canada

Varian Vacuum Technologies
Lexington, MA

Viper/T-Tech Division
Century Manufacturing Company
Minneapolis, MN

Yokagawa Corporation of America
Newnan, GA

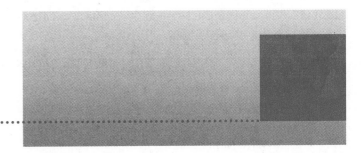

GLOSSARY

A: an abbreviation for ampere.

abbreviation: A shortened form of a word, phrase, title, or name.

absolute: Perfect in quality or nature; complete.

absolute humidity: The measure of the actual measure of water (H_2O) in a given volume of air.

absolute pressure: Pressure measured from absolute zero instead of normal atmospheric pressure.

absolute temperature: Temperature measured on the Rankine and Kelvin thermometers calibrated from absolute zero. The freezing point of water on the Rankine Scale is 492°R (273°K).

absolute zero: The absence of all heat, –459°F (–237.6°C). Also noted as 0° on the Rankine and Kelvin temperature scales. The temperature at which all molecular motion stops.

absorb: To take in, such as a sponge takes in liquid.

absorption: The process of absorbing.

A/C: Abbreviation for air conditioning or air conditioner. Also an abbreviation for alternating current.

accessory: A supplementary item or device, generally nonessential.

access valve: See *service port* and *service valve*.

accumulator: A tank located in the tailpipe to receive the refrigerant that leaves the evaporator. This device is constructed to ensure that no liquid refrigerant enters the compressor.

accumulator-dehydrator: An accumulator that includes a desiccant. See *accumulator* and *desiccant*.

accumulator-drier: See *accumulator-dehydrator*.

actuator: A device that transfers a vacuum or electric signal to a mechanical motion. Typically, an actuator performs an ON/OFF or OPEN/CLOSE function.

adaptive memory: The feature of a computer memory that allows the microprocessor to automatically compensate for changes in the dynamics of the process being controlled. Anything stored in adaptive memory is only lost when power to the computer is interrupted, such as when the battery is disconnected.

additives: Substances that are added, generally in small quantities, to improve or enhance performance.

adequate: Sufficient to satisfy a need or requirement.

adjuster: A means of adjusting a component, such as by linkage, to adapt or conform to new conditions or requirements.

adsorb: To take on, such as a liquid, on the surface of a solid.

affixed: To attach or secure to something, such as a label to a panel.

aftermarket: A term that applies to an accessory or component that has been added to a vehicle generally after the vehicle has been delivered to the customer.

air conditioner: A device used in the control of the temperature, humidity, cleanliness, and movement of air.

air conditioning: The process of adjusting and regulating, by heating or refrigerating, the quality, quantity, temperature, humidity, and the circulation of air in a space or enclosure; to condition the air.

Air-Conditioning and Refrigeration Institute (ARI): An association for the air-conditioning and refrigeration industry. The ARI is very active in research, provides educational materials, and helps to establish standards for the industry.

air delivery system: The component that contains the air ducts, doors, blower, evaporator core, heater core, and controls that delivers air to the interior of the vehicle via the various outlets.

473

air door: A door in the duct system that controls the flow of air in the air conditioner and/or heater.

air inlet valve: A movable door in the plenum blower assembly that permits the selection of outside air or inside air for both heating and cooling systems.

air outlet valve: A movable door in the plenum blower assembly that directs airflow into the heater core or into the ductwork that leads to the evaporator.

air pollution: See *pollution.*

air temperature sensor: A unit consisting of an aspirator, bimetallic sensing element, and a vacuum modulator used to sense in-vehicle temperature.

alignment pins: Index pins are generally used to ensure that two parts are properly mated.

allotropic: A term that applies to elements that appear in more than one form although their atomic composition is the same. For example, the properties of graphite and diamond are the same, the element carbon (C), but their physical properties are entirely different. Both, however, are called an allotropic of carbon (C).

alternate refrigerant: Any substance that is approved to replace an ozone-depleting substance in an air-conditioning or refrigeration system.

alternating current: An electric current, such as that produced by an alternator, that reverses direction in a circuit at regular intervals.

ambient air: Air surrounding an object.

ambient air temperature: See *ambient temperature.*

ambient compressor switch: An electrical switch that energizes the compressor clutch when the outside air temperature is 47°F (8.3°C) or above. Similarly, the switch turns off the compressor when the air temperature drops below 32°F (0°C).

ambient sensor: A thermistor used in automatic temperature control units to sense ambient temperature. Also see *thermistor.*

ambient switch: A switch used to control compressor operation by turning it on or off. The switch is regulated by ambient temperature.

ambient temperature: Temperature of the surrounding air. In air-conditioning work, this term refers to the outside air temperature.

ambient temperature sensor: A sensor that measures the outside air temperature as it enters the evaporator.

American Society of Mechanical Engineers (ASME): An association to which mechanical engineers can belong. Through ASME, members may keep cur-

rent on new technologies and procedures in the engineering field.

ammeter: A meter used to determine the current draw, in amperes, of a circuit or component.

ampere (A): A measure of current.

amplifier: A device used in automatic temperature control units to provide an output voltage that is in proportion to the input voltage from the sensors.

analog: Continuous physical variables, such as voltage or rotation, as opposed to *digital.*

aneroid bellows: An accordion-shaped temperature sensor charged with a small amount of volatile liquid. Temperature change causes the bellows to contract or expand, which, in turn, opens or closes a switch as in a thermostat.

annealed copper: Copper (Cu) that has been heat treated to render it workable; commonly used in refrigeration systems.

antiboil: A substance that may be added to a vehicle cooling system to raise the boiling point of the coolant, thereby preventing loss of coolant due to boilover.

antifreeze: A commercially available additive solution used to increase the boiling temperature and reduce the freezing temperature of engine coolant. A solution of 50 percent water and 50 percent antifreeze is suggested for year-round protection.

appliance: Also known as small appliance. Any device that contains and uses 5 pounds (2.27 kilograms) or less of Class I or Class II substance as a refrigerant, which includes any room air conditioner, refrigerator, freezer, packaged heat pump, dehumidifier, under the counter ice maker, vending machine, and drinking water fountain.

application: The act of putting something to special use; to have a specific purpose.

approved: Considered right or good; to consent, confirm, or sanction, such as to approve HFC-134a for replacing CFC-12 in motor vehicle air-conditioning (MVAC) systems.

ARI: Acronym for Air-Conditioning and Refrigeration Institute.

ASE: Registered trademark of the National Institute for Automotive Service Excellence.

ASHRAE: An abbreviation for American Society of Heating, Refrigeration, and Air-Conditioning Engineers.

ASME: Acronym for American Society of Mechanical Engineers.

aspirator: A device that uses suction to move air, accomplished by a differential in air pressure.

ATC: Abbreviation for automatic temperature control.

ATC servo programmer: A mechanically operated switch to control blower speed whenever the blower switch is in the AUTO position on some vehicle lines.

atmosphere: Air, such as in the air we breathe.

atmospheric ozone: As ultraviolet (UV) rays from the sun reach Earth, they are combined with smog and other pollutants to produce atmospheric ozone. Atmospheric ozone, unlike stratospheric ozone, is considered harmful. Whenever possible it is to be avoided.

atmospheric pollution: See *pollution*.

atmospheric pressure: Air pressure at a given altitude. At sea level, atmospheric pressure is 14.696 psia (101.329 kPa absolute).

atom: The smallest possible particle of matter.

auto: Abbreviation for automatic or automobile.

auto control: See *automatic control*.

automatic: A self-regulating system or device that adjusts to variables of a predetermined condition.

automatic control: A thermostatic dial on the instrument panel that can be set at a comfortable temperature level to automatically control the flow of air.

automatic temperature control (ATC): The name of an air-conditioner control system designed to automatically maintain an in-vehicle temperature and humidity level at a preset level or condition.

auxiliary: A component or device giving assistance or support to the primary component or device, such as a compressor auxiliary shaft oil seal.

auxiliary seal: A seal mounted outside the seal housing to prevent refrigeration oil from entering the clutch assembly.

axial: Pertaining to an axis; a pivot point.

axial compressor: A compressor so designed that the cylinders are arranged parallel to the output shaft.

axial plate: A term often used for "swash plate" or "wobble plate."

azeotrope: A mixture of two or more liquids, such as refrigerants, which, when mixed in precise proportions, behave like a compound when evaporating or condensing.

back idler: A pulley that tightens the drive belt; the pulley rides on the back or flat side of the belt.

back seat (service valve): Turning the valve stem to the left (ccw) as far as possible back seats the valve. The valve outlet to the system is open and the service port is closed.

barb fitting: A fitting that slips inside a hose and is held in place with a gear-type clamp. Ridges (*barbs*) on the fitting prevent the hose from slipping off.

barometer: An instrument used to measure the weight or pressure of the atmosphere.

barometric pressure: Atmospheric pressure as measured with a barometer.

barrier hose: A hose specially constructed with a liner to prevent refrigerant leakage through its walls. Most air-conditioning systems in vehicles manufactured after 1988 have barrier-type hoses.

battery voltage: The voltage of a fully charged battery generally assumed to be 12 volts. Actually, a fully charged motor vehicle lead-acid battery may be 13.2 volts.

bellows: An accordion-type chamber that expands or contracts with temperature changes to create a mechanical controlling action such as in a thermostatic expansion valve.

belt: See *V-belt*, *V-groove belt*, and *serpentine belt*.

belt dressing: A prepared spray solution formulated for use on automotive belts to reduce or eliminate belt noise. Not recommended for serpentine belts.

bi-level: Two levels; generally refers to vent and floor air delivery in a motor vehicle.

bimetal: See *bimetallic*.

bimetallic: Two dissimilar metals fused together; these metals expand (or contract) at different temperatures to cause a bending effect. Bimetallic elements are used in temperature-sensing controls.

bimetallic sensor: A sensor using a bimetallic strip or coil.

bimetallic temperature sensor: See *bimetallic thermostat*.

bimetallic thermostat: A thermostat that uses bimetallic strips instead of a bellows for making or breaking contact points.

bleeding: Slowly releasing pressure in the air-conditioning system by drawing off some liquid or gas.

bleed orifice: A calibrated orifice, such as in a vacuum system, that allows ambient air to enter the system to equalize the vacuum.

blend air: The control of air quality by blending heated and cooled air to the desired temperature.

blend air door: A door in the duct system that controls temperature by blending heated and cooled air.

blend airstream: See *blend air door.*

blend door: A door in the case/duct system that may be positioned so as to mix cooled air with heated air to achieve the desired in-vehicle temperature-humidity conditions.

blower: See *squirrel-cage blower.*

blower circuit: All of the electrical components required for blower speed control.

blower fan: See *squirrel-cage blower* or *fan.*

blower motor: See *motor.*

blower motor relay: See *blower relay.*

blower relay: An electrical device used to control the function or speed of a blower motor.

blower resistor: See *resistor.*

blower speed controller: A solid-state control device that operates the blower motor and compressor clutch on signals from the microprocessor.

blower switch: A dash-mounted device that allows the operator to turn the blower motor on and off and/or control its speed.

blown: A term often used when referring to a defective fuse.

boiling: A rapid change in state of a liquid to a vapor by adding heat, decreasing pressure, or both.

boiling point: The temperature at which a liquid changes to a vapor.

boils: The changing of a substance from a liquid to a vapor by adding heat energy.

bore: A compressor cylinder, or any cylinder hole. The *bore* size is the inside diameter of the hole.

Bowden cable: A wire cable inside a metal or rubber housing used to regulate a valve or control from a remote place.

box socket: A type of socket wrench.

brazing: A high-temperature metal joining process that is satisfactory for units with relatively high internal pressures.

British thermal unit (Btu): A measure of heat energy; one Btu is the amount of heat necessary to raise 1 pound of water 1°F.

Btu: Abbreviation for British thermal unit.

bug screen: A mesh-like covering that is placed over the front of a vehicle to "catch" bugs before they collide with the vehicle and/or condenser. Not generally recommended because of blocking the airflow.

bypass control valve: See *hot gas bypass valve.*

CAA: Acronym for Clean Air Act.

cable: A wire-like device, generally in a sheath or housing, used to impart a mechanical motion between two points.

calibrated: A device, such as a pressure gauge, that has been corrected for accuracy.

calibration: The act of correcting a device, such as a pressure gauge, by comparing it to one of known accuracy, called a *standard.*

calorie: The smallest measure of heat energy. One calorie is the amount of heat energy required to raise 1 gram of water 1°C. There are 252 calories in 1 Btu.

cam: An off-center member of a turning shaft; a lobe.

can tap: a device used to pierce, dispense, and seal small cans of refrigerant.

can tapper: See *can tap.*

can valve: See *can tap.*

capacitor: An electrical device for accumulating and holding a charge of electricity.

capacity: The refrigeration produced, measured in tons or Btu per hour.

capillary: A small tube with a calibrated length and inside diameter used as a metering device.

capillary attraction: The ability of tubular bodies to draw up a fluid.

capillary tube: A tube with a calibrated inside diameter and length used to control the flow of refrigerant. In automotive air-conditioning systems, the tube connecting the remote bulb to the expansion valve or to the thermostat is called the capillary tube.

capture: To attract and/or hold, such as to capture moisture in an air-conditioning system.

carbon: An element; chemical symbol is C.

carbon dioxide: A harmful, colorless, odorless gas formed by respiration, combustion, and decomposition. The chemical symbol is CO_2.

carbon monoxide: A hazardous by-product of burned gasoline. It is odorless and colorless, therefore not easily detected. Inhalation of carbon monoxide (CO) can be fatal.

carbonyl chlorofluoride: A toxic by-product of Refrigerant-12 if allowed to come into contact with an open flame or heated metal. The fumes of carbonyl chlorofluoride (COClF) should be avoided.

carbonyl fluoride: A toxic by-product of Refrigerant-12 if allowed to come into contact with an open flame or heated metal. The fumes of carbonyl fluoride (COF_2) should be avoided.

caution: A warning; careful forethought to warn of possible danger or harm.

CCFOT: Abbreviation for cycling clutch fixed orifice tube.

CCOT: Abbreviation for cycling clutch orifice tube.

CCPS: An acronym for clutch cycling pressure switch.

ccw: Abbreviation for counterclockwise.

Celsius: A metric temperature scale using the freezing point of water as zero. The boiling point of water is 100°C (212°F English).

center-mount components: The installation in a heating and air-conditioning system, whereby the evaporator is mounted in the center of the fire wall on the engine side and the heater core is mounted directly to the rear in the passenger compartment.

centigrade: A term often used to indicate *Celsius*. A term not used in the SI metric system. See *Celsius*.

centimeter: A unit of measure in the SI metric system. One centimeter is equal to 0.3937 inch in English measure.

centrifugal force: A natural tendency of a body to separate when spinning or turning.

certification: The state of being "certified." A document attesting to the truth of something, such as one's knowledge and ability to service motor vehicle air-conditioning (MVAC) systems.

CFC: Abbreviation for chlorofluorocarbons, a man-made compound used in refrigerants.

CFM: Also cfm. Abbreviation for cubic feet per minute. Sometimes expressed as ft_3/min.

CFR: Acronym for Code of Federal Regulations.

change of state: Rearrangement of the molecular structure of matter as it changes between any two of the three physical states: solid, liquid, or gas.

charge: A specific amount of refrigerant or oil by volume or weight.

charge tolerance: The accuracy, plus/minus, required in the amount of refrigerant to be charged into the system.

charging: The act of placing a charge of refrigerant or oil into the air-conditioning system.

charging cylinder: A container with a visual indicator for use where a critical, or exact, amount of refrigerant must be measured.

charging hose: A hose with a small diameter constructed to withstand high pressures; the hose is located between the unit and the manifold set.

charging station: A unit containing a manifold and gauge set, charging cylinder, vacuum pump, and leak detector. This unit is used to service air conditioners.

check relay: See *check valve relay*.

check valve: A device located in the liquid line or inlet to the drier. The valve prevents liquid refrigerant from flowing the opposite way when the unit is shut off. Also, a device used in the vacuum system to ensure adequate vacuum reserve during periods of acceleration and other low-vacuum periods.

check valve relay: An electrical switch used to control a solenoid-operated check valve.

chemical: A substance that has a particular molecular composition, such as HFC-134a refrigerant.

chemical compound: Two or more chemically pure elements blended to form a mixture, such as HFC-134a.

chemical instability: An undesirable condition caused by the presence of contaminants in the refrigeration system.

chlorofluorocarbon: A man-made compound used in refrigerants, such as R-12, more accurately designated CFC-12.

chlorine: A highly irritating greenish-yellow gas found in many Class I and Class II refrigerants. Chlorine (Cl) has been found to be hazardous to the ozone, and refrigerants containing this gas have been or are being phased out of production.

CID: Abbreviation for cubic-inch displacement.

circuit breaker: A bimetallic device used instead of a fuse to protect a circuit.

Cl: An abbreviation for chlorine.

clamps: Mechanical devices used to hold two or more components, generally temporarily.

clamping diode: A diode that is used to protect delicate solid-state electronic circuits against unwanted voltage "spikes."

Class I substance: This term relates to fully halogenated chlorofluorocarbon refrigerants. The production, use, and sale of Class I refrigerants are controlled by the Clean Air Act (CAA).

Class II substance: This term relates to a group consisting of halons that are used in fire extinguishers. Their use, production, and sale are controlled by the Clean Air Act (CAA).

Clean Air Act (CAA): A Title IV amendment, signed into law in 1990 by President Bush, established national policy relative to the reduction and elimination of ozone-depleting substances.

clean: See *purge* and *flush.*

clogged: Obstructed, such as a debris-filled TXV or FOT screen.

clutch: A coupling device that transfers torque from a driving to a driven member when desired.

clutch armature: That part of the clutch attached to the compressor crankshaft that is pulled in when engaged.

clutch coil: See *clutch field.*

clutch cycle time (total): See *cycle clutch time (total).*

clutch cycle pressure switch: A pressure-actuated switch that controls compressor clutch action to prevent evaporator icing.

clutch cycle switch: An electrical switch, pressure or temperature actuated, that cuts off the compressor at a predetermined evaporator temperature.

clutch field: Consists of many windings of wire and is fastened to the front of the compressor. Current applied to the field sets up a magnetic field that pulls the armature in to engage the clutch.

clutch plate: See *clutch armature.*

clutch rotor: That portion of the clutch in which the belt rides. The rotor is freewheeling until the clutch is engaged. On some clutches, the field is found in the rotor and the electrical connection is made by the use of brushes.

cm: Abbreviation for centimeter.

CO: Chemical symbol for carbon monoxide.

Code of Federal Regulations (CFR): Regulations that are generated, published, and enforced by the United States government.

coil: A term often used for clutch field coil.

cold: The absence of heat.

cold leak: A refrigerant leak that is easier to detect when the engine and air-conditioning system have been turned off long enough for all components to have cooled.

comb: See *condenser comb.*

combination valve: Used on some vehicles; an H-valve having a suction throttling valve and expansion valve combined.

comfort: A pleasing and enjoyable environment; the removal of excessive heat, moisture, dust, and pollen from the air.

Comfortron: Another name for an automatic temperature control.

common ground: An area where all ground wires are attached, such as the metal frame and/or body parts of a vehicle.

compatibility: A term used to indicate that a substance may be used with another substance, such as PAG lubricant is compatible with refrigerant HFC-134a.

complete circuit: A circuit without interruption in which electrical current may travel to and from the battery.

components: Parts of mechanical or electrical systems.

compound: A combination of two or more elements, substances, ingredients, or parts.

compound gauge: A gauge that registers both pressure and vacuum (above and below atmospheric pressure); used on the low side of the systems.

compress: To reduce the volume by pressure.

compression ratio: The ratio of maximum to minimum piston chamber volume during one piston stroke.

compressor: A component of the refrigeration system that pumps refrigerant and increases the pressure of the refrigerant vapor.

compressor discharge pressure switch: A pressure-operated electrical switch that opens the compressor clutch circuit during high-pressure conditions.

compressor displacement: A value obtained by multiplying the displacement of the compressor cylinder or cylinders by a given rpm, usually the average engine speed of 30 mph, or 1,750 rpm.

compressor protection switch: An electrical switch installed in the rear head of some compressors to stop the compressor in the event of a loss of refrigerant.

compressor shaft seal: An assembly consisting of springs, snap rings, O-rings, shaft seal, seal sets, and gasket. The shaft seal is mounted on the compressor crankshaft and permits the shaft to be turned without a loss of refrigerant or oil.

concentration: The amount of substance in a unit compared to another substance in the same unit.

condensate: Water taken from the air; the water forms on the exterior surface of the evaporator.

condensation: The process of changing the vapor to a liquid.

condenser: The component of a refrigeration system in which refrigerant vapor is changed to a liquid by the removal of heat. Also, a term often used (incorrectly) when referring to a capacitor.

condenser comb: A comb-like device used to straighten the fins on the evaporator or condenser.

condenser temperature: The temperature at which compressed gas in the condenser changes from a gas to a liquid.

condensing pressure: Head pressure as read from the gauge at the high-side service valve; the pressure from the discharge side of the compressor to the condenser.

conditioned air: Air that is cool, dry, and clean.

conditions: Modes or states of being, such as low or high pressure, and hot or cold temperature.

conduction: The transmission of heat through a solid.

conduction of heat: The ability of a substance to conduct heat.

conductor: A material capable of carrying an electrical current, such as a copper (Cu) wire.

connector: A device used to join or fasten two or more parts, such as electrical wires.

contacts: A term used for electrical points that open and close to break or make a circuit.

container: A receptacle in which something is held or carried.

contaminant: Anything other than refrigerant and refrigeration oil in the system.

contaminated: A term generally used when referring to a refrigerant cylinder or a system that is known to contain foreign substances such as other incompatible or hazardous refrigerants.

control head: The master controls (such as temperature and fan speed) that the driver uses to select the desired system condition.

control panel: A dash-mounted panel that contains the controls for a device, such as the air-conditioning system.

control valve: A term sometimes used for the heater valve or thermostatic expansion valve.

convection: The transfer of heat by the circulation of a vapor or liquid.

conversion: The substitution of one element or component for another as in substituting HFC-134a for CFC-12 in an automotive air-conditioning system. More properly referred to as *retrofit*.

coolant: The fluid, usually a mixture or antifreeze and water, that is circulated through the cooling system to carry away engine heat.

coolant flow valve: A term used for heater valve.

coolant recovery tank: See *expansion tank*.

coolant thermostat: See *thermostat*.

cooling coil: See *evaporator*.

cooling fan controller module: An electronic unit that will cycle the cooling fan on and off in response to signals from other engine sensors.

cooling system: All of the components required to remove heat from the engine. These include the engine water jackets, water pump, radiator, thermostat, pressure cap, and connecting hoses.

Cool Pack®: A trade name used by Harrison Radiator Division of General Motors to describe their hang-on or under-dash (aftermarket) air-conditioning systems.

core: The coolant passages and fins of a radiator or heater found between the two header tanks.

core hole plug: See *core plug*.

core plug: Commonly known as a *freeze plug*, the core plug is a metal cup-shaped disc that is inserted into the engine block to seal holes that were provided to remove casting sand when the block was cast.

corrosion: The decomposition of metal, caused by a chemical action, usually acid.

CO_2: Chemical symbol for carbon dioxide.

cowl air intake: The inlet at the base of the windshield that allows outside air to enter the heating/air-conditioning system.

cracked: A term often used for the midseated position of a service valve. Also see *midpositioned*.

crankcase: See *sump*.

crankshaft: That part of a reciprocating compressor on which the wobble plate or connecting rods are attached to provide for an up-down or to-fro piston action.

crankshaft seal: See *compressor shaft seal*.

crossflow radiator: A radiator in which the coolant flow is from one side to the other, as opposed to a *vertical flow radiator*.

cross mixing: The intentional or unintentional mixing of two or more substances.

cubic feet per minute (cfm): The quantity of air or fluid that will pass a given point in one minute.

cubic-inch displacement: The cylinder volume of a compressor as the piston moves from the bottom of its stroke to the top of its stroke, in cubic inches (in.^{3}).

current: The movement of electrons through a conductor. Current is measured in *amperes*.

current draw: The amount of current, in amperes, required to operate an electrical load device such as a blower motor.

custom: Made to order.

custom system: A deluxe automotive air-conditioning system that uses both inside and outside air.

cutoff switch: An electrical switch that is pressure or temperature operated. The switch is used to interrupt the compressor clutch circuit during certain low- or high-pressure conditions.

cw: Abbreviation for clockwise.

cycle: An event, from start to finish. Also see *refrigeration cycle*.

cycle clutch time (total): Time from the moment the clutch engages until it disengages, then re-engages. Total time is equal to on time plus off time for one cycle.

cycling clutch: A clutch that is turned on and off to control temperature.

cycling clutch fixed orifice tube: An air-conditioning system that has a fixed orifice tube (expansion tube) in which the air temperature is controlled by starting and stopping the compressor with a thermostat or pressure control. See *CCFOT*.

cycling clutch FOT: See *cycling clutch fixed orifice tube*.

cycling clutch orifice tube: See *cycling clutch fixed orifice tube*.

cycling clutch system: An air-conditioning system in which the air temperature is controlled by starting and stopping the compressor with a thermostat or pressure control.

cylinder: A circular tube-like opening in a compressor block or casting in which the piston moves up and down or back and forth; a circular drum used to store refrigerant.

dam: The sealing provisions located between the radiator and condenser to ensure adequate ambient and ram air through both components.

dampened pressure switch: An electrical pressure switch that opens the compressor clutch circuit when the low-side pressure is too low.

DB: Abbreviation for dry bulb.

DC: Abbreviation for direct current.

debris: The fragmented remains of a defective or eroded part; also may refer to trash or other such matter.

decal: A transfer picture, design, or label.

declutching fan: An engine cooling fan mounted on the water pump. A temperature-sensitive device is provided to govern or limit terminal speed.

decomposing: The act of separating into separate elements or components; to break down by rot or decay.

defective: Faulty; something that does not perform or function as it should.

defogger: That part of the heater system designed to clear the windshield of fog haze under certain conditions.

defrost: To remove frost.

defrost door: A small door within the duct system to divert a portion of the delivery air to the windshield.

defroster: That part of the heater system designed to clear heavy frost or light ice from the inside or outside of the windshield.

defroster door: See *defrost door*.

defrost switch: A thermostatic-type switch, sensing evaporator temperature, that turns the compressor off and on to prevent frosting or freezing of the evaporator.

dehumidify: To remove water vapor from the air.

dehydrate: See *purge* and *evacuate*.

dehydrated: A device or system that is dry from which all water has been removed.

dehydrator: See *filter drier*.

dehydrator filter: See *filter drier*.

deice switch: A switch used to control the compressor operation to prevent evaporator freeze-up.

delay relay: See *time-delay relay*.

delta t: A range in temperature from high to low or low to high; often referred to as Δ_t.

density: The weight or mass of a gas, liquid, or solid.

deoxidized: A tubing or metal surface that is free of oxide formations, which are caused by the action of air or other chemicals.

Department of Transportation (DOT): The United States Department of Transportation is a federal agency charged with regulation and control of the shipment of all hazardous materials.

depressing: To press down, such as a Schrader valve core stem.

depressurize: See *discharge*.

desiccant: A drying agent used in refrigeration systems to remove excess moisture. The desiccant is located in the receiver-drier or accumulator-drier.

design working pressure: The maximum allowable working pressure for which a specific system component is designed to work safely.

dew point: The point where air becomes 100 percent saturated with moisture at a given temperature.

diagnosis: The procedure followed to locate the cause of a malfunction.

diagrams: Plans, sketches, drawings, or outlines, such as electrical schematics.

diaphragm: A rubber-like piston or bellows assembly that divides the inner and outer chambers of back-pressure-regulated air-conditioning control devices.

dichlorodifluoromethane: The proper name for Refrigerant-12. *See R-12.*

dieseling: The engine continuing to run when the ignition switch is turned off due to an overheated condition.

digital: Of or pertaining to digits. Generally refers to a meter, such as an ammeter, that displays numbers as opposed to *analog.*

diode: An electrical check valve. Current flows only in one direction through a diode.

direct current (DC): An electric current that flows in one direction only.

discharge air: Conditioned air as it passes through the outlets and enters the passenger compartment.

discharge line: Connects the compressor outlet to the condenser inlet.

discharge pressure: Pressure of the refrigerant being discharged from the compressor; also known as the high-side pressure.

discharge pressure switch: See *compressor discharge pressure switch.*

discharge side: That portion of the refrigeration system under high pressure, extending from the compressor outlet to the thermostatic expansion valve inlet.

discharge stroke: The movement of a piston from the bottom to the top of its stroke.

discharge valve: See *high-side service valve.*

disinfectant: An agent, such as heat, radiation, or a chemical that destroys, neutralizes or inhibits growth of microorganisms.

dispense: To give out or distribute in parts or portions.

displacement: In automotive air conditioning, this term refers to the compressor stroke X bore.

disposable: Designed to be disposed of, such as a virgin refrigerant cylinder.

disposable cylinders: Containers for one-time use in the packaging, transporting, and dispensing of refrigerant. It is a violation of federal law to refill these cylinders, commonly referred to as *DOT-39s.*

dissipation: The elimination of matter, such as water vapor, into the atmosphere, which becomes known as humidity.

distributor: A device used to divide the flow of liquid refrigerant between parallel paths in an evaporator.

diverted: Turned aside from a course or direction.

DIY: Also DIYer. An acronym for *do-it-yourselfer.*

domestic: A term that generally refers to products and goods that are manufactured in the United States.

DOT: An acronym for the U.S. Department of Transportation.

DOT 39: Trade jargon for a disposable refrigerant cylinder. Disposable cylinders should never be refilled or used as compressed air tanks.

double action: To impart motion or exert energy in two directions, generally opposing.

double flare: A flare on the end of a piece of copper tubing or other soft metal; the tubing is folded over to form a double face.

downflow radiator: A radiator in which the coolant flow is from the top tank to the bottom tank, as opposed to a *crossflow radiator.*

downstream: A component or device that is placed after another component, such as the downstream blower of a heater core.

downstream blower: A blower arranged in the duct system so as to pull air through the heater and air-conditioner core(s).

drain cock: Drain provisions that allow the removal of coolant from the radiator or engine block.

drier: A device containing desiccant; a drier is placed in the liquid line to absorb moisture in the system.

drip pan: A shallow pan, located under the evaporator core, used to catch condensation. A drain hose is fastened to the drip pan and extends to the outside to carry off the condensate.

drop in: A term used by some alternate refrigerant manufacturers for their refrigerant as areplacement for

CFC-12, meaning that no system alterations or changes are required. In spite of claims, the automotive industry advises that there is no such thing as a drop-in refrigerant to replace CFC-12.

dry bulb (DB): A term often used when referring to a standard spirit-type thermometer, as opposed to a "wet bulb" thermometer.

drying agent: See *desiccant*.

duct: A tube or passage used to provide a means to transfer air or liquid from one point or place to another.

DVOM: An abbreviation for digital volt-ohmmeter.

Dytel®: A red dye additive included in some DuPont class I refrigerants used for detecting leaks.

EATC: Abbreviation for electronic automatic temperature control.

EEVIR: Abbreviation for evaporator equalizer valves-in-receiver; see *valves-in-receiver*.

effective temperature: The expected or intended temperature.

efficiency: The output of energy divided by the input of energy; may be expressed in percentage or ratio.

electro-drive cooling fan: An electrically controlled engine cooling fan that cycles on and off with the compressor clutch.

electromagnet: A temporary magnet created by passing electrical current through a coil of wire. A clutch coil is a good example of an electromagnet.

electromagnetic: A device consisting of a ferrous-metal core and a coil of wire that produces a magnetic effect when an elecrtic current is applied in the coil.

electromagnetic field: The magnetic force created by an electromagnet.

electronic cycling clutch switch: An electronic switch that prevents the evaporator from freezing by signaling various electronic control devices when the evaporator reaches a predetermined low temperature.

electronic leak detector: An electrically (AC or DC) powered leak detector that emits an audible and/or visual signal when its sensor is passed over a refrigerant leak.

electrovacuum relay: A device that prevents blower motor operation when heat is selected and the engine is not up to operating temperature.

element: Any of the 105 identified chemicals known to make up all matter. Each element has a symbol, an atomic number, and an atomic weight.

eliminates: Getting rid of; removing.

energy: The capacity for performing work.

engine cooling system: See *cooling system*.

engine idle compensator: A thermostatically controlled device on the carburetor that prevents stalling during prolonged hot weather periods while the air conditioner is operated.

engine thermal switch: An electrical switch designed to delay the operation of the system in cool weather to allow time for the engine coolant to warm up.

engine thermostat: A temperature sensitive mechanical device found at the coolant outlet of an engine that expands (opens) or contracts (closes) to control the amount of coolant allowed to leave the engine, based on its temperature.

English scale: A term that refers to the English system of measurement as opposed to the metric system.

environment: The conditions or circumstances that surround one; the surrounding ambient atmosphere.

Enviromental Protection Agency (EPA): An agency of the U.S. government charged with the responsibility of protecting the environment and enforcing the Clean Air Act (CAA) of 1990.

EPA: An acronym for the Environmental Protection Agency.

EPR: See *evaporator pressure regulator*.

equalized: To be equal or in balance.

equalizer line: A small-bore line used to provide a balance of pressure from one point to another, as in a thermostatic expansion valve.

ESTER: A trade term for the synthetic lubricant, polyol ester.

ethylene glycol: See *antifreeze*.

ETR: See *evaporator temperature regulator*.

evacuate: To create a vacuum within a system to remove all trace of air and moisture.

evaporation: The process of changing from a liquid to a vapor.

evaporator: The component of an air-conditioning system that conditions the air.

evaporator coil: Also referred to as "evaporator core," the tube and fin assembly that is located inside the evaporator case where heat is removed from air passing through it.

evaporator control valve: Can refer to any of the several types of evaporator suction pressure control valves or devices used to regulate the evaporator temperature by controlling the evaporator pressure.

evaporator core: The tube and fin assembly located inside the evaporator housing. The refrigerant fluid picks up heat in the evaporator core when it changes into a vapor.

evaporator equalizer valves-in-receiver: See *valves-in-receiver.*

evaporator housing: The cabinet, or case, that contains the evaporator core. Often, the diverter doors, duct outlets, and blower-mounting arrangement are found on the housing.

evaporator pressure control valve: See *evaporator control valve.*

evaporator pressure regulator: A back-pressure-regulated temperature control device used on some vehicles.

evaporator temperature regulator: A temperature-regulated device used by Chrysler Air-Temp to control the evaporator pressure.

exhaust stroke: A term used for discharge stroke.

expansion: The increase in volume of a gas or a liquid as it becomes heated.

expansion plug: See *core plug.*

expansion tank: An auxiliary tank, usually connected to the inlet tank or a radiator, which provides additional storage space for heated coolant. Often called a coolant recovery tank.

expansion tube: A metering device, used at the inlet of some evaporators, to control the flow of liquid refrigerant into the evaporator core. Also see *fixed orifice tube.*

expansion valve: See *thermostatic expansion valve.*

external equalizer: See *equalizer line.*

Fahrenheit: An English thermometer scale using 32°F as the freezing point of water (H_2O), and the boiling point of water (H_2O) as 212°F.

fan: A device that has two or more blades attached to the shaft of a motor. The fan is mounted in the evaporator and causes air to pass over the evaporator. A fan is also a device that has four or more blades, mounted on the water pump, which cause air to pass through the radiator and condenser.

fan clutch: A term used for declutching fan.

fast flushing: The use of a special machine to clean the cooling system by circulating a cleaning solution.

federal regulations: Rules and laws imposed by agencies of the federal government, such as the Environmental Protection Agency (EPA).

feedback: The intended return of a portion of an output signal used to determine condition or position.

feeler gauge: See *nonmagnetic feeler gauge.*

field: A coil with many turns of wire located behind the clutch rotor. Current passing through this coil sets up a magnetic field and causes the clutch to engage.

field coil: See *clutch field* or *electromagnet.*

filler neck: That part of the radiator that has the pressure cap.

filter: A device used with the drier or as a separate unit to remove foreign material from the refrigerant.

filter drier: A device that has a filter to remove foreign material from the refrigerant and a desiccant to remove moisture from the refrigerant.

fin comb: See *condenser comb.*

fins: Thin metal strips in an evaporator, condenser, or radiator found around the tubes to aid in heat transfer.

fittings: Components, generally at the end of a hose or tube, that permit attachment to a component or another hose or tube.

Fitz-All: A can tap designed to be used on screw-top and flat-top disposable refrigerant cans.

fixed orifice tube: A refrigerant metering device, used at the inlet of evaporators, to control the flow of liquid refrigerant allowed to enter the evaporator. See *FOT.*

fixed orifice tube cycling clutch system: An air-conditioning system that has a fixed orifice tube as a metering device and a thermostat controlled cycling clutch as a means of temperature control.

flammable: Easily ignited and/or capable of burning.

flare: A flange or cone-shaped end applied to a piece of tubing to provide a means of fastening to a fitting.

flare-nut wrench: A special semi-open end wrench used to loosen and tighten flare nuts.

flash: The ignition (flash) point of a liquid; a lamp (flashing) on/off.

flash gas: Gas resulting from the instantaneous evaporation or refrigerant in a pressure-reducing device such as an expansion valve or a fixed orifice tube.

flooded: A term used to indicate that the evaporator is receiving more liquid refrigerant than what can be evaporated.

flooding: A condition caused by too much liquid refrigerant being metered into the evaporator.

fluid: A liquid, free of gas or vapor.

fluorescent: Glowing, vivid, bright.

fluorocarbon: Pertains to a group of refrigerants; R-12, for example, is a fluorocarbon.

flush: To remove solid particles such as metal flakes or dirt. Refrigerant passages are purged with a clean dry gas, such as nitrogen (N).

flushing agent: An approved liquid or gas used to flush an air-conditioning system.

flux: A substance used in the joining of metals when heat is applied to promote the fusion of metals.

foaming: The formation of a froth of oil and refrigerant due to the rapid boiling out of the refrigerant dissolved in the oil when the pressure is suddenly reduced.

fogging: A clouding or blurring condition.

foot-pound: A unit of energy required to raise 1 pound a distance of 1 foot.

forced air: Air that is moved mechanically, such as by a fan or blower.

FOT: Abbreviation for fixed orifice tube. See *fixed orifice tube.*

FOTCC: Abbreviation for *fixed orifice tube cycling clutch.*

fractionation: When one, or more, refrigerants of the same blend leak at a faster rate than the other refrigerant(s) in that same blend.

freeze plug: See *core plug.*

freeze protection: Controlling evaporator temperature so that moisture on its surface does not freeze and block the airflow.

freeze-up: Failure of a unit to operate properly due to the formation of ice at the expansion valve.

freezing point: The temperature at which a given liquid solidifies. Water freezes at 32°F (0°C); this value is its freezing point.

Freon®: Registered trademark of E. I. DuPont for its group of refrigerants.

Freon 12®: See *R-12.*

front idler: A groove pulley used in automotive air conditioning as a means of tightening the drive belt. The belt rides in the pulley groove(s).

front-of-dash components: The installation of heating and air-conditioning components that are mounted on the fire wall in the engine compartment.

front seating: Closing off the line leaving the compressor open to the service port fitting. This allows service to the compressor without purging the entire system. Never operate the system with the valves front seated.

frosting back: The appearance of frost on the tailpipe and suction line extending back as far as the compressor.

ft.-lb.: Abbreviation for foot-pound.

fully halogenated: All hydrogen (H) atoms of a hydrocarbon molecule are replaced with chlorine (Cl) or fluorine (F).

functional test: See *performance test.*

fuse: An electrical device used to protect a circuit against accidental overload or unit malfunction.

fusible link: A type of fuse made of a special wire that melts to open a circuit when current draw is excessive.

fusible plug: A plug, generally found in a pressure vessel of an air-conditioning system, that will melt at a predetermined temperature to release system pressure.

fusion: The act of melting.

gas: A state of matter. A vapor that has no particles or droplets of liquid.

gasket: A thin layer of material or composition that is placed between two machined surfaces to provide a leak-proof seal between them.

gauge: A device, generally mechanical, to indicate pressure in English or metric terms.

gauge manifold: See *manifold.*

gauge set: Two or more instruments attached to a manifold and used for measuring or testing pressure.

Genetron 12®: Registered trademark of Allied Chemicals Company (Refrigerant-12).

glazed: The condition of a well-worn drive belt that may result in slipping.

global warming: The gradual warming of the earth's atmosphere due to the greenhouse effect. See *greenhouse effect.*

graduated container: A container, such as a measuring cup, that has incremental marks to indicate the quantity of its contents.

gram: A unit of measure in the metric system. One gram is equal to 0.0363 ounce in the English system.

greenhouse effect: A greenhouse is warmed because glass allows the sun's radiant heat to enter but prevents Earth's radiant heat from leaving. Global warming is caused by some gases in the atmosphere that act like greenhouse glass; hence, the term *greenhouse effect.*

grid: The wire circuit of the rear window defogger.

ground: A term that generally refers to the negative side of the battery and all electrical components.

gurgle method: A method of adding refrigerant, from small cans, to a system without running the engine.

halide leak detector: A device consisting of a tank of acetylene gas, a stove, chimney, and search hose used to detect leaks by visual means.

halogen leak detector: See *electronic leak detector.*

halogens: Refers to any of the five chemical elements—astatine (At), bromine (Br), chlorine (Cl), fluorine (F), and iodine (I)—that may be found in some refrigerants,

hang-on unit: An under-dash aftermarket air conditioner. Also may refer to an aftermarket *built-in* system.

HCFC: An acronym for identifying hydrochlorofluorocarbon refrigerants.

head: That part of a compressor that covers the valve plates and separates the high side from the low side.

header tanks: The top and bottom tanks (downflow) or side tanks (crossflow) of a radiator. The tanks in which coolant is accumulated or received.

headliner: That part of the automobile interior overhead or covering the roof inside. Some early air conditioners had duct work in the headliner.

head pressure: Pressure of the refrigerant from the discharge reed valve through the lines and condenser to the expansion valve orifice.

heat: Energy; any temperature above absolute zero.

heat exchanger: An apparatus in which heat is transferred from one fluid to another, on the principle that heat moves to an object with less heat.

heat intensity: The measurement of heat concentration with a thermometer.

heat load: The *load* imposed on an air conditioner due to ambient temperature, humidity, and other factors that may produce unwanted heat.

heat of fusion: The amount of latent heat, in Btu, required to cause a change of state from a solid to a liquid.

heat of respiration: The heat given off by ripening vegetables or fruits in the conversion of starches and sugars.

heat quantity: The amount of heat as measured on a thermometer. See *British thermal unit.*

heat radiation: The transmission of heat from one substance to another while passing through, but not heating, intervening substances.

heat transfer: A flow of heat based on a principle of physics.

heat transmission: Any flow of heat.

heater core: A water-to-air heat exchanger that provides heat for the passenger compartment.

heater hose: Rubber or composition lines used to move heated coolant to the heater and back to the cooling system.

heater valve: A manual or automatic valve in the heater hose used to open (start) or close (stop) coolant flow to the heater core.

heliarc: The act of joining two pieces of aluminum or stainless steel using a high-frequency electric weld and an inert gas, such as argon. This weld is made electrically while the inert gas is fed around the weld. This gas prevents oxidation by keeping the surrounding air away from the metals being welded.

hex key wrench: A six-sided (hexagon) wrench.

HFC: Abbreviation for hydrofluorocarbon, a man-made compound used in refrigerants, such as R-134a.

Hg: Chemical symbol for mercury. Used to identify a vacuum.

high head: A term used when the head (high-side) pressure of the system is excessive.

high-heat load: Refers to the maximum amount of heat that can be absorbed by CFC-12 as it passes through the evaporator.

high-load conditions: Those instances when the air conditioner must operate continuously at its maximum capacity to provide the cool air required.

high-pressure control: See *high-pressure cutoff switch.*

high-pressure cutoff switch: An electrical switch activated by a predetermined high pressure. The switch opens a circuit during high-pressure periods.

high-pressure lines: The lines from the compressor outlet to the expansion valve inlet; these lines carry high-pressure liquid and gas.

high-pressure relief valve: A mechanical device designed so that it releases the extreme high pressures of the system to the atmosphere.

high-pressure side: That part of the system, from the compressor outlet to the evaporator inlet, that is under high pressure.

high-pressure switch: See *high-pressure cutoff switch.*

high-pressure vapor line: See *discharge line.*

high side: See *discharge side.*

high-side pressure: See *discharge pressure.*

high-side service valve: A device located on the discharge side of the compressor; this valve permits the service technician to check the high-side pressures and perform other necessary operations.

high suction: The low-side pressure is higher than normal due to a malfunction of the system.

high vacuum: A vacuum below 500 microns (0.009 6 psia or 0.66 kPa).

high-vacuum pump: A two-stage vacuum pump that has the capability of pulling below 500 microns (0.009 6 psia or 0.66 kPa). Many vacuum pumps can pull to 25 microns (0.005 psia or 0.003 kPa).

holding fixture: A handy device for holding components, such as a compressor, during disassembly and assembly procedures.

horsepower: A measurement of mechanical power or the rate at which work is done.

hot gas: The condition of the refrigerant as it leaves the compressor until it gives up its heat and condenses.

hot gas bypass line: The line that connects the hot gas bypass valve outlet to the evaporator outlet. Metered hot gas flows through this line.

hot gas bypass valve: A device used to meter hot gas back to the evaporator through the bypass line to prevent condensate from freezing on the core.

hot gas defrosting: The use of high-pressure gas in the evaporator to remove frost.

hot leak: A refrigerant leak that is best detected when the engine and air-conditioning system are at normal operating temperatures.

housing: The cover or container for a component, device, or system.

HP: Abbreviation for horsepower.

humidify: To add moisture to the atmosphere. To increase the relative humidity.

humidity: See *moisture.* Also see *relative humidity.*

HVAC: Abbreviation for Heating, Ventilating, and Air Conditioning. Though applicable, HVAC is a term that is seldom used when referring to the automotive comfort system.

H-valve: An expansion valve with all parts contained within that is used on some Chrysler and Ford vehicles.

hydrocarbon: An organic compound containing only hydrogen (H) and carbon (C).

hydrochloric acid (HCl): A corrosive acid produced when water and CFC-12 are mixed as within an automotive air-conditioning system.

hydrochlorofluorocarbon (HCFC): A group of refrigerants that contain the chlorine (Cl) atom and the hydrogen (H) atom, which causes the chlorine (Cl) atom to dissipate more rapidly in the atmosphere.

hydrofluorocarbon: A man-made compound used in refrigerants, such as HFC-134a.

hydrolyzing action: The corrosive action within the air-conditioning system induced by a weak solution of hydrochloric acid formed by excessive moisture chemically reacting with the refrigerant.

hydrometer: A device used to measure the specific gravity of the coolant to determine its freezing temperature.

hygrometer: A device used to measure relative humidity.

hygroscopic: Readily absorbing and retaining moisture.

hysteresis: The general failure of a device to return to its original condition after the cause of change has been determined and remedied, such as a bellows in a thermostat.

ice melting capacity: Refrigerant equal to the latent heat of fusion of a stated weight of ice at 144 Btu per pound.

ID: Also id. Abbreviation for inside diameter.

ideal humidity: A relative humidity of 45 percent to 50 percent.

ideal temperature: Temperature from 68°–72°F (20°–22.2°C)

idler: A pulley device that keeps the belt whip out of the drive belt of an automotive air conditioner. The idler is used as a means of tightening the belt.

idler eccentric: A device used with the idler pulley as a means of tightening the belt.

illegal: Unlawful.

impeller: A rotating member with fins or blades used to move liquid. The rotating part of a water pump, for example.

imported: A term used to identify goods manufactured in other countries and brought into this country for sale or consumption.

improper: Unsuitable, irregular, or abnormal.

in-car sensor: A thermistor used in automatic temperature control units for sensing the in-car temperature. Also see *thermistor.*

inches of mercury: An English unit of measure when referring to a vacuum; abbreviated in. Hg.

inch-pound: A unit of energy required to raise 1 pound a distance of 1 inch; abbreviated in.-lb.

in-duct sensor: A thermistor used in automatic temperature control units for sensing the in-duct return air temperature. Also see *thermistor*.

inert: An element, such as a gas, that is not readily reactive with other elements.

inflammable: Easily ignited and capable of burning rapidly.

in.-lb.: Abbreviation for inch-pound.

inside diameter: The measure across the inside walls of a tube or pipe at its widest point.

insulate: To isolate or seal off with a nonconductor.

insulation tape: Tape (either rubber or cork) used to wrap refrigeration hoses and lines to prevent condensate drip.

insulators: Materials that are nonconductors of electrical current.

intake stroke: The movement of a piston from top to bottom of its stroke.

integrated: Made a part of.

intentional: Deliberate; on purpose.

interchangeable: Switchable from one system or device to another.

intermittently: At no particular interval or period.

in-vehicle sensor: See *in-car sensor*.

in-vehicle temperature sensor: See *in-car sensor*.

inverted: Turned upside down.

involuntary: Unintentional.

isolate: A technique whereby a component may be separated from the rest of the system; for example, a compressor may be *isolated* from the rest of the system by front-seating the service valves.

Isotron 12®: A trademark of Penn Salt Company for Refrigerant-12.

jumper: A wire used to bypass an electrical control or a device.

junction: A point where two or more components, such as electrical wires or vacuum hoses, are joined.

junction block: A device on which two or more junctions may be found.

kelvin: A thermometer scale using 273°K as the freezing point of water. Absolute zero is the beginning of this temperature scale: 0°K.

kilogram: A unit of measure in the metric system. One kilogram is equal to 2.205 pounds in the English system.

kilopascal: A unit of measure in the metric system. One kilopascal (kPa) is equal to 0.145 pound per square inch (psi) in the English system.

kilopascal absolute: See *kPa absolute*.

kinetic: Refers to motion.

kinetic energy: Refers to energy in motion.

km/h: A metric unit of measure; the abbreviation for kilometers per hour.

kPa: Abbreviation for kilopascal.

kPa absolute: A metric unit of measure for pressure measured from absolute zero.

kPa gauge: A metric unit of measure for pressure measured from atmospheric or sea level pressure.

LAL: Acronym for lowest achievable level.

lamp: A term used for bulb.

latent heat: The amount of heat required to cause a change of state of a substance without changing its temperature.

latent heat of condensation: The quantity of heat given off while changing a substance from a vapor to a liquid.

latent heat of evaporation: The quantity of heat required to change a liquid into a vapor without raising the temperature of the vapor above that of the original liquid.

latent heat of fusion: The amount of heat that must be removed from a liquid to cause it to change to a solid without causing a change of temperature.

latent heat of vaporization: See *latent heat of evaporation*.

lb.: An abbreviation for the English unit of measure; pound.

leak: An unintentional loss of material, such as refrigerant and/or oil.

leak detector: See *halide leak detector* or *halogen leak detector*.

Leak Find: A trade name for a dye solution used to find difficult leaks in the system.

LED: Abbreviation for light-emitting diode.

light-emitting diode: A diode that emits a light when current passes through it. LEDs are available in many colors, such as red, green, and yellow. See *diode*.

liquefier: Same as condenser. See *condenser*.

liquid: A column of fluid without solids or gas pockets.

liquid line: The line connecting the drier outlet with the expansion valve inlet. The line from the condenser outlet to the drier inlet is sometimes called a liquid line.

liter: A metric unit of measure. One liter is equal to 0.2642 gallon in the English system.

load: The required rate of heat removed in a given time.

loop: A complete recirculating path where a substance, such as refrigerant, continuously moves.

lowest achievable level (LAL): A term used to define the lowest amount of emissions possible of CFC and HCFC refrigerants.

low-head pressure: The high-side pressure is lower than normal; less than expected for a given condition.

low-loss fitting: A device designed to close automatically or to be closed manually to prevent refrigerant loss when used at fluid connection points between hoses and service valves, vacuum pump, recovery, or recycle machines.

low pressure: Usually refers to system pressure below normal; less than expected for a given condition.

low-pressure control: See *low-pressure cutoff switch.*

low-pressure cutoff switch: An electrical switch that is activated by a predetermined low pressure. This switch opens a circuit during certain low-pressure periods.

low-pressure line: See *suction line.*

low-pressure side: See *suction side.*

low-pressure switch: See *low-pressure cutoff switch.*

low-pressure vapor line: See *suction line.*

low side: See *suction side.*

low side pressure: The pressure in the low side of the system, from the evaporator inlet to the compressor inlet, as may be noted on the low-side pressure gauge.

low-side service valve: A device located on the suction side of the compressor that allows the service technician to check low-side pressures or perform other necessary service operations.

low-suction pressure: Pressure lower than normal in the suction side of the system due to a malfunction of the unit.

low voltage: Usually refers to voltage below normal; less than expected for a given condition.

lubricant: The new synthetic product PAG and ESTER used with new refrigerants. A term often used to identify an organic mineral-based grease or oil product. New retrofit synthetic lubricants; PAG and ESTER are not mineral based although they are often referred to as *oil.* See *refrigeration lubricant* and *refrigeration oil.*

MACS: Registered trademark for the Mobile Air-Conditioning Society.

magnetic clutch: A coupling device used to turn the compressor on and off electrically.

maintenance: To repair or maintain.

malfunction: To operate improperly or inadequately; not performing to standards.

manifold: A device equipped with a hand shutoff valve. Gauges are connected to the manifold for use in system testing and servicing.

manifold gauge: A calibrated instrument used to measure pressure in the system.

manifold gauge set: A manifold complete with gauges and charging hoses.

manifold vacuum: An unregulated vacuum source at the intake manifold or an engine. See *vacuum.*

master control: The dash-mounted control panel.

matter: Anything that occupies space and has weight.

MAX: An abbreviation for maximum.

mean altitude: 900 feet (274.3 m) is used as the mean, or average, altitude by engineers.

measure: Dimensions, quantity, pressure, or capacity by comparison with a standard.

melting point: The temperature above which a material cannot exist as a solid at a given pressure.

mercury: See *Hg.*

meter: To regulate the flow of a fluid or gas. Also a unit of measure in the metric system. One meter is equal to 39.37 inches in the English system.

metering device: Any device that meters or regulates the flow of a liquid or vapor. See *thermostatic expansion valve, fixed orifice tube,* and *expansion tube.*

metrics: A system of measure.

micron: A unit of measure; 1,000 microns = 1 millimeter = 0.03937 inch.

microprocessor: A solid-state control unit used to control the electronic temperature control. It processes signals received from sensors and then signals appropriate actuators to respond.

midpositioned: The position of a stem-type service valve where all fluid passages are interconnected. Also referred to as *cracked.*

midseated: See *midpositioned.*

mildew: Any variety of fungi, usually whitish, that forms a superficial growth on plants and various organic material.

millimeter: A metric unit of measure. One milliliter is equal to 1/1 000 meter = 0.039 37 inch.

mineral spirits: A petroleum distillate suitable for use as a solvent.

miscible: Capable of being mixed in all proportions. For example, mineral oil is miscible in CFC-12 but is not miscible in HFC-134a.

mm: Abbreviation for millimeter.

Mobile Air-Conditioning Society (MACS): A not-for-profit organization, founded in 1981, for the dissemination and distribution of comprehensive technical information, training, and communications to its members consisting of automotive air-conditioning service shops, installers, distributors, suppliers, and manufacturers in the United States and Canada.

Mobil Gel: A trade name. See *desiccant*.

Mobil Sorbead: A drying agent. See *desiccant*.

mode door: A door in the case/duct system that may be positioned to provide selected in-vehicle temperature and humidity conditions.

modulated vacuum: A vacuum signal regulated to a particular level. See *vacuum*.

modulates: To change and adjust and readjust to certian specific characteristics.

moisture: Droplets of water in the air; humidity, dampness, or wetness.

molecular sieve: A drying agent. See *desiccant*.

molecule: Two, or more, atoms chemically bonded together.

monochlorodifluoromethane: See *Refrigerant-22*.

Montreal Protocol: An agreement signed by representatives of the participating countries who have agreed to reduced CFC and HCFC emissions into the atmosphere by restricting and regulating the manufacture and sale of such products.

motor: An electrical device that produces a continuous turning motion. A motor is used to propel a fan blade or a blower wheel.

mount and drive: Pulleys, mounting plates, belts, and fittings necessary to mount a compressor and clutch assembly on an engine.

muffler: A hollow tubular device used in the discharge line of some air conditioners to minimize the compressor noise transmitted to the inside of the car. Some units use a muffler on the low side as well.

mulitmeter: A meter having several functions.

multiple pass: This term applies to a recovery/recycle unit that removes refrigerant from an air-conditioning system and circulates it through the recovery/recycle unit to remove contaminants before it is pumped into the recovery cylinder. It must be noted that this method of recycling does not remove noncondensables and/or other types of refrigerants.

multiplier: The number by which another number is multiplied.

MVAC: An abbreviation for motor vehicle air conditioning.

MVAC-like appliance: Air-conditioning systems on other types of vehicles.

National Institute for Automotive Service Excellence (ASE): A not-for-profit corporation founded in 1972 to promote and encourage high standards of automotive service and repair through voluntary testing.

National Institute for Standards and Technology (NIST): A national organization that helps to set standards and technology for mobile air-conditioning systems.

needle valve: A valve that has a slender point fitting into a conical seat, used to accurately control the flow of a liquid or gas.

negative: Less than zero; a terminal having an excess of electrons; one of the two poles of a battery representing ground. Often indicated by the minus (–) symbol.

Newton-meter: A unit of measure in the metric system. One Newton-meter is equal to 0.737 ft.-lb. or 8.844 in.-lb.

NIASE: An acronym for the National Institute for Automotive Service Excellence (ASE).

Nichrome wire: Wire made of an alloy of nickel (Ni) and chromium (Cr) that withstands high temperatures. Used for dropping resistors in blower speed controls.

NIST: An acronym for the National Institute for Standards and Technology.

N•m: Abbreviation for Newton-meter.

noisy: Making a sound.

nonmagnetic feeler gauge: Thin strip(s) of metal of calibrated thickness made of nonferrous metals to check air gap of components that may have a magnetic field.

nonmiscible: Not capable of being mixed in any proportion. For example, mineral oil, although miscible in CFC-12 is nonmiscible in HFC-134a.

normal evacuation: An evacuation with only one pump down.

NPT: An abbreviation for national pipe thread.

O: Chemical symbol for oxygen.

OD: Also od. Abbreviation for outside diameter.

ODP: An acronym for ozone depletion potential.

ODW: An acronym for ozone depletion weight.

ohm: a unit of electrical resistance.

ohmmeter: An electrical instrument used to measure the resistance, in ohms, of a circuit or component.

Ohm's laws: The laws relating to the behavior of electricity.

oil: An organic chemical used as a lubricant. A specially formulated oil is used in air-conditioning systems.

oil bleed line: An external line that usually bypasses an expansion valve, evaporator pressure regulator, or bypass valve to ensure positive oil return to the compressor at high compressor speeds and under a low charge or clogged system condition.

oil bleed passage: Internal orifice that bypasses an expansion valve, evaporator pressure regulator, or bypass valve to ensure a positive oil return to the compressor.

oil injection cylinder: A special cylinder that may be used to inject a measured amount of refrigeration oil into the system.

oil injector: See *oil injection cylinder.*

oil seal: A term used for compressor shaft seal.

open: A term generally used to indicate a break in an electrical circuit, such as an open switch.

operational test: See *performance test.*

option: To choose; to have a choice.

orbits: The paths of a body as it rotates around another body, such as the moon's rotation around Earth.

orifice: A small hole; a calibrated opening in a tube or pipe to regulate the flow of a fluid or liquid.

orifice tube: See *expansion tube* or *fixed orifice tube.*

O-ring: A synthetic rubber gasket with a round (O-shaped) cross section.

OSHA: An abbreviation for Occupational Safety and Health Administration.

O₂: Chemical symbol for oxygen, as in the air we breathe.

O₃: The chemical symbol for ozone.

outlet: A term used for discharge line.

outside diameter: The measure across the outside walls of a tube or pipe at its widest point.

overcharge: Indicates that too much refrigerant or refrigeration oil is added to the system.

overcooling: A term indicating the engine does not reach normal operating temperature.

owner's manual: A book, generally supplied with the vehicle, that provides valuable and helpful information to the owner.

oxidize: The formation of a crust on certain metals due to the reaction of the metal, heat, and oxygen.

oxygen: An element (O), as in the air we breathe.

oz: Abbreviation for ounce.

ozone (O₃): An unstable pale-blue gas, with a penetrating odor; it is an allotropic form of oxygen (O) formed usually by a silent electrical discharge in the air.

ozone depletion: The reduction of the ozone layer due to contamination, such as by the release of refrigerants into the atmosphere. See *ozone layer.*

ozone depletion potential (ODP): A term used when referring to ozone depletion weight (ODW).

ozone depletion weight (ODW): A number value assigned to a chemical to represent its ability to destroy stratospheric ozone.

ozone layer: Also called ozonosphere. A layer at a height of about 20 miles (32 kilometers) having a high concentration of ozone.

package tray: Shelf behind the rear seat in a sedan. Trunk-mounted air-conditioner units use ducts through the package tray as the intake and outlet of the unit.

PAFTT: Acronym for the Program for Alternative Fluorocarbon Toxicity Testing.

PAG: An acronym for the synthetic lubricant poly alkaline glycol.

panel: A cover.

parallel circuit: Electrical circuits that run parallel with each other.

parts per million: The unit used to measure the amount of moisture in refrigerant. The maximum (desirable) moisture content is ten parts of moisture to one million parts of refrigerant, or 10 ppm.

Pb: The chemical symbol for lead.

percentage: The parts of a hundred. For example, 25 parts of 100 is 25 percent.

performance: The manner in which something functions or is expected to function.

performance test: Readings of the temperature and pressure under controlled conditions to determine if an air-conditioning system is operating at full efficiency.

personal hygiene: Care of one's person by bathing and wearing personal protective equipment and clothing.

perspiration: A fluid, mostly water, that is excreted through the pores of the skin by the sweat glands.

phosgene gas: A highly toxic gas, carbonyl chloride ($CCOCl_2$). Until recently, it was believed that phosgene gas was produced when CFC-12 came into contact with heated metal or an open flame. It is now known that little or none of this gas is produced in this manner.

photovoltaic: See *photovolotaic diode.*

photovoltaic diode: A device that has a junction of two dissimilar metals that produces an electrical signal proportional to the amount of light that strikes it.

pickup tube: A tube extending from the outlet of the receiver almost to the bottom of the tank to ensure that 100 percent liquid is supplied to the liquid line or metering device.

pilot operated: A small valve used to energize or regulate a large valve.

pilot-operated evaporator pressure regulator: An EPR valve that is regulated by an internal pilot valve pressure.

pintle valve: The ball and seat found inside a thermostatic expansion valve. The pintle is attached to the diaphragm, which causes it to open and close in response to pressure changes.

piston: A cylindrical part that moves up and down or back and forth in a compressor cylinder.

plenum: See *plenum chamber.*

plenum blower assembly: Located on the engine side of the fire wall, this assembly contains air ducts, air valves, and a blower that permits the selection of air from the outside or inside of the car and directs it to the evaporator or to the heater core if desired.

plenum chamber: An area filled with air at a pressure that is slightly higher than the surrounding air pressure, such as the chamber just before the blower motor.

POASTV: See *positive absolute suction throttling valve.*

POA valve: See *positive absolute suction throttling valve.*

POE: An abbreviation for polyol ester.

POEPR: The abbreviation for pilot-operated evaporator pressure regulator.

pollen: An irritant to those who suffer from hay fever or other allergies; the fine, yellowish powder from the anthers of flowers.

poly alkaline glycol (PAG): A synthetic lubricant. PAG is the lubricant of choice in automotive applications for use with HFC-134a refrigerant.

polyol ester (POE): A synthetic oil-like lubricant that is occasionally recommended for use in an HFC-134a system. This lubricant is compatible with both HFC-134a and CFC-12.

positive: One of the two poles of a magnet. One of the two terminals of an electrical circuit or device.

positive absolute suction throttling valve: A suction throttling valve used by Delco Air. This valve has a bronze bellows under a nearly perfect vacuum, which is not affected by atmospheric pressure.

potentiometer: Ses *rheostat*

pounds per square inch absolute: Pressure that is not compensated or adjusted for altitude or other variables.

power module: A component that controls the operation of the blower motor.

power servo: A servo unit used in automatic temperature control that is operated by a vacuum or an electrical signal.

PPM: also ppm. Abbreviation for *parts per million.*

preformed: Preshaped to fit a particular application, such as a heater bypass hose.

prescribed: A rule or guide for a specific application.

pressure: Force per unit of area; the pressure of refrigerant is measured in pounds per square inch.

pressure cap: A radiator cap that increases the pressure of the cooling system and allows higher operating temperatures.

pressure control: A pressure-actuated electrical control.

pressure control valve: A device located between the evaporator outlet and compressor inlet used to control the pressure of the refrigerant in the evaporator.

pressure drop: The difference in pressure between any two points; a pressure drop may be caused by a restriction or friction.

pressure gauge: A device, usually mechanical, used to measure pressure.

pressure line: Although all refrigerant lines are under pressure, the term *pressure line* refers to the discharge line. See *discharge line.*

pressure reduction: A drop of pressure due to a restriction or metering device.

pressure release grill: An air vent that prevents pressure from building up inside the car while the comfort system is operating.

pressure sensing line: See *remote bulb.*

pressure switch: An electrical switch that is actuated by a predetermined low or high pressure. A pressure switch is generally used for system protection.

pressure tester: A device used to pressure test the cooling system and pressure cap to ensure that the systems are not leaking under pressure.

Prestone 12: A trade name of the Union Carbon and Carbide Chemical Company (Refrigerant-12).

primary seal: A seal between the compressor shaft seal and the shaft to prevent the leakage of refrigerant and oil.

probes: Slender, sometimes flexible, devices used to investigate and/or obtain information from a remote location.

program: Instructions for a computer in order for it to do its job. May include fixed and variable data.

Program for Alternative Fluorocarbon Toxicity Testing (PAFTT): A testing agency founded by refrigerant manufacturers for the purpose of studying and testing of new refrigerants to determine toxicity levels.

programmer: That part of an automatic temperature control system that controls the blower speed, air mix doors, and vacuum diaphragms. Also the onboard computer.

prohibit: To forbid, usually by authority.

propane: A flammable gas used in the halide leak detector.

psi: Abbreviation for pounds per square inch.

psia: Abbreviation for pounds per square inch absolute.

psig: Abbreviation for pounds per square inch gauge.

psychrometer: See *sling psychrometer*.

pulldown: Another term for *pumpdown*. See *evacuate*.

pulley: A flat wheel with a V-groove machined around the outer edge; when attached to the drive and driven members, the pulley provides a means of driving the compressor.

pump: The compressor. Also refers to the vacuum pump.

pump down: See *evacuate*.

purge: To remove moisture and/or air from a system or a component by flushing with a dry gas, such as nitrogen (N), to remove all of the refrigerant from the system.

purity test: A static test that may be performed to compare the suspect refrigerant pressure to an appropriate temperature chart to determine its purity.

quantity: A specified or indicated amount or number.

quick coupler: A coupler that allows hoses to be quickly connected and/or disconnected. Most shop air hoses, for example, are equipped with quick couplers.

radial compressor: A space-saving compressor used on small cars.

radiation: The transfer of heat without heating the medium through which it is transmitted.

radiator: A coolant to air-heat exchanger; the device that removes heat from coolant passing through it.

radiator cap: See *pressure cap*.

radiator core: See *core*.

radiator hose: Rubber or synthetic tubes used to carry coolant from the engine to the radiator and from the radiator to the engine.

radiator pressure cap: See *pressure cap*.

radiator pressure tester: See *pressure tester*.

ram air: Air forced through the radiator and condenser coils by the movement of the vehicle or the action of the fan.

Ranco Control: A trade name used when referring to a thermostat. See *thermostat*.

Rankine: A thermometer scale for which the freezing point of water is 492°R. Absolute zero is the beginning of this thermometer scale.

RCD: An acronym for refrigerant containment device.

receiver: A container for the storage of liquid refrigerant.

receiver-dehydrator: A combination container for the storage of liquid refrigerant and a desiccant.

receiver-drier: See *receiver-dehydrator*.

reciprocating: To move back and forth, up and down, or to and fro, such as the actio of a piston.

reciprocating compressor: An air-conditioning compressor in which the pistons move up and down or back and forth.

recirc door: See *recirculate door*.

recirculate door: A door in the plenum that regulates the amount of recirculated airflow.

recirculated: Reused.

reclaim: To reclaim is to process used refrigerant to new product specifications by means that may include distillation. This process requires that a chemical analysis of the refrigerant be performed to determine that appropriate product specifications are met. This term implies the use of equipment for processes and procedures usually available only at a reprocessing facility.

recovery: the recovery of refrigerant is to remove it, in any condition, from a system and to store it in an external container without necessarily testing or processing it in any way.

recovery cylinder: A recovery cylinder for CFC-12 and/or HFC-134a must meet DOT specifications 4BA-300. These cylinders are characterized by a combined liquid/vapor valve located at the top. A dip tube is used to feed liquid refrigerant from the bottom so it can be dispensed without inverting the cylinder. A recovery cylinder should be painted gray with a yellow shoulder.

recovery/recycle systems: A term often used to refer to the circuit inside the recovery unit used to recycle and/or transfer refrigerant from the air-conditioning system to the recovery cylinder.

recovery/recycle unit: A term used to identify the complete unit used to recover and/or recycle refrigerant from the air-conditioning system.

recovery tank: See *expansion tank.*

recycle: To recycle is to clean the refrigerant for reuse by oil separation and pass it through other devices, such as filter driers, to reduce moisture, acidity, and particulate matter. Recycling applies to procedures usually accomplished in the repair shop or at a local service facility.

red dye trace solution: The dye shows the exact location of a leak in the air-conditioning system by depositing a colored film around the leak.

reed valve: Thin leaf of steel located in the valve plate of automotive compressors, serving as suction and discharge valves. The suction valve is located on the bottom of the valve plate and the discharge valve is on top.

refrigerant: The chemical compound used in a refrigeration system to produce the desired cooling.

refrigerant containment device (RCD): A device introduced on some 1994 model car lines to help guard against high pressure, resulting in refrigerant loss by controlling the compressor and/or condenser fan motor.

refrigerant identifier: A device used to determine the purity of refrigerant.

refrigeration: To use an apparatus to cool; keep cool; chill; keep chilled, under controlled conditions, by natural or mechanical means, as an aid to ensure personal safety and comfort; to cool the air by removing some of its heat content.

refrigeration cycle: The complete cycle of the refrigerant back to the starting point, evidenced by temperature and pressure changes.

refrigeration lubricant: A synthetic oil-like lubricant, such as PAG and ESTER, that is formulated for specific use and application in designated refrigeration systems. Also see *refrigeration oil.*

refrigeration oil: A highly refined organic mineral oil free from all contaminants, such as sulfur, moisture, and tars. Also see *refrigeration lubricant.*

reheat principle: A principle used in automotive air-conditioning systems to control in-car relative humidity. The air is first cooled then reheated.

related: To be associated with.

relative humidity: The actual moisture content of the air in relation to the total moisture that the air can hold at a given temperature.

relay: An electrical switch device activated by a low-current source to control a high-current device.

remote bulb: A sensing device connected to the expansion valve by a capillary tube. This device senses the tailpipe temperature and transmits pressure to the expansion valve for its proper operation.

remote sensing bulb: See *remote bulb.*

replace: To put back in its former position.

reserve tank: See *vacuum reserve tank* or *expansion tank.*

resistance: The property of a substance that impedes current and results in the dissipation of power in the form of heat.

resistor: A voltage dropping device, usually wire wound, which provides a means of controlling fan speeds.

restricted: Kept or confined within certian limits.

restriction: A blockage in the air-conditioning system caused by a pinched or crimped line, foreign matter, or moisture freeze-up.

restrictor: An insert fitting or device used to control the flow of refrigerant or refrigeration oil.

retainer ring: A "C-" or "E-" type ring that holds a component in position.

retrofit: To modify equipment that is already in service using parts and/or materials available or made available after the time of original manufacture.

returnable cylinder: See *recovery cylinder.*

reusable cylinder: See *recovery cylinder.*

reverse flush: A method of cleaning an engine and/or radiator by flushing in a direction opposite of normal coolant flow under pressure.

revolutions per minute: The number of times a moving member rotates through 360 degrees in 1 minute.

RH: An abbreviation for relative humidity.

rheostat: A wire-wound variable resistor used to control blower motor speed.

r/min: Also RPM or rpm; abbreviation for revolutions per minute.

room temperature: With reference to the temperature range of 68°F (20°C) to 72°F (22.2°C).

R-134a: Trade term for Refrigerant HFC-134a.

rotary vacuum valve: That part of a vacuum control used to divert a vacuum signal for operation of doors, switches, and/or valves.

rotary vane: A type of positive displacement air-conditioner compressor.

rotate: To turn.

rotor: The rotating or freewheeling portion of a clutch; the belt sides on the rotor.

routed: The careful attention where the wiring harness and hoses are placed (routed) to avoid damage.

RPM: Also, rpm or r/min. Abbreviation for *revolutions per minute.*

R-12: Abbreviation for Refrigerant-12.

R-22: Trade term for Refrigerant HCFC-22.

saddlebags: Air chambers or openings in the left and right front corners of the car body between the kick pads and the exterior of the car. The evaporator is sometimes located in the right saddlebag.

saddle clamp access valve: See *saddle valve.*

saddle valve: A two-part accessory valve that may be clamped around the metal part of a system hose to provide access to the air-conditioning system for service.

SAE: An acronym for the Society of Automotive Engineers.

safety factor: A margin allowed for error in determining safety perimeters, such as the pressure ratings of a vessel.

safety glasses: Eyeglasses with shatterproof lenses worn for eye protection.

safety goggles: Goggles worn over eyeglasses for eye protection.

SATC: An abbreviation for semiautomatic temperature control.

saturated: A condition whereby a medium holds as much moisture (H_2O) as it can at a given temperature and pressure without forming droplets.

saturated desiccant: A desiccant that contains all of the moisture it can hold at a given temperature.

saturated drier: See *saturated desiccant.*

saturated point: The point at which matter must change state at any given temperature and pressure.

saturated temperature: The boiling point of a refrigerant at a particular pressure.

saturated vapor: Saturation indicates that the space holds as much vapor as possible. No further vaporization is possible at this particular temperature.

Saybolt Seconds Universal (SSU): A method used to determine the flow rate of fluids, usually to designate the weight of a lubricant.

scale: A mechanical or electronic device used for determining weight. Also a term relating temperature, pressure, and/or weight.

schematics: Drawings of electrical circuits or systems.

Schrader: A type of service valve.

Schrader valve: A spring-loaded valve similar to a tire valve. The Schrader valve is located inside the service valve fitting and is used on some control devices to hold refrigerant in the system. Special adapters must be used with the gauge hose to allow access to the system.

Scotch yoke: A type of compressor design.

screen: A metal mesh located in the receiver, expansion valve, and compressor inlet to prevent particles of dirt from circulating through the system.

scroll: A type of compressor design.

seat: The surface on which another part rests. Also, to wear to a good fit.

semiautomatic: An automatic device or system that requires manual input or assistance.

semiconductors: Materials that will conduct electrical current in one direction only.

sending unit: That part of a temperature or pressure warning device that triggers or transmits a warning signal to the dash gauge or lamps.

sensible heat: Heat that causes a change in the temperature of a substance, but does not change the state of the substance.

sensor: A temperature-sensitive unit such as a remote bulb or thermistor. See *remote bulb* and *thermistor.*

sensor: An electronic device that is sensitive to light and/or heat.

series circuit: An electrical circuit whereby two or more loads are in series.

serpentine belt: A flat or V-groove belt that winds through all of the engine accessories to drive them off the crankshaft pulley.

service access gauge port adapter: An adapter used to connect the manifold gauge set to the service port of the system in certain applications.

service access gauge port valve: See *service port.*

service hose: A hose that attaches to the manifold and gauge set and to the high and low sides of the system. Service hoses are also used to evacuate and charge the air-conditioning system.

service port: Fitting found on the service valves and some control devices; the manifold-set hoses are connected to this fitting.

service valve: See *high-side (low-side) service valve.*

servo motor: An electric motor used to control a mechanical device such as a heater valve. Also see *vacuum motor.*

setscrew: A screw used to secure a part onto another part, such as a pulley on a shaft.

shaft key: A key used to engage a part onto another part, such as a clutch armature on the compressor crankshaft.

shaft seal: See *compressor shaft seal.*

shaft seal: A term used for compressor shaft seal.

shim: Thin material, usually metal, to space and/or align two mating parts.

short: An electrical term used to indicate an unwanted electrical path, usually to ground.

short cycling: Can be caused by poor air circulation or a maladjusted thermostat. The unit runs for very short periods.

side dash components: The installation of heating and air-conditioning components that have the evaporator mounted on the curb side of the fire wall in the engine compartment and the heater core in the back, in the passenger compartment.

sight glass: A window in the liquid line or in the top of the drier; this window is used to observe the liquid refrigerant flow.

Significant New Alternatives Policy (SNAP): A rule established in 1994, under Section 612, to initiate a program in which the EPA is to evaluate applications for use of substitute chemicals and technology designated to replace ozone depleters in specific uses. SNAP testing will include flammability, chemical toxicity, global warming potential, exposure of workers, consumers, the general public, and aquatic life.

silica gel: A drying agent used in many automotive air conditioners because of its ability to absorb large quantities of water. A desiccant.

silver solder: An alloy containing from 5 to 45 percent of silver. Silver solder melts at 1,120°F (604°C) and flows at 1,145°F (618°C). It is an ideal material for use in refrigeration service.

single pass: This term applies to a recovery unit that removes refrigerant from the air conditioner and passes it through once on its way to the recovery cylinder.

sling psychrometer: A device using two matched mercury-filled thermometers to obtain the relative humidity reading.

slugging: The return of liquid refrigerant or oil to the compressor.

small appliance: See *appliance.*

SNAP: Acronym for Significant New Alternatives Policy.

snap ring pliers: Pliers used for removing and replacing snap rings.

Society of Automotive Engineers (SAE): The society of Automotive Engineers (SAE) is a professional organization of the automotive industry. Founded in 1905 as the Society of Automobile Engineers, the SAE is dedicated to providing technical information and standards to the automotive industry.

soft solder 50/50: A metallic alloy of 50 percent tin (Sn) and 50 percent lead (Pb); used to repair or join ferrous metal parts for temperatures up to 250°F (121°C). No longer recommended for use due to its lead (Pb) content.

soft solder 95/5: A metallic alloy of 95 percent tin (Sn) and 5 percent antimony (Sb); used to repair or join ferrous metal parts for temperatures below 350°F (176°C). Often used for refrigeration service.

solder: A metallic alloy used to unite metals.

solenoid valve: An electromagnetic valve controlled remotely by electrically energizing and de-energizing a coil.

solid: A state of matter that is not liquid nor a gas or vapor.

solid state: Referring to electronics consisting of semiconductor devices and other related nonmechanical components.

Sorbead: A desiccant.

spanner: A special wrench that may be used to hold a clutch armature while removing the crankshaft nut or screw.

specifications: Service information and procedures provided by the manufacturer that must be followed in order for the system to operate properly.

specific heat: The quantity of heat required to change 1 pound of a substance by 1°F.

specialized tools: Tools that have special application, such as for removing a shaft seal from the front head cavity.

spikes: Unwanted electrical surges.

squirrel cage: A blower case designed for use with the squirrel-cage blower.

squirrel-cage blower: A blower wheel designed to provide a large volume of air with a minimum of noise. The blower is more compact than the fan, and air can be directed more efficiently.

SSU: An acronym for Saybolt Seconds Universal.

standard: The normal and expected circumstances and conditions.

standard ton: See *ton of refrigeration.*

starved: A term that refers to an evaporator that has too little refrigerant, resulting in a poor to no cooling effect.

starving: Describes a condition when the evaporator does not get enough refrigerant to properly function.

static pressure: See *static system pressure.*

static system pressure: The pressure of the system when it and the engine are off. Low- and high-side pressure should be near equal at this time.

stethoscope: An instrument used to convey sounds of the engine to the ear of the technician.

stem-type service valve: See *service valve.*

strainers: See *screen.*

stratosphere: An upper portion of the atmosphere that extends 10–30 miles (16–48 km) above the serface of the earth.

stratospheric ozone: Ozone (O_3) found in the stratospheric layer of the atmosphere.

stratospheric ozone layer: A layer above the earth; extending from 6–15 miles (9.7–24.1 km) above Earth's surface, protecting the earth from ultraviolet (UV) rays from the sun.

stroke: The distance a piston travels from its lowest point to its highest point.

STV: See *suction throttling valve.*

subcooler: A section of liquid line used to ensure that only liquid refrigerant is delivered to the expansion valve. This line may be a part of the condenser or may be placed in the drip pan of the evaporator.

subcooling: Cooling of liquid refrigerant below its condensing temperature.

substance: Any form of matter.

suction accumulator: See *accumulator.*

suction accumulator-drier: See *accumulator.*

suction line: The line connecting the evaporator outlet to the compressor inlet.

suction line regulator: See *suction throttling valve* or *evaporator pressure regulator.*

suction manifold: The point where refrigerant enters the compressor.

suction pressure: Compressor inlet pressure. Reflects the pressure of the system on the low side.

suction service valve: See *low-side service valve.*

suction side: That portion of the refrigeration system under low pressure; the suction side extends from the expansion device to the compressor inlet.

suction stroke: A term used for intake stroke.

suction throttling valve: A back-pressure-regulated device that prevents the freeze-up of the evaporator core.

suction throttling valve-POA: See *positive absolute suction throttling valve.*

sump: The bottom part of the compressor that contains oil for lubrication of the moving parts of the compressor. Not all compressors have a sump.

sunload: Heat intensity and/or light intensity produced by the sun.

sunload sensor: A device that senses heat and/or light intensity. See *photovoltaic diode.*

superheat: Adding heat intensity to a gas after the complete evaporation of a liquid.

superheated vapor: Vapor at a temperature higher than its boiling point for a given pressure.

superheat switch: An electrical switch activated by an abnormal temperature-pressure condition (a superheated vapor); used for system protection.

SUVA®: A trade name for DuPont's new generation of ozone-friendly refrigerants. HFC-134a is a SUVA® refrigerant recommended for automotive use.

swaging: A means of shaping soft tubing so that two pieces of the same size of tubing can be joined without the use of a fitting. The inside diameter of one tube is increased to accept the outside diameter of the other tube.

swash plate: A mechanical system that is used for pumping, having an angle plate attached to a center shaft, and pistons that are attached to the plate along the axis of the shaft. As the shaft is rotated, the pistons move in and out of the cylinders, producing suction and pressure.

swash plate compressor: A compressor in which the pistons are driven by an offset (swash) plate affixed to the main shaft, such as the Delco Air six-cylinder compressor.

sweat: The use of a soft solder to join two pieces of tubing or fittings using heat.

sweat fitting: A fitting designed to be used in sweating.

sweeping: See *purge*.

synthetic: A substance or material that is not natural or genuine, such as plastic used for leather.

system: All of the components and lines that make up an air-conditioning system.

system-dependent recovery system: Refrigerant recovery system that relies on system components, such as the compressor, to remove the refrigerant from the system.

tailpipe: The outlet pipe from the evaporator to the compressor. See *suction line*.

tank: See *header tank* and *expansion tank*.

Taps All Valve: see *Fitz-All Valve*.

technician: Any person who performs maintenance, service, or repair that could reasonably be expected to release Class I or Class II substances into the atmosphere. This includes, but is not limited to, installers, employees, service personnel, and owners.

temper: To modify; to bring to the desired consistency, hardness, texture or other desired physical state.

temperature: Heat intensity measured on a thermometer.

temperature gauge: A dash-mounted device that indicates engine temperature.

temperature glide: A range of evaporating or condensing temperature for a given pressure.

temperature gradient: A condition whereby an area of high temperature is placed into contact with an area of low temperature, causing heat to flow.

temperature indicator: See *temperature gauge*. May also be COLD and HOT lamps to warn of overcooling or overheating of engine coolant.

temperature-regulated valve: See *hot gas bypass valve*.

temperature sending unit: See *sending unit*.

temperature sensing bulb: See *remote bulb*.

terminal: A part that forms the end, such as to an electrical wire.

ternary: Having three parts, elements, or divisions.

tetrafluoroethane: Chemical name for the *ozone-friendly* refrigerant commonly known as HFC-134a or, more simply, R-134a. Its chemical symbol is CH_2FCF_3.

TEV: Abbreviation for thermostatic expansion valve.

theory: Systematically organized knowledge.

thermal: Of, caused by, or pertaining to heat.

thermal delay fuse: A device used with the *compressor protection switch* that heats and blows a fuse to stop compressor action during abnormal operation.

thermal fuse: A temperature-sensitive fuse link designed so that it melts at a certain temperature and opens a circuit.

thermal limiter: An electrical or mechanical device used to control the intensity or quantity of heat. A device similar to a fuse that opens at 300°F (149°C).

thermistor: A temperature-sensing resistor that has the ability to change values with changing temperature.

thermostat: A device used to cycle the clutch to control the rate of refrigerant flow as a means of temperature control. The driver has control over the temperature desired.

thermostatic clutch control: See *thermostat*.

thermostatic deicing switch: A thermostat that prevents evaporator icing up due to low-temperature conditions.

thermostatic expansion valve: The component of a refrigeration system that regulates the rate of flow of refrigerant into the evaporator as governed by the action of the remote bulb sensing tailpipe temperatures.

thermostatic switch: See *thermostat*.

third gauge: A low-side gauge used to check pressure drop across a suction pressure regulator, at the compressor inlet, on some vehicles.

threshold limit value (TLV): The percentage, generally given in parts per million (ppm), of refrigeration in atmospheric air above which a human can become drowsy or have a loss of concentration.

throttle: A valve that regulates the flow of fluid.

throttling valve: See *suction throttling valve* and *evaporator pressure regulator*.

time delay: The intentional delay of an action or function, based on a predetermined fixed or variable time.

time-delay relay: An electrical switch device that provides a time delay before closing (or opening).

tinning: Coating two surfaces to be joined with solder.

TLV: Acronym for threshold limit value.

ton of refrigeration: The effect of melting 1 ton of ice in 24 hours. 1 ton equals cooling 12,000 Btu per hour.

torque: A turning force; for example, the force required to seal a connection; measured in (English) foot-pounds (ft.-lb.) or inch-pounds (in.-lb.); (metric) Newton-meters (N•m).

torque wrench: A wrench calibrated in a manner to determine torque, in in.-lb. or ft.-lb. (N•m) of a bolt or nut.

total heat load: The amount of heat to be removed or added, based on all conditions.

Totaltest tester: A device that may be used to check for moisture and acid in refrigerants.

trace: A colored dye (suitable for use in a refrigeration system) introduced to the system to detect leaks.

transducer: A vacuum valve used to transfer the electrical signal from the amplifier into a vacuum signal. This vacuum signal regulates the power servo unit in automatic temperature control units.

triple evacuation: A process of evacuation that involves three pump downs and two system purges with an inert gas, such as dry nitrogen (N).

trunk unit: An automotive air-conditioning evaporator that mounts in the trunk compartment and is ducted through the package tray.

TXV: Abbreviation for thermostatic expansion valve.

typical: An example of the qualities, traits, or nomenclature of a component or system.

Ucon: A trade name for refrigerant.

UL: An acronym for Underwriters Laboratories.

ultrasonic: Frequencies at a range above that of the human ear.

ultraviolet (UV): That part of the electromagnetic spectrum, emitted by the sun, that lies between visible violet light and x rays.

ultraviolet (UV) radiation: The invisible rays from the sun that have damaging effects on the earth. It is ultraviolet (UV) radiation that causes sunburns.

undercharge: A system that is short of refrigerant; this condition results in improper cooling.

under-dash unit: See *hang-on unit.*

Underwriters Laboratories (UL): An independent laboratory that performs standards testing on devices, systems, and materials for compliance and/or approval.

unloading solenoid: An electrically controlled valve for operating the throttling valve or bypass valve in some applications.

upstream blower: A blower arranged in the duct system so as to push air through the heater and/or air-conditioner core(s).

UV: An acronym for ultraviolet.

V: Abbreviation for volt.

vacuum: Any pressure below atmospheric pressure.

vacuum check relay: A mechanical air-operated device that checks (closes off) a vacuum line to a pot whenever the manifold vacuum pressure falls below the applied vacuum pressure.

vacuum check valve: An air-operated mechanical device that checks (closes) a vacuum line to the vacuum reserve tank whenever the manifold vacuum pressure falls below the reserve vacuum pressure.

vacuum diverter valve: See *vacuum motor.*

vacuum hose: See *vacuum line.*

vacuum line: A rubber tube used to transmit a vacuum signal from one component to another.

vacuum modulator: A component of the air temperature sensor that regulates the amount of vacuum signal that is applied to the servo motor.

vacuum motor: A device designed to provide mechanical control by the use of a vacuum.

vacuum pot: See *vacuum motor.*

vacuum power unit: A device for operating the doors and valves of an air conditioner using a vacuum as a source of power.

vacuum programmer: A device with a bleed valve that changes vacuum pressure by bleeding more or less air, thereby controlling the vacuum signal.

vacuum pump: A mechanical device used to evacuate the refrigeration system to rid it of excess moisture and air.

vacuum reserve tank: A container used to store reserve (engine) vacuum pressure.

vacuum selector valve: A vacuum valve that controls the vacuum motors, which, in turn, operate the airflow control doors.

vacuum tank: See *vacuum reserve tank.*

valence ring: The outer ring of an atom.

valve plate: A plate containing suction and/or discharge valves located under the compressor heads.

valves-in-receiver (VIR): An assembly containing the expansion valve, suction throttling valve, desiccant, and receiver.

vapor: See *gas.*

vaporization: The act of evaporating. Changing from a liquid to a vapor.

vapor lines: Lines used to carry refrigerant gas or vapor.

variable: Subject to change; not fixed.

variable displacement: To change the displacement of a compressor by changing the stroke of the piston(s).

V-belt: A rubber-like continuous loop placed between the engine crankshaft pulley and accessories to transfer rotary motion of the crankshaft to the accessories.

VDOT: Abbreviation for variable displacement orifice tube.

ventilation: The act of supplying fresh air to an enclosed space, such as the inside of an automobile.

ventilation: To admit and/or recirculate fresh or conditioned air.

venturi: A tube-like device that contains a restriction to create a negative pressure.

venturi effect: The reduction in pressure and increase in speed as gases pass through a restricted orifice such as in an aspirator.

vertical flow radiator: See *downflow radiator.*

V-groove belt: See *serpentine belt.*

VIR: Abbreviation for *valves-in-receiver.*

viscose: See *viscosity.*

viscose friction: Friction between moving parts, such as in a compressor, resulting from using an oil that is too heavy or thick.

viscosity: The thickness of a liquid or its resistance to flow.

VOC: Acronym for volatile organic compound.

volatile liquid: A liquid that evaporates readily to become a vapor.

volatile organic compound (VOC): A term that relates to chemicals that are negligibly photochemically reactive.

volt (V): A unit of measure of electrical force.

voltmeter: A device used to measure volt(s).

V-pulley: Used in automotive applications to drive the accessories, such as a water pump, generator, alternator, power steering, and air-conditioner compressor.

washers: Flat disks made of metal, plastic, or other matter, used to relieve friction, prevent leakage, or distribute pressure.

water control valve: A mechanically operated or vacuum-operated shutoff valve that stops the flow of hot water to the heater.

water jacket: Passages in an engine block and head that allow coolant to flow through the engine for carrying away its heat.

water pump: A device, usually belt driven, that provides a means of circulating coolant through the engine and cooling system.

water valve: See *water control valve.*

watt: A unit of electrical power; volts times amps equal watts.

wet bulb: A device, such as a thermometer, that has a wet sock over its sensing element.

wet bulb temperature: The ambient temperature measured with a wet bulb thermometer.

wide-open throttle cutout relay: A relay that cuts power to the compressor clutch during heavy acceleration.

windchill: See *windchill factor.*

windchill factor: The temperature of wind-less air that would have the same effect on exposed human skin as a given combination of wind speed and temperature.

windings: The small enamel-covered wires that make coils, such as the relay, compressor clutch coil, and motor field coils.

wobble plate: A term often used for "swash plate"

wobble plate compressor: See *swash plate compressor.*

woodruff key: An index key that prevents a pulley from turning on a shaft.

WOT: Abbreviation for wide-open throttle.

zener diode: A diode with properties that prevent current flow up to a given voltage and allow current flow above a given voltage.

INDEX